Y0-BEB-379

CONCORDIA UNIVERSITY CHICAGO
3 4211 00188 4546

ANNUAL REVIEW OF EARTH AND PLANETARY SCIENCES

ANNUAL REVIEW OF EARTH AND PLANETARY SCIENCES

VOLUME 12, 1984

GEORGE W. WETHERILL, *Editor*
Carnegie Institution of Washington

ARDEN L. ALBEE, *Associate Editor*
California Institute of Technology

FRANCIS G. STEHLI, *Associate Editor*
University of Oklahoma

ANNUAL REVIEWS INC. 4139 EL CAMINO WAY PALO ALTO, CALIFORNIA 94306 USA

ANNUAL REVIEWS INC.
Palo Alto, California, USA

International Standard Serial Number: 0084-6597
International Standard Book Number: 0-8243-2012-3
Library of Congress Catalog Card Number 72-82137

TYPESET BY A.U.P. TYPESETTERS (GLASGOW) LTD., SCOTLAND
PRINTED AND BOUND IN THE UNITED STATES OF AMERICA

Annual Review of Earth and Planetary Sciences
Volume 12, 1984

CONTENTS

SOME RELATED ARTICLES IN OTHER *ANNUAL REVIEWS*

From the *Annual Review of Astronomy and Astrophysics*, Volume 21 (1983)

Herbig-Haro Objects, Richard D. Schwartz

Variations in Solar Luminosity, Gordon Newkirk, Jr.

From the *Annual Review of Ecology and Systematics*, Volume 14 (1983)

Stasis in Progress: The Empirical Basis of Macroevolution, Jeffrey S. Levinton

From the *Annual Review of Materials Science*, Volume 13 (1983)

Ceramic Materials for the Immobilization of Nuclear Waste, D. R. Clarke

From the *Annual Review of Microbiology*, Volume 37 (1983)

Evolutionary Relationships in Vibrio *and* Photobacterium: *A Basis for a Natural Classification*, Paul Baumann, Linda Baumann, Marilyn J. Woolkalis, and Sookie S. Bang

Evolution of Wall-less Prokaryotes, Jack Maniloff

From the *Annual Review of Nuclear and Particle Science*, Volume 33 (1983)

Cosmic-Ray Record in Solar System Matter, R. C. Reedy, J. R. Arnold, and D. Lal

Ann. Rev. Earth Planet. Sci. 1984. 12: 1–10

THE GREENING OF STRATIGRAPHY 1933–1983

L. L. Sloss

Department of Geological Sciences, Northwestern University, Evanston, Illinois 60201

Readers should be aware that anyone who writes a kind of autobiographical prefatory chapter for a volume such as this must be an egotist. Self-discipline has never permitted me to think of myself as an egotist but, by accepting this assignment, I am clearly out of the closet that may have protected me in the past. Now the problem becomes one of avoidance of fatuous egotism; I am sure that I can rely on friends to judge.

No special significance attaches to the year 1933 except that 50 years seems a span quite long enough to cover in reminiscence. In 1933 I was past midstream as a Stanford undergraduate, and stratigraphy was something you read about in textbooks and in assigned references to US Geological Survey (USGS) Professional Papers—a concept of order and succession not subject to observation and verification in the adjacent California coastal ranges or, without a degree of blind faith, in the mountains of western Nevada where Stanford took its students for field exercises. I was identified as a student of paleontology, and an indifferent example at that; would-be paleontologists were placed under the benign wing of S. W. Muller and the more exacting regime of H. G. Schenck. Here, I dutifully memorized the Tertiary stratigraphic column for this and that region of California and learned to identify a few hundred "index" fossils, mostly representing early Mesozoic and Cenozoic times and commonly recognized as much by position in a tray or drawer as by morphologic detail. None of this effort conferred much of a feeling of intellectual reward, but lip service to the rules gained one admittance to the seminars run by Muller, Schenck, and their graduate students. These sessions were lively, contentious, and exciting; the subject was stratigraphic classification at a time when this seemingly stupefying subject was being revivified as heretics arose to challenge the principles imposed by nineteenth century dogma—ancient

0084-6597/84/0515-0001$02.00

positions newly supported by the recently published "Stratigraphic Code." The Code, reflecting conventional wisdom, demanded equity and coincidence of boundaries among groupings of strata defined by lithologic, paleontologic, and chronologic criteria; Muller, Schenck, and company made it clear that successions of strata were divisible into units defined by lithologic character (or by interruptions in lithologic continuity), while quite different stratal units were defined only by their ages. This was pretty heady stuff because it flew in the face of contemporary belief in an Earth history punctuated by global "revolutions" and lesser "disturbances" marked in the rock record by discontinuities that served to discriminate systems, series, groups, and formations, each representing a position in the rank ordering of segments of geologic time (period, epoch, age, and stage).

At Chicago, my next station on the road to maturity, I continued to be identified as a protopaleontologist and thus became a supplicant at the very active feet of Carey Croneis. Carey was a magnificent teacher, but he tended to let his research students sink or swim on their own; his interests then revolved around neglected groups of fossil organisms (worm jaws, immature crinoids, etc), while students were offered the riches of the Hall Collection from which to choose subjects for paleontologic investigation. Stratigraphy, except in a confined descriptive sense, was something Croneis had already mastered during his fruitful years in Arkansas; furthermore, the task at Chicago was relegated to R. T. Chamberlin. Rollin taught his brand of stratigraphy in an historical-geology context heavily weighted by the concepts expressed in the 1905 editions of the texts by Chamberlin père and Salisbury. Mostly, I remember a spirited defense of the Planetesimal Hypothesis, a spirited refutation of the contrary views of Jeans and Jeffreys, and long lists of stratigraphic names from every part of North America. These lists were not to be taken lightly because they would inevitably occupy positions of prominence at the dreaded predoctoral oral examination.

Yes, stratigraphy as presented in the graduate schools and journals of learned societies of the mid-1930s could charitably be described as uninspiring. At Chicago, inspiration was available from two sources. One of these bubbled from the students themselves; a fairly rough lot by present standards, they were men (very few women) returned to the halls of learning by the collapse of their jobs in the oil camps of Venezuela and the copper mines of Chile. Given a supply of grapefruit juice diluted with lab alcohol, impromptu seminars would explore the failings of contemporary stratigraphic concepts. The second source rose from the basement of Rosenwald Hall where Francis Pettijohn, a partially reclaimed student of the Archean, and Bill Krumbein, a born-again refugee from a career in business, were separately and jointly examining the materials of stratigraphy, sediments and sedimentary rocks, as objects of quantitative study and statistical analysis. Here was the breakthrough toward release from the mnemonic exercises of classical stratigraphy; I became infatuated with numbers (although I remained functionally illiterate as a mathematician), and my enthusiasm for paleontologic taxonomy expired even as I was completing a dissertation on Devonian corals.

Seven years after leaving high school, degree in hand but very moist behind the ears, I was transported to Butte where a niche was available in a three-man department at the Montana School of Mines. Mines was a great place; there were no committee chores or other administrative duties because these were all efficiently subsumed by the office of the president, freeing the faculty to conduct teaching and research without pretensions of academic democracy. The State Bureau of Mines and Geology was the research arm of the School; the Bureau received an annual budget of about $25,000, out of which the Instructor in Paleontology and Stratigraphy was given a per diem allowance of $4, a graduate-student assistant, and a pickup truck, all to be applied to a project of choice in the summer months. My immediate predecessor, Harold Scott, had done important things in straightening out certain late Mississippian strata, but no one had worried much about the underlying early Mississippian limestones since the mapping of the Three Forks and Little Belt Mountain folios decades earlier. So, the better parts of three glorious field seasons were spent climbing about in some of the world's greatest scenery—it is a well-known fact that stratified rocks, especially those that include cliff-forming carbonate masses, develop scenery superior to any produced by the random erosion of granites and metamorphic massifs; further, the water for morning coffee has mineral substance, and the trout are more virile.

Enjoying the topographic delights was one thing, but nothing in my education had prepared me to make the right observations on carbonates. They could not be disaggregated and passed through sieves; they defied analysis in terms of Stokesian settling velocities; and when we looked at thin sections, we saw nothing that transmitted useful information. Still, it was clear that numbers must be generated if progress were to be made. So, tens of sections were measured and divided into units defined by bedding thickness and gross texture. Thousands of samples were collected and these were crushed and digested to yield insoluble residues. The thickness data could be assembled to form isopach maps, a rather far-out data-integration device for the times, and the residues became useful as quantitative checks of correlation from one measured section to another.

In my zeal to wring more information from limestones than my primitive observations produced, I persuaded a colleague (S. R. B. Cooke) to try spectrochemical analysis as a means of quantifying the differences among carbonates. Cooke had made a grating spectrograph with his own skilled hands, and he and I put in uncounted hours grinding samples, loading measured amounts into graphite electrodes, igniting these one-by-one, and deriving percentages of major elements (Fe, Mg, Al, Sr) using external standards and a densitometer. The results proved "interesting," but the cost effectiveness was so low as to turn Cooke back to mineral dressing and me to the pursuit of other avenues of quantitative truth.

Meanwhile, strictly qualitative stratigraphic knowledge was beginning to surface—the kind of knowledge that my younger associates, partly in disparagement and partly in sheer jealousy, call "lore." Example: throughout its outcrop area, the top of the Madison Limestone is scarred by great depressions and channels filled with red mud and sand. Clearly some major karst-forming episode followed the close of Madison deposition, and this event had no relationship to classically recognized chronology. Example: over most of Montana Mississippian, strata rest without evidence of major discontinuity on late Devonian beds. But in south-central Montana and south into Wyoming, Madison rocks succeed Ordovician (Bighorn), whereas both Ordovician and Silurian are unknown over major areas of Montana. Example: my beloved Madison is in contact with marine Jurassic on the Sweetgrass Arch of northwestern Montana (resulting in the Kevin-Sunburst oil field and the Dempsey-Gibbons prizefight of 1923), while to the south the Jurassic oversteps younger and younger beds to lie on Triassic at the Wyoming line. All of this was common knowledge to generations of older geologists, but none found the relationships of sufficient concern to display them in map form or discuss their significance. I began to accumulate both concern and map representations.

The number of examples and types of stratigraphic maps was growing. I have mentioned isopach maps on which lines of equal thickness describe the geometry and distribution of stratigraphic units. Tentative stabs were being made to differentiate areas of differing *facies* defined by petrology (e.g., evaporite facies) or by interpretations based on thickness, color, paleontology, assumed provenance, etc (e.g. nonmarine facies, geosynclinal facies). The major difficulty lay in defining the reasons why a line separating facies "A" from facies "B" should follow a certain course on a map. Meanwhile, Midcontinent workers (notably A. I. Levorsen) were identifying regional unconformities and describing these by drafting paleogeographic maps (later, and more accurately, to be called subcrop maps) showing the distribution of stratigraphic units formerly exposed at ancient erosion surfaces.

These were times, the late-1930s and early-1940s, when major advances in descriptive stratigraphy and in stratigraphic concept were being made by petroleum geologists, often in at least partial isolation from the academic community. Indeed, a measure of mutual distrust served to intensify the separation into two camps. Academics had little opportunity to work with subsurface data, preferring to draw interpretations from exposed rocks, while, perforce, stratigraphy in the oil patch relied on the glimpses of rocks and fossils brought up by the drill. One group had a wealth of observations, but the wealth was unevenly distributed by irrelevant accidents of erosion and exposure; the other had data from drilling over vast regions but, especially after rotary drilling replaced cable tools, the derived stratigraphic information was held suspect by academic unbelievers. After all, the proportion of the sedimentary section represented by cores recovered in any critical exploratory drill hole was always very small (and taken with a view toward the solution of engineering, rather than stratigraphic, problems). In the absence of cores, subsurface stratigraphy had to rely on a mishmash of cuttings returned to the surface in the drilling mud along with cavings from thousands of feet up the hole. The kind of mature biostratigraphy based on micropaleontology that integrated surface and subsurface observations in California and the Gulf Coast was not available in the Rocky Mountains and Great Plains. To make matters worse, drilling commonly penetrated units of strata unrepresented, or at least unexposed, in outcrop, serving to further separate surface and subsurface workers and concepts.

Nevertheless, trying to be a stratigrapher in Montana forced one to delve into the seemingly occult world of the subsurface. The center of the state and its southern border are decorated by isolated uplifts that expose the Paleozoic rocks that were my playground; leaping across Judith Gap from the Little Belts to the Big Snowies, or from the Beartooths to the Pryors, was not an impossible assignment, but interpolating over a distance of about 100 miles between the Big Snowies and the Little Rockies, for example, raised bewildering questions unanswerable without recourse to the subsurface. Stretching east to the Black Hills or, much worse, to the unplumbed depths of the Williston Basin created even greater challenges. The necessity of treating regional stratigraphy as a related succession of time-dependent three-dimensional problems was becoming apparent, and any relevant data set, even the feared subsurface, had to be applied. Fortunately for me, two developments of the early-1940s accelerated my stratigraphic education.

The Montana Bureau was an informal repository for cores and samples of wildcat wells; these and the systematic surface samples collected for insoluble-residue studies brought experienced company men to Butte, and I was exposed to repeated short courses in how to wring information from rotary cuttings. (Max Littlefield of Gulf was the greatest of teachers of practical stratigraphy in a generation that included little-known giants.) At the same time, geophysical logs displaying the electrical and radioactive properties of strata and fluids in bore holes were becoming routinely available, providing uninterrupted records of stratigraphic successions and making correlation from well to well a vastly easier exercise.

It became apparent to me, as it must have to other workers who had better things to do than give talks and write papers, that the character or facies of a unit of strata as well as part of the reasons for its distribution and thickness were governed by the tectonic framework of the big region represented by Montana, "tectonic framework" being a buzz term for the geography of basins and arches. Another dominant control over distribution and thickness was exercised by regional unconformities that dictated the degree to which units were preserved or were thinned or eliminated by erosion. The unconformities, in turn, were obvious creatures of the tectonic framework, cutting deeply into underlying strata on arches, less effective but often traceable across basins. Again, these were observations common to anyone with experience in three-dimensional regional stratigraphy, but no one had introduced such findings in a classical stratigraphic education.

A little sober consideration led to conclusions that were all apparent long before but which came upon me and my classical background as startling cultural events. For example, the Montana stratigraphic succession, and that of adjacent states and provinces, is divisible into packages, each bounded by a regional unconformity representing uplift with respect to baselevel. Further, each package is witness to a different tectonic framework that, if understood, clarifies the preserved thickness and distribution of individual units. Thus, Ordovician and Silurian beds are missing on the Sweetgrass Arch and its Paleozoic antecedents, but Ordovician is preserved on the eastern flank of the Arch (as in the Little Rockies) and Silurian is present in abundance in Manitoba. Everywhere the beveled edges of Ordovician and Silurian rocks are covered by overstepping Devonian beds (early Mississippian as the Wyoming line and the Black Hills are approached). Clearly, the Devonian marched to a different tectonic drummer, although much of the *status quo ante* was restored prior to early Jurassic erosion... and so on, as is now made clear in most elementary texts.

Conventional theory and practice of the time placed western Montana firmly within the borders of the Cordilleran geosyncline; therefore, it was reasonable for a newly arrived student of the rocks to expect significant differences in thicknesses and lithologies as the line separating "stable interior" and "geosyncline" was crossed. In actuality, most stratigraphic units display no significant change in character from central to western Montana, and westward shifts in the rates of change of thickness are more easily explained by stacked thrust sheets than by transition from one depositional-tectonic province to another. Here again my training failed me, but I cannot claim to have enunciated anything useful; before I left Montana, Hans Stille's "miogeosyncline" began to be discussed and I was not motivated to pursue the matter further than that.

In 1946–1947 the Golden Years in Montana came to their inevitable end, and I was abruptly transported to Serendipity City, aka Northwestern University. Here were Bill Krumbein, newly returned to academe after war service with the USGS and Gulf Oil, and Ed Dapples, a product of the combined efforts of the coal industry and W. H. Twenhofel. Krumbein had lost none of his zeal for quantitative expression of geologic facts and factors, Dapples represented integrity in terms of hard-nosed sedimentary petrology and petrography of detrital sediments, and I was supposed to supply a level of competence in carbonates and a general background of accumulated stratigraphic lore. These were the times when Paul Krynine was establishing a geosynclinal cycle based on the composition and texture of sandstones—"orthoquartzites" and "low-" and "high-rank greywackes" complemented by the color of tourmaline grains and the presence/absence of strained quartz. Krynine had, for better or worse, exposed the whole topic of sedimentary tectonics and we were swept along, erecting our own classifications and tectonic interpretations illustrated by complex trigonal and tetrahedral diagrams, while Francis Pettijohn joined the action, producing yet another classification or two while having the good sense to decry classifications in general. The exercise of tectonic synthesis by means of the polarizing microscope was enormous good fun, highly educational and stimulating, and relatively harmless (except to the Internal Revenue Service, which lost a tax suit because Krynine referred to the St. Peter Sandstone as a quartzite).

Much more valuable, from my point of view, were joint efforts to develop a methodology for quantitative facies mapping. Dapples and I would never have gone far in facies mapping without Krumbein's genius with numbers, but we all contributed to procedures that made it possible to integrate hundreds of bits of data and produce stratigraphic maps from which soundly based interpretations could be drawn. The technology transfer to students was easy, and sufficient numbers were attracted to permit facies mapping of many parts of the stratigraphic column in many areas of the US and Canada; nothing, not reading nor outcrop and subsurface work, was nearly as useful to me in the accumulation of stratigraphic lore as the supervision of a decade of students involved in facies studies.

While all this was going on, the stratigraphic packages and their bounding unconformities identified in Montana were pursued around the whole of the interior of North America and found to be nearly universally applicable. The packages, now termed *sequences*, were given North

American Indian names (Sauk, Tippecanoe, etc) so that there would be no confusion with system/period names (Cambrian, Ordovician, etc) of European heritage. Both facies mapping and the sequence concept were far enough along so that when a symposium, "Sedimentary Facies in Geologic History," was held at the 1948 El Paso meeting of the Geological Society of America, I presented a paper on behalf of my colleagues covering both topics. This at one stroke violated and verified Bill Krumbein's often stated "one paper–one idea" principle. The facies methodology attracted a lot of favorable attention, but the sequence concept was either buried or treated roughly. (One highly qualified formal discussant of the symposium referred to sequences and the related concepts as "wholly unwarranted," "specious," and tending to "confuse rather than illuminate.")

Fifteen years of further work, largely contributed by students, demonstrated the applicability of stratigraphic sequences, and by the dawn of the plate-tectonic age the empirical framework for the North American interior was firmly established. The need for one major change became apparent between 1948 and 1963. Originally, the boundary between the Kaskaskia and Absaroka sequences was placed between middle and late Mississippian strata in recognition of marked changes initiated at the base of Chester sedimentation in Montana and Illinois. There were problems of precise definition, however, and these problems became intractable as the "geosynclinal" margins of North America were approached. Peter Vail, working in eastern Tennessee, showed the continuity of deposition and the artificiality of trying to isolate late Mississippian strata on physical grounds. Vail found a much more meaningful discontinuity just below the biostratigraphically defined base of the Pennsylvanian and convinced me that the base of the Absaroka should be redefined.

Meanwhile, Marshall Kay had built on Stille's sparsely published ideas, and it became clear that the stratigraphic sequences that concerned me were creatures of the craton and commonly were extendable into the "miogeosynclines." The latter term was always difficult to apply because the tectonic state represented required a volcanically endowed "eugeosyncline" to form the outboard limits but no genetic or other interdependence of "mio-" and "eu-" was apparent. Much of this dilemma evaporated when Robert Dietz showed that the "eugeosynclines" are oceanic in origin and tectonically emplaced with respect to "miogeosynclines," which are shelf-prism sediments at continental margins.

The isolationist dreams of a North American stratigraphic structure and organization and a taxonomy based on aboriginal tribal names were dissipated by the new paradigms of the 1960s. If cratonic sequences were for real, they should be applicable on all the cratons of Pangaea. They are, of course, but recognition waited on the kind of stratigraphic integration of surface and subsurface observations that had long been available in North America; equally important, as long as Europeans emphasized that only biostratigraphy is "real" stratigraphy, no tie across Iapetus and the Atlantic was clear to those who published their ideas widely. Fortunately, growing energy demands resulted in accelerated drilling and the publication of regional isopach-facies studies. Among the latter the major eye-opener was the Soviet Academy's *Atlas of the Russian Platform*; here, for all who would but look, were the North American sequences, their bounding unconformities, everything.

In the late-1960s and through the 1970s stratigraphic observation was enormously enhanced by developments emerging from the digital recording of multichannel exploration-seismic data. A number of stratigraphers in what is now Exxon's research arm—Vail, R. M. Mitchum, J. B. Sangree, and less intensively, R. G. Todd—were exposed to sequence philosophy and practice in their formative years; this group was preadapted to recognizing unconformity-bounded units among the simulations of stratigraphic cross sections produced by reflection profiling. The wealth and clarity of the data are such that stratigraphic relationships, within the finite limits of the resolving power of the method, can be observed in great detail as two-dimensional continua, rather than as series of discrete points demanding interpolation. Thus, the stratigraphic geometry of unconformities (truncation, overstep, onlap, etc.) is exposed to observation and analysis in almost embarrassing profusion, leading to the identification of many more regionally significant stratal packages than previously recognized. Add to this the near-global coverage provided by multinational exploration efforts and the realization that many unconformities, enclosing many sequence-like successions, are traceable from region to region and from continent to continent, and it becomes obvious that the original six named sequences are only part of the story.

It might seem reasonable to retire from the action, taking with me the Indian tribes and abandoning the scene to the seismic stratigraphers and their sequences identified by cold-blooded, unromantic letter and number. Not just yet, if you please! I will not long defend Absaroka (late Paleozoic–early Mesozoic) and Tejas (Eocene and younger)—they are marked by internal changes and complexities deserving recognition in nomenclature. The others, however, represent more than pride of authorship. All four are radically different from Absaroka and Tejas, and each is unified by individualized tectonic modes and geographies and by idiosyncratic petrologies. That is why, for example, experienced stratigraphic folklorists can visit anybody's craton, stop at a road cut, and say "this is Cambrian" or "this is almost certainly Cretaceous" without benefit of body fossils or a mass spectrometer. What is going on?

The folks with the greatest volume of data, the seismic types, are deeply impressed by the apparent global synchrony of stratigraphic patterns clearly related to the freeboard of continental margins and the rate of change of sea levels with respect to sea bottoms. They find that they sleep well when they place their faith in eustatic sea levels and dream pleasant dreams when glacial controls on eustatics can be invoked. Certain geophysicists, especially those adept at numerical modeling, note that the rate of subsidence of passive margins relative to sea level mimics the decline curve of thermal contraction; hence, all is related to heating, as at a midocean ridge, and subsequent cooling. There are variants of the thermal theories, but none can be made to operate at the wide spectrum of frequencies exhibited by cratonic stratigraphy.

I don't have any answers but I can say, and often do, that no unified theory has yet been developed to encompass the total field of observations and lore that must be satisfied. Most relevant cratonic data reside in sedimentary basins; therefore, a student of Phanerozoic cratons is, perforce, a student of basins. Many cratonic basins with a legible history of the past 180 million years can be shown to have a subsidence history remarkably similar to that of passive margins, but no heat episode is evidenced so thermal contraction is awkward. Older sedimentary basins can be shown to have subsided irregularly, but with a degree of global synchrony, over hundreds of millions of years; further, changes in the mode and rate of subsidence coincide with the times of the bounding unconformities of the Indian-tribe sequences. In other words, whatever controls basin subsidence also governs the freeboard of cratons and their tectonic modes.

It is my unabashedly parochial opinion that cratons, those seemingly inert blocks, are blessed with outstanding questions of considerable significance, and that such questions will not approach resolution without full consideration of the hard data and less well-defined lore that constitute the muscle and heart of stratigraphy.

Ann. Rev. Earth Planet. Sci. 1984. 12: 11–37

DOUBLE-DIFFUSIVE CONVECTION DUE TO CRYSTALLIZATION IN MAGMAS

Herbert E. Huppert

Department of Applied Mathematics and Theoretical Physics, University of Cambridge, Cambridge CB3 9EW, England

R. Stephen J. Sparks

Department of Earth Sciences, University of Cambridge, Cambridge CB2 3EQ, England

1. INTRODUCTION

An important development in the understanding of the fluid dynamics of convection has been the recognition that heat- and mass-transfer processes in multicomponent systems are often fundamentally different to those in the more familiar one-component systems. In a system containing two or more properties that have different molecular diffusivities and opposing effects on the vertical density gradient, a wide range of novel and complex convective phenomena can occur. In the last few years fluid dynamicists and geologists have recognized that such convection, known by the general term of *double-diffusive convection*, can occur in magmas and in other fluids of geophysical interest. The implications of this form of convection are far-reaching and are likely to revolutionize our perceptions of many geophysical processes.

Theoretical and experimental analyses of double-diffusive convection were first developed in relation to the oceans, specifically with a view to explaining several different kinds of layering (Huppert & Turner 1981a). It is also apparent that double-diffusive convection is likely to occur in many other systems of geophysical interest. Much attention has been given in particular to the notion that double-diffusive convection can play an important role in the differentiation of magmas. Some of the most

0084–6597/84/0515–0011$02.00

important effects in magmas are due to the crystallization induced by cooling.

This article reviews the theoretical and experimental developments in the study of convective phenomena in both multicomponent fluids and crystallizing fluid systems. We concentrate on convection in magmas because of the recent blossoming of research in this area. We also indicate other types of fluids where double-diffusive and compositional effects are likely to be important.

2. CONVECTION IN MULTICOMPONENT FLUIDS

Most of the early studies of the fluid dynamics of double-diffusive convection were developed with oceanographic applications in mind (Huppert & Turner 1981a), for which the two relevant properties are heat and salt. These two properties have different molecular diffusivities ($D_{HEAT} \approx 10^{-3}$ cm^2 s^{-1}; $D_{SALT} \approx 10^{-5}$ cm^2 s^{-1}). In many natural situations the gradients of temperature and salinity are such that one of the properties has a destabilizing effect (i.e. leads to convection), although the system has an overall apparently stable density stratification. In these circumstances, horizontal layering inevitably occurs. Layering developed in this way has attracted the attention of petrologists because of the resemblance of these structures to layering observed in igneous intrusions (McBirney & Noyes 1979). This is arguably a superficial resemblance, but the physical and chemical properties of silicate melts are such that we consider double-diffusive convection certain to occur. Recent detailed studies of volcanic rocks have also demonstrated that many magma chambers contain stable compositional gradients but unstable temperature gradients (Hildreth 1981), which is a situation where double-diffusive effects can be expected.

Before illustrating the principles of double-diffusive convection with some examples, we briefly review the physics of thermal convection in a one-component system so that the differences will be more apparent. If a layer of uniform fluid is heated from below, the lowermost parts of the layer heat up, expand, and become buoyant. The buoyant fluid parcels rise. If the heating is sufficiently vigorous, these parcels cause circulation throughout the fluid layer, which is recognized as convection. The pattern of circulation and the heat transfer across the fluid can generally be predicted given the physical properties of the fluid and the applied temperature field. The dimensionless parameters characterizing the motion are the thermal Rayleigh number and the Prandtl number, defined as

$$\mathrm{Ra}_T = g\alpha\Delta T d^3/\kappa_T \nu \qquad (1)$$

and

$$Pr = \nu/\kappa_T, \tag{2}$$

where g is the gravitational acceleration, α the thermal expansion coefficient, ΔT the temperature difference across the layer, d the fluid layer thickness, κ_T the thermal diffusivity, and ν the kinematic viscosity. The Rayleigh number is the ratio of the buoyancy force driving convection to the viscous force resisting fluid movement, and the Prandtl number is the ratio of the diffusivities of momentum to heat. Convective motions are initiated when Ra_T is of the order 10^3. The exact value depends on the specific boundary conditions but is independent of the Prandtl number. The convective motion becomes turbulent when Ra_T exceeds a value of approximately 10^6 for the large Prandtl numbers that characterize magmas.

Several authors have calculated typical thermal Rayleigh numbers in magma chambers (Shaw 1965, Bartlett 1969, Rice 1981). These calculations yield Rayleigh numbers generally well in excess of 10^6 and range up to 10^{23} for large basaltic chambers. Even in a 1-km-deep magma chamber containing a relatively high-viscosity rhyolitic melt, the calculated Rayleigh number can still be 10^9. Such high values imply that the vast majority of magma chambers are in vigorous turbulent convection. This application of purely thermal convection theory has two implications. First, crystal settling would be inhibited because the convective velocities in the interior of the chamber are orders of magnitude larger than the settling velocities of crystals (Sparks et al 1984). Second, because of the convection, no vertical property gradients could survive in the interior of a magma chamber. Of course, these problems do not necessarily arise, because in many cases the application of purely thermal convection is inappropriate, as is described below.

Two specific concepts of double-diffusive convection are illustrated in Figure 1, which shows part of a laboratory experiment using aqueous solutions of K_2CO_3 and KNO_3. In the experiment, described in detail by Huppert et al (1982b), a tank measuring 40 × 20 × 30 cm high was filled with aqueous K_2CO_3 in such a way that there was a linear density gradient with depth, resulting in a stable stratification. The temperature of the K_2CO_3 solution was initially uniform at 11°C. An intrusion of a hot (65°C), relatively heavy, dyed KNO_3 solution was rapidly introduced at the base of the tank. Because of its density, the KNO_3 solution ponded at the base of the tank to form a separate layer, as marked by the dye. We return in a later section to the behavior of the lower layer; in this section, we are concerned only with the early behavior of the K_2CO_3 solution.

Figure 1 shows the experiment 20 minutes after the introduction of the KNO_3 layer. We observe that the stably stratified K_2CO_3 layer has been transformed into two different series of layers. These are a consequence of double-diffusive convection. There are two scales of layers produced, and we consider their origin in turn. The finer-scale layers in the upper part of the K_2CO_3 solution are the result of sidewall heating and were produced, before the introduction of the hot KNO_3 solution, by the following mechanism. The laboratory temperature was approximately 20°C, and heat was transferred into the tank. At any given level, fluid close to the sidewall becomes hotter and therefore lighter. As heating proceeds, the thermal boundary layer thickens and produces upward flow. However, because of the small diffusivity of K_2CO_3, the upward-flowing fluid retains its K_2CO_3 content from its level of origin, while moving upward into a decreasing K_2CO_3 concentration and therefore a decreasing density. The upward flow is thus constrained by the density stratification and eventually reaches a level of equal density, where it is diverted sideways. This creates a local circulation that progressively drives a layer of circulating fluid into the

Figure 1 Shadowgraph view of an experiment in which an upper layer of compositionally stratified K_2CO_3 solution is heated from below by a hot layer of KNO_3 solution. Double-diffusive layering occurs in the K_2CO_3 solution, seen here 20 minutes after the experiment began. Details are given in Huppert et al (1982b).

interior of the tank. Such a series of layers cannot occur in a one-component system.

The formation of the layers depends critically on the contrast in diffusivities of K_2CO_3 and heat. It is because the diffusivity of K_2CO_3 is much less than the diffusivity of heat that the hot fluid rising along the sidewall retains almost all its original content of K_2CO_3. The heat transferred from the room is only capable of changing the density slightly. In contrast, the background vertical density variations due to compositional effects are relatively large. Consequently, the fluid can only rise a short distance before coming to a level where the densities are equal. The spacing of layers formed in this way is dependent on the temperature difference across the walls of the tank, the vertical density gradient, and the physical properties of the fluid. The same type of layering would form if the exterior were cold and the compositionally graded fluid in the tank were hot. The only difference between the two cases would be the sense of circulation (see Figures 3 and 5 in Huppert & Turner 1980).

In Figure 1, layering of a larger scale is observed in the lower regions of the K_2CO_3 solution immediately above the hot KNO_3 layer. These layers are the consequence of the heat transfer to the compositional gradient from the KNO_3 layer below. The temperature at the base of the K_2CO_3 solution increases, forming a thermal boundary layer that grows with time. Eventually, the base of the K_2CO_3 solution is heated sufficiently to cause parcels of buoyant fluid to rise and initiate convection. However, in contrast to the one-component case, a parcel of fluid retains its K_2CO_3 content as it rises (due to its low diffusivity). In an ambient density gradient, a parcel of fluid can only rise a short distance before it reaches a level where the ambient density has the same value as that of the parcel. A layer of circulation is formed up to this level. This convecting layer grows while transferring some heat to the fluid above. With time this causes another convecting layer to form, and then another and another. In this way the potential energy available from the thermal input drives convection, despite the stable stratification in the K_2CO_3 solution. The spacing of these layers is greater than that of the layers formed from sidewall heating because the thermal energy is greater from the hot KNO_3 layer, thus providing more potential energy to create thicker layers.

The phenomena illustrated in Figure 1 are just two of many different examples of double-diffusive structures (Huppert & Turner 1981a), some of which are considered later. However, the principles discussed in this section are general: variations in properties with different diffusivities have a profound influence on convection. In the case of silicate melts, quite small changes in composition can produce substantial density changes. Furthermore, the major chemical components in silicate melts have very small diffusivities (typically in the range 10^{-6} to 10^{-10} cm^2 s^{-1}), whereas the thermal diffusivity ranges approximately from 10^{-2} to 10^{-3} cm^2 s^{-1}. Magmas are thus suitable fluids for the formation of double-diffusive structures.

The theoretical and experimental understanding of double-diffusive convection has been recently reviewed by Turner (1979, Chapter 8) and Huppert & Turner (1981a). The reader is referred to these articles for more detailed and comprehensive treatments. We now briefly review aspects of this work that are particularly relevant to geological systems. In a two-component system containing two properties with different molecular diffusivities, four different situations can be envisaged. In practice the two properties of interest are temperature (T) and composition (S). The four different possibilities are shown in Figure 2. One of these situations is absolutely stable (Figure 2*a*); in this case the temperature increases upward and the heavy chemical components are concentrated toward the base of the system. Another situation is unstable (Figure 2*b*); in this case the temperature decreases upward and the heavy chemical components are

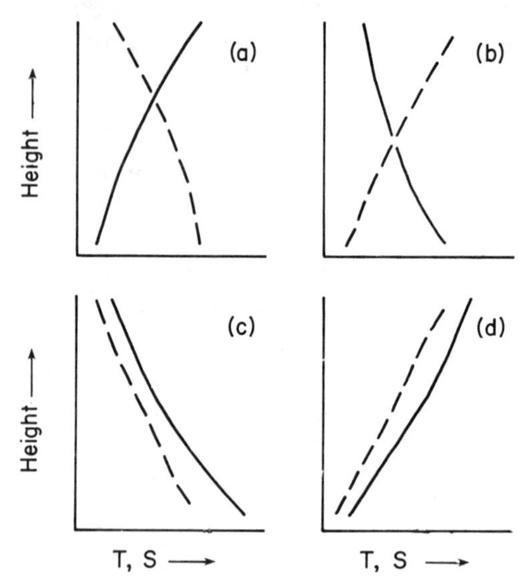

Figure 2 Sketches of the four different kinds of temperature and chemical composition that can occur through a fluid layer. Temperature (T) variation with depth in the layer is shown as a solid line, and the variation in the distribution of heavy chemical components (S) is shown as a broken line. (*a*) Both properties have stable distributions; (*b*) both properties have unstable distributions; (*c*) temperature distribution is destabilizing, and compositional distribution is stabilizing; (*d*) compositional distribution is destabilizing, and temperature distribution is stabilizing.

concentrated toward the top. The resulting convection in this situation is similar to that in a one-component system, with the compositional distribution enhancing the convection. The Rayleigh-number criterion can be used to predict when convection can begin. Equation (1) is modified to include the influence of compositional variations in density, such that

$$\mathrm{Ra}_e = \mathrm{Ra}_T + g\beta\Delta S d^3/\kappa_S \nu, \tag{3}$$

where β is the coefficient of expansion due to compositional change, $\rho_0\beta\Delta S$ is the (positive) difference in density due to compositional changes across a fluid layer of mean density ρ_0 and thickness d, and κ_S is the compositional diffusivity.

The two other possibilities of temperature and compositional gradients (Figures 2*c*, *d*) constitute the conditions for double-diffusive convection. In both cases the overall density is stable. In Figure 2*c* the system is unstable with respect to temperature (the temperature increases downward), but stable with respect to composition, since the heavy components are concentrated downward. This situation was illustrated in the experiment described earlier (Figure 1). It is known as the "diffusive" regime, and a series of horizontal layers can readily form in this situation. We believe it is the geologically most important situation, because there are many circumstances where hot magma can lie beneath colder magma of lower density. Some of these circumstances are discussed in the sections that follow.

The fourth situation (Figure 2*d*) involves a stable temperature distribution (temperature increasing with height) with an unstable compositional profile in which heavy components are concentrated upward. In this situation the phenomenon of "fingering" occurs, in which fluids from different levels of the system interpenetrate in long, thin convection cells (see Figures 1 and 3 in Huppert & Turner 1981a). Adjacent cells transfer their heat in the manner of a heat exchanger, with the energy for the motion coming from the unstable composition field. While such motions definitely occur in the ocean, where they are called "salt-fingers" and have been photographed at a depth of 1500 m in the North Atlantic (Williams 1974), so far this situation does not appear to be geologically important. This is because rather unusual circumstances have to be invoked to get hot, compositionally dense magma to overlie colder, compositionally less-dense magma. However, Kantha (1980) has suggested that some columnar jointing in lavas can be produced in this way, but there is little evidence to favor this explanation over contraction jointing.

At this point we outline some of the theoretical considerations governing double-diffusive convection. In a one-component system the motion can be specified by the thermal Rayleigh number (Equation 1) and the Prandtl number (Equation 2). In a two-component system two additional parameters are required, such as the compositional Rayleigh number[1]

$$\mathrm{Ra_S} = g\beta\Delta S d^3/\kappa_T \nu \tag{4}$$

and the ratio of the diffusivities

$$\tau = \kappa_S/\kappa_T, \tag{5}$$

which lies between 0 and 1. Frequently, it is convenient to use the ratio of the compositional to the thermal Rayleigh number

$$Q = \beta\Delta S/\alpha\Delta T \tag{6}$$

in place of the compositional Rayleigh number.

Linear stability theory applied to a single layer of fluid indicates that infinitesimal motions are initiated for $\mathrm{Pr} \gg 1$ and $\tau \ll 1$ when $\mathrm{Ra_T} - \mathrm{Ra_S}$ exceeds a critical value of order 10^3. However, nonlinear convection is known to be possible at lower values of $\mathrm{Ra_T} - \mathrm{Ra_S}$ (Huppert & Moore 1976), and Proctor (1981) has shown that in the limit of very small τ, nonlinear convection is possible if $\mathrm{Ra_T}$ alone exceeds a critical value of order 10^3. The minimum Rayleigh number needed to maintain a series of layers still remains to be found; but all laboratory experiments indicate that convecting layers easily form in double-diffusive systems.

Because much of the development of double-diffusive convection has been related to the oceans, many of the intuitive and quantitative results obtained are for parameters appropriate for salty water. Considerable caution is required in applying results in a quantitative way to geological systems. The Prandtl number and ratio of the diffusivities are different in the two cases, and the quantitative influences of these two parameters have not been evaluated for all possible situations. For example, the thickness of the thinner layers in Figure 1 is given by (Huppert & Turner 1980)

$$h = 0.65\Delta_h \rho \bigg/ \frac{d\rho}{dz}, \tag{7}$$

where $\Delta_h \rho$ is the horizontal density difference between fluid at the wall and in the far interior, and $d\rho/dz$ is the vertical density gradient. The constant of proportionality should vary little, if at all, with variations in Pr and τ. However, the mean thickness $\bar{h}$ of the lowest layers in Figure 1 is given by (Huppert & Linden 1979)

$$\bar{h} \propto \kappa_T^{1/2}(g\, d\rho/dz)^{-1/4}, \tag{8}$$

[1] We use κ_T, rather than κ_S, in the denominator for convenience and historical reasons; one could use the single parameter $\mathrm{Ra_S}/\tau$, which would have κ_S in the denominator.

where the constant of proportionality for NaCl in water is 51. This value is likely to be strongly dependent on Pr and τ, and its value for magmatic values of Pr and τ is as yet undetermined.

3. CONVECTION IN CRYSTALLIZING FLUIDS

In the previous section we showed that once compositional and thermal gradients are established, double-diffusive effects can occur in any multicomponent fluid, but we left aside the question of how suitable gradients can be generated in geophysically interesting fluids. An important mechanism in the case of magmas is crystallization, which generally causes much greater changes in melt density than are caused by associated temperature changes (McBirney 1980, Sparks & Huppert 1984). Following studies by Chen & Turner (1980) and McBirney (1980), it has become evident that crystallization in a fluid can cause a wide range of novel and interesting convective effects, including the generation of compositional and thermal gradients in initially homogeneous fluids. The convective effects of crystallization are of fundamental importance to the understanding of magma genesis and the behavior of the Earth's core (Gubbins et al 1979, Loper & Roberts 1983), as well as being important in many industrial processes [see Vol. 2, No. 4 of *Physicochemical Hydrodynamics*, edited by Hurle & Jakeman (1981)].

The basic principles of convection due to crystallization are simple. Crystals growing from a multicomponent, saturated fluid selectively deplete fluid adjacent to the growing interface in either light or heavy components. The composition, and therefore the density of the fluid, is changed and convection can occur.

The effects of crystallization on convection have been studied using aqueous solutions of simple salts (Chen & Turner 1980, McBirney 1980, Turner & Gustafson 1981, Huppert & Turner, 1981b, Huppert et al 1982b, Kerr & Turner 1982). The solubility of most salts in water increases with temperature. Thus, cooling a saturated solution (for example, KNO_3) will cause crystallization and a decrease in the density of the residual solution (Huppert & Turner 1981b). The decrease in fluid density is a consequence of the removal of heavy components to form crystals. Purely thermal contributions to density, which would tend to increase the density of a cooling solution, are here dominated by the compositional contributions. In some systems, crystallization can cause density to increase in residual fluids as a result of selective removal of light components. Examples include the formation of ice from some aqueous solutions and the geologically important case of plagioclase feldspar crystallizing from basaltic magma (Sparks & Huppert 1984).

Laboratory Experiments

The study of convective phenomena in crystallizing solutions is a new field of fluid mechanics. Some of the physical and chemical effects are currently being elucidated by laboratory experiments on aqueous solutions. One type of experiment has involved crystallization of an initially homogeneous solution by cooling at a boundary. In the case of cooling along a vertical wall, a number of experimental studies have shown that when crystallization takes place on the vertical wall, so that heavy components are removed from the fluid, a boundary-layer flow of light solute-depleted fluid is generated (McBirney 1980, Turner 1980, Turner & Gustafson 1981). Light fluid accumulates at the top of the container to form a layer that is thermally and compositionally stratified. This stratified region grows at the expense of a convecting, well-mixed lower layer (Figure 3). Layering is formed in the stratified upper fluid by various double-diffusive processes, such as internally generated instabilities, cooling across a sidewall, and heating by the underlying homogeneous fluid. Eventually the whole container can become stratified and layered. The structure of the boundary layer is strongly influenced by double-diffusive effects because the low

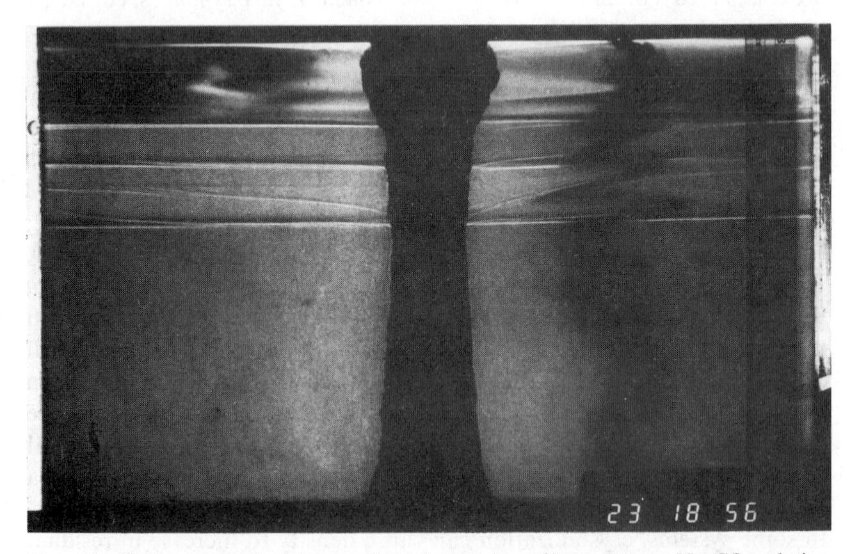

Figure 3 Shadowgraph view of the cooling of a container of homogeneous Na_2CO_3 solution at a vertical central pipe. After 23 hours 19 minutes, crystals have grown around the pipe, and a layer of compositionally stratified and depleted solution occupies the upper part of the tank. The temperature gradient imposed across the stratified region has resulted in double-diffusive layers forming. Details are given in Turner & Gustafson (1981).

diffusivity of chemical components enables the light, solute-depleted fluid generated by crystallization to retain its composition.

When crystallization occurs by cooling the floor of a container, convection can also occur, even though the temperature gradient is stable. Once crystals form they are observed to generate thin streams of uprising depleted fluid that drive convective circulation (Chen & Turner 1980). Such experiments reveal some interesting features of compositional convection driven by crystallization. The vertical streamers are typically very thin (<0.1 cm), and the flow is usually laminar. Consequently, such convection produces rapid vertical exchange but rather slow lateral mixing.

A variety of experiments have been performed to investigate additional interesting effects. For example, Kerr & Turner (1982) showed that the crystal layers in the solidified product resulted from fluid layers in a container, although the compositions and thicknesses of the fluid and crystal layers were different from one another. Tait et al (1984) have shown that compositional convection can occur during crystallization in a porous bed of glass balls. We have also carried out as yet unpublished experiments (with J. S. Turner) on crystallization along tilted boundaries. Compositional gradients are formed as a consequence, because light fluid rising from different positions along the boundary mixes with different amounts of overlying fluid.

A different kind of experiment involves replenishment of a container with fluid of another, dissimilar composition and temperature. These experiments were designed to investigate fluid-dynamic aspects of open-system magma chambers, for which there is an influx of primitive magma into a chamber containing more-differentiated magma (see Huppert & Sparks 1980a,b). In the experiments (Huppert & Turner 1981b), a layer of hot KNO_3 solution was emplaced beneath a layer of cold and lower-density $NaNO_3$ solution. A sharp double-diffusive interface was formed between the layers, across which heat was transferred. Convective cooling of the lower layer resulted in crystals of KNO_3 growing in the interior and on the bottom of the container. Compositional convection augmented the effects of thermal convection, maintaining a well-mixed lower layer. The density of the residual fluid decreased with time and eventually reached the same density as the overlying $NaNO_3$ solution. Thereafter, the residual fluid ascended rapidly into the overlying layer and mixed thoroughly with it.

This experiment illustrates some important principles, which are relevant to a number of geological situations. Hot fluid can lie stably beneath cold fluid if its composition is such that it is denser. Such a situation can occur in nature when primitive dense magma from the mantle is emplaced into a chamber containing more-differentiated magma (Sparks et al 1980, Huppert & Sparks 1980a,b), when hot brines or sulfide-bearing hydrothermal solutions are discharged onto the seafloor (Turner & Gustafson 1978, Solomon & Walshe 1979), or when a layer of higher-temperature mantle of one composition underlies a mantle layer of another composition. In the case of magmas and brines, the cooling of such hot fluid layers can lead to crystallization or precipitation, respectively, and allow instabilities to develop, causing the fluid layers to mix. An important principle related to igneous processes is that fluid layers of different composition can coexist in the same container and evolve as chemically independent systems. We note that no crystals could have formed in the experiment described had the fluids been initially mixed together.

Huppert et al (1982b) modified this experiment in two ways. In the first variation, the hot KNO_3 solution was introduced beneath compositionally stratified K_2CO_3 solution, as already described (Figure 1). The lower layer of KNO_3 solution (see Figure 1) will eventually overturn owing to crystallization, but because of the vertical density gradient, it only mixes with the lower parts of the stratified fluid. In the second variation, hot KNO_3 solution was emplaced continuously and slowly into a chamber of cool $NaNO_3$ solution. Quench crystallization occurred around the inlet orifice. This released light residual fluid into the container, and a mound of interlocking crystals was formed.

A further modification of the replenishment studies has been to investigate the effects of viscosity variations (Huppert et al 1983). In these experiments the viscosity and density of the upper layer in the two-layer system were systematically varied by using combinations of glycerin, sugar, and water. When the viscosity ratio was large (upper : lower viscosity = 2700 : 1), crystallization took place along the interface, and light residual KNO_3 solution was continuously swept up into overlying viscous fluid (Figure 4). No abrupt overturn took place, and release of residual fluid began immediately. This behavior is believed to be due to viscous coupling at the interface, which enables the residual fluid from the lower layer to be incorporated into thermal plumes in the viscous fluid.

As plumes of light fluid ascended into the residual fluid, further crystallization of KNO_3 occurred. These crystals were initially carried up in the plumes but eventually fell out. The plumes ascended to the top of the tank and deposited a layer of light KNO_3 solution on top of the glycerin. The experiments indicate that viscosity differences could play an important role in the fluid dynamics of crystallizing systems, with interfacial effects becoming significant. They also illustrate the difficulty of mixing together fluids of strongly contrasted viscosity.

Theoretical Aspects

Much of the research on convection in crystallizing systems has been concerned with identifying new phenomena, and many of the processes observed in the experiments have yet to be characterized quantitatively. There have been only limited developments in the theoretical understanding of convection associated with crystal growth. Loper & Roberts (1983) have developed a theoretical model, which has yet to be tested by experiment. In industrial contexts, where the main interest is in eliminating convective effects (see Hurle & Jakeman 1981, Coriell et al 1980, Coriell & Sekerka 1981, Schaefer & Coriell 1982), some theoretical work has been completed concerning instabilities along single faces of crystals.

Major differences between compositional convection associated with growing crystals and thermal convection are the length scales of the flows and the critical conditions for instability. Some insight into these differences can be ascertained by first considering the development of a compositional boundary layer next to a growing crystal face. In a quiescent liquid, rejection or incorporation of solute results in an exponential concentration profile in a thin film adjacent to the growing crystal (Coriell et al 1980, Coriell & Sekerka 1981). This film has a characteristic thickness $\delta = \kappa_S/V$, where κ_S is the diffusion coefficient, and V is the growth velocity of the crystal face. An approximate approach to determining the stability of a

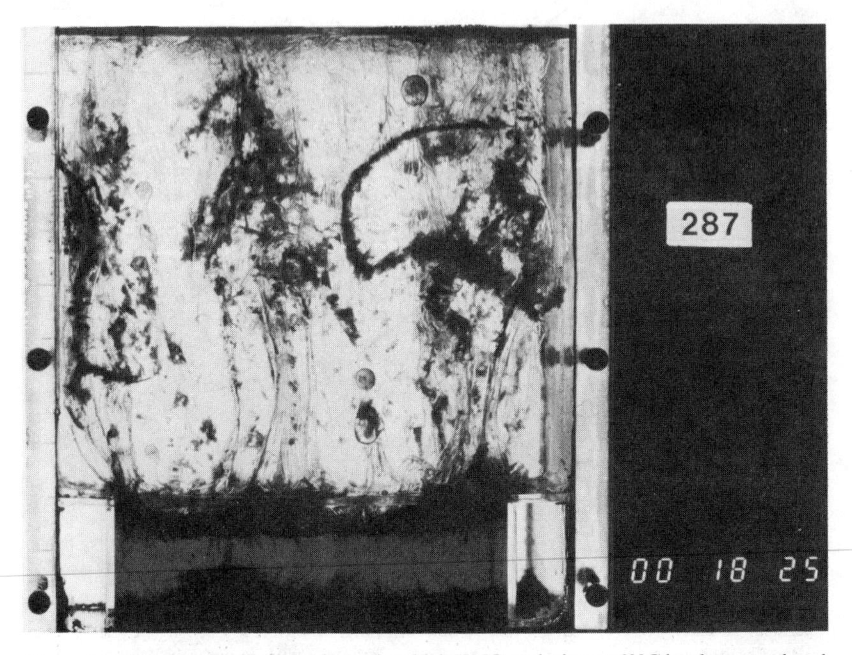

Figure 4 Photograph of an experiment in which KNO_3 solution at 60°C has been emplaced beneath glycerin at 11°C. Blobs and plumes of residual KNO_3 solution are generated at the interface and rise continuously into the glycerin. Crystals of KNO_3 form in the plumes as they cool, and the crystals, often in the form of a long chain, fall back. Residual KNO_3 solution forms a layer above the glycerin.

compositional film is to substitute the film thickness into the compositional Rayleigh number, which can be re-expressed as

$$\mathrm{Ra_S} = \frac{g \Delta \rho \kappa_S^2}{V^3 \mu}. \tag{9}$$

For a horizontal film above a crystal face (g vertical), the classic Rayleigh criterion would be $\mathrm{Ra_S} > 10^3$.

Figure 5 shows fields of stable films ($\mathrm{Ra_S} < 10^3$) and unstable films ($\mathrm{Ra_S} > 10^3$), for fluids of viscosities 10^2 and 10^6 poise and a value of $\Delta\rho = 0.01$ g cm^{-3}, on a plot of diffusion coefficient against crystal growth velocity. Growth velocities in the range 0.1–10 cm yr^{-1} are typical of slowly cooled magma chambers. Also shown are lines of constant κ_S/V. The diagrams illustrate that in many magmatic situations compositional convection should occur as crystallization proceeds.

A similar diagram can be constructed for the laboratory experiments

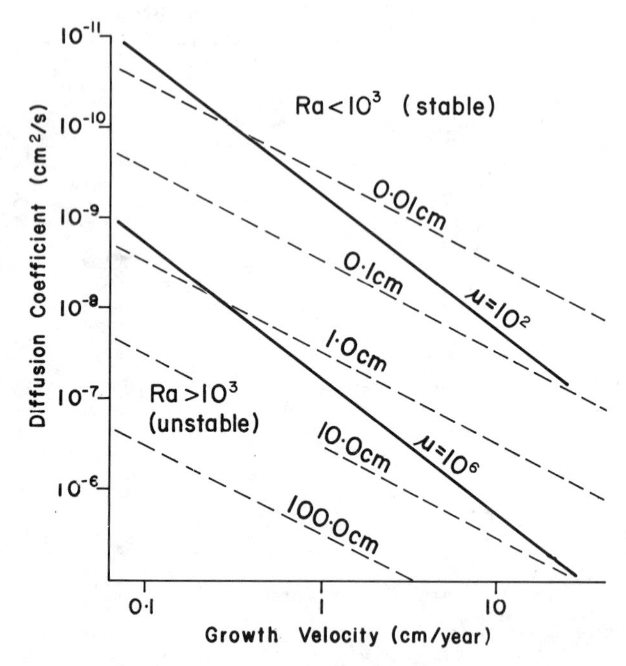

Figure 5 The solid lines in this diagram mark the value of $\mathrm{Ra_{(film)}} = 10^3$ as a function of chemical diffusivity and crystal growth velocity for magma with low viscosity ($\mu = 10^2$ poise) and with high viscosity ($\mu = 10^6$ poise). A value of $\Delta\rho = 0.01$ g cm^{-3} in Equation (9) is assumed. These lines separate conditions of stability ($\mathrm{Ra} < 10^3$) from conditions where compositional convection away from a horizontal crystal face is expected. The dashed lines are lines of constant film thickness (κ_S/V).

with aqueous solutions (Figure 6), where crystal growth rates are in the range 10^{-3} to 10^{-5} cm s^{-1}, viscosities are of the order 10^{-2} poise, and diffusivities are of the order 10^{-5} cm^2 s^{-1}. In all the experimental studies referred to previously, horizontal crystal faces should generate unstable compositional boundary layers. From this point of view, a KNO_3 crystal face growing at 10^{-4} cm s^{-1} ($D = 10^{-5}$ cm^2 s^{-1} and $\mu = 0.01$ poise) would be dynamically similar to an anhydrous crystal growing at 0.66 cm yr^{-1} in a wet rhyolitic magma ($D \approx 10^{-7}$ cm^2 s^{-1} and $\mu = 10^6$ poise). In the laboratory experiments the compositional film would become unstable at $\delta \approx 0.01$ cm. This is broadly consistent with the plumes that are observed to rise from growing crystals, which are certainly considerably thinner than a millimeter. We note that a horizontal crystal face represents the most stable configuration and thus the worst case for assessing whether convection will occur. For vertical crystal faces, no stable condition exists. Consequently, there will always be some tendency for the compositional film to detach from the crystal surface.

Unfortunately, an analysis of the dynamic development of a compositional boundary layer next to growing crystals has not yet been completed. There are already, however, some indications of the nature of such boundary layers and the problems likely to be encountered with the

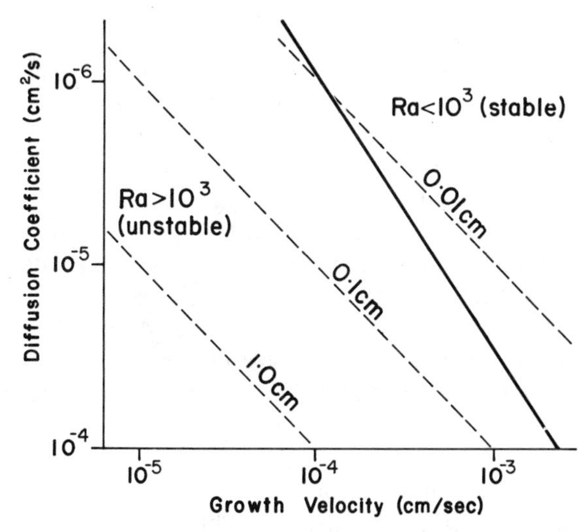

Figure 6 The solid line marks the position of conditions where $Ra_{(film)} = 10^3$ on a plot of chemical diffusivity vs crystal growth velocity. The fluid in this case has a viscosity of 0.01 poise, a typical value for the aqueous solutions used in the laboratory experiments. The value of $\Delta\rho$ in Equation (9) is taken as 0.01 g cm^{-3}. The laboratory experiments all fall in the field of $Ra > 10^3$.

analysis. The stagnant-film model described above and experimental observations indicate that the flows will be thin. In both the laboratory experiments and in magmas, the flow will be laminar. One problem is that the boundary layers are often much thinner than the associated crystals. Consequently the flow will be complicated by the geometrical irregularities of the surrounding solid boundaries.

These features suggest some interesting possibilities for experimental studies of silicate systems. Because of the small length scales necessary for compositional convection, these effects should be noticeable in small capsules, including some currently used by experimental petrologists. Dowty & Berkebile (1982), for example, have reported strong gradients in melt composition during crystallization experiments in capsules of 1 cm diameter. Although the authors attribute the gradients to crystal settling, there is no evidence to favor this interpretation over one involving convection. Kirkpatrick et al (1981) have also identified compositional gradients adjacent to olivine crystals grown in a melt of diopside composition. The effects of compositional convection should be identifiable in silicate melts, and suitably designed experiments would provide a most interesting approach.

4. CONTROLS ON MAGMA DENSITY

From the previous discussions on convection it is clearly important to establish the relative contributions of composition and temperature to the density of magmas, particularly during crystallization and melting processes. There have been a large number of determinations of the densities of silicate melts, and it is now possible to predict the density of a melt of given composition and temperature with an accuracy generally better than 1% (Nelson & Carmichael 1979, Bottinga et al 1982). These studies reveal that small changes in melt composition can have large effects on density.

Recent studies (McBirney 1980, Sparks et al 1980, Stolper & Walker 1980, Sparks & Huppert 1984) have shown that during fractional crystallization of basaltic magmas the compositional changes during an increment of fractionation are generally much larger than associated thermal effects. The quantitative aspects were considered by Sparks & Huppert (1984), who introduced a parameter called the *fractionation density*, which is defined as the density of the fluid components being selectively removed by crystallization. The fractionation density is thus

$$\rho_c = \frac{\hat{M}_c}{\hat{V}_c}, \tag{10}$$

where $\hat{M}_c$ is the gram formula weight, and $\hat{V}_c$ is the partial molar volume of the components being removed from the melt into crystals. When ρ_c is greater than the melt density, the residual melt decreases in density. When ρ_c is less than the melt density, the residual melt increases in density.

Figure 7 plots the fractionation density against crystal composition of some common minerals formed during fractionation of basaltic magma. In the plot, the composition of the crystals is expressed in terms of the two main end members of the solid solution series (for example, A = forsterite and B = fayalite for olivine). The upper and lower density lines for each mineral represent temperatures of 1200 and 1400°C, respectively. The typical ranges of densities for basaltic melts are shown. The diagram demonstrates that in general the effects on density of fractionating minerals are much greater than are those of temperature.

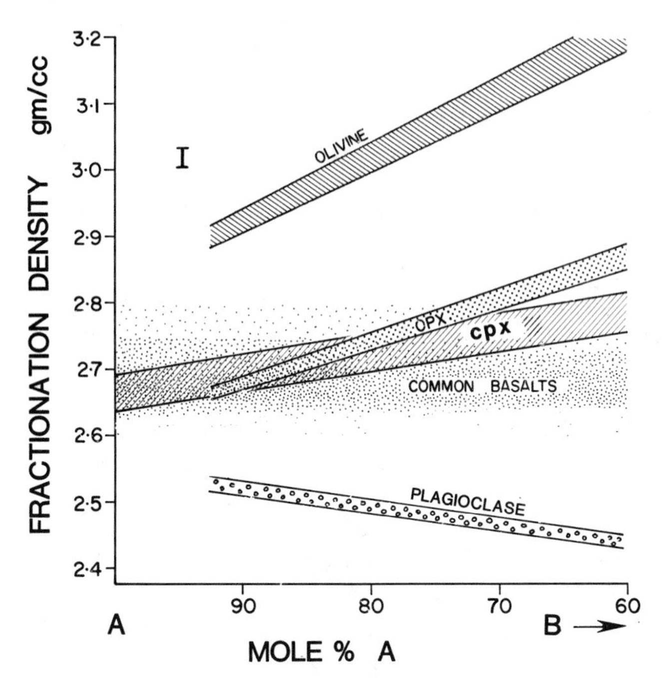

Figure 7 Fractionation densities of olivine, orthopyroxene, clinopyroxene, and plagioclase plotted against composition. The components A and B are separately defined for each mineral: olivine, $A = Mg_2SiO_4$ and $B = Fe_2SiO_4$; orthopyroxene, $A = MgSiO_3$ and $B = FeSiO_3$; clinopyroxene, $A = CaMgSi_2O_6$ and $B = CaFeSi_3O_8$; plagioclase, $A = CaAl_2Si_2O_8$ and $B = NaAlSi_3O_8$. For each mineral, fractionation densities have been calculated at 1200°C (*upper line*) and at 1400°C (*lower line*). The density range of common basaltic magmas is indicated, and the 2σ error bar is shown in the top left-hand corner for typical density estimates.

Figure 8 shows a schematic plot of density vs a convenient compositional parameter (Mg/Mg + Fe) for basaltic magmas related to one another by fractional crystallization. The general shape of the curve for density vs composition is considered characteristic of dry basaltic systems and can be interpreted with the aid of Figure 7. For high values of (Mg/Mg + Fe) a high-temperature primitive basalt (marked *P* in Figure 8) fractionates olivine and/or pyroxenes. The fractionation density of mafic minerals is greater than basaltic melt, and $\beta\Delta S/\alpha\Delta T \gg 1$. Consequently, the residual melts decrease in density despite the temperature decrease. Once plagioclase joins other mafic phases in cotectic assemblages, the fractionation density can often become less than the basalt melt density. For example, olivine gabbro assemblages have fractionation densities in the range 2.62–2.64 g cm^{-3} (Sparks & Huppert 1984). Thus, the density of residual melts can increase, causing a density minimum marked *A* in Figure 8. Although both temperature and compositional changes contribute to the increase in density, compositional effects are usually dominant. When other dense phases join in (for example, magnetite), or the ferromagnesian phases become very iron-rich, the fractionation density can become larger than the melt density, causing a decrease in density and a density maximum *B*.

Some important principles are illustrated in these diagrams. First, compositional effects usually dominate over thermal effects in determining melt density. The same statement will be true of other processes involving crystal-melt interaction, such as partial melting and contamination. Thus melts can decrease in density during fractionation despite thermal effects. Second, when new phases enter during fractionation, there can be important changes in slope, including density maxima and minima. Although only basaltic systems have so far been considered in detail by this approach, we feel that these principles will apply to all magmatic systems. They imply that compositional effects dominate thermal effects on density during major igneous processes and that both double-diffusive and compositional effects will be important in convection.

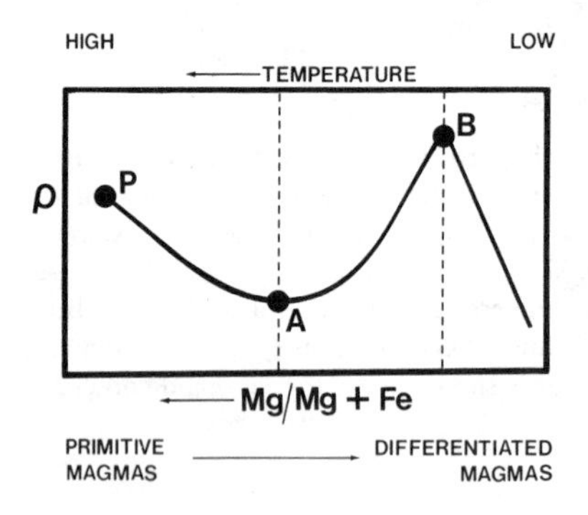

Figure 8 Schematic relationship between magma density and (Mg/Mg + Fe) for basaltic melts related to one another by fractional crystallization.

5. APPLICATIONS TO MAGMA GENESIS

Much of the recent interest in double-diffusive convection has focused on its role in the evolution of igneous rocks. In only a few years an extensive literature has developed. We briefly summarize the main ideas and applications that have emerged. Double-diffusive effects probably play a role in all major igneous processes.

Crystal Fractionation

There is now a general consensus among petrologists that the bulk of magmas are affected by crystal fractionation during their evolution. This process requires solid and liquid phases to be separated. The major mechanism for fractionation was thought until recently to be crystal settling (Bowen 1928, Wager & Brown 1967). There is, however, a good deal of evidence to suggest that crystal settling is an inadequate explanation. Some of the geological situations cited to support settling have been found suspect (see Campbell 1978, McBirney & Noyes 1979, Hildreth 1979). Theoretical arguments indicate that settling is opposed by convection in large magma chambers and that the fluid motions can keep crystals in suspension (Bartlett 1969, Sparks et al 1983).

Separation of liquid from crystals by compositional convection provides an alternative mechanism for fractionation (McBirney 1980). We have proposed the name *convective fractionation* for the process (Sparks et al 1984). This term [also used by Rice (1981)] embraces a wide variety of convective phenomena caused by crystallization, as already described from laboratory experiments on aqueous solutions. Below we summarize the main concepts and some speculations, which provide an alternative to settling. We do not have the space here to provide a detailed case for the new views, but instead we recommend that the reader consult the appropriate references.

In *closed-system* fractionation of magma, much of the crystallization will occur along the margins of the chamber, where the steepest temperature gradients and largest undercoolings occur and nuclei already exist (see Campbell 1978, McBirney & Noyes 1979). Crystals grown on the margins or in thin viscous boundary layers change the composition and therefore the density of the magma *locally*, and this melt can be convected away from its point of origin. The effects of this convection will depend strongly on the physical properties of the magma, on the fractionation density of the crystallizing phases, and on the chamber geometry (McBirney 1980, Sparks & Huppert 1984, Sparks et al 1984). In many circumstances, compositional and thermal gradients can be generated that are appropriate for the development of double-diffusive layering.

Some consequences and implications of this model for closed-system evolution of magmas are as follows:

1. Compositional and thermal gradients can be set up in magma chambers containing initially homogeneous magma. Gradients can be such that double-diffusive layering can develop. Eruption of parts of a chamber can produce volcanic rock sequences that are zoned and contain compositional discontinuities.
2. Once compositional gradients are established in a chamber, crystallization along the margins will lead to cumulates varying in composition with position. In mafic intrusions, rhythmic layering parallel to the margins will be discordant to phase layering parallel to gravitational stratification in the fluid (see Casey & Karson 1981, Wilson & Larsen 1982, Sparks et al 1984). Double-diffusive layering in the magma could also influence crystal layering in mafic intrusions (Kerr & Turner 1982, Irvine et al 1982).
3. Sidewall crystallization influences immediately only a small proportion of the total magma body, because selective removal of components occurs locally and affects only the magma in the thin compositional boundary layer. This can result in highly fractionated magmas accumulating at the top, or bottom, of a chamber without requiring large amounts of crystallization. Crystal-settling models often require large fractions of the magma chamber to crystallize in order to form highly differentiated melts (see Sparks et al 1984). Highly differentiated melts can form at the top of a chamber at an early stage by sidewall crystallization.
4. Once compositional gradients are established, double-diffusive layers can be formed, and each layer can subsequently evolve as a chemically independent system.
5. Compositional convection can occur in porous media. Tait et al (1984) have proposed that convection will play a major role in postcumulus crystallization. For example, they suggest that adcumulates can form when intercumulus crystallization produces light residual fluid that rises through the crystal pile and is continuously replaced by heavy magma from the overlying magma chamber.
6. The crystals that form cumulates and those that form phenocrysts in magmas may have different origins. Phenocrysts can form internally in a magma chamber at double-diffusive interfaces, where undercoolings and supersaturations are greatest. They can also form during the mixing of magmas (see below), by erosion of cumulates, and as residual crystals from the source region (restite). Once crystals of diverse origins are suspended in the magma, further growth can occur. [Sparks et al (1984) provide a fuller discussion of the origins of phenocrysts.] Phenocrysts, therefore, do not

necessarily provide accurate information on the proportions and compositions of fractionating phases.

Magma-Chamber Replenishment

There is now a great deal of evidence to support the idea that many magma chambers are open systems that are periodically or continuously replenished by new magma from depth. This is now perceived by many petrologists as important in controlling many petrological and geochemical features of igneous rocks (see, for example, Walker et al 1979, Eichelberger 1980, Huppert & Sparks 1980b, O'Hara & Matthews 1981, Sakuyama 1981). Double-diffusive phenomena are certain to occur during replenishment, because the compositions, temperatures, and densities of the new magma and of the magma already resident in the chamber are generally different. These compositional differences arise because the resident magma is more differentiated than the new magma.

A number of interesting convective processes due to crystallization and double-diffusive effects can occur during magma-chamber replenishment. Two fundamental situations can be considered, depending on the sign of the density difference between the new magma and the resident magma. In nearly all cases the incoming magma will have a higher temperature than the resident magma. The density contrast will, however, generally be attributable to the compositional difference between the magmas.

When the new magma is denser than the resident magma, the new magma will pond at the base of the chamber and form a layer with a double-diffusive interface beneath the overlying magma. This situation has been analyzed in some detail for the case of picritic magma emplaced into a basaltic magma chamber (Sparks et al 1980, Smewing 1981, Huppert & Sparks 1980a,b). The important features of the analysis are that heat is exchanged across the double-diffusive interface and that crystallization occurs in the basal magma layer. When the density of the residual melt has become equal to the density of the overlying magma, rapid mixing will occur between the magmas. The basic features of the model were demonstrated in laboratory experiments (Huppert & Turner 1981b, Huppert et al 1982b).

Input of dense mafic magma from the mantle is undoubtedly a fundamental feature of many magmatic plumbing systems. Overturn will not necessarily occur, particularly if the resident magma has a low density; for example, if basalt is emplaced beneath rhyolite, overturn cannot take place unless some other factor such as exsolution of abundant volatiles in the basaltic magma occurs (Eichelberger 1980, Huppert et al 1982a). Rice (1981) has postulated that when basalt is emplaced beneath rhyolite, layering develops at the interface and eventually overturn occurs. There is no experimental or theoretical justification for this model. Overturn only happens when the densities of adjacent magma layers become equal, and it is not clear how this is accomplished in Rice's model.

When the new magma is lighter than the resident magma, mixing can take place immediately. This situation can arise in the important case of replenishment of basaltic magma chambers with primitive magma, which is lighter than highly fractionated basaltic magma (Sparks et al 1980). The light magma will rise to form a turbulent plume, provided the Reynolds number is sufficiently large, resulting in mixing between the magmas. The hybrid magma will then rise to the top of the chamber to produce compositional stratification. Alternatively, if the Reynolds number is small, the new magma will rise without significant mixing to form a layer at the top of the chamber.

From the foregoing discussion it is evident that both compositional stratification and magma mixing can result from replenishment. Repeated replenishment can cause compositional stratification and thermal gradients suitable for double-diffusive convection. At the same time, crystallization can cause magma layers to change and mix with one another. These ideas have been used to explain the compositional uniformity of mid-ocean ridge basalts (Huppert & Sparks 1980a), the origin of cyclic layering in ultramafic intrusions (Huppert & Sparks 1980b), and the origin of chemical diversity in composite intrusions (Marshall & Sparks 1983). They have also been used to interpret mixing phenomena in calc-alkaline volcanic rocks (Eichelberger 1980, Huppert et al 1982a).

Compositional Zoning

The origin of compositional zonation in magma chambers has attracted considerable attention (Hildreth 1979, 1981). We have already indicated that crystallization along the margins of a chamber and replenishment are two mechanisms for generating zonation. Two other mechanisms could also be significant in nature. First, fractional partial melting can generate a succession of magmas of increasing density and temperature. If successive melts are fed into a chamber, zonation could result. Second, contamination of the magma along the walls of a chamber can produce zonation. For example, melting of the granitic walls can generate boundary-layer flows of low-density rhyolite that can rise to the top of the chamber. If the melting involves assimilation at the margins, the contaminated magma can differ in density from the uncontaminated magma and lead to compositional convection and stratification.

Much has been made in particular of the striking trace-element gradients to be found in some high-silica rhyolite magma chambers (Hildreth 1979,

1981). Diffusive effects have been invoked to explain these gradients, and a rather complicated model, termed *convection-driven thermogravitational diffusion*, has been invoked to explain these gradients (Shaw et al 1976, Hildreth 1981). The model involves ascent of water-rich boundary layers along the sides of the chamber to form a stratified region at the top. The compositional gradients are then considered to originate by Soret diffusion due to the temperature gradient imposed across the stratified region and to the volatile fluxing from beneath. The boundary layers are considered to form by diffusion of water into the magma from outside.

Although Soret diffusion should not be dismissed as a differentiation mechanism (Walker & De Long 1982, Schott 1983), there is no clear evidence to support its operation on a large scale. While it may be that Soret effects could become significant in a stratified magma chamber containing convecting double-diffusive layers (Schott 1983), experiments on Soret differentiation in rhyolitic melts produced gradients opposite in sign to those observed in nature (Lesher et al 1982). We suggest that fractional crystallization by sidewall accretion and convection coupled with volatile fluxing is a simpler explanation for the gradients in high-silica rhyolites. The principal objection to the role of crystal fractionation (Hildreth 1981) is that large amounts of crystallization are required to generate the rhyolitic magmas with extreme trace-element concentrations. This problem does not arise in a sidewall crystallization model, as highly fractionated fluids can be generated in the boundary-layer flows. A fuller discussion of these issues can be found in the articles by Hildreth (1981), Michael (1983), Schott (1983), and Sparks et al (1984).

Magmatic Ore Deposits

The formation of some types of magmatic ore deposits could be influenced by double-diffusive effects. Chromitites appear to form during replenishment of magma chambers, and mixing processes can be invoked to account for their formation (Irvine 1977, Huppert & Sparks 1980b, Sparks & Huppert 1984). Campbell et al (1983) have suggested that platinum sulfide horizons in layered intrusions are generated when primitive magma is emplaced into a magma chamber. Mixing of large volumes of the resident magma with the new magma leads to the precipitation of platinum sulfides.

In silicic systems, highly differentiated magma, enriched in volatiles and incompatible elements, can accumulate at the top of a chamber by sidewall crystallization. The evolution of a vapor phase from such highly evolved magma could lead to mineralization events. Tin mineralization in the roof zones of some granitoids (see, for example, Groves & McCarthy 1978) might provide an example of this process. A new aspect of the mineralization envisaged here is that "late-stage" fluids may be formed early in the evolution of some magma chambers. Mineralization of the roof zones could occur from an evolved stratified region at the top of the chamber while a substantial fraction of the chamber still contains undifferentiated magma.

6. OTHER GEOPHYSICAL APPLICATIONS

There are other important examples of natural multicomponent fluids in which double-diffusive effects can be anticipated. The solidification of the inner core of the Earth from the molten outer core involves growth of metallic iron-nickel crystals and rejection of light components such as silicon. It has been argued that compositional convection resulting from crystallization will be much more important than thermal convection in driving fluid motions in the outer core and generating the Earth's magnetic field (Gubbins et al 1979, Loper & Roberts 1983).

Many hydrothermal fluids cool and crystallize as they ascend to the Earth's surface, and double-diffusive effects should be expected. The discharge of hot brines onto the seafloor can be cited as one situation where double-diffusive phenomena occur (Turner & Gustafson 1978). Pore fluids in sedimentary rocks vary in both composition and temperature and could develop gradients appropriate for double-diffusive convection. Evaporites also involve crystallization from complex aqueous solutions, and double-diffusive phenomena should occur.

Perhaps the largest-scale example of double-diffusive convection is to be found in the dynamic behavior of the Earth's mantle. Richter (1979) has suggested that layering in the mantle may have originated by the heating of a chemical gradient from below. Compositional variations occur in the oceanic lithosphere, and when subduction occurs, cold dense parts of the lithosphere can become lighter than the surrounding mantle when they are heated (see Ringwood 1983). Partial melting could also play a role analogous to crystallization in magmas, because large-density changes occur when melting takes place.

Many of the early developments in the study of double-diffusive convection were motivated by oceanographical considerations. There is a large primary literature and a number of reviews devoted to this application, many of which are cited in Huppert & Turner (1981a). It would upset the balance of this article to discuss the oceanographic situation in detail. It suffices to say that double-diffusive effects of both the diffusive type (Figure 2*c*) and the fingering type (Figure 2*d*) are known to play a large role in many different parts of the world's oceans. The influence is felt over sizes ranging from the small scale of naturally occurring salt-fingers to the scale of fronts. In all cases, double-diffusive effects are undoubtedly important, but accurate quantitative determinations of the extent to which these phenomena affect oceanic mixing are not yet available. Once known, the influence of such mixing will have to be assessed in comparison with internal wave breaking, turbulence, and other effects.

7. CONCLUSIONS

Multicomponent fluid systems have the potential for developing double-diffusive convection. In the case of magmas and probably many other geophysically significant fluids, changes in chemical composition have much greater effects on density than associated temperature changes. In addition, natural fluids cool as they approach the Earth's surface. Thus, unstable temperature gradients and stable compositional gradients should be common. Crystallization of multicomponent fluids also leads to a variety of important convection effects. In particular, stable compositional gradients can often be formed in initially homogeneous fluid systems. Experimental and theoretical evidence suggests that double-diffusive convection should be important in magmas. The application of these principles to a number of geological situations has already led to new insights into the origin of igneous rocks.

ACKNOWLEDGMENTS

We wish to thank I. H. Campbell, A. R. McBirney, S. A. Morse, J. S. Turner, and J. A. Whitehead for helpful comments on an earlier version of this review.

Literature Cited

Bartlett, R. W. 1969. Magma convection, temperature distribution, and differentiation. *Am. J. Sci.* 267:1067–82

Bottinga, Y., Weill, D. F., Richet, P. 1982. Density calculations for silicate liquids. I. Revised method for aluminosilicate compositions. *Geochim. Cosmochim. Acta* 46: 909–20

Bowen, N. L. 1928. *The Evolution of Igneous Rocks.* Princeton, N.J.: Princeton Univ. Press. 332 pp.

Campbell, I. H. 1978. Some problems with the cumulus theory. *Lithos* 11:311–21

Campbell, I. H., Naldrett, A. J., Barnes, S. J. 1983. A model for the origin of the platinum-rich sulfide horizons in the Bushveld and Stillwater Complexes. *J. Petrol.* 24:133–65

Casey, J. F., Karson, J. A. 1981. Magma chamber profiles from the Bay of Islands ophiolite complex. *Nature* 292:295–301

Chen, C. F., Turner, J. S. 1980. Crystallization in a double-diffusive system. *J. Geophys. Res.* 85:2573–93

Coriell, S. R., Sekerka, R. F. 1981. Effect of convective flow on morphological stability. *PhysicoChem. Hydrodyn.* 2: 281–95

Coriell, S. R., Cordes, M. R., Boettinger, W. J., Sekerka, R. F. 1980. Convective and interfacial instabilities during unidirectional solidification of a binary alloy. *J. Cryst. Growth* 49:13–28

Dowty, E., Berkebile, C. A. 1982. Differentiation and diffusion in laboratory charges in basaltic composition during melting experiments. *Am. Mineral.* 27:900–6

Eichelberger, J. C. 1980. Vesiculation of mafic magma during replenishment of silicic magma reservoirs. *Nature* 288:446–50

Groves, D. I., McCarthy, T. S. 1978. Fractional crystallization and the origin of tin deposits in granitoids. *Miner. Deposita* 13: 11–26

Gubbins, D., Masters, T. G., Jacobs, J. A. 1979. Thermal evolution of the Earth's core. *Geophys. J. R. Astron. Soc.* 59:57–100

Hildreth, W. 1979. The Bishop Tuff:

Evidence for the origin of compositional zonation in silicic magma chambers. *Geol. Soc. Am. Spec. Pap. No. 180*, pp. 43–75

Hildreth, W. 1981. Gradients in silicic magma chambers: implications for lithospheric magmatism. *J. Geophys. Res.* 86: 10153–92

Huppert, H. E., Linden, P. F. 1979. On heating a stable salinity gradient from below. *J. Fluid Mech.* 95:431–64

Huppert, H. E., Moore, D. R. 1976. Nonlinear double-diffusive convection. *J. Fluid Mech.* 78:821–54

Huppert, H. E., Sparks, R. S. J. 1980a. Restrictions on the compositions of mid-ocean ridge basalts: a fluid dynamical investigation. *Nature* 286:46–48

Huppert, H. E., Sparks, R. S. J. 1980b. The fluid dynamics of a basaltic magma chamber replenished by influx of hot, dense ultrabasic magma. *Contrib. Mineral. Petrol.* 75:279–89

Huppert, H. E., Turner, J. S. 1980. Ice blocks melting into a salinity gradient. *J. Fluid Mech.* 100:367–84

Huppert, H. E., Turner, J. S. 1981a. Double-diffusive convection. *J. Fluid Mech.* 106: 299–329

Huppert, H. E., Turner, J. S. 1981b. A laboratory model of a replenished magma chamber. *Earth Planet. Sci. Lett.* 54:144–72

Huppert, H. E., Sparks, R. S. J., Turner, J. S. 1982a. Effects of volatiles on mixing in calcalkaline magma systems. *Nature* 297: 554–57

Huppert, H. E., Turner, J. S., Sparks, R. S. J. 1982b. Replenished magma chambers: effects of compositional zonation and input rates. *Earth Planet. Sci. Lett.* 57:345–57

Huppert, H. E., Sparks, R. S. J., Turner, J. S. 1983. Laboratory investigations of viscous effects in replenished magma chambers. *Earth Planet. Sci. Lett.* 65: In press

Hurle, D. T. J., Jakeman, E. 1981. *Physico-Chem. Hydrodyn.* 2:237–44

Irvine, T. N. 1977. Origin of chromitite layers in the Muskox intrusion and other layered intrusions: a new interpretation. *Geology* 5:273–77

Irvine, T. N., Keith, D. W., Todd, S. G. 1982. Formation of the Stillwater J-M Reef by concurrent mixing in a stratified magma body. *Carnegie Inst. Washington Yearb.* 81:286–94

Kantha, L. H. 1980. A note on the effect of viscosity on double-diffusive processes. *J. Geophys. Res.* 85:4398–4404

Kerr, R., Turner, J. S. 1982. Layered convection and crystal layers in multicomponent systems. *Nature* 298:231–33

Kirkpatrick, R. J., Kuo, L. C., Melchior, J. 1981. Crystal growth in incongruently-melting compositions: programmed cooling experiments with diopside. *Am. Mineral* 66:223–41

Lesher, L. E., Walker, D., Candela, P., Jays, J. F. 1982. Soret fractionation of natural silicate melts of intermediate to silicic composition. *Geol. Soc. Am. Abstr. with Programs* 14:545

Loper, D. E., Roberts, P. H. 1983. Compositional convection and the gravitationally powered dynamo. In *Stellar and Planetary Magnetism*, ed. A. M. Soward, pp. 297–327. Gordon & Breach

Marshall, L., Sparks, R. S. J. 1984. Origin of some mixed magma and net-veined ring intrusions. *J. Geol. Soc. London.* In press

McBirney, A. R. 1980. Mixing and unmixing of magmas. *J. Volcanol. Geotherm. Res.* 7:357–71

McBirney, A. R. Noyes, R. M. 1979. Crystallization and layering of the Skaergaard intrusion. *J. Petrol.* 20:487–554

Michael, P. J. 1983. Chemical differentiation of the Bishop Tuff and other high-silica magmas through crystallization processes. *Geology* 11:313–34

Nelson, S. A., Carmichael, I. S. E. 1979. Partial molar volume of oxide components in silicate liquids. *Contrib. Mineral. Petrol.* 71:117–24

O'Hara, M. J., Matthews, R. E. 1981. Geochemical evolution in an advancing, periodically replenished, periodically tapped, continuously fractionated magma chamber. *J. Geol. Soc. London* 138:237–78

Proctor, M. R. E. 1981. Steady sub-critical thermohaline convection. *J. Fluid Mech.* 105:507–21

Rice, A. R. 1981. Convective fractionation: A mechanism to provide cryptic zoning (macrosegregation), layering, crescumulates, banded tuffs and explosive volcanism in igneous processes. *J. Geophys. Res.* 86: 405–17

Richter, F. M. 1979. Focal mechanisms and seismic energy release of deep and intermediate earthquakes in the Tonga-Kermadec region and their bearing on the depth and extent of mantle flow. *J. Geophys. Res.* 84:6783–95

Ringwood, A. E. 1983. Phase transformations and differentiation in subducted lithosphere; implications for mantle dynamics, basalt petrogenesis and crustal evolution. *J. Geol.* In press

Sakuyama, M. 1981. Petrological study of the Myoko and Kurohima volcanoes, Japan: crystallization sequence and evidence for magma mixing. *J. Petrol.* 22:553–83

Schaefer, R. J., Coriell, S. R. 1982. Convective and interfacial instabilities during solidification of succinonitrile containing ethanol. In *Materials Processing in the Reduced Gravity Environment of Space*, ed. G. E. Rindone, pp. 479–89. New York: Elsevier. 676 pp.

Schott, J. 1983. Thermal diffusion and magmatic differentiation: a new look at an old problem. *Bull. Mineral.* 106:247–62

Shaw, H. R. 1965. Comments on viscosity, crystal settling, and convection in granitic magmas. *Am. J. Sci.* 263:120–52

Shaw, H. R., Smith, R. L., Hildreth, W. 1976. Thermogravitational mechanisms for chemical variations in zoned magma chambers. *Geol. Soc. Am. Abstr. with Programs* 8:1102

Smewing, J. D. 1981. Mixing characteristics and compositional differences in mantle-derived melts beneath spreading axes: evidence from cyclically layered rocks in the ophiolite of North Oman. *J. Geophys. Res.* 86:2645–60

Solomon, M., Walshe, J. L. 1979. The formation of massive sulfide deposits on the sea-floor. *Econ. Geol.* 74:797–813

Sparks, R. S. J., Huppert, H. E. 1984. Density changes during the fractional crystallization of basaltic magmas: fluid dynamic implications. *Contrib. Mineral. Petrol.* In press

Sparks, R. S. J., Meyer, P., Sigurdsson, H. 1980. Density variation amongst mid-ocean ridge basalts: implications for magma mixing and the scarcity of primitive basalts. *Earth Planet. Sci. Lett.* 46: 419–30

Sparks, R. S. J., Huppert, H. E., Turner, J. S. 1984. The fluid dynamics of evolving magma chambers. *Philos. Trans. R. Soc. London.* In press

Stolper, E., Walker, D. 1980. Melt density and the average composition of basalt. *Contrib. Mineral. Petrol.* 74:7–12

Tait, S. R., Huppert, H. E., Sparks, R. S. J. 1984. The role of compositional convection in the formation of adcumulate rocks. *Lithos.* Submitted for publication

Turner, J. S. 1979. *Buoyancy Effects in Fluids.* Cambridge Univ. Press. 368 pp.

Turner, J. S. 1980. A fluid-dynamical model of differentiation and layering in magma chambers. *Nature* 285:213–15

Turner, J. S., Gustafson, L. B. 1978. The flow of hot saline solutions from vents in the sea floor: some implications for exhalative massive sulfide and other ore deposits. *Econ. Geol.* 73:1082–1100

Turner, J. S., Gustafson, L. B. 1981. Fluid motions and compositional gradients produced by crystallization or melting at vertical boundaries. *J. Volcanol. Geotherm. Res.* 11:93–125

Wager, L. R., Brown, G. M. 1967. *Layered Igneous Rocks.* San Francisco: Freeman. 588 pp.

Walker, D., De Long, S. E. 1982. Soret separation of mid-ocean ridge basalt magma. *Contrib. Mineral. Petrol.* 79:231–40

Walker, D., Shibata, T., De Long, S. E. 1979. Abyssal tholeiites from the Oceanographer Fracture Zone. II. Phase equilibria and mixing. *Contrib. Mineral. Petrol.* 70:111–25

Williams, A. J. 1974. Salt fingers observed in the Mediterranean outflow. *Science* 185: 941–43

Wilson, J. R., Larsen, S. B. 1982. Discordant layering relations in the Fongen-Hyllingen basic intrusion. *Nature* 299:625–26

Ann. Rev. Earth Planet. Sci. 1984. 12: 39–59

APPLICATIONS OF ACCELERATOR MASS SPECTROMETRY

Louis Brown

Department of Terrestrial Magnetism, Carnegie Institution of Washington, Washington, DC 20015

INTRODUCTION

Radioactive isotopes may generally be called rare compared with the stable isotopes of the same element without fear of misusing that adjective. Certainly their observation with a conventional mass spectrometer can, with few exceptions, be ruled out. Of course, it is with radiation detectors, not mass spectrometers, that they are observed, and their rarity has not kept them from finding generous application as tracers or clocks. A few such nuclides did not find application because of lifetimes so long as to preclude their observation through decay, except in special cases. Three such isotopes, ^{10}Be (1.5×10^6 yr half-life), ^{26}Al (7.2×10^5 yr), and ^{36}Cl (3.0×10^5 yr), are now routinely observable using accelerator mass spectrometry and are finding increasing use in earth and planetary sciences. Their use for radioactive dating, an idea that figured prominently in the papers reporting the first experiments, has been a small part of subsequent work, but their use as tracers has developed rapidly, with the result that this review covers areas unsuspected in 1977. The application to radiocarbon has proved arduous owing to the need to determine isotopic ratios to a precision of 1%, a restriction that has seldom hampered the method when applied to the other isotopes, which are generally measured to accuracies of 5 to 10%. The precision desired for radiocarbon, implying an age uncertainty of 83 yr, although trivial by the standards of low-energy mass spectrometry, has only recently been achieved for sustained periods of machine operation. The technique has been applied to research on the atmosphere, the cosmic-ray history found in Arctic ice, manganese nodules,

0084–6597/84/0515–0039$02.00

pelagic sediments, the evolution of soil and its erosion, the formation of continental margins, the dynamics of aquifers, the origin of tektites and impact glasses, the nature of island-arc volcanism, the dating of earthquakes, and the irradiation history of lunar and meteoritic material. The six years following the invention of the method have seen activity in refining the experimental methods, but the fundamentals are still adequately described in Litherland's (1980) review.

THE ISOTOPES AND THEIR PRODUCTION

None of the isotopes considered here remain from the synthesis of the primordial matter of the solar system, nor are they daughter products of radioactive primordial matter. All are produced by cosmic rays incident on the atmosphere, the Earth's and Moon's crust, and extraterrestrial matter. Carbon-14 and ^{36}Cl have an additional source, which may prove to be of consequence for the earth sciences: the production of the isotope in the atmospheric bomb tests circa 1960. Beryllium-10 and ^{36}Cl are also produced in the Earth's crust from natural radioactivity, from cosmic-ray neutrons, and from cosmic-ray-muon-produced neutrons, which imposes observational limits at some level.

An observer initiating a study of cosmic-ray-produced radioisotopes will invariably turn to the review of Lal & Peters (1967), a valuable study despite its age [a defect that can be amended by reference to Reedy et al (1983)]. Lal & Peters summarize the first calculations of production rates for ^{10}Be, ^{26}Al, and ^{36}Cl. These calculations depend on an accurate knowledge of production cross sections, and as Yiou & Raisbeck (1972) have emphasized, they especially depend on the values of the as yet unmeasured neutron-induced reactions. These reactions dominate production because the nucleon flux in the atmosphere is composed mostly of neutrons. The production of isotopes depends on the geomagnetic latitude because of the shielding that results from the Earth's magnetic field, a dependence calculated by Lal (1962) and reproduced in Lal & Peters (1967).

Several investigators have reported measurements of the production rate of ^{10}Be based on observations of rain, polar ice, and sediments. The results of these calculations and observations are given in Table 1; it is obvious that the agreement is bad. Table 2 collects data on ^{26}Al/^{10}Be ratios. Primary cosmic rays contain ^{10}Be, but the flux is some two orders of magnitude lower than what is produced in the atmosphere. Cosmic dust is not, as was one time thought, an important source of ^{26}Al, according to McCorkell et al (1967).

An important source of ^{36}Cl is the pulse of about 85 kg of the isotope injected into the atmosphere by seven marine nuclear explosions in 1956–

Table 1 Calculations and measurements of ^{10}Be deposition rates for production in the atmosphere in units of atoms cm^{-2} s^{-1}

Investigator	Kind of data	Local deposition (raw data)	Global deposition (latitude correction)
Amin et al (1972, 1975)	Sediments	—	0.018
Reyss et al (1981)	Calculated global	—	0.021
Raisbeck et al (1979b, 1981c)	French rain	0.055	0.042
Stensland et al (1983)	Illinois rain	0.040	0.025
McCorkell et al (1967)	Greenland ice	0.020	0.060
Beer et al (1983)	Greenland ice	not reported	0.016
Tanaka & Inoue (1979, 1980, Tanaka et al 1977)	7 Pacific sediment cores	—	0.033
Somayajulu (1977)	7 Pacific sediment cores	—	0.015
Finkel et al (1977)	5 Pacific-Indian sediment cores	—	0.020
Raisbeck et al (1981c)	5 Indian sediment cores	—	0.060
Wahlen et al (1983)	New York lake core	0.015	0.011

58. The first observations of ^{36}Cl in rain by Schaeffer et al (1960) were primarily bomb produced and were recognized as such. The pulse was measured by Elmore et al (1982) in a 100-m Greenland ice core, where the concentrations reached a peak in deposits during the late-1950s that was two orders of magnitude greater than that produced by cosmic rays.

Table 2 Calculations and measurements of $^{26}Al/^{10}Be$[a]

Investigator	Kind of data	Ratio
Lal & Peters (1967)	Calculation	0.0031
Reyss et al (1981)	Calculation	0.0052
Raisbeck et al (1983a)	Air filters	0.0038
McCorkell et al (1967)	Greenland ice	0.0084
Tanaka et al (1968)	Pacific sediments	<0.014
Reyss et al (1976)	Pacific sediments	0.0084
Reyss & Yokoyama (1976)	Manganese nodules	0.0030
Guichard et al (1978)	Manganese nodules	0.0067
Raisbeck et al (personal communication)	Manganese nodules	0.00036

[a] In comparing these values with original literature, bear in mind that they are the ratios of atoms, not of activities, and that before 1972 the accepted half-life of ^{10}Be was 2.7×10^6 yr rather than 1.5×10^6 yr (see Yiou & Raisbeck 1972).

The ^{36}Cl bomb pulse was also observed in natural water by Bentley et al (1982). During their study they evaluated the ^{36}Cl content of 12 prebomb water samples and deduced an average deposition rate for the isotope of 0.0016 ± 0.0004 atoms cm^{-2} s^{-1}. No similar pulse has been observed for ^{10}Be. Aluminum-26 cannot be synthesized by thermal neutron absorption.

No isotopes with mass greater than 40 are produced in the atmosphere in quantities sufficient to occupy us here, owing to the lack of sufficiently heavy target atoms. Heavier isotopes are produced by direct action of cosmic rays on solid matter, such as meteorites, lunar rocks and regolith, and terrestrial rocks and soil at high altitude. For details of these calculations, see Yokoyama et al (1977) and references therein.

PRINCIPLES OF APPLICATION

The efficiency with which the ion source and accelerator make use of the supplied sample material is the most important parameter in the design of an experiment. The efficiency of sample use is the number of atoms of the element that can be transmitted through the machine and detected divided by the amount loaded into the ion source. This quantity is seldom measured directly by experimentalists, but they are nevertheless constrained by its value. The efficiency of forming the ions required by the accelerator varies according to the ion desired and the construction of the source. C, Al, and Cl are best obtained as negative atomic ions and Be as the oxide ion BeO^-. Nearly all the work done so far has employed ion sources of the Middleton (1977, 1983) design that make use of Cs both as an electron donor and as a sputter ion to form copious numbers of atoms or molecules to receive the electron. The overall efficiency is the product of the ionizer efficiency, the accelerator transmission, and the detector efficiency. As an example, assume one wishes to measure a ratio of 5×10^{-15} with an observation of 100 atoms of ^{10}Be, assuming an overall efficiency of 10^{-3}. This requires about 1 mg of BeO, with no allowance for counting losses resulting from beam warm-up and switching. Completion of the measurement in an hour means a $^{9}Be^{3+}$ current of about 3 μA.

The chemical form of the sample must provide a high degree of ionization and must be in solid form, and for ^{10}Be and ^{36}Cl the chemical suppression of B and S to levels low enough for the detector to reject the isobars ^{10}B and ^{36}S is necessary. The lifetimes of the ions $^{14}N^-$ and $^{26}Mg^-$ are so short that none reach the accelerator high-voltage terminal and hence do not interfere in ^{14}C and ^{26}Al work. This is the reason $^{26}AlO^-$ cannot be used, as $^{26}MgO^-$ would overwhelm the detector with ^{26}Mg, regardless of the chemical purity of the load, because Mg has a lower atomic number and hence lower stopping power than Al; this is the opposite relationship to the one between Be and B. Much experimentation is still needed to determine the best loading for ^{14}C, the choice depending to some extent on the chemical composition of the original sample: graphite and metal carbides are favored, carbonates are not. All measurements of ^{10}Be have used the method of isotope dilution, a consequence of the rarity of ^{9}Be in nearly all natural materials. The natural isotopic ratios of ^{14}C, ^{26}Al, and ^{36}Cl are so small in terrestrial samples that isotope dilution finds limited application.

APPLICATIONS

The Atmosphere and Its Precipitation

About 75% of the production of ^{10}Be takes place in the stratosphere, where its residence time is of the order of a year, conforming to times generally ascribed to aerosols. Raisbeck et al (1981a) have measured its concentration in both stratospheric and tropospheric air, thereby beginning the study of this isotope's wanderings at the source. At latitude 65°N, airborne filters exposed at an altitude of 10.7 km yielded a concentration of 7×10^6 atoms m^{-3} (STP), increasing to 1.3×10^7 at 19.2 km. Tropospheric concentrations are more than two orders of magnitude lower, as expected from the much shorter residence time for tropospheric aerosols. After Middleton succeeded in producing copious beams of Al^- from Al_2O_3, Raisbeck et al (1983a) measured the concentration of ^{26}Al at altitudes of 15.0 and 19.2 km, obtaining $^{26}Al/^{10}Be$ ratios of 0.0040 and 0.0033, respectively, to be compared with Lal & Peter's calculated value of 0.0031. There is no report of a ^{36}Cl measurement from the atmosphere.

Aerosols are generally thought to be incident on the Earth's surface in rain and snow, so Raisbeck et al (1979b, 1981c) began collecting precipitation in an open container at Orsay in 1978 and have reported concentrations of 2.5×10^4 and 2.0×10^4 atoms g^{-1} for that year and 1980. Stensland et al (1983) measured the concentration in individual rainstorms, sampling with a wet-only collector, which was opened only during the rainfall; despite this precaution, analysis for various ions showed the presence of soil, which has been observed to have ^{10}Be concentrations from 10^7 to 10^9 atoms g^{-1}. There is an evident danger of including recycled ^{10}Be when sampling rain at land stations, especially if the collector remains continuously open. Finkel et al (1980) report a measurement of 1800 atoms g^{-1} of ^{36}Cl in rainwater at La Jolla, California, and Bentley et al (1982) report 2800 atoms g^{-1} in rainfall at Tucson, Arizona. There are no reports of ^{26}Al measurements in rain, but McCorkell et al (1967) found a concentration of 175 atoms g^{-1} in Greenland ice.

The invention of the accelerator method immediately gave rise to proposals to measure the concentrations of ^{10}Be and ^{36}Cl in deep cores of polar ice, the frozen records of past precipitations. Raisbeck et al (1978, 1981b) began with Antarctic ice obtained from Terres Australes et Antarctiques Françaises, culminating in measurements on a 906-m core taken from Dome C (74°39′S, 124°10′E). A joint American-Danish-Swiss coring in Greenland at Dye 3 (65°11′N, 43°50′W) produced a 1930-m sample, for which Beer et al (1983) reported the ^{10}Be. The immense superiority of the accelerator method can best be seen by comparing these measurements, each of which required a kilogram of ice, with those of McCorkell et al (1967), each of which required more than a thousand tons.

Of particular interest for several lines of study was the threefold increase in ^{10}Be concentrations found by the Orsay group for ice precipitated between 10^4 and 3×10^4 years ago, the time of the most recent ice age. Virtually the same effect is reported for the Greenland core. Oxygen-18 shows a corresponding decrease of about 0.6‰ for both cores, indicating a colder climate. Interpretations of these important data draw on change of climate, such as lowered precipitation or altered circulation patterns or both, or change of the production rate. The data of Dome C show a 2σ increase in ^{10}Be concentration at the time of the Maunder sunspot minimum from 1645 to 1715 A.D., which took place at the time of the "Little Ice Age"; this supports the interpretation that enhanced production was a characteristic of the ice age. There are, however, meteorological reasons for favoring lower precipitation as the reason.

Beer et al (1983) were also able to observe the effect of the 11-yr solar cycle on ^{10}Be production. This effect is masked for ^{14}C because of the slow exchange rate of atmospheric and oceanic CO_2, but is only slightly attenuated by the 1-yr residence time of ^{10}Be in the atmosphere, for which variations of about 60% were observed with a 1.5-yr phase lag.

Nishiizumi et al (1979, 1981, 1982b, 1983a) have measured the concentrations of ^{10}Be and ^{36}Cl in Antarctic ice near Allan Hills (76°45′S, 159°E). For ^{10}Be in modern ice they measure concentrations in the same range as those at Dome C. The ratio of $^{10}Be/^{36}Cl$ for five samples of old but undated ice gave 15.8 ± 2.8, with scatter only slightly more than the analytical error. This will certainly allow dating of ancient ice and has been employed once to ascertain relative ages over a spread of 4×10^5 yr.

The Deep Oceans and Their Deposits

It was in the pelagic ooze that Arnold (1956) first observed cosmic-ray-produced terrestrial ^{10}Be. The motivation for seeking it there is the extremely low sedimentation rates in the deep oceans, rates measured in mm kyr^{-1}, and concentrations were found in excess of 10^9 atoms g^{-1}. By processing kilograms of ooze, investigators could secure enough of the isotope to observe a few thousand decays per day with β-counters, and an extensive investigation of the distribution of ^{10}Be in the top few meters developed over the next two decades. Of particular note are papers by Amin et al (1975), Inoue & Tanaka (1976), Tanaka & Inoue (1979, 1980), Tanaka et al (1977), Somayajulu (1977), and Finkel et al (1977). Investigators using accelerators, who have been attracted to totally new kinds of application, have not yet approached the extent of data reported in these papers. Aluminum-26 was also sought in the deep oceans, but nearly a decade passed from the first report of its measurements in pelagic sediments by Tanaka et al (1968) until Reyss et al (1976) reported an observation that yielded an $^{26}Al/^{10}Be$ ratio in some conformity with what had been observed in Greenland ice and by calculation (see Table 2 for details).

There are advantages from various points of view in considering the dynamics of ^{10}Be in the deep ocean separately from that of the continents and their margins. The isotope's residence there can be conveniently studied in three reservoirs: a surface layer about 100-m thick, a deep-ocean region extending to the bottom, and a region of pelagic sediment. An additional, somewhat special reservoir is the manganese nodules and crusts found on the floor of some parts of the deep oceans.

In an early exploitation of the enhanced sensitivity of the accelerator method, Raisbeck et al (1979a, 1980) measured the concentration of ^{10}Be in three samples of surface ocean water, with an average of 750 atoms g^{-1}, and two samples taken at 800 and 3140 m, with an average of 2200 atoms g^{-1}. They note that the low concentration in the surface layer corresponds to similar observations of certain trace elements, which they attribute to particle scavanging; this leads to a residence time of about 16 yr. The deep-ocean value led to a value for the maximum residence time there of 630 yr, assuming a global deposition rate (to which residence time is inversely proportional) of 0.042 atoms cm^{-2} s^{-1}. Kusakabe et al (1982) have measured concentrations at three depths below the surface layer in the Drake Passage, for which they obtained an average of 1400 atoms g^{-1}, a value that may be low because of storage losses to the walls of polyethylene jugs in which the samples had remained unacidified for more than 10 yr. Krishnaswami et al (1982) measured the ^{10}Be concentration at 4100 m at the GEOSECS station (29°38′N, 121°29′W) to be 6100 atoms g^{-1}, which gives a deep-ocean residence time of 4100 yr, assuming a deposition rate of 0.018 atoms cm^{-2} s^{-1}. No measurements of ^{26}Al or ^{36}Cl in ocean water have been reported.

The ocean currents acting for periods equal to residence times would seem to ensure that the latitude effect in the deposition rate is averaged out or greatly altered by the time the ^{10}Be and ^{26}Al reach the bottom. This is borne out by examining the deposition rates obtained from individual deep-ocean cores of Tanaka & Inoue (1979), which have scatter that can be reconciled with a latitude effect only through the exercise of a healthy imagination (although Arctic Ocean concentrations are down an order of magnitude). Raisbeck et al (1981c) have suggested that because of particle scavanging the concentration of pelagic sediments is constant to a first approximation. An examination of the concentrations of the top few centimeters of 25 deep-ocean cores yields an average value of $4.9 \pm 2.0 \times 10^9$ atoms g^{-1} for dry sediment. If this hypothesis is true, then the deposition rate obtained for a given core will be proportional to the sedimentation rate modulated by local effects. Tanaka & Inoue (1979) demonstrate through extensive measurements of concentration as a function of depth that over the past 2.5 Myr the production rate has not deviated from the mean by more than 10% when averaged over periods of 0.2 Myr, nor by more than 30% when averaged over 0.1-Myr periods. With sedmentation as the clock, they also show the isotope's half-life to be 1.5 Myr.

Manganese nodules are ferromanganese oxide concretions that litter much of the deep-ocean floor. Since their discovery more than a century ago they have attracted the attention of generations of scientists without having disclosed the secret of their origins, although the mystery of how they escape burial by sedimentation 10^3 times faster than the nodule accretion rate may have been solved by Paul's (1976) photographs showing benthic organisms removing the sediment cover. Perhaps it is the inelegance of this explanation that has driven investigators to question the slow accretion rate determined from measurements of ^{10}Be as a function of depth. Somayajulu (1967) first observed the isotope in nodules, and Turekian et al (1979) first applied the accelerator method to them.

Two papers report measured ^{10}Be concentration profiles as deep as 20 mm in a total of five nodules or crusts, a feat that would be out of the question using β-counting. The results found in these two papers (Krishnaswami et al 1982, Ku et al 1982) effectively summarize what is known about ^{10}Be in nodules. Surface concentrations of these five objects, taken from widely separated locations in both the Atlantic and Pacific, fall between 3×10^{10} and 7×10^{10} atoms g^{-1}, with the natural ratios of $^{10}Be/^{9}Be$ between 0.9×10^{-7} and 1.5×10^{-7}. Growth rates agree whether taken from the concentration of ^{10}Be or from the natural isotopic ratio and fall between 1.4 and 2.0 mm Myr^{-1}. There is a suggestion of a faster accretion rate ending 7 to 9 Myr ago. Krishnaswami et al (1982) and Krishnaswami & Cochran (1978) review previous studies, discuss alternate interpretations for slow accretion, and consider what confirmation the Th- and U-decay series offer. The evidence indicates that the slow accretion rate is real.

It required almost a decade after the first observation of ^{10}Be in nodules before ^{26}Al was observed in one by Reyss & Yokoyama (1976). The observed ratio of ^{26}Al/^{10}Be is in rough agreement with values obtained elsewhere (see Table 2 for details). The growth rate for ^{26}Al agrees with that obtained from ^{10}Be, as best one can say for the three datum points obtained by Guichard et al (1978). However, recent measurements by Raisbeck, Yiou, Klein, Middleton & Somayajulu (1982, personal communication) using the accelerator method have found the ^{26}Al/^{10}Be ratio for two nodules to be 0.00029 for Aries 15D and 0.00043 for Aries 12D, values that differ more than an order of magnitude from the atmospheric value and that strongly disagree with the results of Guichard et al.

The Continents and Their Margins

Unlike the deep oceans, polar ice, and various extraterrestrial objects, the continents had almost no investigations of the concentrations of the long-lived rare isotopes prior to the introduction of the accelerator method, with the obvious exception of ^{14}C, which bears a different relationship to the subject at hand than ^{10}Be, ^{26}Al, and ^{36}Cl. The reason for this neglect is, of course, the extremely low concentrations that preclude observation by means of radioactive counting. Moeller & Wagener (1967) did in fact observe ^{10}Be extracted from huge amounts of soil using β-counters, but the task was so arduous that no systematic study of its residence and migration resulted. Other observations of ^{10}Be have had to await the arrival of the accelerator method, but they are now well advanced in soils; in river, lake, estuary, and continental-margin sediments; and in organic matter. Chlorine-36 has been shown to be a splendid isotope for dating and tracing the movement of ground water.

Upon landing, ^{10}Be encounters primarily soil, rock, and vegetation. The fundamental question that the earth scientist must answer is, where does it go? The answer will occupy many of us, one hopes usefully, for many years. The elements from which the answer will be formed can now be imperfectly perceived. A fraction, in some localities a large fraction, penetrates meters of soil during periods measured in units of 10^5 yr. A lesser fraction is incorporated in sediments and in eroded material in transport. Related to the inventories in soils and sediments is the uncertain fraction residing in organic matter; here, the interest centers on ^{10}Be in petroleum. A small fraction, one that has not yet been observed directly, must be dissolved in ground and river water. Central to any discussion of ^{10}Be in soils and sediments is the measure of the element's soil-to-water distribution coefficient. Laboratory measurements with ^{7}Be by Robbins & Eadie (1982) showed that typical soil components had volume soil-to-water coefficients in the range of 10^5 to 10^6. Measures & Edmonds (1982) have measured the dissolved elemental Be in Yangtze and Amazon river waters. If one compares their data with the Be that must be in the sediment loads, the result is consistent with what Robbins & Eadie find.

In the first examinations of ^{10}Be in soils, Pavich et al (1983) have found soils for which all or certainly very much of the incident ^{10}Be can be accounted, whereas Monaghan et al (1983) have found soils for which very little of the incident ^{10}Be can be accounted. Pavich et al examined soil profiles for six sites in Virginia and Maryland, among which were three dated terraces along the Rappahannock River. The concentrations could be reproduced approximately by assuming transport downward by dissolved ^{10}Be with soil-to-water distribution coefficient of 4×10^5. In soils with a more complex history and composition, they noted a correlation of the isotope and the clay content and a deficiency of the isotope in horizons having high organic content, a deficiency also found for extractable Al, presumably because of the similar chemistry of Be and Al. Monaghan et al examined terraces in Mendocino County, California, and found concentrations whose maximum values were much lower than those along the Rappahannock, with the attendant consequence that the inventories of ^{10}Be were much less than what might be expected from ages assigned to the terraces geologically. They give a mean residence time for the Mendocino terraces of only 10^4 yr and attribute the loss to mobilization of Be by unspecified organic components that also mobilize the extractable Al, the same explanation offered by Pavich et al for the depletions in organic-rich horizons.

If the observed, very large soil-to-water distribution coefficients hold during erosion, and doubts in this regard find their origin primarily in the uncertain role played by organic soil components, then sediments will be the next large continental reservoir of ^{10}Be. Measurements by Brown et al (1981) have shown the concentrations from fine-grained samples of a given river system near its mouth to be relatively constant even 100 km seaward. Sediments in the margins of five continents have shown concentrations restricted to values between 2 and 4×10^8 atoms g^{-1}. These values are more than an order of magnitude lower than pelagic sediments and have much less variation than soils, which have been found with values as low as 2×10^7 and as high as 2×10^9 atoms g^{-1}.

Beryllium-10 was first studied in lake sediments by Raisbeck et al (1981c) in Lakes Keilambete (Australia) and Winderemere (England), both of which had concentrations in excess of 10^9 atoms g^{-1}. The deposition rates were an order of magnitude larger than determinations of the atmospheric rate, which they attributed to erosional transport. Later, Wahlen et al (1983) examined a core of Green Lake (New York) sufficiently deep to allow measurement of ^{10}Be deposition rates at times before the advent of

European settlement. Erosion effects are evident during and after the eighteenth century, but five measurements from the period 1200 to 1490 A.D. yield a deposition rate of 0.015 ± 0.002 atoms $cm^{-2} s^{-1}$, a rate so low as to suggest that the ^{10}Be content of that portion of the sediment resulted only from atmospheric deposition with no erosional component. The period that followed, corresponding to the time of the Maunder minimum (1645–1715 A.D.), shows a value three times higher with great scatter. Lundberg et al (1983) have examined sediments and nearby soil in the Maurice River-Union Lake system of New Jersey. They measured the ^{10}Be concentrations for different particle sizes and found the millimeter-sized silicates to have little ^{10}Be (of the order of 10^7 atoms g^{-1}).

Most remarkable about the study of Lundberg et al, however, was the discovery of extraordinarily high ^{10}Be concentrations, values reaching 1.5×10^{10} atoms g^{-1}, in organic matter handpicked out of the sediments. High concentrations are also found in peat. This reverse side of the coin, in which certain kinds of organic matter retain the isotope with an even greater tenacity than clay, in sharp contrast to other soil organics that appear to mobilize it, suggests an extensive and very likely useful field of research. The report by Yiou et al (1983) of concentrations of 4 to 9×10^6 atoms g^{-1} in what is thought to be Plio-Pleistocene petroleum with no observable concentration in much older oil, although suggested independently by B. Peters, in retrospect follows directly from the isotope's incorporation in the lake sediment organics.

A preliminary report[1] of $^{36}Cl/Cl$ ratios measured from samples of halite taken from a 930-m sediment core from Searles Lake, California, shows a correlation of radiometric with magnetostratigraphic ages for five or six datum points.

Specialists in groundwater have long considered ^{36}Cl, introduced by rain, to be the best isotope for determining the age and dynamics of aquifers. For a review of techniques used in this field, see Davis & Bentley (1982). Favorable properties of ^{36}Cl are its half-life of 3.0×10^5 yr, which matches well the times to be measured, and its chemical characteristics, which cause it to remain in solution and neither react with nor be dissolved from the matrix of the aquifer. Properties of Cl that complicate interpretation of the measurement of $^{36}Cl/Cl$ ratios are the variability of the amount of Cl in atmospheric precipitation and the large cross section for $^{35}Cl(n,\gamma)^{36}Cl$. The former depends on the rigors of Cl transport by the atmosphere from the ocean; the latter determines the lower limit of application allowed by the U and Th in the aquifer. Chloride can leak into the aquifer and

[1] F. M. Phillips, G. I. Smith, H. W. Bentley, D. Elmore, and H. E. Gove, personal communication.

complicate interpretations, as can oceanic intrusions, which have been observed to reduce the $^{36}Cl/Cl$ values. Bentley & Davis (1981) used ^{36}Cl to study two aquifers thought to have simple hydrology. One of them, the Fox Hills-Hell Creek Aquifer in southwestern North Dakota, yielded concentrations (not isotopic ratios) that correlated with ages calculated from hydrodynamic principles. The second, the Carrizo Sand Aquifer in south-central Texas, proved to be more difficult to understand.

A terrestrial application of ^{10}Be that pertains to deep oceans and continents is its use as a tracer in studying the source of magmas in island-arc volcanism. These aligned and regularly spaced volcanoes are caused, as are the associated offshore trenches, by a lithospheric plate that is being subducted at speeds measured in centimeters per year. One particular question addressed by geochemists is whether the magmas are derived from the downgoing plate, its sedimentary cover, or the mantle wedge just above the plate. Answers based on the examination of the chemical and isotopic composition of the lavas have not convinced those who do not wish to believe.

I. S. Sacks and F. Tera proposed looking for ^{10}Be in the lavas of these volcanoes, reasoning that if it were found in them at a significantly higher concentration than in non-island-arc lavas, such as from rift, hotspot, and seamount volcanoes and flood basalts, then one could conclude that some of the subducted pelagic sediment, which contains high concentrations, had been incorporated in the magmas. The problem for the experimentalist is to convince himself that the ^{10}Be was not introduced through some mechanism other than subduction, such as surface contamination by rain, mixture of the magma with soil or sediment, or in situ production by cosmic rays of radioactivity. A satisfactory answer must eventually come from the examination of a large number of samples. Brown et al (1982, Brown 1983) have measured the ^{10}Be concentration in 49 lavas: 23 Aleutian and Central American samples have concentrations of a few million atoms per gram, a factor of ten greater than what has been found in the 10 non-island-arc lavas; and 16 samples from the South America, Japan, and Mariana arcs show a variable concentration, with only a few in the range of the Aleutian–Central American samples and most lower. Although one cannot say that the question of contamination has been laid to rest, there is growing confidence that the observed ^{10}Be was introduced from the sediment covering subducted plates. It is not possible to calculate from the data what fraction of the lava is derived from sediment because such calculations require a knowledge of how the ^{10}Be is tapped by the volcanoes, but the data do allow one to reason that about 2% of the isotope that is subducted along the entire front is ejected by the volcanic chain.

The Solar System

Our understanding of the nature of the solar system has grown at a truly bewildering rate during the past three decades, leaving us not only with clever ideas as to the origin of the Earth, the Sun, and planetary neighbors but also with an appreciation of it as a dynamic entity. Just as we no longer see the Earth as a static thing enlivened by a thin skin of biological and geological activity, neither do we see the solar system only as a gigantic clockwork unwaveringly instructing us in the validity of the mechanical laws. Wherever we look, we find change or evidence of past change. Much of this has resulted from the intense examination of the extraterrestrial material that has come into our hands: meteorites, lunar rocks and regolith, and cosmic dust and spherules. The accelerator method has been effectively employed in studying all of these substances and has made use of all four isotopes: ^{10}Be, ^{14}C, ^{26}Al, and ^{36}Cl.

Meteorities have been studied for clues about the solar system ever since their identity as stones that fall from the sky was finally conceded by European scientists in the late eighteenth century, but it is through the study of their isotopic composition, both stable and radioactive, that their messages have been read. It is clearly out of place here to sketch meteorite research in any general form; the reader is referred to the books by Mason (1962) and Wasson (1974). One aspect, however, must be addressed in order to appreciate the contribution of the accelerator method. Prior to striking the Earth, a meteorite is subjected to irradiation by cosmic rays, which results in the continual production of various isotopes. The concentrations of various radioactivities allow one to determine an *exposure age*. This age, generally interpreted as being the period between the meteorite's breakoff from some larger parent body, where it had been shielded from cosmic rays, and its landing on Earth, where it is again shielded, is observed to range from 2×10^4 to 10^9 yr. For a summary of the present knowledge of exposure ages, see Nishiizumi et al (1980). The time between its fall to Earth and its discovery, called the *terrestrial age*, can be determined by the decrease in radioactivity. All in all, this model succeeds in organizing much data in an orderly manner. Meteorites sufficiently small so that shielding is negligible attain saturated values for the concentrations of radioisotopes with much shorter half-lives than the exposure ages. Saturation values, useful as references in determining terrestrial ages, are given in Table 3.

The accelerator method has had an especially important application to meteorites harvested in remarkable regions of Antarctica, where objects stored for hundreds of thousands of years are brought up by glacial ice flowing against a mountain barrier and are eventually exposed by wind

Table 3 Saturation values for stone meteorites and bulk lunar surface (0 to 50 g cm^{-2})

Isotope	Meteorite	Concentration (dpm kg^{-1})	(atoms g^{-1})	Observer
^{10}Be	Bruderheim	25	2.9×10^{10}	Pal et al (1983)
^{14}C	Bruderheim	50	2.2×10^{8}	Fireman (1983)
^{26}Al	L-chondrite composite	59	3.2×10^{10}	Evans et al (1979)
^{36}Cl	Average	23	5.2×10^{9}	Nishiizumi et al (1983a)
Lunar samples				
^{10}Be	Apollo 15 core	13	1.5×10^{10}	Nishiizumi et al (1983b)
^{14}C	73221	22	9.6×10^{7}	Fireman et al (1977)
^{26}Al	Apollo 15 core	44	2.4×10^{10}	Nishiizumi et al (1983c)
^{36}Cl	Apollo 15 core	11	2.5×10^{9}	Nishiizumi et al (1983b)

ablation of the ice. The reader should find Marvin's (1983) description of the region and of the activities going on there instructive. The preservation over long periods, particularly of the stone meteorites that otherwise weather quickly, gives us a unique collection of objects having a much greater spread of terrestrial ages than was heretofore available. Prior to the Antarctic discovery, no stone meteorite had been found with an age greater than 3×10^4 yr, and most of those collected had ages falling within the past 200 yr.

Nishiizumi et al (1979, 1981, 1983a) measured the ^{36}Cl content of 26 Antarctic meteorites, and this first application of the accelerator method showed that ALHA 77002, a chondrite of class L5, has a terrestrial age of 7×10^5 yr, more than an order of magnitude older than any stone ever found before. The continuation of this work, which also includes the interpretation of extensive measurements of ^{53}Mn and incorporates ^{26}Al measurements by Evans et al (1979), found Antarctic meteorites with terrestrial ages between 10^4 and 7×10^5 yr. Meteorites with widely varying terrestrial ages were found near one another on the same ice surface. No clear understanding of the relationship of meteorite age and ice age has developed. The usual complications and disputes about such matters as multistage irradiation histories, shielding effects, and common falls need not concern us here. One small controversy does touch the subject, however, because it is the discrepancy between ^{36}Cl terrestrial ages and the ^{14}C ages determined by Fireman (1983) with both the accelerator method and β-counting. The radiocarbon ages determined for three of five meteorites for which there are

^{36}Cl ages resulted in serious disagreement. One suspects that the ^{14}C suffers from a systematic upper limit imposed by contamination, although Fireman sets the limit that results from the continuing production by cosmic rays at 5.5×10^4 yr, more than 10^4 yr greater than the age determinations in question.

Pal et al (1983) have used measurements of the ^{10}Be concentration in 22 samples from saturated or near-saturated stone meteorites including mineral separates from Bruderheim to relate production rate to chemical composition. They conclude that the production cross section from Mg is twice that of O, contrary to general prejudice. This group also addresses a problem posed by the production of the stable isotope ^{21}Ne, which has low primordial abundance relative to the other Ne isotopes but which is enhanced in meteorites owing to production by cosmic rays. Discrepancies in the production rate had brought forth suggestions of the inconstancy of cosmic rays, but correlation of ^{10}Be with ^{21}Ne reported by Moniot et al (1982) do not support this.

The production of radioisotopes as a function of depth in lunar soils interested early investigators of the Apollo samples, but only ^{26}Al and ^{53}Mn measurements were reported for any significant depth. Of the long-lived nuclides, ^{26}Al is the most easily detected by radiation because it emits a characteristic γ-ray and a β^+ and can be observed in low-level counting equipment with more assurance than the other pure and weak β-emitters. Manganese-53 is very sensitive to analysis by neutron activation. In a series of papers, Nishiizumi et al (1982a, 1983b,c) have examined profiles in the Apollo 15 long core of ^{10}Be, ^{26}Al, ^{36}Cl, and ^{53}Mn, all but the last done with the accelerator method and all done to an accuracy far exceeding previous work. The results confirmed the theoretical calculations of Reedy & Arnold (1972). Saturation values for bulk samples of the lunar surface are given in Table 3, where comparison can be made with similar values for stone meteorites. The accelerator method allows measurement of these isotopes in individual lunar grains, and preliminary studies by Klein et al (1983) on ^{26}Al have disclosed variations of as much as a factor of two in neighboring submillimeter-sized grains. A discussion of the interpretation of these observations is best postponed. The reports by Raisbeck et al (1983b) of ^{10}Be and ^{26}Al in deep-sea stony spherules imply that they have an extraterrestrial origin and have apparently been irradiated as small bodies in space.

Tektites are glass objects found scattered about certain regions of the Earth that have attracted man's attention since prehistory, presenting him with numerous puzzles and apparent contradictions and causing spirited controversy. Two books that summarize what is known while giving opposite interpretations of origin are Barnes & Barnes (1973) and O'Keefe

(1976). Since the Apollo flights, chemical and isotopic evidence has accumulated favoring the view that tektites are composed of terrestrial matter that has been melted, degassed, and propelled out of the atmosphere by a large impacting object. A group from Rutgers University (Pal et al 1982) examined 7 Australasian tektites for their ^{10}Be content, finding concentrations of about 10^8 atoms g^{-1}, which they interpreted as evidence for their origin in sediments. Subsequently, Tera et al (1983) measured the concentration in 12 tektites from Australia and 13 from Southeast Asia and found an average concentration of 1.5×10^8 atoms g^{-1} for the former group and 1.0×10^8 for the latter. Beyond the confirmation of the Rutgers work, the study showed that each subset had a scatter in values of only 15%, perhaps the two tightest distributions of natural ^{10}Be data yet observed. This supports the sediment interpretation, pointing specifically to continental-margin sediments. The uniformity of the data would seem to rule out soils, as observed soil profiles have more scatter in just 3 m. An examination of an australite by Raisbeck et al (1983c) showed ^{26}Al at the level expected for production on the Earth's surface, making a lunar origin again untenable.

Traditional Radiocarbon Dating

There is little need to point out the contributions of radiocarbon dating to archaeology and holocene geology. These successes provided the motivation for developing the accelerator method. The initial demonstrations of ^{14}C-counting were followed by refinement of existing ion optics or design of new machines to provide satisfactory reproducibility. These efforts have led to various laboratories approaching the 1% level generally agreed upon as satisfactory for the first stage.

The origin of man has always been the central problem of archaeology, and a subproblem is when and where anatomically modern man appeared. The scarcity of hominid fossils retards progress, so the discovery of a nearly complete female skeleton in 1972 south of San Francisco Bay was an important event, but the subsequent dating of this Sunnyvale skeleton by Bada & Helfman (1975), using the amino-acid technique, at an age of 7×10^4 yr made it truly sensational, as it placed the appearance of man in America six to ten times earlier than had been thought. The claims were subject to lively dispute based on other methods that gave much lower ages, but the matter has not been resolved to the satisfaction of all parties. Taylor et al (1983) extracted organic fractions specific to bone collagen from four pieces of the skeleton. These fractions, which do not exchange with groundwater bicarbonate, are very small, and only one of the four was large enough for decay counting; the other three were successfully examined using the accelerator method and gave values that averaged 4370 yr, almost identical to the value obtained by decay counting the fourth sample. An analogous controversy about an infant skeleton found near Taber, Alberta, has found a similar resolution by R. M. Brown et al (1983), who have assigned it an age of 4100 yr.

In research directed toward a more practical goal, Tucker et al (1983) determined the age of milligram flecks of charcoal found by trenching the Wasatch fault near Salt Lake City, along which as much as 10 m of displacement has taken place during the past 1.2×10^4 yr. These bits of charcoal, the result of forest or range fires, were deposited sometime after a scarp-producing earthquake. The three samples gave an average age of 8530 yr with a scatter consistent with their 600-yr experimental uncertainty, so they probably date the same event.

CONCLUSIONS

The six years since the introduction of accelerator mass spectrometry have seen the rapid development that characterizes a successful new instrumental method. It has attracted scholars with interests ranging from the time variation of cosmic rays to the evolution of soils. It now reflects the efforts of those who were struggling with the experimental difficulties of the method before 1977 and of those who previously never dreamed that their research would involve a particle accelerator. Its extension to radiocarbon dating, foremost in the minds of the inventors, has proved so difficult that the first reports of application to routine problems are only now appearing, but the difficulty in reproducing isotopic standards to accuracies of 1% or less, that omnipresent embarrassment of the experimentalists, has had almost no effect on the application of the method to ^{10}Be, ^{26}Al, and ^{36}Cl. One reason is easily illustrated by Table 4, which lists observed isotopic concentrations for a variety of natural materials, where variations of seven orders of magnitude are found. The 5 and 10% accuracies generally reported and regretted by observers are trivial when viewed on such scales. Based on accuracy relative to the span of available data, accelerator mass spectrometry with ^{10}Be, ^{26}Al, and ^{36}Cl is superior to Nd mass spectrometry with its much admired precision of a few parts in 10^5. Constrained as it is by a data span of less than three orders of magnitude and rivaled by highly sophisticated β-counting techniques, accelerator mass spectrometry for ^{14}C has required equipment development superior to the needs of the other isotopes. There is every indication that this has succeeded. One can expect the results to be increasingly evident soon in the traditional radiocarbon field.

Table 4 also demonstrates the wide use, almost entirely as a tracer, to which ^{10}Be has been put. One need not look far to learn the reason : a high

Table 4 Observed natural concentrations in atoms g^{-1}

Substance [a]	^{10}Be	^{26}Al	^{36}Cl
Stratospheric air	10^4	30	—
Rain	2×10^4	—	2000
Ocean surface	750	—	—
Deep ocean	4000	—	—
Pelagic sediments	5×10^9	4×10^7	—
Manganese nodule surface	5×10^{10}	2×10^8	—
Greenland ice (present)	10^4	—	4000
Antarctic ice (present)	4×10^4	—	4000
Continental-margin sediments	3×10^8	—	—
Soils	2×10^7 to 2×10^9	—	—
Groundwater	—	—	5000 to 5×10^4
Hard wood	5×10^6	—	—
Pleistocene petroleum	5×10^6	—	—
Organics in sediment	2×10^{10}	—	—
Island-arc basalts	5×10^6	—	—
Non-island-arc basalts	2×10^5	—	—
Stone meteorite	3×10^{10}	3×10^{10}	7×10^9
Lunar surface bulk	1.5×10^{10}	2.4×10^{10}	2.5×10^9
Australasian tektite	10^8	$<2 \times 10^7$	

[a] These values are given only as examples. There is often a significant range of values. Refer to appropriate sections of the text for elaboration.

atmospheric production rate and a low abundance of the stable isotope. In none of the investigations reported has dilution of ^{10}Be by stable Be been a limit. But what is said about ^{10}Be can be inverted when applied to ^{26}Al. There are few places on Earth where one does not find the radioisotope overwhelmed by the stable. The restrictions on ^{26}Al are not fundamental, however, and may be overcome by advances in technique. This would open the possibility of absolute dating with the ^{10}Be-^{26}Al pair, proposed as a consequence of their similar chemistry more than 20 years ago by Lal (1962). Chlorine-36 has proved its use in the chronology of meteorites, polar ice, and groundwater. One need not be foolishly optimistic to expect greatly extended fields of application in the next few years for all these nuclides.

Literature Cited

Amin, B. S., Biswas, S., Lal, D., Somayajulu, B. L. K. 1972. Radiochemical measurements of ^{10}Be and ^{7}Be formation cross sections in oxygen by 135 and 550 MeV protons. *Nucl. Phys. A* 195:311–20

Amin, B. S., Lal, D., Somayajulu, B. L. K. 1975. Chronology of marine sediments using the ^{10}Be method: Intercomparison with other methods. *Geochim. Cosmochim. Acta* 39:1187–92

Arnold, J. R. 1956. Beryllium-10 produced by cosmic rays. *Science* 124:584–85

Bada, J. L., Helfman, P. M. 1975. Amino acid racemization dating of fossil bones. *World Archaeol.* 7:160–73

Barnes, V. E., Barnes, M. A., eds. 1973. *Tektites. Benchmark Papers in Geology Series.* Stroudsburg, Pa: Hutchinson & Ross. 445 pp.

Beer, J., Andre, M., Oeschger, H., Stauffer, B., Balzer, R., et al. 1983. Temporal ^{10}Be variations in ice. *Radiocarbon* 25:269–78

Bentley, H. W., Davis, S. N. 1981. Applications of AMS to hydrology. *Proc. Symp. Accel. Mass Spectrom.*, Argonne, Ill: Argonne Natl. Lab. 458 pp.

Bentley, H. W., Phillips, F. M., Davis, S. N., Gifford, S., Elmore, D., et al. 1982. Thermonuclear ^{36}Cl pulse in natural water. *Nature* 300:737–40

Brown, L. 1983. Beryllium-10 in island-arc volcanism. *Eos, Trans. Am. Geophys. Union* 64:284 (Abstr.)

Brown, L., Sacks, I. S., Tera, F., Klein, J., Middleton, R. 1981. Beryllium-10 in continental sediments. *Earth Planet. Sci. Lett.* 55:370–76

Brown, L., Klein, J., Middleton, R., Sacks, I. S., Tera, F. 1982. ^{10}Be in island-arc volcanoes and implications for subduction. *Nature* 299:718–20

Brown, R. M., Andrews, H. R., Ball, G. C., Burn, N., Imahori, Y., et al. 1983. Accelerator ^{14}C dating of the "Taber Child." *Can. J. Archaeol.* In press

Davis, S. N., Bentley, H. W. 1982. Dating groundwater. *ACS Symp. Ser.* 176:187–221

Elmore, D., Tubbs, L. E., Newman, D., Ma, X. Z., Finkel, R., et al. 1982. ^{36}Cl bomb pulse measured in a shallow ice core from Dye 3, Greenland. *Nature* 300:735–37

Evans, J. C., Rancitelli, L. A., Reeves, J. H. 1979. ^{26}Al content of Antarctic meteorites: implications for terrestrial ages and bombardment history. *Proc. Lunar Planet. Sci. Conf., 10th*, pp. 1061–72

Finkel, R., Krishnaswami, S., Clark, D. L. 1977. ^{10}Be in arctic ocean sediments. *Earth Planet. Sci. Lett.* 35:199–204

Finkel, R. C., Nishiizumi, K., Elmore, D., Ferraro, R. D., Gove, H. E. 1980. ^{36}Cl in polar ice, rainwater and seawater. *Geophys. Res. Lett.* 7:983–86

Fireman, E. L. 1983. Carbon-14 ages of antarctic meteorites. *Lunar Planet. Sci. XIV*, pp. 195–96 (Abstr.)

Fireman, E. L., DeFelice, J., D'Amico, J. 1977. ^{14}C in lunar soil: temperature-release and grain-size dependence. *Proc. Lunar Sci. Conf., 8th*, pp. 3749–54

Guichard, F., Reyss, J.-L., Yokoyama, Y. 1978. Growth rate of manganese nodule measured with ^{10}Be and ^{26}Al. *Nature* 272:155–56

Inoue, T., Tanaka, S. 1976. Be-10 in marine sediments. *Earth Planet. Sci. Lett.* 29:155–60

Klein, J., Middleton, R., Raisbeck, G. M., Yiou, F., Langevin, Y. 1983. ^{26}Al measurements in individual lunar grains. *Lunar Planet. Sci. XIV*, pp. 375–76 (Abstr.)

Krishnaswami, S., Cochran, J. K. 1978. Uranium and thorium series nuclides in oriented ferromanganese nodules: growth rates, turnover times and nuclide behavior. *Earth Planet. Sci. Lett.* 4:45–62

Krishnaswami, S., Mangini, A., Thomas, J. H., Sharma, P., Cochran, J. K., et al. 1982. ^{10}Be and Th isotopes in manganese nodules and adjacent sediments: nodule growth histories and nuclide behavior. *Earth Planet. Sci. Lett.* 59:217–34

Ku, T. L., Kusakabe, M., Nelson, D. E., Southon, J. R., Korteling, R. G., et al. 1982. Constancy of oceanic deposition of ^{10}Be as recorded in manganese crusts. *Nature* 299:240–42

Kusakabe, M., Ku, T. L., Vogel, J., Southon, J. R., Nelson, D. E., et al. 1982. Beryllium-10 profiles in seawater. *Nature* 299:712–14

Lal, D. 1962. Cosmic ray produced radionuclides in the sea. *J. Oceanogr. Soc. Jpn.* 18:600–14

Lal, D., Peters, B. 1967. Cosmic-ray-produced radioactivity on the earth. *Handb. Physik XLVI* 2:551–612

Litherland, A. E. 1980. Ultrasensitive mass spectrometry with accelerators. *Ann. Rev. Nucl. Part. Sci.* 30:437–73

Lundberg, L., Ticich, T., Herzog, G. F., Hughes, T., Ashley, G., et al. 1983. ^{10}Be and Be in the Maurice River-Union Lake system of southern New Jersey. *J. Geophys. Res.* 88:4498–4504

Marvin, U. B. 1983. Extraterrestrials have landed on Antarctica. *New Sci.* 97:710–15

Mason, B. 1962. *Meteorites.* New York: Wiley

McCorkell, R., Fireman, E. L., Langway, C. C. Jr. 1967. Aluminum-26 and beryllium-10 in Greenland ice. *Science* 158:1690–92

Measures, C. I., Edmonds, J. M. 1982. Beryllium in the water column of the central North Pacific. *Nature* 297:51–53

Middleton, R. 1977. A survey of negative ions from a cesium sputter source. *Nucl. Instrum. Methods* 144:373–99

Middleton, R. 1983. A versatile high intensity negative ion source. *Nucl. Instrum. Methods.* 214:139–50

Moeller, P., Wagener, K. 1967. Dating soil layers by ^{10}Be. In *Radioactive Dating and Methods of Low-Level Counting*, pp. 744. Vienna: IAEA

Monaghan, M. C., Krishnaswami, S., Thomas, J. H. 1983. ^{10}Be concentration and the long term fate of particle-reactive nuclides in five soil profiles from California. *Earth Planet. Sci. Lett.* In press

Moniot, R. K., Kruse, T. H., Savin, W., Tuniz, C., Milazzo, T., et al. 1982. Beryllium-10 contents of stony meteorites and neon-21 production rate. *Lunar Planet. Sci. Conf. XIII*, pp. 536–37 (Abstr.)

Nishiizumi, K., Arnold, J. R., Elmore, D., Ferraro, R., Gove, H. E., et al. 1979. Measurements of ^{36}Cl in Antarctic meteorites and Antarctic ice using a Van de Graaff accelerator. *Earth Planet. Sci. Lett.* 45:285–92

Nishiizumi, K., Regnier, S., Marti, K. 1980. Cosmic ray exposure ages of chondrites, pre-irradiation and constancy of cosmic ray flux in the past. *Earth Planet. Sci. Lett.* 50:156–70

Nishiizumi, K., Murrell, M. T., Arnold, J. R., Elmore, D., Ferraro, R., et al. 1981. Cosmic ray produced ^{36}Cl and ^{53}Mn in Allan Hills-77 meteorites. *Earth Planet. Sci. Lett.* 52:31–38

Nishiizumi, K., Arnold, J. R., Elmore, D., Tubbs, L. E., Cole, G., et al. 1982a. Measurements of cosmic ray produced ^{53}Mn and ^{10}Be in lunar core. *Lunar Planet. Sci. XIII*, pp. 396–97 (Abstr.)

Nishiizumi, K., Arnold, J. R., Klein, J., Middleton, R. 1982b. ^{10}Be and other radionuclides in Antarctic meteorites and in associated ice. *Meteoritics* 17:260–61

Nishiizumi, K., Arnold, J. R., Elmore, D., Ma, X., Newman, D., et al. 1983a. ^{36}Cl and ^{53}Mn in Antarctic meteorites and ^{10}Be-^{36}Cl dating of Antarctic ice. *Earth Planet. Sci. Lett.* 62:407–17

Nishiizumi, K., Arnold, J. R., Elmore, D., Ma, X. Z., Gove, H. E., et al. 1983b. ^{36}Cl depth profile in Apollo 15 drill core. *Lunar Planet. Sci. XIV*, pp. 558–59 (Abstr.)

Nishiizumi, K., Arnold, J. R., Klein, J., Middleton, R. 1983c. Measurements of ^{26}Al in Apollo 15 drill core using accelerator mass spectrometry. *Lunar Planet. Sci. XIV*, pp. 560–61 (Abstr.)

O'Keefe, J. A. 1976. *Tektites and Their Origin.* New York: Elsevier. 254 pp.

Pal, D. K., Tuniz, C., Moniot, R. K., Kruse, T. H., Herzog, G. F. 1982. ^{10}Be in Australasian tektites: evidence for a sedimentary precursor. *Science* 218:787–89

Pal, D. K., Tuniz, C., Moniot, R. K., Savin, W., Kruse, T. H., et al. 1983. Composition dependence of the beryllium-10 production rate in stony meteorites. *Lunar Planet. Sci. XIV*, pp. 588–89 (Abstr.)

Paul, A. Z. 1976. Deep-sea bottom photographs show that benthic organisms remove sediment cover from manganese nodules. *Nature* 263:50–51

Pavich, M. J., Brown, L., Klein, J., Middleton, R. 1983. Beryllium-10 accumulation in a soil chronosequence. *Earth Planet. Sci. Lett.* In press

Raisbeck, G. M., Yiou, F., Fruneau, M., Lieuvin, M., Loiseaux, J. M. 1978. Measurement of ^{10}Be in 1,000- and 5,000-year-old Antarctic ice. *Nature* 275:731–32

Raisbeck, G. M., Yiou, F., Fruneau, M., Loiseaux, J. M., Lieuvin, M. 1979a. ^{10}Be concentration and residence time in the ocean surface layer. *Earth Planet. Sci. Lett.* 43:237–40

Raisbeck, G. M., Yiou, F., Fruneau, M., Loiseaux, J. M., Lieuvin, M., et al. 1979b. Deposition rate and seasonal variations in precipitation of cosmogenic ^{10}Be. *Nature* 282:279–80

Raisbeck, G. M., Yiou, F., Fruneau, M., Loiseaux, J. M., Lieuvin, M., et al. 1980. ^{10}Be concentration and residence time in the deep ocean. *Earth Planet. Sci. Lett.* 51:275–78

Raisbeck, G. M., Yiou, F., Fruneau, M., Loiseaux, J. M., Lieuvin, M., et al. 1981a. Cosmogenic ^{10}Be/^{7}Be as a probe of atmospheric transport processes. *Geophys. Res. Lett.* 8:1015–18

Raisbeck, G. M., Yiou, F., Fruneau, M., Loiseaux, J. M., Lieuvin, M., et al. 1981b. Cosmogenic ^{10}Be concentration in Antarctic ice during the past 30,000 years. *Nature* 292:825–26

Raisbeck, G. M., Yiou, F., Lieuvin, M., Ravel, J. C., Fruneau, M., et al. 1981c. ^{10}Be in the environment: some recent results and their applications. *Proc. Symp. Accel. Mass Spectrom.* Argonne, Ill: Argonne Natl. Lab. 458 pp.

Raisbeck, G. M., Yiou, F., Klein, J., Middleton, R. 1983a. Accelerator mass spectrometry measurement of cosmogenic ^{26}Al in terrestrial and extraterrestrial matter. *Nature* 301:690–92

Raisbeck, G. M., Yiou, F., Klein, J., Middleton, R., Yamakoshi, Y., et al. 1983b. ^{26}Al and ^{10}Be in deep sea stony spherules; evidence for small parent bodies. *Lunar Planet. Sci. XIV*, pp. 622–623 (Abstr.)

Raisbeck, G. M., Yiou, F., Klein, J., Middleton, R. 1983c. ^{26}Al/^{10}Be in an Australian tektite; further evidence for a terrestrial origin. *Eos, Trans. Am. Geophys. Union* 64:284 (Abstr.)

Reedy, R. C., Arnold, J. R. 1972. Interaction of solar and galactic cosmic-ray particles with the moon. *J. Geophys. Res.* 77:537–55

Reedy, R. C., Arnold, J. R., Lal, D. 1983. Cosmic-ray record in solar system matter. *Science* 219:127–35

Reyss, J.-L., Yokoyama, Y. 1976. Aluminum-26 in a manganese nodule. *Nature* 262: 203–4

Reyss, J.-L., Yokoyama, Y., Tanaka, S. 1976. Aluminum-26 in deep-sea sediment. *Science* 193:1119–20

Reyss, J.-L., Yokoyama, Y., Guichard, F.

1981. Production cross sections of ^{26}Al, ^{22}Na, ^{7}Be from argon and of ^{10}Be-^{7}Be from nitrogen: implications for production rates in the atmosphere. *Earth Planet. Sci. Lett.* 53:203–10

Robbins, J. A., Eadie, B. J. 1982. Beryllium-7: A tracer of seasonal particle transport processes in Lake Michigan. *Eos, Trans. Am. Geophys. Union* 63:957 (Abstr.)

Schaeffer, O. A., Thompson, S. O., Lark, N. L. 1960. Chlorine-36 radioactivity in rain. *J. Geophys. Res.* 65:4013–16

Somayajulu, B. L. K. 1967. Beryllium-10 in a manganese nodule. *Science* 156:1219–20

Somayajulu, B. L. K. 1977. Analysis of causes for the beryllium-10 variations in deep sea sediments. *Geochim. Cosmochim. Acta* 41: 909–13

Stensland, G. J., Brown, L., Klein, J., Middleton, R. 1983. Beryllium-10 in rain. *Eos, Trans. Am. Geophys. Union* 64:283 (Abstr.)

Tanaka, S., Inoue, T. 1979. ^{10}Be dating of North Pacific sediment cores up to 2.5 million years B.P. *Earth Planet. Sci. Lett.* 45:181–87

Tanaka, S., Inoue, T. 1980. ^{10}Be evidence for geochemical events in the North Pacific during the Pliocene. *Earth Planet. Sci. Lett.* 49:34–38

Tanaka, S., Sakamoto, K., Takagi, J., Tsuchimoto, M. 1968. Aluminum-26 and beryllium-10 in marine sediments. *Science* 160:1348–49

Tanaka, S., Inoue, T., Imamura, M. 1977. The ^{10}Be method of dating marine sediments —comparison with the paleomagnetic method. *Earth Planet. Sci. Lett.* 37:55–60

Taylor, R. E., Payen, L. A., Gerow, B., Donahue, D. J., Zabel, T. H., et al. 1983. Middle holocene age of the Sunnyvale human skelton. *Science* 220:1271–73

Tera, F., Middleton, R., Klein, J., Brown, L. 1983. Beryllium-10 in tektites. *Eos, Trans. Am. Geophys. Union* 64:284 (Abstr.)

Tucker, A. B., Woefli, W., Bonani, G., Suter, M. 1983. Earthquake dating: An application of carbon-14 atom counting. *Science* 219:1320–21

Turekian, K. K., Cochran, J. K., Krishnaswami, S., Lanford, W. A., Parker, P. D., et al. 1979. The measurement of ^{10}Be in manganese nodules using a tandem Van de Graaff accelerator. *Geophys. Res. Lett.* 6:417–20

Wahlen, M., Kothari, B., Elmore, D., Banerjee, S. K., Leskee, W. 1983. ^{10}Be in lake systems. *Eos, Trans. Am. Geophys. Union* 64:282 (Abstr.)

Wasson, J. T. 1974. *Meteorites.* Berlin: Springer-Verlag

Yiou, F., Raisbeck, G. M. 1972. Half-life of ^{10}Be. *Phys. Rev. Lett.* 29:372–75

Yiou, F., Raisbeck, G. M., Klein, J., Middleton, R. 1983. ^{10}Be in crude petroleum. *Eos, Trans. Am. Geophys. Union* 64:284 (Abstr.)

Yokoyama, Y., Reyss, J.-L., Guichard, F. 1977. Production of radionuclides by cosmic rays at mountain altitudes. *Earth Planet. Sci. Lett.* 36:44–50

Ann. Rev. Earth Planet. Sci. 1984. 12 : 61–82

OCEANOGRAPHY FROM SPACE

Robert H. Stewart

Scripps Institution of Oceanography, University of California at San Diego, La Jolla, California 92093 and Jet Propulsion Laboratory, California Institute of Technology, Pasadena, California 91109

INTRODUCTION

For centuries oceanographers have studied the seas from ships. Yet an understanding of the ocean, even in limited regions, often depends on an understanding of processes over ocean basins, and even over the entire ocean in some instances. The "El Niño" phenomenon off the coast of Peru is now known to depend on processes operating across the equatorial Pacific, weather over North America depends in part on other processes in the Pacific, and the extent of sea ice around Antarctica depends in part on the dynamics of currents in the Southern Ocean. Yet until recently, we have lacked the global perspective demanded by studies of the ocean, its interaction with the atmosphere and cryosphere, and its influence on the biosphere.

Now, within the past five years, satellites, particularly those flown by the National Aeronautics and Space Administration (NASA) and the National Oceanic and Atmospheric Administration (NOAA), have begun to provide the first worldwide views of the ocean. From these satellites have come global maps of the variability of ocean currents, surface wind speed, surface temperature, wave height, water vapor in the atmosphere, and percent cloud cover. When these are added to regional studies that promise to expand into global charts of incoming solar radiance, oceanic productivity, and perhaps even surface evaporation, it is clear that we are on the threshold of being able to study the ocean and atmosphere as a complete system, in contrast with the short-term regional studies of the past.

In what follows, I discuss the instruments in general and specific measurements now being made from space.

0084–6597/84/0515–0061$02.00

SATELLITE INSTRUMENTS

Over the past decade, NASA and NOAA have developed and flown many instruments that can observe the sea. Some were developed primarily to measure the atmosphere or the land and were only later shown to be useful for oceanography. Others, particularly those on *Seasat*, were designed specifically for oceanography.

The instruments can be divided into two classes (Table 1): (*a*) passive instruments (radiometers) that observe natural radiation, and (*b*) active instruments that illuminate the sea and observe the reflected radiation. The instruments can be further subdivided into the wavelength or frequency bands used: (*a*) light, (*b*) infrared, and (*c*) super-high-frequency (SHF) radio waves (Stewart 1982). The first two bands are usually observed by combined visible and infrared instruments, while the third requires completely different instruments. All use frequencies that propagate relatively easily through the atmosphere, i.e. the atmospheric windows (Table 2).

Each band has particular advantages. Optical radiation is easily focused into an image with resolutions of a few kilometers. But clouds obscure the surface, and even clear air is not completely transparent. Radio signals are barely influenced by the atmosphere and readily propagate through clouds, but they cannot be easily focused to produce an image. Except for the synthetic-aperture radar, radio instruments tend to produce images with resolutions of 50–100 km.

OBSERVATIONS OF OCEAN SURFACE TEMPERATURE AND HEATING

Three different instruments can measure sea-surface temperatures from space: (*a*) infrared radiometers, (*b*) multifrequency atmospheric sounders, and (*c*) SHF radiometers. All observe radiation emitted by the sea surface, which increases as surface temperature increases, but each has unique advantages. The infrared radiometers produce high-resolution images of the sea by observing in 1–3 bands through cloud-free atmospheres. Because clouds, aerosols, and water vapor all emit or reflect radiation that obscures the signal from the sea surface, they are the primary sources of error, with clouds dominating. Of great concern are clouds too small to be clearly resolved by the instrument. Fortunately, several procedures have been developed that detect the presence of such clouds and so reduce their influence (Henderson-Sellers 1982, Coakley & Bretherton 1982). Water vapor and aerosols are also sources of concern. Usually radiance in two (3.7 and 10.5 μm) or three (3.7, 10.5, and 11 μm) wavelength bands are combined

Table 1 Spaceborne instruments for viewing the sea[a]

Active	
Visible light (lasers)	Radio signals (radars)
	ALT 1973, 1975–78 SAR 1978 SCAT 1973, 1978

Passive		
Visible light (photographs) (radiometers)	Infrared radiation (radiometers)	Radio waves (radiometers)
AVCS 1964–72 MSS 1972– RBV 1972– CZCS 1978– TM 1982–	HRIR 1964–70 SR[b] 1970–72 VHRR[b] 1972–78 AVHRR[b] 1978– VISSR[b] 1974– OLS[b] 1976– HCMR 1978–81 HIRS 1978–	ESMR 1972–76 NEMS 1972 SCAMS 1975–78 SMMR 1978–

[a] Abbreviations:
ALT = Altimeter
AVCS = Advanced Vidicon Camera Systems
AVHRR = Advanced Very High Resolution Radiometer
CZCS = Coastal Zone Color Scanner
ESMR = Electrically Scanned Microwave Radiometer
HCMR = Heat Capacity Mapping Radiometer
HIRS = High Resolution Infrared Radiation Sounder
HRIR = High Resolution Infrared Radiometer
MSS = Multispectral Scanner
NEMS = Nimbus-E Microwave Scatterometer
OLS = Optical Line Scanner
RBV = Return Beam Vidicon
SAR = Synthetic-Aperture Radar
SCAMS = Scanning Microwave Spectrometer
SCAT = Scatterometer
SMMR = Scanning Multifrequency Microwave Radiometer
SR = Scanning Radiometer
TM = Thematic Mapper
VHRR = Very High Resolution Radiometer
VISSR = Visible and Infrared Spin Scan Radiometer
[b] Also view with visible light.

to reduce the atmospheric errors. The influence of water vapor is greatest in the 11-μm band and least at 3.7 μm. Aerosols scatter sunlight and are most important at 3.7 μm. By using nighttime observations to avoid sun glint, data from several bands to reduce the influence of humidity, and techniques to detect clouds, infrared radiometers can measure sea-surface tempera-

Table 2 The primary atmospheric windows

Radiation band	Variability due to	Transmittance
0.4–0.7 μm	ozone, aerosols, clouds	0.6–0.8
2.0–2.5 μm	water vapor, aerosols, clouds	0.9
3.5–4.0 μm	water vapor, clouds	0.6–0.9
8.0–9.0 μm	water vapor, ozone, clouds	0.5–0.9
10–13 μm	water vapor, clouds	0.2–0.9
25–40 GHz	water vapor, rain	0.8–0.9
0.1–20 GHz	water vapor, rain	0.9–0.99

tures with an accuracy of better than 1°C (Bernstein 1982a). Alternatively, radiosonde measurements of water vapor can be used to correct the satellite measurements in some regions with similar accuracy (Maul 1981). Such carefully selected or corrected observations of high accuracy are now used to produce monthly maps of surface temperature on a one-degree grid covering the world's oceans. Because infrared radiometers have a resolution of about 1 km and a sensitivity of about 0.1°C, they are also widely used to map the subtle features in the surface-temperature field mostly due to surface currents. Global studies of these features are leading to an improved understanding of ocean dynamics, particularly on scales of a few hundred to a thousand kilometers (Legeckis 1978, 1979).

Because clouds are so common, techniques to correct for their influence may be more useful than techniques that merely detect their presence. Multifrequency atmospheric sounders that observe infrared radiance are most useful for this task. Measurements of radiance in 20–24 bands are used to estimate surface temperature, percent cloud cover, and temperature profiles in the atmosphere. Thus, data from the Tiros Operational Vertical Sounder carried on the NOAA meteorological satellites have been used to calculate surface temperatures with an accuracy of around 1°C when averaged over areas of a few hundred kilometers on a side (Figure 1).

Lastly, microwave radiometers can observe the sea even through persistent clouds. The measurement of sea temperature is more difficult in SHF bands than in the infrared bands because sea water has a low emissivity, typically around 0.35–0.40 at frequencies below 20 GHz, and so the sea has a brightness temperature of between 120 K and 140 K, depending on incidence angle and polarization. The brightness temperature is converted to thermodynamic temperature by multiplying by the reciprocal of emissivity. This magnifies the influence of small errors; and atmospheric clouds and water vapor, wind speed, and errors in the instrument itself must all be considered. Present estimates of the accuracy of surface-temperature observations made by the *Seasat* SMMR (see Table 1 for a list of abbreviations) lead to values of around $\pm 1.5°C$ globally, with somewhat better accuracy in midlatitudes (Hofer et al 1981, Lipes 1982, Njoku & Hofer 1981). This accuracy requires some restrictions on the operation of the instrument: (*a*) the oceanic areas must be at least 300 km from land, or else the influence of the relatively hot land viewed through the sidelobes of the antenna adds additional error that is difficult to remove; (*b*) sun glint must be avoided, including certain sun angles at the spacecraft for the SMMR on *Nimbus 7*; and (*c*) the radiometer must be out of sight of terrestrial transmitters operating at the same frequency as the radiometer. In some extreme instances this requires that the land be below the horizon seen by the satellite.

At present, the accuracies of the three different types of instruments are being compared globally to determine their relative accuracy and sources of systematic error. These observations will then be compared with surface observations in selected regions made from ships. For studies of climate it is particularly important to know whether or not the maps of surface temperature have long-wavelength, long-period errors similar to the natural variability of temperature (Figure 2).

Closely allied with the techniques used to measure surface temperature are those used to calculate incoming shortwave solar radiation at the sea surface. This is the primary source of surface heating, particularly in the

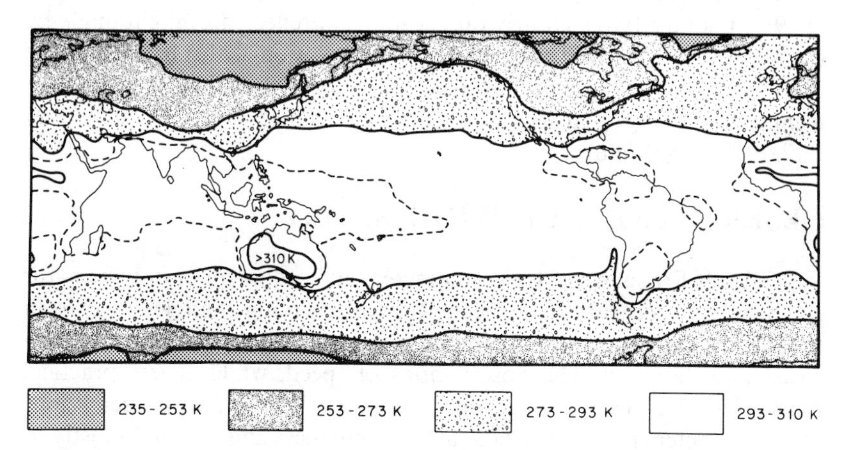

Figure 1 Mean surface temperature of the Earth during January 1979 calculated using data from the Tiros Operational Vertical Sounder on the NOAA meteorological satellites. The original map has an accuracy of $\pm 1°C$ and a resolution of $2° \times 5°$, but for clarity only a few contours of temperature are shown here. The dashed contours denote areas with temperature of 300 K (from an original figure by M. T. Chahine of the Jet Propulsion Laboratory and J. Susskind of the Goddard Space Flight Center).

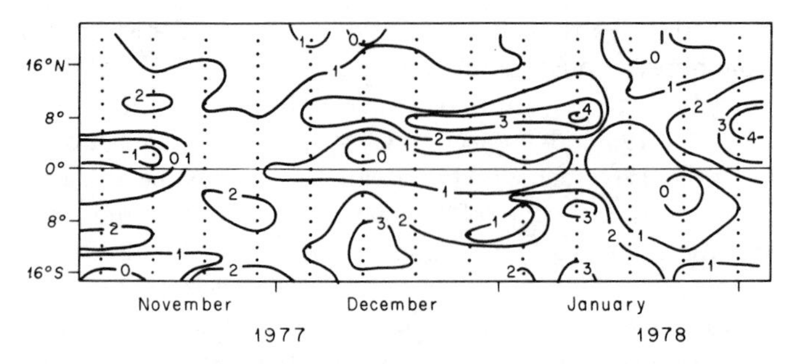

Figure 2 The nature of the space and time scales of errors resulting from atmospheric influences on satellite measurements of surface temperature. The plot shows surface temperature (°C) measured by air expendable bathythermographs minus satellite measurements of sea temperature along 150°W. Note that the difference is significant, and that it has spatial and temporal scales similar to naturally occurring variations in sea-surface temperature (after Barnett et al 1979).

tropics, and the measurements are important for understanding the surface heat balance. The incoming radiation is accurately estimated from visible observations. Essentially, almost all radiation not reflected back out to space by clouds is absorbed by the sea. The relatively small amounts of radiation absorbed by clouds and the atmosphere can be estimated from the amount of cloud cover (itself estimated from the visible observations) and from knowledge of the average properties of the atmosphere (Gautier et al 1980, Gautier 1981). Hourly observations of reflected sunlight made by the VISSR can be used to estimate the insolation with an accuracy of around 9% despite the lack of an accurate calibration of the sensor, and the technique is being used to map incoming radiance over the Pacific and Indian Oceans (Figure 3).

OBSERVATIONS OF WIND SPEED

Super-High-Frequency (SHF) radiometers and radars observing the nadir can be used to measure wind speed. Radars observing at angles away from the vertical and observations of cloud motion can be used to measure both speed and direction. The observations of speed, while of less practical importance, are easier to make, but all the observations are indirect. Radar and radiometers measure ocean-surface roughness and respond mostly to small wind-produced wavelets that influence the emissivity and reflectivity of the surface, so surface wavelets must be related to wind velocity. Sequential pictures of cloud position give cloud motion due to winds a few hundred meters above the sea, so the velocity must be extrapolated to the surface.

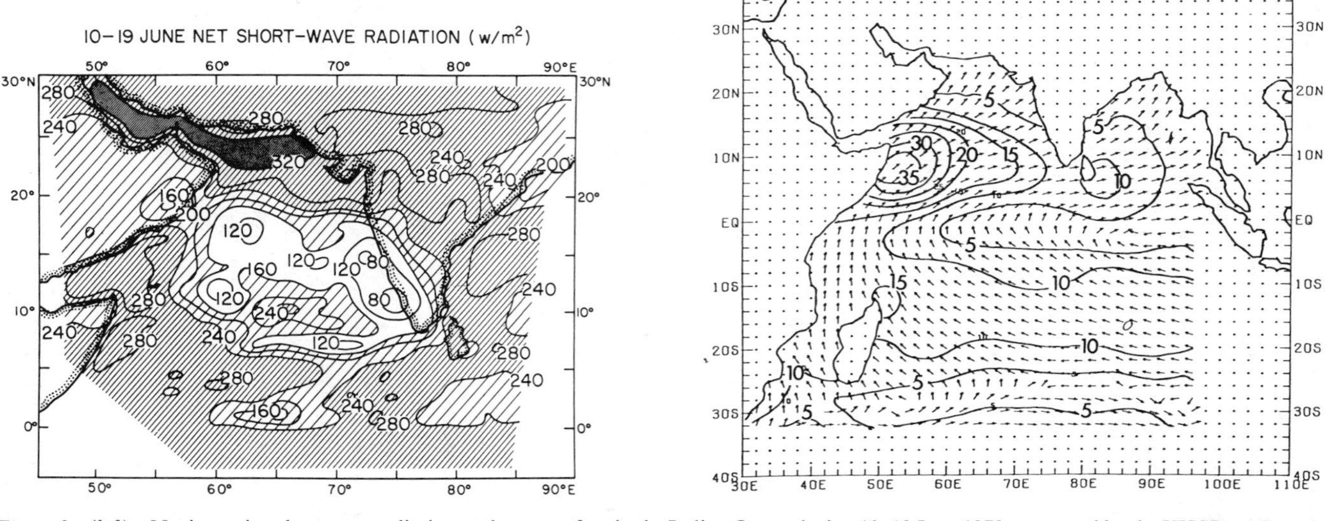

Figure 3 (*left*) Net incoming shortwave radiation at the sea surface in the Indian Ocean during 10–19 June 1979, measured by the VISSR on *Goes-1* (from C. Gautier, Scripps Institution of Oceanography). (*right*) Average wind stress for the same time and place calculated from measurements of low-level cloud velocities by the same instrument (from Wylie & Hinton 1982b).

The variation of emissivity with wind speed is large at frequencies above 3 GHz and is due to both waves and foam from breaking waves. The influence of waves dominates at angles well away from vertical. Thus, waves produce the signal observed by the SMMRs on *Seasat* and *Nimbus* 7, which view the sea at an incidence angle of 50°. At vertical incidence the influence of surface roughness is less, and foam is more important.

Radars transmit radio signals and observe the strength of the signal reflected from the sea surface. The reflectivity is due to small waves, and is a function of radio wavelength, polarization, incidence angle, and azimuth angle relative to the wind. Jones & Schroeder (1978), summarizing a large body of observations, showed that reflectivity increased with wind speed at incidence angles greater than 10–20°, that it decreased with wind speed for angles near vertical, and that the variation was greatest at higher frequencies and for horizontally polarized radiation. However, the reflectance is only weakly dependent on direction relative to the wind. Thus, very precise observations of radar backscatter at several azimuth angles are required to measure both wind speed and direction.

Preliminary assessments of radiometer observations of wind speed made by the *Seasat* SMMR indicate that wind was measured from space with an accuracy of ± 2.0 m s^{-1}; however, improvements are expected (Njoku 1982, Cardone et al 1983). A similar accuracy was achieved using radar measurements made at vertical by the *Seasat* altimeter (Brown et al 1981, Fedor & Brown 1982), and a slightly better accuracy of 1.6 m s^{-1} was obtained using the *Seasat* scatterometer, a radar that measured at incidence angles of 20–50° (Pierson 1981, Jones et al 1982).

Radar observations of wind direction are more difficult, requiring special radar geometries such as that of the *Seasat* scatterometer. This instrument observed the reflectivity of the sea at two azimuth angles, and produced estimates of wind direction that had a twofold ambiguity; that is, the instrument produced two to four possible wind directions at each point on the surface (Figure 4). Although one of these directions was within $\pm 18°$ of the true wind direction on average, the problem remains to find automatically which of the four vectors represents the true wind (Wurtele et al 1982). As a result, *Seasat* scatterometer measurements of wind have not yet been processed to produce vector wind fields, although work has begun to do so for a limited set of data, primarily to test various techniques for removing ambiguities and to produce a set of data useful for testing numerical weather predictions.

Measurements of wind speed made over the Pacific by the *Seasat* altimeter, scatterometer, and SHF radiometer have been compared with ship observations from the same region, and systematic and important differences were observed, casting doubt on earlier estimates of accuracy.

Several processes should be involved, but their relative importance is not yet known. The radio-frequency properties of the surface, particularly the radar reflectivity, depend on the height of short waves on the sea surface. The height, in turn, depends on a balance between the input of energy by the wind and other wave components (via wave-wave interactions) and the loss of energy through viscosity, surface films, and other waves (again through wave-wave interactions).

Measurements of radar reflectivity as a function of the height of longer waves on the surface, with wind speed constant, indicate that wave-wave interactions are perhaps only of minor importance. Thus, water viscosity and surface films, both of which damp short waves, and boundary-layer stability, which influences the input of energy to waves, must singly or together cause errors in radar measurements of wind speed. For *Seasat*, the influence of stability has been crudely accounted for, and work on estimating the influence of surface tension and viscosity is now underway.

Surface films are less well understood, and their influence has only been

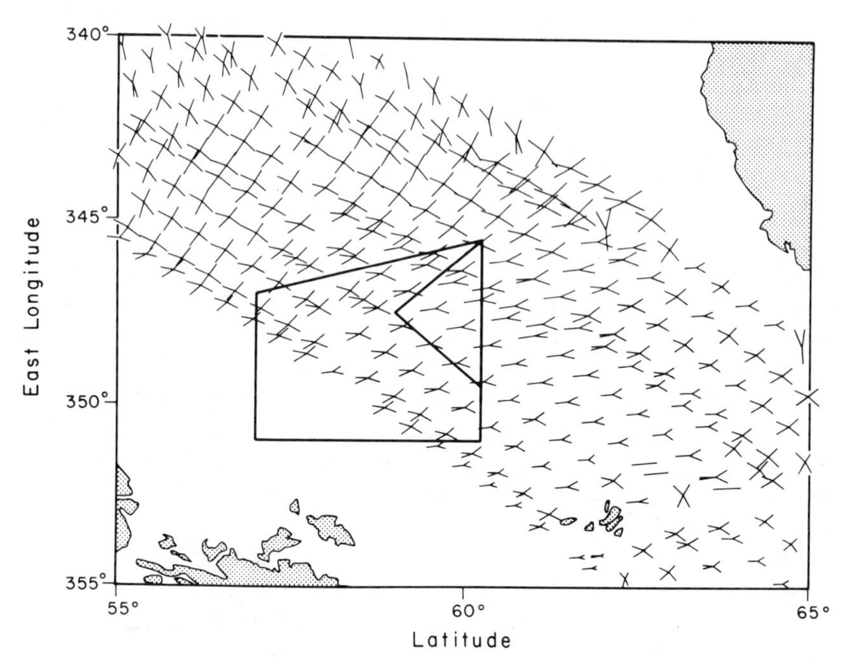

Figure 4 *Seasat* scatterometer observations of wind direction are ambiguous. In general, four directions satisfy the radar observations, but in some areas the ambiguity degenerates to two or three directions. Future scatterometers should have only a 180° ambiguity. The solid polygon is the area of the JASIN experiment used to calibrate the scatterometer (from Wurtele et al 1982).

crudely examined. Weak films are ubiquitous (Huhnerfuss et al 1977), and since they strongly effect the damping of short waves (Garrett 1967), they cannot be ignored. To better understand these errors, further experiments are now being planned that will use radars on aircraft to observe selected areas of the sea at the same time that the surface properties of the sea are measured independently by other means.

The measurement of surface winds via cloud motion is no easier than measurement via radar scatter. Over much of the ocean, and particularly in the tropics, small low-level cumulus clouds are common, and their position can be determined every thirty minutes using visible-light images from the VISSR on geostationary satellites. The wind at cloud level is calculated from the displacement of the cloud images. The measurement is indirect because the wind at height is faster than the wind near the surface, and because the wind direction rotates in the lower levels of the atmosphere. The change in speed can be 2–5 m s^{-1}, and the rotation can be as large as 35° depending on atmospheric conditions (Wylie & Hinton 1982a, Halpern 1979). Usually, the extrapolation from cloud levels to the surface is based on regional studies of the boundary layer over the ocean, so the technique is not yet truly global. Despite this, cloud motion is routinely used to measure oceanic winds (Figure 3), and at present (there being no scatterometers in space) the technique provides the only measurements of wind velocity from space (Green et al 1975, Wylie & Hinton 1982b).

OBSERVATIONS OF CURRENTS

Surface geostrophic currents produce variations in sea level with wavelengths ranging from 40 km to the dimensions of the ocean basins, and with periods ranging from weeks to centuries or longer. The variations are small, at most 1–2 m in height, but those with wavelengths of 40–3000 km and periods of 3–30 days were easily measured from space using a precise radar altimeter on *Seasat*. Such a radar sends out a radio pulse and measures its height by noting the time required for the pulse to travel to the surface and back. If the satellite's orbit (position) is also known, the measurement gives the height of sea level.

Measurements of time-variable currents require only observations with a precision of a few centimeters made repeatedly along an exactly recurring satellite track. Measurements of mean time-averaged currents (those produced by the general circulation of the ocean) are much more difficult because regional differences in gravity also produce variations in sea level. Thus, measurements of mean currents not only require a knowledge of the height of the altimeter relative to the center of the Earth (the vertical component of the satellite ephemeris) and a measurement of the height of the altimeter above the sea surface, but also a knowledge of the geoid over distances of 1000–10,000 km, all with an accuracy of better than ±5–10 cm and with no correlated, systematic errors so that many independent observations of the surface may be averaged together to obtain a mean sea level with an accuracy of 1–5 cm.

The accuracies required of measurements of oceanic topography, and hence surface currents, are stringent but possible. Errors are contributed by surface waves, which distort the reflected altimeter pulse; air molecules and water vapor, which delay the pulse in the troposphere; free electrons in the ionosphere, which further delay the pulse; and the instrument itself. Taken together, these can amount to an inaccuracy of 10–20 cm under typical conditions. Fortunately, these sources of error can be reduced to acceptable levels. The ionospheric electron content is routinely measured by various nations, and can be measured directly using a dual-frequency altimeter. Water vapor is measured by SHF radiometers, air pressure is provided by conventional surface observations and by surface weather analyses, and wave height is independently measured by the altimeter itself (Table 3). Only uncertainties in the knowledge of the geoid and the satellite ephemeris remain as important unsolved problems.

Using present techniques, orbits can be calculated with an uncertainty of ±40 cm in the height of the satellite (Table 4). Expected improvements in knowledge of Earth's gravity field, and new methods of continuously tracking satellites using coherent radio signals, should make possible calculations of the height of a satellite with an accuracy of ±5–15 cm. At the same time, accurate orbits will further refine our knowledge of gravity, and may allow the long-wavelength components of the geoid to be calculated with an accuracy of ±5–10 cm. In addition, special gravity-measuring

Table 3 Typical altimeter errors

Source of error	Uncorrected (*Seasat*) (cm)	Corrected (*Seasat*) (cm)	Corrected (*Topex*) (cm)	Decorrelation distance (km)
Geoid	10^4	10–200	10–50	3
Orbits	5×10^5	30–50	5–10	>10,000
Ionosphere	0.2–20	0.2–5	1.0	20
Mass of air	230	0.7	0.7	1000
Water vapor	6–30	2.0	1.2	50
Electromagnetic bias[a]	4	2	2.0	200–1000
Altimeter (noise)	5	5	1.5	20
Altimeter (calibration)	50	5–10	1.5	∞

[a] Assumes 2-m waveheight.

satellite systems, such as the Geomagnetic Research Mission, should give the shorter-wavelength components, also with good accuracy.

Satellite altimeters, flying on *Skylab*, *Geos-3*, and *Seasat*, have observed the sea with ever-increasing precision and accuracy. The first observations showed the geoid, which has undulations 100 times larger than those due to ocean currents (Figure 5); observations from *Geos-3* and *Seasat* have been combined to map the oceanic geoid with an accuracy of ± 1 m over a grid of roughly 100–300 km. Of course, the map has small errors due to permanent ocean currents, but these are small enough to be neglected. The observations of the geoid are now being used to study the thickness and strength of the oceanic lithosphere (Watts 1979) and to map the bathymetry of poorly surveyed regions of the ocean (Dixon et al 1983, Dixon & Parke 1983).

The *Geos-3* altimeter was the first to map time-variable ocean currents, but it could only observe the strongest currents. The improved accuracy and precision of the *Seasat* altimeter, together with the exactly recurring orbit, allowed the observation of much weaker currents. The observations of variable currents have been summarized by Cheney & Marsh (1981), who have published a global map of the variability of surface velocity; and the mean (time-averaged) currents have been investigated by Tai & Wunsch (1983), who have mapped the long-wavelength components of the mean currents in the Pacific.

Future satellites, such as the proposed *Topex* Mission, are expected to produce measurements of the oceanic topography accurate enough to make major contributions to oceanography. The goals of *Topex* are to map the surface topography of the ocean over entire ocean basins for several years, to integrate these measurements with subsurface measurements and models of the ocean's density field in order to determine the general circulation of the ocean and its variability, and then to use this information to understand the nature of ocean dynamics, to calculate the heat transported by the oceans and the interaction of currents with waves, and to test the ability to predict circulation from forcing by winds (*Topex* Science Working Group 1981).

Table 4 Accuracy of geodetic satellite orbits

Satellite	Inclination (°)	Height (km)	Area/Mass ($m^2\ kg^{-1}$)	Radial Orbit error (m)
Geos-3	115	840	4.1×10^{-3}	1.0–2.0
Lageos	110	5900	0.7×10^{-3}	0.3–0.4
Starlett	50	800–1100	1.0×10^{-3}	1.5–2.0
Seasat	108	800	11.4×10^{-3}	0.4–0.5

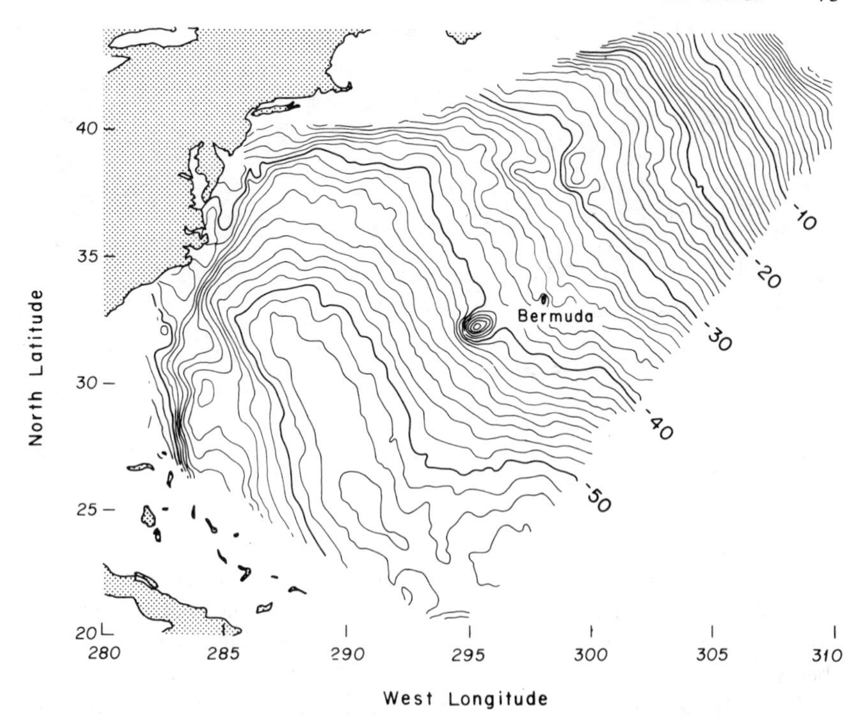

Figure 5 The mean sea surface in the northwest Atlantic produced from satellite altimeter measurements. The surface is determined almost entirely by the local geoid. Similar maps, with somewhat less spatial resolution, now exist for all the oceans. Contours given in meters (figure from J. C. Marsh, Goddard Space Flight Center).

To make the required measurements, the satellite will have a dual-frequency altimeter, with an accuracy of ± 2 cm, and a SHF radiometer to measure atmospheric water vapor. Both laser and radio-frequency tracking will be used to help determine the orbit within an accuracy of ± 5–15 cm.

Although our primary interest in this section has been to describe measurements of ocean currents, it is worth noting that altimeters also measure wind speed and wave height. The wave measurements are of practical importance, and the altimeters on *Geos-3* and *Seasat* have produced the first global maps of wave height (Figure 6).

OBSERVATIONS OF OCEAN COLOR

Both phytoplankton and sediments change the color of the ocean, and these changes can be seen from space. Their observation requires sensitive measurements of radiance in several narrow bands of visible light (usually

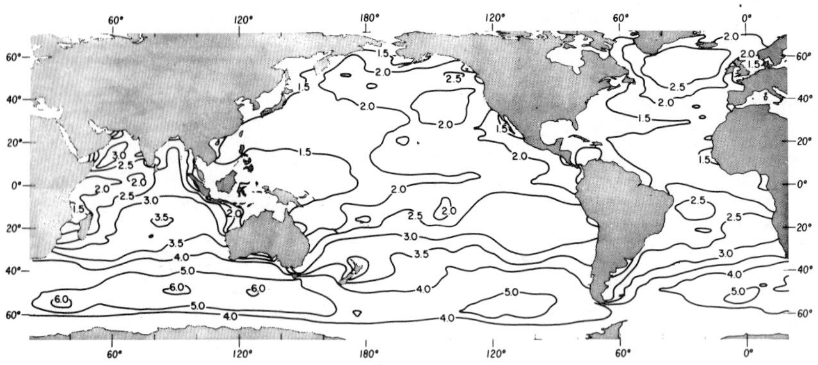

Figure 6 Average significant wave height (in meters) measured between 7 July and 10 October by the *Seasat* altimeter (from Chelton et al 1981).

blue, green, yellow, orange, and red), together with an accurate estimate of the light contributed by the atmosphere. Sensitive measurements are possible with modern multispectral scanners, but estimates of air light are difficult to make, and techniques for obtaining this information are still being evaluated.

Roughly 90% of the upwelling radiance observed from space comes from the atmosphere and must be removed to enable visible-light radiometers to see the sea surface clearly. Usually, it is assumed that the ocean emits no red light, so all upwelling red light comes from the air. The red radiance is then extrapolated to shorter wavelengths using known spectral properties of aerosols and Rayleigh-scattered light. The extrapolated values are subtracted from the observed upwelling radiance to obtain the fraction of the radiance that comes from the sea (Gordon & Clark 1980). The calculated values of radiance from the sea are then examined to determine if the original assumption of no upwelling red light is correct: some very turbid waters have a reddish hue and must be avoided. Alternatively, very clear waters with no measurable chlorophyll have well-known optical properties, and the signal from such water can be subtracted from the signal observed in space to yield the aerosol signal (Gordon et al 1983).

The calculated values of the radiance from the sea surface in several wavelength bands are combined to calculate ocean color and hence the concentration of chlorophyll-like pigments in the upper layers of the sea, and this concentration is then converted to productivity (Smith & Baker 1978). Alternatively, the radiance can be used to calculate the amount of suspended sediments in the water.

The first observations of chlorophyll from space were made using the

Multispectral Scanner (MSS) on *Landsat*. However, the images were not at the best wavelengths, nor were they sufficiently sensitive to small changes in radiance. As a result, they could not be used to make accurate measurements of oceanic productivity. Later, a much-improved instrument especially designed to view ocean color, the Coastal Zone Color Scanner (CZCS), was flown on *Nimbus* 7 (Hovis et al 1980). This instrument is now being used to study the distribution of chlorophyll in the ocean and its relationship to currents, insolation, and types of water mass. As a byproduct of this work comes accurate measurements of aerosol types and concentrations in the atmosphere. Observations of ocean color are expected to continue into the future with data from a new instrument, the Thematic Mapper (TM), on the recently launched *Landsat 4*. Although not specifically designed for oceanography, the bands at which it observes were chosen so that considerable oceanographic information can be obtained (Solomonson et al 1980).

OBSERVATIONS OF SEA ICE

The extent of sea ice, its thickness and composition, and whether it is first year or multiyear are important questions for practical purposes as well as for studies of climate. Fortunately, many means of observing ice are available: (*a*) Visible and infrared instruments can peer through holes in clouds, especially in the spring and autumn, and produce pictures (images) of the surface. From these it is possible to observe ice extent and ice leads, and to calculate some ice properties using infrared radiation (McClain 1974). (*b*) Super-high-frequency radiometers can measure variations in surface emissivity and be used to distinguish among first-year ice, multiyear ice, snow, and open water, depending on the region and season (Gloersen et al 1978). Similar observations of continental glaciers can be used to estimate ice grain size and hence snow-accumulation rate (Zwally 1977). (*c*) Radar reflectivity is a function of ice properties and surface roughness, and detailed maps of reflectivity having a resolution of 25 m produced by the *Seasat* SAR can be used to distinguish some ice types in the Arctic (Fu & Holt 1982). (*d*) Radar altimeters can note the position of the edge of an ice cover with an accuracy of a few kilometers.

The ease with which ice can be observed, and the great contrast between open water and ice, has allowed various instruments to observe ice over many years. The SHF radiometers (ESMR, SCAMS, NEMS, and SMMR) on the *Nimbus* satellites have been used to observe ice extent and type since 1972 (Figure 7). Later, the *Seasat* SAR produced many very detailed images of ice over a period of a few months, which are now being used to study ice dynamics. Less extensive, but still useful, observations have been made

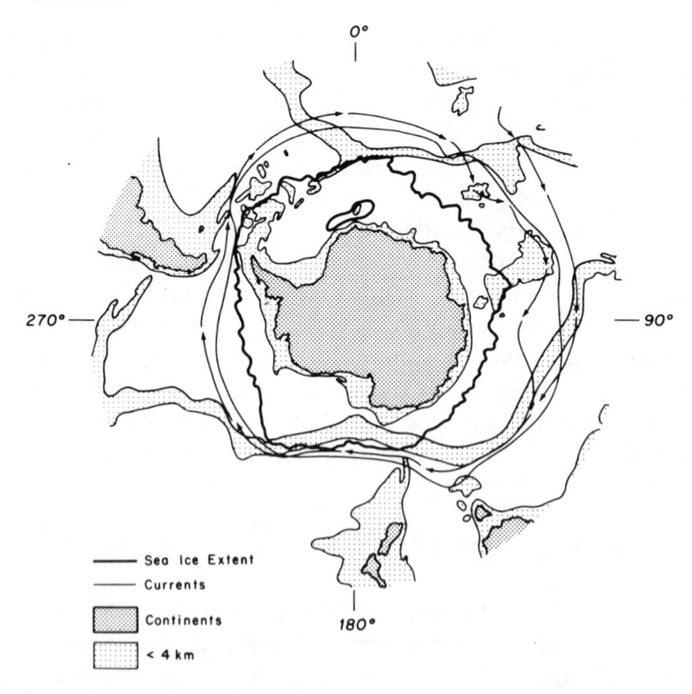

Figure 7 Distribution of Antarctic sea ice on 31 August 1974. The boundary is the 15% sea-ice contour derived from brightness temperature observed by the *Nimbus-5* ESMR. Note the open water near 0°—the Weddel Sea Polynya. Arrows show the approximate position of the Antarctic Circumpolar Current. Underwater ridges appear to deflect the current, and this, in turn, influences the ice boundary (from data in Zwally et al 1976).

using the visible and infrared scanners (VHRR, AVHRR, and MSS). These show ice motion, leads, and the position of large icebergs, one such being tracked for years as it drifted near Antarctica (Figure 8).

ANALYSIS OF DATA

The usefulness of satellite data depends critically on their accuracy and availability, both of which present difficult problems. The accuracy must be maintained from the time data are collected by a particular instrument, throughout their handling and processing, and within the data archives. Once processed to useful units, the calibrated data must be further processed and reduced in volume to yield maps that are accurately documented and easy to obtain in convenient form. The great volume of satellite data, the need to combine data from different instruments and satellites, and the fact that the data users are seldom involved in the design

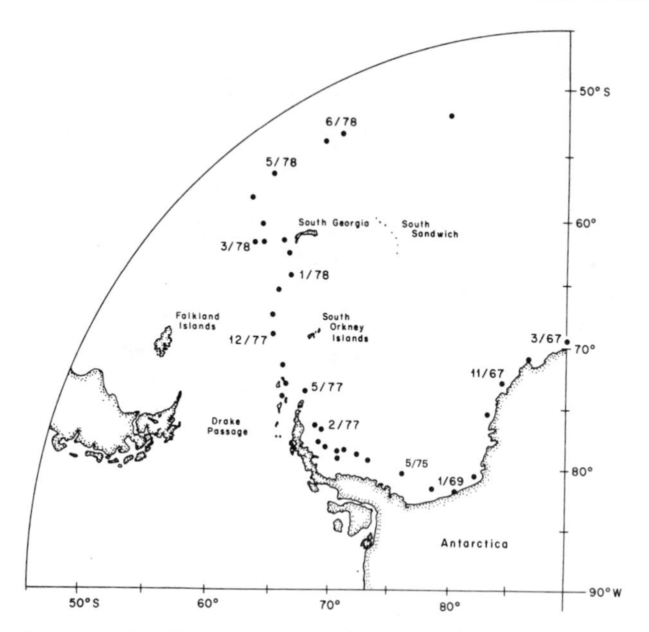

Figure 8 Positions of Trolltunga, a very large tabular iceberg, observed over a period of nearly eleven years by *Landsats* and meteorological satellites (data from McClain 1978).

of instruments or the collection of data make the problems particularly acute. Solutions to these problems are just now being sought.

The accuracy of any particular instrument tends to be documented and published. Examples include the analyses of the accuracy of the *Seasat* instruments and observations (Bernstein 1982b, Weissman 1980), and the instruments on the later *Nimbus* and *Landsat* satellites. However, some widely used instruments have been either uncalibrated (such as the VIS on the early *Geos* satellites) or only poorly calibrated.

Continuous calibration using both calibration on the spacecraft (such as using cold space as a reference temperature) and special calibration areas on the sea surface is rare. Yet both types of calibration are essential if data from a series of instruments are to be used to piece together a long record of observations. For example, many years of SHF radiometric observations of water vapor exist from a series of instruments carried on *Nimbus* satellites, but there was no continuous calibration that would have allowed all these data to be easily combined into a decade-long record of the change in water vapor in the atmosphere over the tropical Pacific.

The large data sets produced by satellites raise other problems. The data are difficult to handle, store, condense, and distribute. Yet most users require easy access to greatly reduced sets of data in convenient form, such as weekly or monthly maps of surface wind speed. In this context, present archives are not very useful. It is difficult for users to know what satellite data exist, to determine the accuracy of these data, or to find and use particular data sets.

The problems associated with present archives have led to proposals that they be replaced by information archives (Space Sciences Board 1982). Such archives would deal primarily with information contained in the satellite data, rather than with the bulky unprocessed data. An essential feature of the archive would be a scientific staff who use the satellite data for their own work, who produce reduced data sets, and who, in a sense, lend credibility to the data. In addition, the archive must include means for documenting, finding, storing, and distributing data electronically. Several prototype archives are being developed by NASA, and a Pilot Ocean Data System used exclusively for oceanographic data is now operating at the Jet Propulsion Laboratory.

The Pilot Ocean Data System contains most of the data from *Seasat*, some *Nimbus* data, conventional oceanographic data used to calibrate or verify the satellite data, and abstracts of papers describing the data, data processing, and calibration. The satellite data are in geophysical units located as a function of time, latitude, and longitude. These original data have been further reduced by oceanographers at the laboratory to produce weekly and monthly maps of variables such as wind speed or sea-surface temperature. These are the source of some of the illustrations in this review. Both original and processed data, together with appropriate documentation and references to the scientific literature, can be called up and displayed by any user, whether local or remote, with a computer terminal. Further processing of data is possible, and this leads to yet greater reduction in the volume of data.

Although only a pilot archive, the system is growing, and it is expected to be a model for future, larger archives. All programs used to manipulate the data are written in standard form and can be easily installed on other computers. Many oceanographers are contributing their time and ideas, and through their help the system is producing the type of information required for further studies of the ocean.

THE EMERGING GLOBAL PERSPECTIVE

The development of many different spaceborne instruments, the ability of different types of instruments to measure the same quantity, and the development of efficient means to analyze, combine, and compare observations from space have given oceanography a new global perspective. By intercomparing global observations, it is now possible to evaluate systematic or regional errors in the observations, and then to explore the relationships among the many phenomena viewed from space. The latter is by far the more exciting prospect, but the former commands our attention at the present.

The intercomparisons of observations of the same phenomena made by different instruments lead to improved knowledge of instrument errors, and to estimates of error that often exceed those predicted separately for each instrument. For example, D. B. Chelton (unpublished manuscript) has examined measurements of wind speed in the tropical Pacific during July, August, and September 1978 by ships and the *Seasat* altimeter and scatterometer, and he found systematic differences that were considerably greater than expected. Similarly, others are examining global measurements of sea-surface temperature made by the AVHRR, SMMR, and HIRS instruments during various periods in 1979. If formal errors are correct, all should agree within $\pm 1°C$, but no one will be very surprised if they do not.

Despite these lingering doubts about the accuracy of particular instruments, the global data sets now being distributed and analyzed are already producing new insights into the oceans (Legeckis 1978). And the combination of measurements from different instruments deepens these insights. Early measurements of the patterns of sea-surface temperature made by the VHRR and AVHRR infrared radiometers clearly showed differences in ocean circulation along the eastern and western boundaries of the oceans. When combined with measurements of ocean color from the CZCS, these data are beginning to show interactions of currents and oceanic productivity (Figure 9). Studies now underway are relating temperature to productivity to the distribution of commercial fish stocks. Ultimately, we expect this work will produce a better understanding of fisheries and a better foundation for the management of fish stocks.

Other important studies are also in progress. Ocean currents are driven primarily by the wind, although solar heating and evaporation play a role in some areas. Yet the statistics of ocean variability, the distribution of currents, and their response to the wind are only poorly understood; nor are we able to use numerical models of the ocean circulation to predict even the present state of the ocean, let alone its variability. Satellite data are essential to the solution of these problems.

Altimeters can easily measure the variability of ocean currents, particularly motion that is too brief or too small to be resolved by models of ocean circulation. After the statistics of this motion have been observed, it will then be possible to improve the models. The *Seasat* altimeter has already mapped the global distribution of ocean variability, but the mission was too short to obtain the statistics required to improve models of oceanic circulation. The next altimeter satellites, particularly *Topex*, are expected to provide this information.

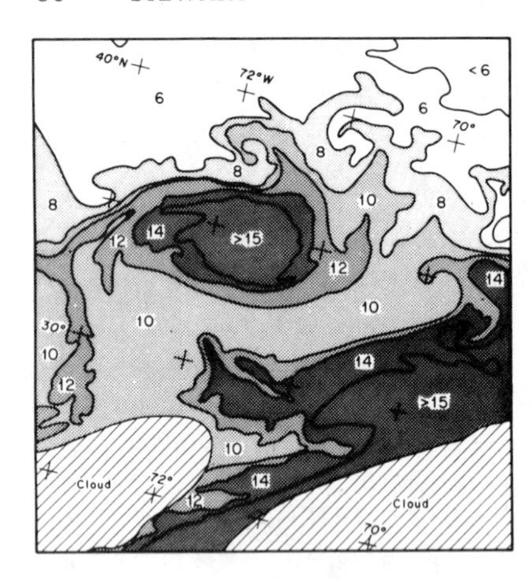

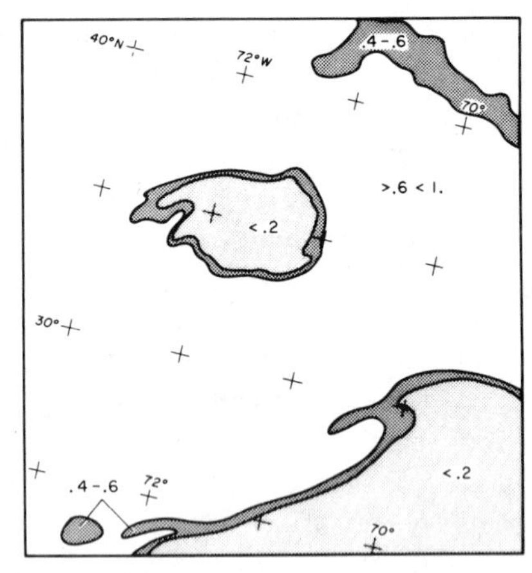

Figure 9 (*left*) The pattern of sea-surface temperature in the Gulf Stream on 1 May 1982 in °C observed by the AVHRR on the *Noaa*-7 meteorological satellite. (*right*) The chlorophyll pigment concentration in mg cm^{-2} in the same area observed by the CZCS on *Nimbus* 7. The images were received and processed by Otis Brown of the Rosenstiel School of Marine and Atmospheric Science of the University of Miami, and the figures here are redrawn directly from facsimiles sent via the *ATS* satellite to ships at sea participating in the "Rings" experiment.

Scatterometers are the only source of the global surface wind data required by the numerical models of ocean circulation. (Since currents in one region can result from wind forcing in a distant region, it is clear that global measurements are required.) The *Seasat* scatterometer demonstrated the ability to obtain the required data, but again the observations were for too brief a time to be very useful. Thus, in the future, scatterometers must be flown along with altimeters in order to understand the relationship between winds and currents.

These studies of ocean dynamics, to be successful, require good conventional measurements, particularly measurements of the density distribution within the ocean, information that cannot be obtained from space, and in situ measurements of currents and winds, information necessary to check the accuracy of the satellite observations of these fields. Here too, satellites make important contributions. Most oceanographic ships rely on satellite navigation, for this allows them to set subsurface instruments, such as current meters, and then return months later to retrieve the instruments and their data. Satellites also track the position of surface buoys that transmit back surface temperature and pressure. This system, called ARGOS, is very effective, and over a period of years has tracked many buoys and provided essential weather data from remote ocean areas. Newer buoys are now drogued to follow surface currents, and a small number of buoys are now being tracked to obtain more information about ocean circulation (Richardson 1983). In addition, satellites, particularly the *ATS-6*, transmit data to ships at sea, and these help guide ships into regions where measurements should be made (cf. Figure 9).

ACKNOWLEDGMENT

This work was supported by NASA through a contract to the Jet Propulsion Laboratory of the California Institute of Technology.

Literature Cited

Barnett, T. P., Patzert, S. C., Webb, W. C., Bean, B. R. 1979. Climatological usefulness of satellite determined sea-surface temperature in the tropical Pacific. *Bull. Am. Meteorol. Soc.* 60(3): 197–205

Bernstein, R. L. 1982a. Sea surface temperature estimation using the *NOAA 6* satellite advanced very high resolution radiometer. *J. Geophys. Res.* 87(C12): 9455–65

Bernstein, R. L. 1982b. *Seasat* special issue I: Geophysical evaluation. *J. Geophys. Res.* 87(C5): 3173–3438

Brown, G. S., Stanley, H. R., Roy, N. A. 1981. The wind-speed measurement capability of spaceborne radar altimeters. *IEEE Trans. Oceanic Eng.* 6(2): 59–63

Cardone, V., Chester, T., Lipes, R. 1983. Evaluation of *Seasat* SMMR wind speed measurements. *J. Geophys. Res.* 88(C3): 1709–26

Chelton, D. B., Hussey, K. J., Parke, M. E. 1981. Global satellite measurements of water vapor, wind speed and wave height. *Nature* 294(5841): 529–32

Cheney, R. E., Marsh, J. G. 1981. Global mesoscale variability from *Seasat* collinear altimeter data. *Eos, Trans. Am. Geophys. Union* 62(17): 298 (Abstr.)

Coakley, J. A., Bretherton, F. P. 1982. Cloud cover from high resolution scanner data: Detecting and allowing for partially filled fields of view. *J. Geophys. Res.* 87(C7): 4917–32

Dixon, T. H., Parke, M. E. 1983. Bathymetry estimates in the southern oceans from *Seasat*. *Nature* 304: 406–11

Dixon, T. H., Naraghi, M., McNutt, M. K., Smith, S. M. 1983. Bathymetric prediction from *Seasat* altimeter data. *J. Geophys. Res.* 88(C3): 1563–71

Fedor, L. S., Brown, G. S. 1982. Waveheight and wind speed measurements from the *Seasat* altimeter. *J. Geophys. Res.* 87(C5): 3254–60

Fu, L. L., Holt, B. 1982. *Seasat* views oceans and sea ice with synthetic-aperture radar. *Jet Propul. Lab. Rep. 81–120*, Pasadena, Calif. 200 pp.

Garrett, W. D. 1967. Damping of capillary waves at the air-sea interface by organic surface-active material. *J. Mar. Res.* 25(3): 279–91

Gautier, C. 1981. Daily shortwave energy budget over the ocean from geostationary satellite measurements. In *Oceanography From Space*, ed. J. F. R. Gower, pp. 201–6. New York: Plenum

Gautier, C., Diak, G., Masse, S. 1980. A simple physical model to estimate incident solar radiation at the surface from *Goes* satellite data. *J. Appl. Meteorol.* 19(8): 1005–12

Gloersen, P., Zwally, H. J., Chang, A. T. C., Hall, D. K., Campbell, W. J., Ramseier, R. O. 1978. Time-dependence of sea-ice concentration and multiyear ice fraction in the Arctic Basin. *Boundary-Layer Meteorol.* 13: 339–59

Gordon, H. R., Clark, D. K. 1980. Atmospheric effects in the remote sensing of phytoplankton pigments. *Boundary-Layer Meteorol.* 18: 299–314

Gordon, H. R., Clark, D. K., Brown, J. W., Brown, O. B., Evans, R. H., Broenkow, W. W. 1983. Phytoplankton pigment concentrations in the Middle Atlantic Bight: comparison of ship determinations and CZCS estimates. *Appl. Opt.* 22(1): 20–36

Green, R., Hughes, G., Novak, C., Schreitz, R. 1975. The automatic extraction of wind estimates from VISSR data. In *Central Processing and Analysis of Geostationary Satellite Data*, ed. C. L. Bristor, pp. 94–110. *NOAA/NESS Tech. Mem. 64*

Halpern, D. 1979. Surface wind measurements and low-level cloud motion vectors near the Intertropical Convergence Zone in the Central Pacific Ocean from November 1977 to March 1978. *Mon. Weather Rev.* 107: 1525–34

Henderson-Sellers, A. 1982. De-fogging cloud determination algorithms. *Nature* 298(5873): 419–20

Hofer, R., Njoku, E. G., Waters, J. W. 1981. Microwave radiometric measurements of sea surface temperature from the *Seasat* satellite: First results. *Science* 212:1385–87

Hovis, W. A., Clark, D. K., Anderson, F., Austin, R. W., Wilson, W. H., et al. 1980. *Nimbus 7* Coastal Zone Color Scanner: System description and initial imagery. *Science* 210(4465):60–63

Huhnerfuss, J., Walter, W., Kruspe, G. 1977. On the variability of surface tension with mean wind speed. *J. Phys. Oceanogr.* 7(4):567–71

Jones, W. L., Schroeder, L. C. 1978. Radar backscatter from the ocean: dependence on surface friction velocity. *Boundary-Layer Meteorol.* 13:133–49

Jones, W. L., Schroeder, L. C., Boggs, D. H., Bracalente, E. M., Brown, R. A., et al. 1982. The *Seasat-A* satellite scatterometer: The geophysical evaluation of remotely sensed wind vectors over the ocean. *J. Geophys. Res.* 87(C5):3297–3317

Legeckis, R. 1978. A survey of worldwide sea surface temperature fronts detected by environmental satellites. *J. Geophys. Res.* 83(C9):4501–22

Legeckis, R. V. 1979. Satellite observations of the influence of bottom topography in the seaward deflection of the Gulf Stream off Charleston, South Carolina. *J. Phys. Oceanogr.* 9(3):483–97

Lipes, R. G. 1982. Description of *Seasat* radiometer status and results. *J. Geophys. Res.* 87(C5):3385–95

Maul, G. A. 1981. Application of *Goes* visible-infrared data to quantifying meso-scale ocean surface temperatures. *J. Geophys. Res.* 86(C9):8007–21

McClain, E. P. 1974. Environmental Earth satellites for oceanographic-meteorological studies of the Bering Sea. In *Oceanography of the Bering Sea*, ed. D. W. Hood, E. J. Kelley, pp. 579–93. Fairbanks: Univ. Alaska

McClain, E. P. 1978. Eleven year chronicle of one of the world's most gigantic icebergs. *Mar. Weather Log* 22(5):328–33

Njoku, E. G. 1982. Passive microwave remote sensing of the Earth from space—a review. *Proc. IEEE* 70(7):728–50

Njoku, E. G., Hofer, R. 1981. *Seasat* SMMR observations of ocean surface temperature and wind speed in the North Pacific. In *Oceanography From Space*, ed. J. F. R. Gower, pp. 673–81. New York: Plenum

Pierson, W. J. 1981. Winds over the ocean as measured by the scatterometer on *Seasat*. In *Oceanography From Space*, ed. J. F. R. Gower, pp. 563–71. New York: Plenum

Richardson, P. L. 1983. Eddy kinetic energy in the North Atlantic from surface drifters. *J. Geophys. Res.* 88(C3):1563–71

Smith, R. C., Baker, K. S. 1978. The bio-optical state of ocean waters and remote sensing. *Limnol. Oceanogr.* 23(2):247–59

Solomonson, V. V., Smith, L. P., Park, A. B., Webb, U. C., Lynch, T. J. 1980. An overview of progress in the design and implementation of *Landsat-D* systems. *IEEE Trans. Geosci. Remote Sensing* 18(2):137–46

Space Sciences Board. 1982. *Data Management and Computation: Issues and Recommendations*, Vol. 1. Washington DC: US Natl. Res. Counc. 167 pp.

Stewart, R. H. 1982. Satellite oceanography: The instruments. *Oceanus* 24(3):66–74

Tai, C. K., Wunsch, C. 1983. Absolute measurement of the dynamic topography of the Pacific Ocean by satellite altimetry. *Nature* 301(5899):408–10

Topex Science Working Group. 1981. Satellite altimetric measurements of the ocean. *NASA Rep. 400-111*, Jet Propul. Lab., Pasadena, Calif. 78 pp.

Watts, A. B. 1979. On geoid heights derived from *Geos-3* altimeter data along the Hawaiian-Emperor Seamount chain. *J. Geophys. Res.* 84(B8):3817–26

Weissman, D. E. 1980. Special issue on *Seasat* sensors. *IEEE J. Oceanic Eng.* 5(2):71–180

Wurtele, M. G., Woiceshyn, P. M., Peteherych, S., Borowski, M., Appleby, W. S. 1982. Wind direction alias removal studies of *Seasat* scatterometer-derived wind fields. *J. Geophys. Res.* 87(C5):3365–77

Wylie, D. P., Hinton, B. B. 1982a. A comparison of cloud motion and ship wind observations over the Indian Ocean for the year of FGGE. *Boundary-Layer Meteorol.* 23:197–208

Wylie, D. P., Hinton, B. B. 1982b. The wind stress patterns over the Indian Ocean during the summer monsoon of 1979. *J. Phys. Oceanogr.* 12:186–99

Zwally, H. J. 1977. Microwave emissivity and accumulation rate of polar ice. *J. Glaciol.* 18(79):195–215

Zwally, H. J., Wilheit, T. T., Gloersen, P., Mueller, J. L. 1976. Characteristics of Antarctic sea ice as determined by satellite-borne microwave imagers. *Proc. Symp. Meteorol. Obs. Space: Their Contrib. First GARP Exp.*, pp. 94–97. Boulder, Colo: Nat. Cent. Atmos. Res.

Ann. Rev. Earth Planet. Sci. 1984. 12: 83–106

THE HISTORY OF WATER ON MARS[1]

Steven W. Squyres

Theoretical Studies Branch, NASA Ames Research Center, Moffett Field, California 94035

INTRODUCTION

Ever since Schiaparelli's description of the now infamous "canals" on Mars, the subject of water on that planet has been of considerable interest, both within and outside the scientific community. While the canals of Schiaparelli and Lowell were optical illusions, our exploration of Mars to date has shown conclusively that H_2O, often in the liquid state, has played a major role in shaping the appearance of the martian surface. With the close of the Viking mission, it is now an appropriate time to attempt a synthesis of this subject. This paper briefly summarizes what is known about the history of water on Mars. First we consider the long-term evolution of martian water. This discussion includes the geomorphic evidence for water on Mars, the chemical evidence concerning the total amount of water that has been outgassed, and the various ways in which water on Mars can be stored or lost. This is followed by a discussion of water cycles that operate over time scales of tens or hundreds of thousands of years, driven by climatic changes that appear to be related to variations in the orbital elements of Mars.

LONG-TERM EVOLUTION OF WATER ON MARS

Geomorphic Evidence for Water

CHANNELS One of the most startling and important discoveries of the Mariner 9 mission to Mars was widespread evidence for modification of the martian surface by the action of liquid water (McCauley et al 1972, Masursky 1973, Milton 1973). This evidence comes in the form of channels, which are often remarkably similar to terrestrial stream, river, and flood features. Martian channels have a wide range of morphologies, but those that show the strongest evidence for a fluvial origin can be loosely grouped in two categories: *outflow channels* and *runoff channels* (Sharp & Malin 1975). Outflow channels arise fully developed from spatially limited, often clearly bounded source regions. Runoff channels, however, possess tributary systems and are much more similar to conventional terrestrial drainage systems. Channels in a third category, often termed *fretted channels* (Sharp 1973, Sharp & Malin 1975), are less clearly of simple fluvial origin, and they are discussed below in the section on ground ice features.

Outflow channels Outflow channels (Figure 1) are concentrated in the equatorial regions of Mars. They are most common along the northern lowland/southern highland boundary, where they arise from source regions in the highlands and debouch onto the lowland plains. The source regions generally show very complex topography that earns them the name "chaotic terrain." These regions are depressed below their surroundings and commonly are bounded by steep, inward-facing escarpments. Their floors show very complex topography consisting of irregular hummocks, tilted slump blocks, and downdropped remnants of the original highland plateau. The appearance strongly suggests removal of subsurface material

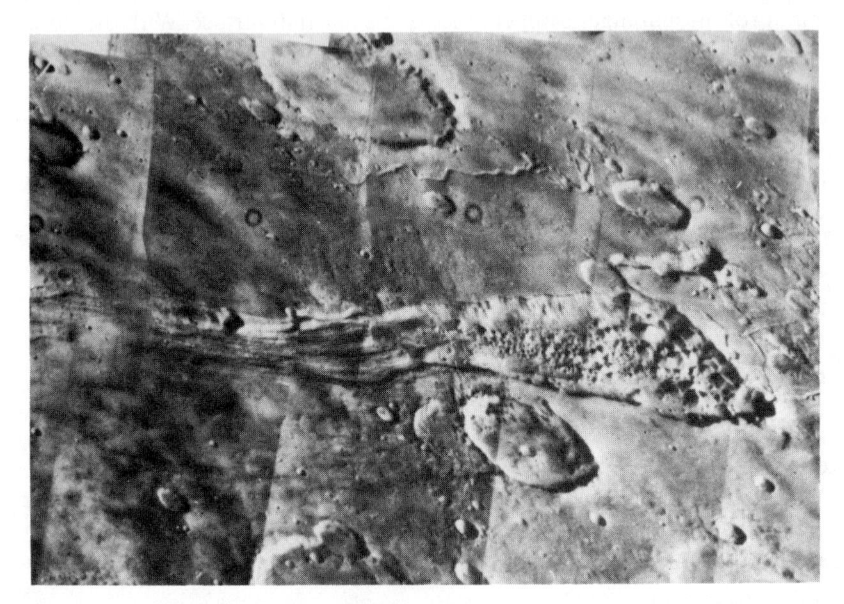

Figure 1 An outflow channel 20 km in width arising from a region of chaotic terrain (−1° lat, 42° long). The channel flows eastward into Simud Vallis.

and widespread collapse of topography. The channels arise fully born from these chaotic regions and may extend for many hundreds of kilometers. In some cases they are narrow and deeply incised, while in others they are broad and anastomosing. Their debouchments are much less distinct than their sources. In fact, they often simply fade into obscurity at their downstream ends. There are no clearly demonstrated regions of sedimentation at their debouchments.

Although clearly the result of fluid flow, outflow channels bear only the most superficial similarity to terrestrial rivers. They are much more similar to the types of features formed by catastrophic floods on Earth (Milton 1973, Baker & Milton 1974, Baker & Kochel 1979). The classic locality for massive flood features is the Channeled Scabland of eastern Washington state (Bretz 1923, 1969), formed by catastrophic breakouts of ice-dammed Lake Missoula and by the overflow of Lake Bonneville. The channels of the Channeled Scabland and of Mars have a great many features in common. These include (Baker & Milton 1974, Baker & Kochel 1979) regional anastomosis, indistinct fluid sink relationships, erosion of diverse rock types hundreds of kilometers from the source, residual uplands separating the channels and streamlined by flow, flow constrictions and expansions, bar complexes below expansion points, generally high width/depth ratios, low sinuosity, differential erosion controlled by lithology, longitudinal grooves, inner channels, cataracts, pendant-shaped streamlined obstacles, and scour marks around many obstacles. The most significant difference is that martian features are consistently several times larger than their proposed terrestrial counterparts. Caution must always be exercised when making arguments regarding the geology of other planets based on terrestrial analogy, but to many workers the evidence seems overwhelming that martian outflow channels were formed by catastrophic release of liquid water from the chaotic terrain.

The magnitudes of the floods implied are enormous. For example, Carr (1979) estimates a peak discharge of 7×10^6 to 5×10^8 m^3 s^{-1} for a flood originating in Juventae Chasma and extending across Lunae Planum. As in terrestrial floods, the principal erosion mechanisms would probably be very-large-scale turbulence, streamlining, and cavitation (Baker 1979). In addition, the martian climatic conditions may have favored scouring and erosion by ice blocks formed in the flow by evaporative cooling.

A variety of mechanisms have been suggested for triggering large floods on Mars. McCauley et al (1972) and Masursky et al (1977) suggested volcanic eruptions beneath glaciers as a possible mechanism. There is, however, no evidence for either massive former ice sheets or for volcanic activity at the locations of the channel sources. Milton (1974) suggested that floods could have resulted from pressure release dissociation of large amounts of subsurface CO_2-H_2O clathrate. This mechanism suffers from some thermodynamic shortcomings, however, including required surface temperatures in the range 0–10°C and termination of the reaction if more than 1% clathrate is present in the host rock. Several authors have suggested release by geothermal warming of subsurface ice in the chaotic terrain (McCauley et al 1972, Sharp & Malin 1975). This is an attractive possibility in many respects, although it too suffers from the difficulty that the channel source regions are not clearly associated with any volcanic features.

One problem in explaining outflow channels as flood features is that the volume of material removed from the chaotic terrain appears insufficient in some cases to produce the required flood discharges. A solution to this problem was proposed by Carr (1979). He suggested that water was released at high pressure from an extensive groundwater system. Much of the early martian crust may have been highly brecciated and porous due to meteorite impacts, so that a great deal of water could be stored beneath the surface. With cold surface temperatures, this subsurface aquifer would be capped by an impermeable permafrost layer. Thickening of the permafrost or tectonic warping of the surface (for example, formation of the Tharsis bulge) would have created very high pore pressures, particularly in low-lying areas. Carr proposed that channels formed when the pore pressure exceeded lithostatic pressure, causing hydraulic breakout and discharge of the aquifer. The total source region could therefore have been significantly larger than just the chaotic terrain where the breakout took place. This hypothesis also has the advantage that it requires no juxtaposition with volcanic regions.

The catastrophic flood hypothesis is widely, but not uniformly, accepted, and a variety of other models have been proposed. Nummedal (1978, 1980) has suggested that the channels were formed by subsurface liquefaction of clay-rich material in the chaotic terrain and release of highly mobile debris flows. A factor that could contribute to generation of a debris flow would be high water pore pressures, and there is clearly a continuum of possibilities between the moist debris flow envisioned by Nummedal and the debris-laden flood envisioned by Carr, Baker, and others. A more radical proposal is that of Lucchitta et al (1981), who suggest that the outflow channels may have been carved by enormous glaciers. They point out that ice streams are capable of producing many of the morphologic features found in the channels. The principal shortcoming of the glacial hypothesis is that the chaotic terrain, which clearly results from removal of subsurface material, is an unlikely source for large ice streams. Ice streams of the type and size cited by Lucchitta et al as possible terrestrial analogues are found at the perimeter of large ice sheets such as those in Greenland and Antarctica.

Other possibilities that have been investigated for outflow channel formation include eolian erosion (Cutts & Blasius 1981), lava erosion (Carr 1974), and flow of liquid alkanes (Yung & Pinto 1978). It seems very likely that at least eolian erosion has played a role in the subsequent modification of outflow channels (Nummedal 1981). The evidence is fairly convincing, however, that rapid flow of large amounts of liquid water, probably with substantial entrained debris and ice, was responsible for outflow channel formation. If this hypothesis is correct, then the magnitudes of discharges implied are sufficient for the features to have formed under the present martian climatic conditions (e.g. Wallace & Sagan 1979).

Using the density of superimposed impact craters, it is possible to determine the relative ages of the outflow channels (Malin 1976, Masursky et al 1977). As would be expected, the channels are younger than the cratered highlands from which they arise. Based on Mariner 9 images, Malin (1976) found the outflow channels to have crater ages older than the oldest plains of the Tharsis region. The absolute age derived depends on the model of the impact flux used. Using the flux model of Soderblom et al (1974), Malin concluded that the outflow channels are ancient features that date from the period of heavy bombardment some 4 billion years ago. Using Viking data, Masursky et al (1977) obtained crater ages somewhat younger, although the youngest ages they suggest are still 2.5–1.0 billion years, again using the flux model of Soderblom et al (1974). They find clear evidence for variations in age among the channels. It should be pointed out that use of a more recent flux model derived by Neukum & Wise (1976) puts all channeling events in the range of 3.5 to 4.0 billion years ago. There is obviously considerable uncertainty in the absolute ages, but it can be stated that the outflow channels formed over an interval of time subsequent to formation of the southern highlands but apparently concentrated fairly early in martian history.

Runoff channels (*valley systems*) A second type of channel apparently caused by flow of liquid is the runoff channel (Sharp & Malin 1975). These are more similar to terrestrial drainage systems, consisting of narrow, often sinuous valleys with tributary systems (Figure 2). It is very likely that runoff channels are not true channels, in the sense of being open conduits that were once filled with moving fluid. In fact, Viking resolution is insufficient to show any unquestionable evidence for fluid flow, such as streamlined obstacles, bars, or interior channels. The runoff channels are therefore more correctly termed "valley systems."

Although they possess tributary systems reminiscent of terrestrial drainage networks, the valley systems are in many ways markedly different from typical terrestrial features (Pieri 1980). They commonly have steep, clifflike walls with talus slopes, flat floors, and rounded amphitheater terminations to tributaries. The tributaries do not show typical dendritic patterns, but instead show parallelism and a lack of competition for undissected intervalley terrain. In fact, there is generally no evidence for runoff erosion on intervalley surfaces. These observations strongly suggest that rainfall was not responsible for carving most valley systems. The landforms exhibited by the valley systems are more typical of those formed by sapping and runoff of subsurface fluid.

It is not clear that all valley systems originated by sapping. Some, such as Nirgal, have a single, very long main valley with only a few short, stubby tributaries. In such instances, the main valley may be an endogenic feature (e.g. a collapsed lava tube or a graben) that has undergone minor fluvial modification. In fact, it has even been suggested that martian channels are primarily endogenic structural features (Schumm 1974). Also, a few valley systems have features consistent with formation by rainfall. For example,

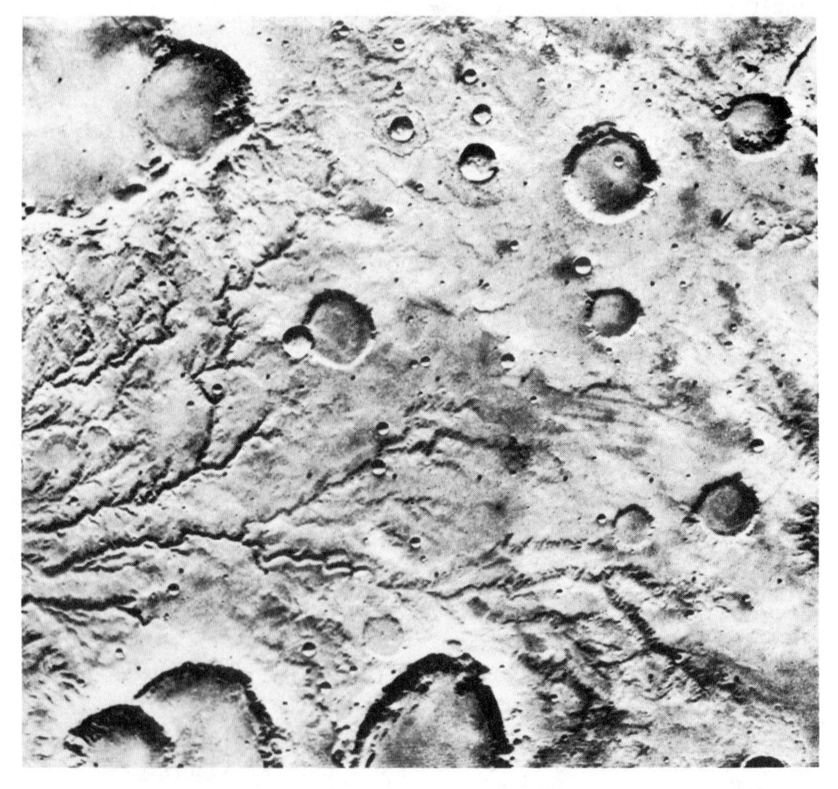

Figure 2 Typical runoff channels. The scale across the image is 265 km (−23° lat, 10° long).

the system in Figure 3 shows very complete dissection, with no broad intervalley surfaces. While the majority of valley systems seem to have developed by sapping, there are a few instances for which precipitation cannot be ruled out.

All of the valley systems on Mars are found in the ancient cratered highlands. They are found in all parts of the cratered highlands, although there is an indication of concentration in equatorial regions (Pieri 1976, 1980). The density of superimposed impact craters is substantially higher than that for the outflow channels and indicates an age as old as the cratered highlands themselves. The formation of valley systems or runoff channels on Mars was limited to the earliest part of the recorded geologic history, probably more than 4 billion years ago.

A particularly important aspect of the valley systems is that their formation by flow of liquid water is almost certain. The fluid discharges implied are quite modest, so that it is unlikely that they could have formed

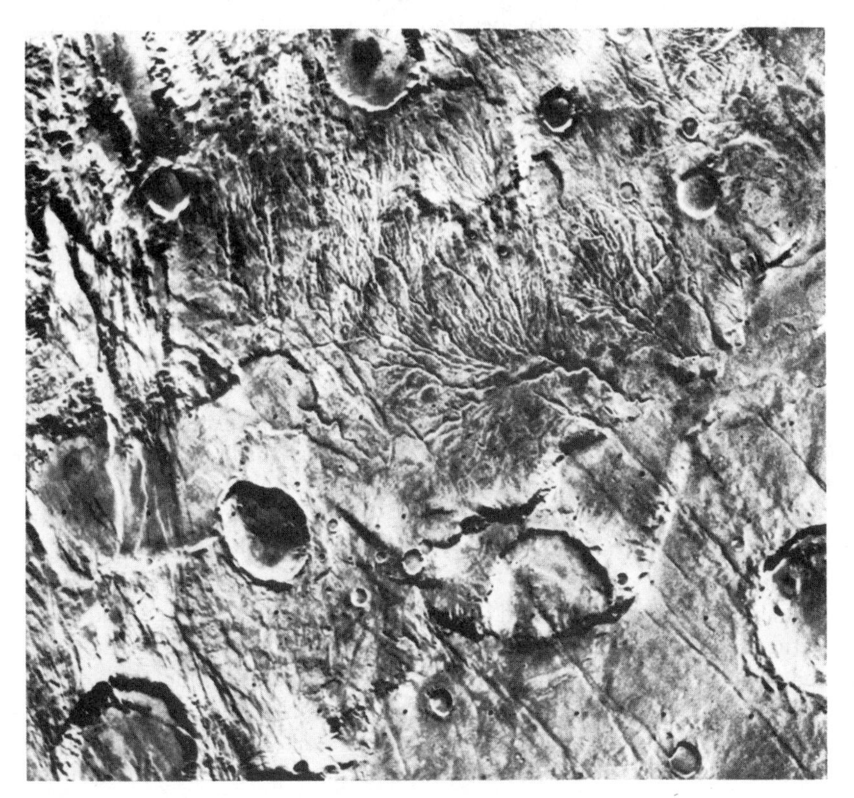

Figure 3 An atypical runoff channel with morphology perhaps consistent with formation by precipitation. The scale across the image is 250 km (−48° lat, 98° long).

under the present climatic conditions. While it is presently possible for water to flow for considerable distances on Mars if protected by a thick ice layer (Wallace & Sagan 1979), it is likely that the very shallow tributaries at the heads of the networks would become choked with ice under the present atmosphere, cutting off the main channel. Formation by sapping also of course implies the presence of subsurface liquid water at very shallow depths, requiring fairly high surface temperatures. The valley systems therefore present strong evidence that the surface atmospheric temperature and pressure very early in Mars' history were significantly higher than they are today. The lack of valley systems in terrains younger than about 4.0–3.5 billion years old suggests that this more clement era did not extend past the earliest part of martian history.

GROUND ICE Additional evidence for subsurface H_2O on Mars is provided by a class of geomorphic features that suggest the presence of ground ice. It is worth noting that subsurface ice cannot presently exist in equilibrium with the atmosphere at latitudes lower than about 40° (Fanale 1976). Any ground ice below these latitudes must, therefore, be sealed off from the atmosphere by burial. Smoluchowski (1968) has shown, however, that under stable climatic conditions a layer of fine-grained debris as thin as a few tens of meters could act as an effective diffusion barrier over time scales comparable to the age of Mars, so sealing from the atmosphere is unlikely to present a major problem.

Fretted terrain A number of types of features on Mars appear to owe their formation to removal of subsurface ice. One is the chaotic terrain already mentioned. Another is fretted terrain (Figure 4). This was first described by Sharp (1973) as "smooth, flat, lowland areas separated from a cratered upland by abrupt escarpments of complex planemetric configuration." The landforms produced range from isolated buttes and mesas to large plateau regions divided by narrow, flat-floored valleys (the "fretted channels"). Fretted terrain is found primarily along the northern lowland/southern highland boundary. Beginning at a break in the upland surface, such as a crater or fault, Sharp proposed that escarpment recession occurred, leaving behind a smooth lowland. The escarpment recession may have resulted from evaporation of exposed ground ice or emergence of groundwater. Either would imply a large amount of H_2O in the original highland material. A major problem is how the debris was removed from the scene, as there is no clear evidence of fluvial transport associated with fretted terrain. One possibility is that the debris was removed by eolian activity, which requires that the debris be of transportable size. There is, of course, abundant evidence for extensive eolian erosion elsewhere on Mars.

Figure 4 A region of fretted terrain in Deuteronilus Mensae (45° lat, 335° long). The mosaic is about 900 km across. Note the lobate debris aprons at the bases of many of the escarpments.

Thermokarst Aside from the major areas of fretted terrain, where escarpment heights may reach 1–2 km, there are a number of other regions where smaller-scale collapse has occurred, creating tablelands with scalloped edges, and small closed depressions (Carr & Schaber 1977). These features are very similar to "thermokarst" features formed by melting of subsurface ice at high latitudes on the Earth. Again, caution must be exercised when invoking terrestrial analogues for martian features, but the quantity of ground ice in thermokarst regions on Earth is typically 80–90% of the total volume of the deposits (Washburn 1973).

Lobate debris aprons A type of landform that implies very large amounts of interstitial ice is the lobate debris apron (Squyres 1978; see also Figure 4). As the name implies, these are debris aprons, lobate in plan view and distinctly convex in cross section, that are common at the bases of escarpments in the latitudes 30–50° in both hemispheres (Squyres 1979a). The convex profiles strongly suggest slow creep or flow of the debris, as do abundant flow lineations. It is probable that these features owe their appearance to gradual flow by creep of interstitial ice, much like terrestrial rock glaciers (Wahrhaftig & Cox 1959). The marked contrast between these very mobile flows and the steep scarps from which they arise (many in fretted terrain) suggests that they contain proportionally more ice than is present in the highland material. Their prevalence in the mid-latitudes, which undergo the largest amount of seasonal deposition of atmospheric H_2O (e.g. Leovy 1973, Jones et al 1979), suggests an atmospheric source for the additional ice. They are generally limited in extent and must contain only a small fraction of the total subsurface ice on Mars.

Rampart craters Many impact craters on Mars possess unusual ejecta deposits consisting of overlapping lobes of debris apparently fluidized at impact (Figure 5). These have commonly been referred to as "rampart

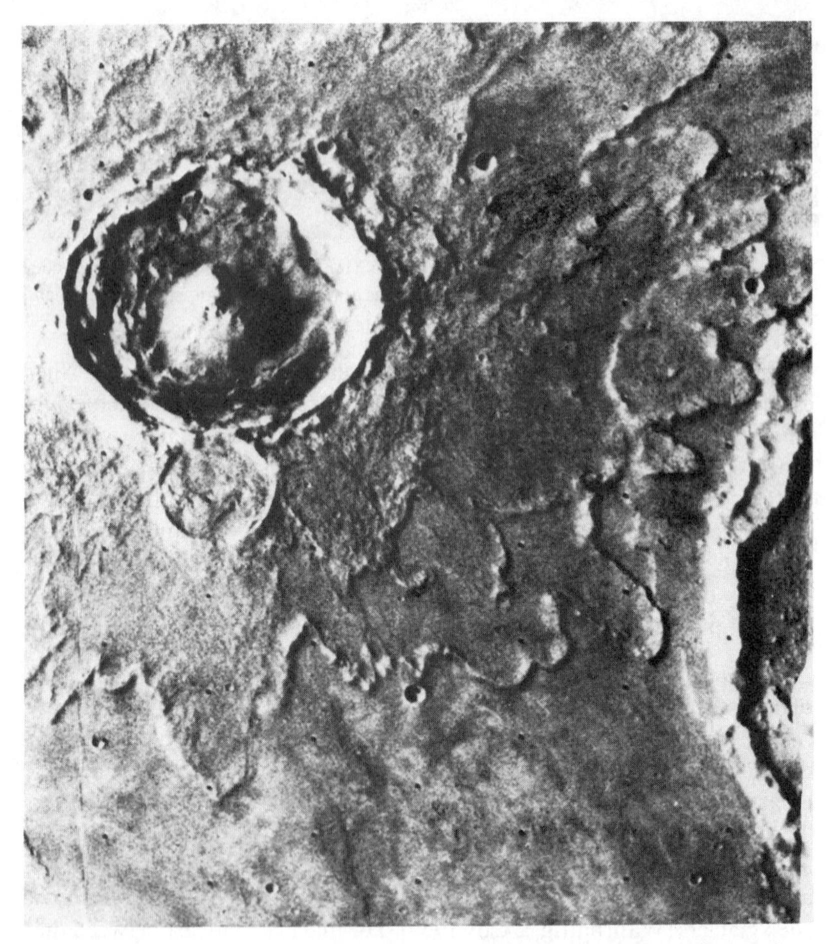

Figure 5 The impact crater Yuty (22° lat, 34° long), showing several sheets of lobed ejecta. The crater is 19 km in diameter.

craters" (McCauley 1973, Carr et al 1977a). There are a number of instances where outward flow of the debris has been deflected around obstacles not much higher than the flow thickness, indicating that little vertical compaction has taken place since the flow. This suggests that the flow was not gas supported, but instead involved lubrication by interstitial liquid, probably water. Rampart craters occur in most geologic units on Mars (Allen 1978) and at virtually all latitudes. The extent of deposits tends to be greatest at low altitudes and high latitudes (Mouginis-Mark 1979). This effect may reflect proportionately higher H_2O content at these locales, as cratering experiments in viscous media show an increase in flow extent with decreasing viscosity (Gault & Greeley 1978). It is not clear whether the water involved in the flow was in a liquid or solid state prior to the impact event. In an impact into ice-rich soil, ice near the impact will be vaporized and ice far from the impact point will remain solid, with only a thin transition zone being converted to liquid. It may be that the thickness of this zone is insufficient to account for the volume of the ejecta. If so, rampart craters may provide evidence for subsurface liquid water at the time of the impact.

Patterned ground Finally, there are a number of features on Mars that resemble "patterned ground" common at high latitudes on Earth. Patterned ground on Earth forms as a result of repeated diurnal and seasonal freezing and thawing of ice-rich soil, causing movement of material and segregation by ice content or sorting by particle size (Washburn 1973). The resultant patterns include circles, stripes, and networks of polygons. The most common types of patterns on Mars are networks of polygonal fractures. These could perhaps be frost-related, although their scale is one to two orders of magnitude larger than that of terrestrial patterns. The larger scale could be related to much longer period climatic fluctuations on Mars (discussed below) and consequently higher thermal skin depths (Helfenstein 1980). A major problem with the hypothesis that these features are formed by frost action is that the temperatures where they are found are currently substantially lower than the freezing point of water. Diurnal or seasonal thawing would require substantially warmer temperatures than are possible at the present or in the recent past. These features are found primarily in the younger northern plains of Mars, however, which apparently formed after the warm period early in martian history. Alternatively, the features could be cooling fractures in lava or extensional tectonic features (Pechmann 1980).

To summarize, there is substantial geomorphic evidence for the former existence of large amounts of H_2O in the surface material of Mars, extending to depths of 1–2 km. Some of this water has been removed to form the many collapse and sapping features observed, but some must still remain. How the regolith was originally charged with water is not clear. Volcanic outgassing and precipitation during a time of higher atmospheric temperatures and pressures is an obvious possibility, but the geologic evidence for precipitation is scanty. It is likely that little of the geomorphic evidence of the very early outgassing has been preserved through the subsequent geologic activity. Our understanding of the geology of Mars is not sufficient to make accurate estimates of Mars' H_2O content based on the geomorphic evidence alone.

Chemical Evidence for Water

Potentially useful data for estimating the total amount of H_2O on Mars have come from compositional studies of the martian atmosphere. Noble gas abundances have long been thought to be particularly important, because once a noble gas such as argon is released into the martian atmosphere, it is effectively there for good. It cannot chemically react with other atmospheric constituents or with surface materials, and the mass is sufficiently high that atmospheric escape is negligible. It has commonly been assumed that the relative abundances of volatiles outgassed by Mars are the same as those for the Earth, so that if the abundance of a primordial noble gas not subject to loss is known, the initial abundance of another volatile subject to loss may be inferred. As is seen, however, recent data from Venus cast considerable doubt on the validity of this assumption.

The first attempt at an in situ measurement of the noble gas composition of the martian atmosphere was made by the failed Soviet lander Mars 6 (Istomin & Grechnev 1976). From engineering telemetry received during the descent, it was deduced that the martian atmosphere had a large noble gas component, with subsequent laboratory duplication of instrument performance suggesting about 35% argon by volume. This figure implied that the ratio of the mass of outgassed argon to the mass of the planet was roughly equal for Mars and the Earth. The enormous amount of outgassed H_2O implied for Mars (about 10^5 g cm^{-2}) seemed inconsistent with reasonable ideas of the capacity of the regolith and prompted some fairly ingenious schemes for hiding large amounts of water.

The situation was of course resolved shortly thereafter by the successful operation of the Viking landers. The Viking mass spectrometers showed that the martian atmosphere at the planet's surface is only 1.6% Ar (Owen & Biemann 1976, Biemann et al 1976, Owen et al 1976, Owen et al 1977). The total abundance of nonradiogenic ^{36}Ar, relative to the planet's mass, is 0.0075 the terrestrial value. Using this number alone to deduce the amount of H_2O outgassing on Mars, one arrives at a figure of 10^3 g cm^{-2}, or a layer 10 m deep spread over the surface of the planet.

More recent results from Venus suggest very strongly that the similitude assumption is not a valid one. Results from the Pioneer Venus large probe and bus mass spectrometers indicate a ^{36}Ar abundance there that, normalized to the planet's mass, is roughly two orders of magnitude larger than the terrestrial value (Hoffman et al 1980). While there is evidence that Venus formerly had more water than is there presently (Donahue et al 1982), it is fairly clear that it never had a hundred times the amount on the Earth. There is therefore no reason to expect similar $^{36}Ar/H_2O$ ratios for the Earth and Mars. It has been proposed instead that the steep ^{36}Ar gradient observed from Venus to Earth to Mars reflects the decrease in nebular pressure (and hence ^{36}Ar partial pressure) in a near-isothermal solar nebula at the time of volatile incorporation (Pollack & Black 1979). Alternatively, it could have resulted from preferential solar wind implantation near the Sun in the solar nebula (Wetherill 1981). The intense interest in the ^{36}Ar abundance on Mars for use in estimating H_2O outgassing may have been unwarranted, then, as the Pioneer Venus data suggest very strongly that the assumption upon which it was based was flawed.

Fortunately, some other methods exist for estimating the amount of H_2O outgassed from Mars. Rasool & LeSergeant (1977) deduced the outgassed H_2O on Mars from the measured ^{36}Ar abundance and the $^{36}Ar/H$ ratio of ordinary chondrites, obtaining a figure of 600 g cm^{-2}. This might be a more reasonable way of using argon, if the ordinary chondrites formed at a position close to that of Mars in the solar nebula. [The validity of using ordinary chondrites in this manner has been disputed, however (Bogard & Gibson 1978).] Pollack & Black (1979) made a calculation based on the assumption that the K/H_2O ratio is similar for Mars and the Earth. From the measured amount of ^{40}Ar in the atmosphere (which is derived from radioactive decay of ^{40}K), they inferred outgassing of Mars sufficient to produce roughly 10^4 g cm^{-2} of water.

McElroy et al (1976, 1977) have considered the implications of the abundances of nitrogen. The present atmosphere of Mars is 2.7% N_2, with a $^{15}N/^{14}N$ ratio of 0.0060, or 1.7 times the terrestrial value. They interpret this enrichment as indicating that Mars had a much larger nitrogen atmosphere in the past. If we consider only atmospheric escape, the ^{15}N enrichment implies an original atmospheric nitrogen partial pressure of at least 1.3 mbar. (The present atmosphere of 95% CO_2 has a surface pressure of $\sim$7 mbar.) A great deal of nitrogen may have been lost from the atmosphere by incorporation of HNO_2 and HNO_3 into surface materials. Depending on the reaction rates chosen, the initial N_2 outgassing may have been as much as 30 mbar. Assuming a N/H_2O ratio like the Earth's, McElroy et al deduce outgassing of up to 1.3×10^4 g cm^{-2} of H_2O on Mars. This assumption is more likely to be correct than the argon similitude assumption, as N and H_2O were probably incorporated into planetesimals in a similar manner (within chemical compounds), whereas Ar and H_2O almost certainly were not (Pollack & Black 1979). In order to have reached the present low nitrogen concentration, McElroy et al conclude that N_2 outgassing (and, by inference, H_2O outgassing) was concentrated very early in martian history.

There are obviously very large uncertainties in these various techniques for estimating Mars' H_2O content. The uncertainty in the Mars-meteorite similitude assumption and the difficulties involved in estimating how much nitrogen has been lost by reaction with the surface are particularly troublesome. Nevertheless, it seems likely that early in its history Mars underwent H_2O outgassing sufficient to produce a layer of water something like 10 to 100 m deep over the surface of the planet.

Loss and Storage of Water

Where has this water gone? Clearly, much of it is frozen in the regolith, as is demonstrated by the geomorphic evidence. In fact, it is probable that ground ice constitutes the largest present sink of water on Mars. The amount in the regolith is less than it once was, however, as is shown by the abundance of features created by removal of subsurface ice or water. There are several other possible fates for water on Mars. It may be present in the current atmosphere, lost from the atmosphere to space, chemically locked in silicates, physically adsorbed on regolith grains, or trapped as ice in the polar deposits.

ATMOSPHERIC WATER A complete discussion of the behavior of water vapor over diurnal and seasonal time scales in the present martian atmosphere is beyond the scope of this article. Compared to other sinks, however, the water content of the atmosphere is quite small. It has long been known from Earth-based observations that water vapor in the martian atmosphere undergoes significant amounts of transport annually, reaching a maximum in each hemisphere after summer solstice (e.g. Barker 1976). The Viking orbiter mapped the distribution of water in the atmosphere and its variation with time (Farmer et al 1977, Jakosky & Farmer 1982). Planet-wide, the highest concentrations occur near the north polar region just after summer solstice, but even there the largest column abundance measured is only about 100 precipitable microns. The maximum total amount of water observed in the martian atmosphere at any time during the year is equivalent to just 1.3 km^3 of ice. A typical water vapor content of the atmosphere near the Viking landing sites is 0.03% (Owen et al 1977).

LOSS TO SPACE Models of the photochemical processes that take place in the martian atmosphere suggest that the H_2O content is relatively constant from the surface up to an altitude of about 40 km, but that it drops off very abruptly above that level due to photochemical decomposition (McElroy et al 1977). Initial decomposition takes place by photolysis, producing OH and H, and also by reaction with excited oxygen, producing OH + OH. These products undergo transport and a variety of other reactions, the net result of which is that H and O escape at a rate equivalent to a water loss of 6×10^7 molecules $cm^{-2}\ s^{-1}$. If this escape rate is simply assumed constant over the history of Mars, it would amount to a total loss of roughly 250 g cm^{-2}, which is probably a fairly small fraction of the total H_2O outgassed. The former loss rate may have been somewhat higher, however, if the atmosphere early in Mars' history was warmer and contained more H_2O than it does at present.

CHEMICAL INCORPORATION IN THE REGOLITH The reddish color of Mars has long suggested that rocks at the surface have undergone a great deal of chemical alteration, and that the surface material may contain a significant amount of chemically bound water. An elemental analysis of martian surface materials was carried out by the Viking lander X-ray fluorescence experiment (Clark et al 1977). Surface fines were acquired from both landing sites to a depth of 25 cm beneath the surface (Baird et al 1977). The elemental compositions of the samples were nearly identical at both sites, suggesting that all of the material acquired was part of a thin, fine-grained blanket of eolian debris that covers much of the planet's surface. The composition is unlike that of any single known mineral or rock type and apparently represents a mixture of minerals. The data can be fit well by a mixture of iron-rich smectite clays (e.g. montmorillonite

$$Mg_{0.3}Al_{1.70}Si_4O_{10}(OH)_2 \cdot Ca_{0.15},$$

saponite

$$Mg_3Al_{0.5}Si_{3.5}O_{10}(OH)_2 \cdot Ca_{0.25},$$

and nontronite

$$Fe_2Al_{0.5}Si_{3.5}O_{10}(OH)_2 \cdot Ca_{0.25}),$$

with smaller amounts of carbonates, iron oxides (such as maghemite), and sulfates (Toulmin et al 1977, Baird et al 1982). Besides the hydroxyl (OH) water contained in the smectites, such clays can also contain substantial amounts of water sandwiched between the layers of the crystal structure. This interlamellar water is readily lost in a dry environment, however, and it is not clear that a great deal of it could be trapped in martian clays. The hydroxyl water is much more stable, however, and may be an important sink of martian H_2O. Heating of martian fines produced little water below 200°C but generated H_2O vapor equivalent to 0.1–1.0 wt % at temperatures up to 500°C (Biemann et al 1977). This behavior is consistent with dehydroxylization of minerals containing OH.

While a mixture dominated by smectites provides a good fit to the X-ray fluorescence data, this solution is by no means unique, and other assemblages may do as well. Specifically, palagonite, the low-temperature hydrothermal alteration product of basaltic glass, may be a significant component of the martian fines. Detailed reflectance spectra of areas similar to the Viking landing sites appear to be incompatible with large amounts of smectites, but they are consistent with a composition dominated by palagonites (Singer 1982). Like smectites, palagonites may contain substantial amounts of water.

Smectite clays commonly form on Earth as a result of weathering of mafic rocks. The efficiency of weathering of crystalline rocks in the present martian environment is quite low, however (Gooding & Keil 1978), and this seems to be an unlikely source for the abundant martian fines. Both smectites and palagonite can form as a result of hydrothermal alteration of volcanic materials. Basaltic glasses erupted in a water-rich environment can be rapidly hydrated to form palagonite, which may yield smectites upon cooling and devitrification. On the Earth this takes place commonly at submarine volcanic vents (Bonatti 1965) but also occurs as a result of subglacial volcanic eruptions (Walker & Blake 1966). It is reasonable to expect that basaltic eruptions into or over ice-rich permafrost could also produce large amounts of smectites or palagonites, trapping some of the water released by heating of the ground ice. There is, of course, widespread evidence for basaltic vulcanism on Mars (e.g. Carr et al 1977b), and the evidence for large amounts of ground ice has been discussed above. It is inevitable that there has been a great deal of basalt-ground ice interaction, and this interaction may have produced significant quantities of smectite clays or palagonites that lock up water. Another source of water-rich fines may be impact-induced hydrothermal alteration related to cratering events (Newsom 1980), but the amount produced by this method is probably small relative to other sources (Allen et al 1982).

ADSORPTION Water can also be physically adsorbed onto surface grains. Adsorption is a reversible process, so water stored in this manner is available for exchange with the atmosphere and can play a role in any short-term climatic fluctuations. Based on experimental and theoretical considerations, it has been estimated that at the very most 3×10^3 g cm^{-2} of water could be adsorbed on all the free surfaces of a regolith 2 km thick (Fanale & Cannon 1974). The actual amount of adsorbed water is probably substantially less. Adsorption may be a very important form of storage for CO_2 that is not in the atmosphere, however.

SHORT-TERM CYCLES IN CLIMATE AND WATER TRANSPORT

Geomorphic Evidence: The Polar Deposits

One other possible fate for water on Mars is that it can become part of the polar deposits. The martian polar regions exhibit a very complex stratigraphy but from the simplest standpoint can be considered to consist of three major units. From base to top, these are the *layered deposits*, the *perennial ice*, and the *seasonal frost cap*.

LAYERED DEPOSITS The layered deposits were first recognized in Mariner 9 images of Mars (Murray et al 1972, Soderblom et al 1973, Cutts 1973). They are found at both poles and extend equatorward to 85–80° latitude. They consist of a thick sequence of thin, horizontal layers. Individual layers observed range in thickness from about 10 to 50 m (Dzurisin & Blasius 1975, Blasius et al 1982), but it is possible that still finer layering exists beyond the resolution limit of orbital images. Individual layers commonly are laterally continuous for hundreds of kilometers. The total thickness of the deposits is difficult to estimate accurately but may be 1–2 km in the south and 4–6 km in the north (Dzurisin & Blasius 1975).

The surface of the layered deposits is free of impact craters, indicating a very young age. The surface is not uniformly smooth, however, but is transected by a roughly spiral pattern of deep arcuate troughs that expose the edges of the layers on their walls. The slopes of the troughs are quite gentle, only about 1–8° (Blasius et al 1982). The troughs are typically 10–20 km wide and range in length from a few tens of km to a maximum of roughly 500 km.

PERENNIAL ICE The layered deposits are overlain by the perennial ice. As its name implies, this is a deposit of ice that remains in place throughout the martian year. At the north pole, the perennial ice reaches almost to the perimeter of the layered deposits, while in the south it covers a smaller area. At both poles the coverage of the layered deposits is determined by slope angle. Even very close to the pole the equatorward-facing slopes of the troughs are kept free of perennial ice by the higher insolation there, while the poleward-facing slopes and the flats separating the troughs are ice covered.

The composition of the perennial ice was long a subject of debate (e.g. Leighton & Murray 1966, Ingersoll 1974). The debate was at least partially settled when the Viking orbiter infrared thermal mapper measured late-summer temperatures over the perennial ice at both poles. At the north pole, typical surface temperatures are 205 K (Kieffer et al 1976). This is substantially higher than 148 K, the saturation temperature of CO_2 at the mean martian surface pressure of 6.1 mbar. Even when we take into account the possibility of existence in a clathrate, the temperatures are inconsistent with the presence of solid CO_2 on the surface. The northern perennial ice, therefore, is H_2O, in accord with large, near-saturation amounts of atmospheric water vapor detected at the north pole during the summer (Farmer et al 1977). The albedo of the ice is ~43%, indicating an admixture of a small amount of dust or other darker debris. At the south pole the situation is more complex (Kieffer 1979). Late-summer temperatures there during the first year of Viking observations were significantly colder and were consistent with the presence of CO_2 frost at the surface. In addition, the infrared brightness temperatures at the infrared thermal mapper's four detection wavelengths showed a spectral pattern best matched by a mixture of CO_2 frost and small amounts of dust. The preservation of CO_2 frost at the surface during the summer may be in part due to screening of solar radiation by atmospheric dust, as global dust storms occur during the southern summer. It is likely that the perennial CO_2 frost at the south pole is quite thin and underlain by trapped H_2O ice like that at the north pole.

The thickness of the perennial ice deposits is not well determined. Atmospheric water vapor detections above the northern perennial ice suggest a thickness of at least 1 m, based on the inferred thermal inertia to the depth of penetration of the seasonal thermal wave (Davies et al 1977). Considering the lack of topographic expression and the sensitivity of the ice to subtle changes in slope, it seems unlikely that its thickness exceeds that of the individual layers in the layered deposits. Because the perennial ice covers no more than 1% of the martian surface, the total amount of H_2O stored there probably does not substantially exceed 10 g cm^{-2}, globally averaged, and could be much less.

SEASONAL FROST CAP During the winter at each pole, the perennial ice and layered deposits are completely covered by a seasonal frost cap. This extends from the pole down to a latitude of 45–40°. It was long suspected that the seasonal frost was composed of a thin layer of solid CO_2 condensed from the atmosphere. This suspicion was confirmed by Mariner 7 infrared radiometer results, which showed frost temperatures of ~150 K (Neugebauer et al 1971).

Layered Deposit Formation and Climatic Change

A number of models have been proposed to account for the appearance of the layered deposits. The deposition process was first described by Cutts (1973). He noted that a substantial fraction of the atmospheric CO_2 on Mars is displaced into the seasonal frost cap each winter. He suggested that this poleward flux of CO_2 could entrain dust particles put into suspension by planet-wide dust storms. Entrained particles would then become embedded in the seasonal cap and be left behind when the cap evaporates in summer. Cutts proposed that the layered deposits were formed during an earlier era and were presently undergoing eolian erosion to form the troughs. Later models (Howard 1978, Cutts et al 1979, Squyres 1979b, Toon et al 1980, Cutts & Lewis 1982) have retained this fundamental idea of dust deposition but have recognized the importance of the perennial ice and have taken a more uniformitarian approach to the timing of deposition and erosion. The role of the perennial ice in forming the layered deposits is implied by the limited extent of the layers. Seasonal dust deposition will take place at all latitudes covered by the seasonal caps, and indeed there is evidence in the Viking orbiter images for dust mantling down to the mid-latitudes. The layered deposits are limited to within 5–10° of the poles, however, which at least in the north coincides with the present limit of the perennial ice. It seems, then, that only dust that is deposited onto the perennial ice is incorporated into the layered deposits. The more recent models have viewed the layered deposits as a mixture of dust and water ice. The layers may be deposited when dust is seasonally placed into the atmosphere by global dust storms and H_2O accumulates at the polar cold traps, transported there from lower latitudes by the atmosphere. They may erode when exposed to high insolation, causing sublimation of the ice, and to winds, causing deflation of the dust. The ice content of the deposits is not known, but values as high as 85% have been suggested (Toon et al 1980). The layered deposits could therefore contain H_2O equivalent to several times 10^3 g cm^{-2}, globally averaged, and may be one of the largest current sinks of water on Mars.

The fine layering in the deposits suggests some type of periodicity in the deposition and erosion processes, causing alternating periods of deposition and erosion, or perhaps simply varying the dust/ice ratio in a cyclic manner. Because most models suggest that layer formation requires the presence of perennial ice and injection of dust into the atmosphere, the appearance of the layers may indicate cyclic fluctuations in perennial ice extent, atmospheric dust loading, or both.

Mars undergoes a number of variations in its orbital motion and axial orientation that may result in cyclic variations in climate. The variations consist of oscillations in obliquity (the tilt of the rotation axis with respect to the orbit plane), oscillations in orbital eccentricity, and precession of the equinoxes. All of these variations are driven by gravitational interactions with other planets and the Sun (Ward 1973, 1974). The various oscillations exhibit a number of periodicities in the range 10^5–10^6 yr. The most important from the climatic standpoint is the obliquity cycle. The obliquity varies with a period of 1.2×10^5 yr, modulated by a longer cycle of

1.2×10^6 yr. The excursions can be quite large, from a minimum of 15° to a maximum of 35°. (The present value is about 25°.) At high obliquity the annual mean insolation at the poles is substantially greater than at low obliquity (Murray et al 1973, Ward 1974). These variations in insolation may drive large fluctuations in atmospheric CO_2 pressure. For example, Toon et al (1980) calculated that at low obliquity the mean atmospheric pressure would be less than 1 mbar, while at maximum obliquity desorption of CO_2 from the regolith at high latitudes could drive it to perhaps 20 mbar.

Large variations in pressure can affect polar deposition in at least two ways. First, when the atmospheric pressure is at its lowest, it will probably be insufficient to initiate dust storms, preventing atmospheric dust loading. Second, the extent of the perennial ice will be strongly influenced by the amount of polar insolation. At times like the present, when the obliquity is of the intermediate value, the variations in eccentricity and the equinoctial precession may also be important in controlling the extent of perennial ice and the timing and severity of dust storms. The precise details of the interplay between the orbital and axial element fluctuations and the formation of polar layers are not yet fully understood, but a connection seems very likely. It is noteworthy that the oscillations experienced by Mars are similar to, but significantly larger than, oscillations suggested to have driven the Earth's ice ages (e.g. Imbrie 1982).

SUMMARY

There is abundant evidence that water has played a major role in the history of Mars. Geomorphic evidence for the action of liquid water includes outflow channels probably formed by catastrophic release of large floods and valley systems formed by runoff during an early period of warmer temperatures and higher atmospheric pressure. A variety of landforms indicate that significant amounts of H_2O may reside beneath the surface as ground ice. The isotopic composition of the present martian atmosphere is difficult to interpret, but it suggests that Mars underwent outgassing early in its history that would be sufficient to produce a layer of water something like 10–100 m deep over the surface of the planet. Besides ground ice, other possible sinks for water include storage in the present atmosphere, loss to space by dissociation and escape, chemical incorporation in silicates, adsorption on regolith grains, and trapping in the polar deposits. The polar layered deposits may contain much of the present H_2O ice on Mars and exhibit a complex stratigraphy probably related to climatic oscillations induced by variations in Mars' orbital and axial elements. It must be stressed that there are still very large uncertainties in the relative quantities of H_2O in the various sinks.

One important point concerning water on Mars is that Mars does not appear to have ever had a hydrological cycle like that on Earth. Although there is abundant evidence for flow of water across the surface, there is no evidence for extended periods of precipitation or continuous flow of surface streams. Instead, the formation of Mars' fluvial features took place as a sequence of brief, isolated events, with flow at each site ceasing forever once the supply of water in the source area was exhausted.

Much remains to be learned about the history of water on Mars. Particularly important questions concern the present distribution of ground ice, the precise mineralogy of the abundant martian dust, and the dust/ice ratio of the polar layered deposits. In an exhaustive recent study, NASA's Solar System Exploration Committee has recommended a program of planetary missions through the year 2000 (Morrison & Hinners 1983). One of the highest priority missions recommended is a Mars Geochemistry/Climatology Orbiter that would directly address these issues. Such a mission is the logical next step in the study of water on our sister planet.

ACKNOWLEDGMENTS

I am grateful to M. H. Carr, J. B. Pollack, R. T. Reynolds, and C. P. McKay for helpful discussions and comments. This work was supported by a National Research Council postdoctoral fellowship.

Literature Cited

Allen, C. C. 1978. Areal distribution of rampart craters on Mars. *NASA Tech. Memo. TM 79729*, pp. 160–61

Allen, C. C., Gooding, J. L., Keil, K. 1982. Hydrothermally altered impact melt rock and breccia: contributions to the soil of Mars. *J. Geophys. Res.* 87:10083–10101

Baird, A. K., Castro, A. J., Clark, B. C., Toulmin, P., Rose, H., et al. 1977. The Viking X-ray fluorescence experiment: sampling strategies and laboratory simulations. *J. Geophys. Res.* 82:4595–4624

Baird, A. K., Weldon, R. J., Tsusaki, D. M., Schnabel, L., Candelaria, M. P. 1982. Chemical composition of martian fines. *J. Geophys. Res.* 87:10059–67

Baker, V. R. 1979. Erosional processes in channelized water flows on Mars. *J. Geophys. Res.* 84:7985–93

Baker, V. R., Kochel, R. C. 1979. Martian channel morphology: Maja and Kasei Valles. *J. Geophys. Res.* 84:7961–83

Baker, V. R., Milton, D. J. 1974. Erosion by catastrophic floods on Mars and Earth. *Icarus* 23:27–41

Barker, E. S. 1976. Martian atmospheric water vapor observations: 1972–1974 apparition. *Icarus* 28:247–68

Biemann, K., Owen, T., Rushneck, D. R., Lafleur, A. L., Howarth, D. W. 1976. The atmosphere of Mars at the surface: isotope ratios and upper limits on noble gases. *Science* 194:76–78

Biemann, K., Oro, J., Toulmin, P., Orgel, L. E., Nier, A. O., et al. 1977. The search for organic substances and inorganic volatile compounds in the surface of Mars. *J. Geophys. Res.* 82:4641–58

Blasius, K. R., Cutts, J. A., Howard, A. D. 1982. Topography and stratigraphy of martian polar layered deposits. *Icarus* 50:140–60

Bogard, D. D., Gibson, E. K. 1978. The origin and relative abundances of C, N, and the noble gases in the terrestrial planets and meteorites. *Nature* 271:150–53

Bonatti, E. 1965. Palagonite, hyaloclastites, and alteration of volcanic glass in the ocean. *Bull. Volcanol.* 28:257–69

Bretz, J. H. 1923. The Channeled Scablands of

the Columbia Plateau. *J. Geol.* 31 : 617–49
Bretz, J. H. 1969. The Lake Missoula floods and the Channeled Scabland. *J. Geol.* 77 : 505–43
Carr, M. H. 1974. The role of lava erosion in the formation of lunar rilles and martian channels. *Icarus* 22 : 1–23
Carr, M. H. 1979. Formation of martian flood features by release of water from confined aquifers. *J. Geophys. Res.* 84 : 2995–3007
Carr, M. H., Schaber, G. G. 1977. Martian permafrost features. *J. Geophys. Res.* 82 : 4039–54
Carr, M. H., Crumpler, L. S., Cutts, J. A., Greeley, R., Guest, J. E., Masursky, H. 1977a. Martian impact craters and emplacement of ejecta by surface flow. *J. Geophys. Res.* 82 : 4055–65
Carr, M. H., Greeley, R., Blasius, K. R., Guest, J. E., Murray, J. B. 1977b. Some martian volcanic features as viewed from the Viking orbiters. *J. Geophys. Res.* 82 : 3985–4015
Clark, B. C., Baird, A. K., Rose, H. J., Toulmin, P., Christian, R. P., et al. 1977. The Viking X-ray fluorescence experiment: analytical methods and early results. *J. Geophys. Res.* 82 : 4577–94
Cutts, J. A. 1973. Nature and origin of layered deposits of the martian polar regions. *J. Geophys. Res.* 78 : 4231–49
Cutts, J. A., Blasius, K. R. 1981. Origin of martian outflow channels: the eolian hypothesis. *J. Geophys. Res.* 86 : 5075–5102
Cutts, J. A., Lewis, B. H. 1982. Models of climatic cycles recorded in martian polar layered deposits. *Icarus* 50 : 216–44
Cutts, J. A., Blasius, K. R., Roberts, W. J. 1979. Evolution of martian polar landscapes: Interplay of long term variations in perennial ice cover and dust storm intensity. *J. Geophys. Res.* 84 : 2975–94
Davies, D. W., Farmer, C. B., LaPorte, D. D. 1977. Behavior of volatiles in Mars polar areas: a model incorporating new experimental data. *J. Geophys. Res.* 84 : 3815–22
Donahue, T. M., Hoffman, J. H., Hodges, R. R., Watson, A. J. 1982. Venus was wet: A measurement of the ratio of D to H. *Science* 216 : 630–33
Dzurisin, D., Blasius, K. R. 1975. Topography of the polar layered deposits of Mars. *J. Geophys. Res.* 80 : 3286–3306
Fanale, F. P. 1976. Martian volatiles: their degassing history and geochemical fate. *Icarus* 28 : 179–202
Fanale, F. P., Cannon, W. A. 1974. Exchange of adsorbed H_2O and CO_2 between the regolith and atmosphere of Mars caused by changes in surface insolation. *J. Geophys. Res.* 79 : 3397–402
Farmer, C. B., Davies, D. W., Holland, A. L., LaPorte, D. D., Doms, P. E. 1977. Mars: water vapor observations from the Viking orbiters. *J. Geophys. Res.* 82 : 4225–48
Gault, D. E., Greeley, R. 1978. Exploratory experiments of impact craters formed in viscous-liquid targets: analogs for martian rampart craters? *Icarus* 34 : 486–95
Gooding, J. L., Keil, K. 1978. Alteration of glass as a possible source of clay minerals on Mars. *Geophys. Res. Lett.* 5 : 727–30
Helfenstein, P. 1980. Martian fractured terrain: Possible consequences of ice-heaving. *NASA Tech. Memo. TM 82385*, pp. 373–74
Hoffman, J. H., Hodges, R. R., Donahue, T. M., McElroy, M. B. 1980. Composition of the Venus lower atmosphere from the Pioneer Venus mass spectrometer. *J. Geophys. Res.* 85 : 7882–90
Howard, A. D. 1978. Origin of the stepped topography of the martian poles. *Icarus* 34 : 581–99
Imbrie, J. 1982. Astronomical theory of the Pleistocene ice ages: a brief historical review. *Icarus* 50 : 408–22
Ingersoll, A. P. 1974. Mars: the case against permanent CO_2 frost caps. *J. Geophys. Res.* 79 : 3403–10
Istomin, V. G., Grechnev, R. V. 1976. Argon in the martian atmosphere: evidence from the Mars 6 descent module. *Icarus* 28 : 155–58
Jakosky, B. M., Farmer, C. B. 1982. The seasonal and global behavior of water vapor in the Mars atmosphere: complete global results of the Viking atmospheric water detector experiment. *J. Geophys. Res.* 87 : 2999–3019
Jones, K. L., Arvidson, R. E., Guiness, E. A., Bragg, S. L., Wall, S. D., et al. 1979. One Mars year: Viking lander imaging observations. *Science* 204 : 799–806
Kieffer, H. H. 1979. Mars south polar spring and summer temperatures: a residual CO_2 frost. *J. Geophys. Res.* 84 : 8263–88
Kieffer, H. H., Chase, S. C., Martin, T. Z., Miner, E. D., Palluconi, F. D. 1976. Martian north pole summer temperatures: dirty water ice. *Science* 194 : 1341–44
Leighton, R. B., Murray, B. C. 1966. Behavior of carbon dioxide and other volatiles on Mars. *Science* 153 : 136–44
Leovy, C. B. 1973. Exchange of water vapor between the atmosphere and surface of Mars. *Icarus* 18 : 120–25
Lucchitta, B. K., Anderson, D. M., Shoji, H. 1981. Did ice streams carve martian outflow channels? *Nature* 290 : 759–63
Malin, M. C. 1976. Age of martian channels. *J. Geophys. Res.* 81 : 4825–45
Masursky, H. 1973. An overview of geological results from Mariner 9. *J. Geophys. Res.* 78 : 4009–30

Masursky, H., Boyce, J. M., Dial, A. L., Schaber, G. G., Strobell, M. E. 1977. Classification and time of formation of martian channels based on Viking data. *J. Geophys. Res.* 82:4016–38

McCauley, J. F. 1973. Mariner 9 evidence for wind erosion in the equatorial and mid-latitude regions of Mars. *J. Geophys. Res.* 78:4123–37

McCauley, J. F., Carr, M. H., Cutts, J. A., Hartmann, W. K., Masursky, H., et al. 1972. Preliminary Mariner 9 report on the geology of Mars. *Icarus* 17:289–327

McElroy, M. B., Yung, Y. L., Nier, A. O. 1976. Isotopic composition of nitrogen: implications for the past history of Mars' atmosphere. *Science* 194:70–72

McElroy, M. B., Kong, T. Y., Yung, Y. L. 1977. Photochemistry and evolution of Mars' atmosphere: a Viking perspective. *J. Geophys. Res.* 82:4379–88

Milton, D. J. 1973. Water and processes of degradation in the martian landscape. *J. Geophys. Res.* 78:4037–47

Milton, D. J. 1974. Carbon dioxide hydrate and floods on Mars. *Science* 183:654–56

Morrison, D., Hinners, N. W. 1983. A program for planetary exploration. *Science* 220:561–67

Mouginis-Mark, P. 1979. Martian fluidized crater morphology: variations with crater size, latitude, altitude, and target material. *J. Geophys. Res.* 84:8011–22

Murray, B. C., Soderblom, L. A., Cutts, J. A., Sharp, R. P., Milton, D. J., Leighton, R. B. 1972. Geological framework of the south polar region of Mars. *Icarus* 17:328–45

Murray, B. C., Ward, W. R., Yeung, S. C. 1973. Periodic insolation variations on Mars. *Science* 180:638–40

Neugebauer, G., Munch, G., Kieffer, H. H., Chase, S. C., Miner, E. D. 1971. Mariner 1969 infrared radiometer results: temperature and thermal properties of the martian surface. *Astron. J.* 76:719–27

Neukum, G., Wise, D. U. 1976. Mars: a standard crater curve and possible new time scale. *Science* 194:1381–87

Newsom, H. E. 1980. Hydrothermal alteration of impact sheets with implications for Mars. *Icarus* 44:207–16

Nummedal, D. 1978. The role of liquefaction in channel development on Mars. *NASA Tech. Memo. TM 97929*, pp. 257–58

Nummedal, D. 1980. Debris flows and debris avalanches in the large martian channels. *NASA Tech. Memo. TM 81776*, pp. 289–91

Nummedal, D. 1981. Wind-modification of the Chryse channels. *NASA Tech. Memo. TM 84211*, pp. 229–31

Owen, T., Biemann, K. 1976. Composition of the atmosphere at the surface of Mars: detection of Argon-36 and preliminary analyses. *Science* 193:801–3

Owen, T., Biemann, K., Rushneck, D. R., Biller, J. E., Howarth, D. W., Lafleur, A. L. 1976. The atmosphere of Mars: detection of krypton and xenon. *Science* 194:1293–95

Owen, T., Biemann, K., Rushneck, D. R., Biller, J. E., Howarth, D. W., Lafleur, A. L. 1977. The composition of the atmosphere at the surface of Mars. *J. Geophys. Res.* 82:4635–39

Pechmann, J. C. 1980. The origin of polygonal troughs on the northern plains of Mars. *Icarus* 42:185–210

Pieri, D. 1976. Martian channels: distribution of small channels on the martian surface. *Icarus* 27:25–50

Pieri, D. 1980. Martian valleys: morphology, distribution, age, and origin. *Science* 210: 895–97

Pollack, J. B., Black, D. C. 1979. Implications of the gas compositional measurements of Pioneer Venus for the origin of planetary atmospheres. *Science* 205:56–59

Rasool, S. I., LeSergeant, L. 1977. Implications of the Viking results for volatile outgassing from Earth and Mars. *Nature* 266:822–23

Schumm, S. A. 1974. Structural origin of large martian channels. *Icarus* 22:371–84

Sharp, R. P. 1973. Mars: fretted and chaotic terrain. *J. Geophys. Res.* 78:4073–83

Sharp, R. P., Malin, M. C. 1975. Channels on Mars. *Geol. Soc. Am. Bull.* 86:593–609

Singer, R. B. 1982. Spectral evidence for the mineralogy of high-albedo soils and dust on Mars. *J. Geophys. Res.* 87:10159–68

Smoluchowski, R. 1968. Mars: Retention of ice. *Science* 159:1348–50

Soderblom, L. A., Malin, M. C., Cutts, J. A., Murray, B. C. 1973. Mariner 9 observations of the surface of Mars in the north polar region. *J. Geophys. Res.* 78:4197–210

Soderblom, L. A., West, R. A., Herman, B. M., Condit, C. D. 1974. Martian planetwide crater distribution, implications for geologic history and surface processes. *Icarus* 22:239–63

Squyres, S. W. 1978. Martian fretted terrain: flow of erosional debris. *Icarus* 34:600–13

Squyres, S. W. 1979a. The distribution of lobate debris aprons and similar flows on Mars. *J. Geophys. Res.* 84:8087–96

Squyres, S. W. 1979b. The evolution of dust deposits in the martian north polar region. *Icarus* 40:244–61

Toon, O. B., Pollack, J. B., Ward, W., Burns, J. A., Bilski, K. 1980. The astronomical theory of climatic change on Mars. *Icarus* 44:552–607

Toulmin, P., Baird, A. K., Clark, B. C., Keil,

K., Rose, H. J., et al. 1977. Geochemical and mineralogical interpretations of the Viking inorganic chemical results. *J. Geophys. Res.* 82:4625–34

Wahrhaftig, C., Cox, A. 1959. Rock glaciers in the Alaska Range. *Geol. Soc. Am. Bull.* 70:383–436

Wallace, D., Sagan, C. 1979. Evaporation of ice in planetary atmospheres: ice-covered rivers on Mars. *Icarus* 39:385–400

Walker, G. P. L., Blake, D. H. 1966. The formation of a palagonite breccia mass beneath a valley glacier in Iceland. *Q. J. Geol. Soc. London* 122:45–61

Ward, W. R. 1973. Large-scale variations in the obliquity of Mars. *Science* 181:260–62

Ward, W. R. 1974. Climatic variations on Mars. I. Astronomical theory of insolation. *J. Geophys. Res.* 79:3375–86

Washburn, A. L. 1973. *Periglacial Processes and Environments.* New York: St. Martin's. 320 pp.

Wetherill, G. W. 1981. Solar wind origin of ^{36}Ar on Venus. *Icarus* 46:70–80

Yung, Y. L., Pinto, J. P. 1978. Primitive atmosphere and implications for the formation of channels on Mars. *Nature* 273:730–32

Ann. Rev. Earth Planet. Sci. 1984. 12: 107–31

THE ORIGIN OF ALLOCHTHONOUS TERRANES: Perspectives on the Growth and Shaping of Continents[1]

Elizabeth R. Schermer, David G. Howell, and David L. Jones

US Geological Survey, Menlo Park, California 94025

INTRODUCTION

In recent years new theories of continental growth have focused on the idea that the crust of the North American continent has grown through the accretion of discrete allochthonous fragments of oceanic and continental material at active margins. Though the processes by which materials are added to continental margins are still poorly understood, we can at least identify the distinct accreted blocks that compose the new margins. This identification and characterization is a major step in understanding the growth and shaping of continents.

The early division of continents into cratonal, miogeosynclinal, and eugeosynclinal areas offered a framework for theories of continental growth by peripheral accretion. Various forms of growth through collision of continental blocks and subduction-related accretion, volcanism, and tectonism (e.g. Dewey & Bird 1970) were proposed in later plate-tectonics models. More detailed work revealed complexities in active margins that could not be explained by assuming that the present spatial relations of tectonic domains within the margins imply original genetic and geographic relations. A growing body of geological and geophysical evidence demonstrates that large translational and rotational displacements have occurred within and between tectonic provinces of the eugeosynclinal belt. A natural outgrowth of plate-tectonics theories—the concept of terrane analysis—recognizes that the diverse fragments of vastly differing geologic histories cannot be assumed to have genetic relationships in time and space and therefore are "suspect." These fragments are termed *suspect tectonostratigraphic terranes* (Figure 1).

The terrane concept, as now applied to the North American Cordillera, originated in the Klamath Mountains of northern California as a direct outgrowth of the investigations of W. P. Irwin (1960). He first subdivided the Klamaths into four arcuate belts, each representing a unique package of rocks characterized by particular ages and lithic assemblages. This fourfold subdivision was later modified (Irwin 1972) when he subdivided one belt into three "subbelts," which he termed "terranes." The term "terrane," as used by Irwin (1972, p. C103), "refers to an association of geologic features, such as stratigraphic formations, intrusive rocks, mineral deposits, and tectonic history, some or all of which lend a distinguishing character to a particular tract of rocks and which differ from those of an adjacent terrane." In the same year, the terrane concept was applied to rocks in southern Alaska by Berg et al (1972), who recognized several terranes as discrete, fault-bounded tectonic units. The Alaskan terranes were later described in more detail by Berg et al (1978) in the first terrane map prepared for the northern part of the North American Cordillera. The definition used on this map emphasized the fault-bounded nature of terranes and also introduced the concept of composite terranes formed by preaccretionary amalgamation (see below). These pioneering efforts have now become standardized, as explained in Jones et al (1983) and below, and as portrayed on more recent terrane maps of the Cordillera (Jones et al 1981, Blake et al 1982a, Campa & Coney 1983, Silberling et al 1983).

One of the earliest indications that terranes might be far travelled came from work by Monger & Ross (1971), who recognized tectonically displaced, fault-bounded faunal provinces in the Canadian Cordillera that had more affinity to the Tethyan realm than to North American realms. Many later workers have identified other displaced terranes of oceanic

---

Figure 1 Highly generalized map of circum-Pacific tectonostratigraphic terranes. Explanation of symbols: (1) Archean cratonal blocks; (2) Proterozoic orogenic belts and cratonal blocks consolidated during the Proterozoic; (3) Paleozoic orogenic belts, composed of terranes accreted during the Paleozoic; (4) Mesozoic-Cenozoic orogenic belts and accreted terranes; (5) continental fragments; (6) remnant volcanic arcs; (7) oceanic islands and seamounts, including hotspot tracks; (8) mixed origin—oceanic with anomalously thick, continental-like structure. Abbreviations of terranes and localities mentioned in text (clockwise from Alaska): P = Peninsular; D = Dillinger; N = Nixon Fork; C = Chulitna; MN = Manley; W = Wrangellia; A = Alexander; CC = Cache Creek; S = Stikine; M = Methow; F = Franciscan; GB = Guyana and Brazilian shields; AR = Arequipa; B = eastern Brazilian shield; V = Sierra de la Ventana; NE = New England fold belt; LF = Lachlan fold belt; MA = Malaysia; YG = Yangtze; SK = Sino-Korea; T = Tarim; J = Japan; SA = Sikhote Alin.

SHIRSHOV RIDGE
BOWERS RIDGE
HESS RISE
HAWAIIAN RIDGE
EMPEROR SEAMOUNTS
SEA OF OKHOTSK
SHATSKY RISE
MID-PACIFIC MOUNTAINS
LINE ISLANDS RIDGE
CAROLINE RIDGE
ONTONG JAVA PLATEAU
PALAU-KYUSHU RIDGE
QUEENSLAND PLATEAU
LORD HOWE RISE
CAMPBELL PLATEAU
EXMOUTH PLATEAU
CARNEGIE RIDGE
GALAPAGOS RISE
NAZCA RIDGE
Figure 4
1 2 3 4 5 6 7 8

rocks containing exotic faunal communities throughout the Cordillera of western North America (Monger 1977, Nestel 1980, Nichols & Silberling 1979, Tozer 1982). Early paleomagnetic work seemed to support the idea that several terranes of the Cordillera were exotic (e.g. Irving & Yole 1972, Packer & Stone 1972, 1974), but the interpretations of the data were not widely accepted until 1977, when combined geophysical, stratigraphic, and paleobiogeographic data gave strong support for thousands of kilometers of displacement of the Wrangellia terrane of Alaska (Hillhouse 1977, Jones et al 1977). Since that time, increasing numbers of paleomagnetic studies have provided evidence for large amounts of translation and rotation for terranes throughout western North America (e.g. Beck & Cox 1979, Alvarez et al 1980, Champion et al 1980, and many others).

It is now becoming apparent that terrane accretion is not limited to North America, but has occurred since at least Proterozoic time throughout the circum-Pacific region. This review briefly summarizes the principles of terrane analysis as discussed in Coney et al (1980) and Jones et al (1983) and presents an application of terrane analysis for the circum-Pacific region. Additional background material for terranes of western North America can be found in these two articles.

TERRANE ANALYSIS

Terrane Definition

A tectonostratigraphic terrane is a fault-bounded geologic entity of regional extent that is characterized by a geologic history different from that of neighboring terranes (Coney et al 1980, Jones et al 1983). Terranes can be classified as stratigraphic, metamorphic, disrupted, or composite. Stratigraphic terranes are composed of coherent sedimentary and igneous sequences. They may include a complicated array of lithofacies reflecting one or more depositional environments, e.g. continental, oceanic, and/or island-arc basins. Metamorphic terranes are represented by fault-bounded blocks that have a regional penetrative metamorphic fabric that obscures and is more distinctive than original lithotypes. Disrupted terranes are characterized by blocks of heterogeneous lithology and age that are set in a matrix of foliated graywacke or serpentinite. Composite terranes consist of two or more terranes that amalgamated prior to their accretion to a continental margin.

TERRANE BOUNDARIES Terranes are by definition fault-bounded, implying that there is at least some movement between adjacent terranes. Commonly the boundaries are suture zones or cryptic faults that are inferred because of the juxtaposition of differing lithologies. Ophiolite belts and blueschist are good evidence for the location of a subduction zone and closure of an ocean basin. However, a terrane boundary need not be marked by ophiolite or ophiolitic mélange; in fact, most terrane boundaries are not, indicating that they probably represent (*a*) accretion via strike-slip or overprint by later consolidation and dispersion processes (Figure 2; see below); (*b*) closure of only relatively small marginal basins rather than entire oceans; and/or (*c*) most of the oceanic crust is either subducted or not exposed due to the structural complexities intrinsic to accreting margins. Worldwide, ocean basins have been destroyed several times during the Phanerozoic, yet ophiolite is only rarely exposed in continental masses. This fact, combined with the observation that terranes are commonly detached from their substrata, suggests that even the closure of an entire ocean basin leaves scant evidence of oceanic crust. Some terranes appear intact with their basement rocks [e.g. Wrangellia on an arc basement and Chulitna on an ophiolite basement (Figure 1; Jones et al 1977, 1980)], but many more are flakes of sedimentary and volcanic strata that occur either as fault-bounded nappes or as coherent blocks embedded in a matrix of foliated graywacke. With rare exceptions, terranes are not equivalent to microplates, as the latter term implies discrete lithospheric blocks.

Plate Tectonics and Terrane Analysis

The fact that petrotectonic assemblages, such as volcanic arcs, forearc basins, and subduction complexes, may be extensive along a continental margin does not in itself prove that these assemblages formed in their present position. Detailed examination of tectonic and stratigraphic relations and paleomagnetic and paleobiogeographic indicators, as has been done for terranes throughout the western North American Cordillera, has demonstrated that parts of these assemblages may be allochthonous. Therefore, one must be cautious in applying plate tectonics models that assume that tectonic domains that are presently spatially juxtaposed are

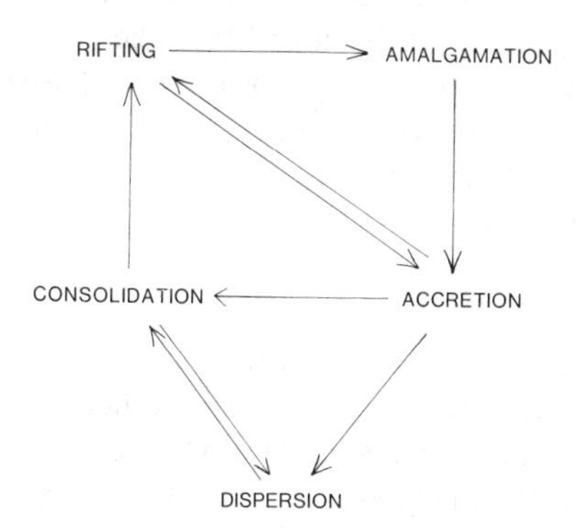

Figure 2 Diagram showing how the various kinematic states of terranes are affected by the complex interrelationship of movement processes, resulting in either the breakup or the welding together of terranes and continental masses.

genetically related. Separation of fact from interpretation is essential here; one does not define a terrane because it fits or does not fit a model but because of its distinctive stratigraphy and geologic history. The genetic linkages and paleogeographic history must be determined by rigorous analysis of geologic and geophysical data. Where tectonostratigraphic sequences are not unequivocally correlative, we take a conservative stance and treat them as separate terranes. This taxonomic splitting permits a variety of palinspastic interpretations and reconstructions once pertinent data are gathered. The skepticism about genetic relationships that is inherent in terrane analysis allows for more objective collection and interpretation of data, as facies relationships must be proven before they serve as the basis for a model.

Unfortunately, the inappropriate application of many popular models to the specifics of orogenic belts has led to error and confusion about these highly mobile, diverse regions. In several ancient orogenes, the typical model of arc, forearc, and subduction complex belts arranged in parallel along an active margin fails when examined in detail. For example, in New Zealand, the tectonic belts from west to east consist of (*a*) middle to late Paleozoic sialic crust and continental strata of the Tuhua terrane, (*b*) Permian to Mesozoic ensimatic island-arc rocks of the Hokonui terrane, (*c*) Paleozoic and Mesozoic(?) volcaniclastic graywacke of the Caples terrane, and (*d*) Permian to Cretaceous quartzo-feldspathic graywacke of the Torlesse terrane (Figure 3; Howell 1980). The continentally derived strata of the Torlesse terrane are spatially separated from their expected source area (the Tuhua terrane) by a coeval, entirely ensimatic assemblage. Similarly, in California, Jurassic quartzo-feldspathic strata of the Franciscan assemblage are separated from their expected continental source to the east by coeval ensimatic igneous and volcaniclastic rocks at the base of the Great Valley sequence (Figure 4; Blake & Jones 1981).

Because of the dynamics of environments within active continental margins and because a given region may pass from one plate tectonic regime to another in both time and space, a plethora of lithofacies may occur within any given terrane. The key to terrane analysis is to distinguish between diverse lithologic packages that are linked in time and space and lithofacies units that are genetically unrelated and thereby reflect juxtapositioning of two or more allochthonous terranes.

Tectonostratigraphic Analysis of Terranes

The stratigraphy of a terrane is probably the most important key to unraveling its origin; however, this stratigraphy commonly reflects a diverse geologic history. This situation is largely a result of both the tectonic and eustatic dynamics along continental margins and on islands and

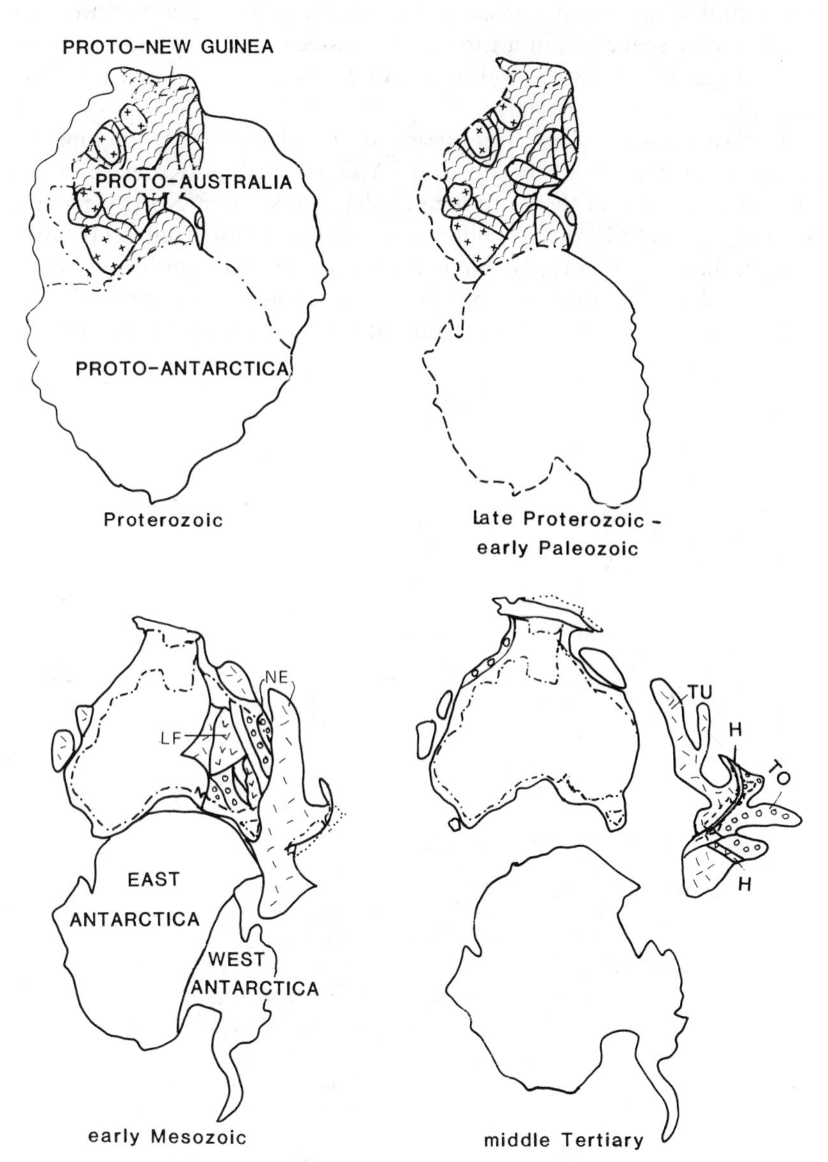

Figure 3 Schematic representation of accretion and dispersion events in Antarctica, Australia, and New Zealand. NE = New England fold belt; LF = Lachlan fold belt; TU = Tuhua terrane; TO = Torlesse terrane; H = Hokonui terrane. Open circles indicate terrane composed of continental margin strata; other symbols as in Figure 1.

seamounts within ocean basins. In the context of plate tectonics we view stratigraphic sequences in a three-part classification: (*a*) passive or trailing margin sequences, (*b*) consuming or leading margin sequences, and (*c*) transform margin sequences.

Passive margin facies are preserved in allochthonous continental fragments of the size of the Australia, Yangtze, or Sino-Korea terranes; smaller fragments of the Cordilleran collage of western North America, such as the Nixon Fork and Dillinger terranes, may also exhibit a passive-margin history (Figure 1). Nonetheless, within this cordillera, many terranes show characteristics of rifting and subsidence. For example, the Triassic portion of the Chulitna terrane of Alaska is composed of basalt and

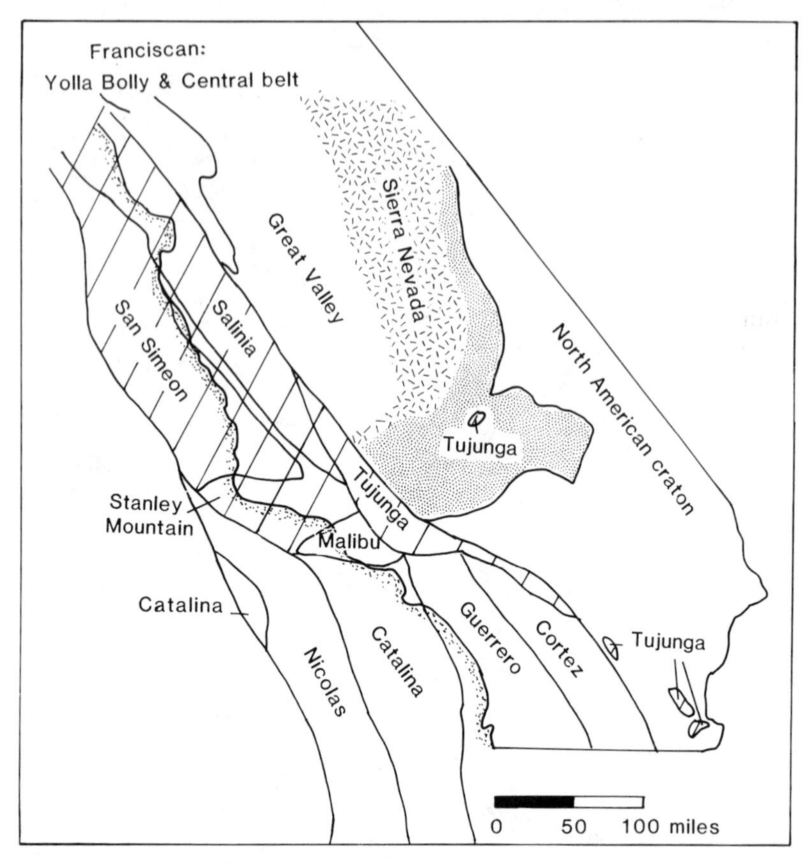

Figure 4 Terranes of southern California. Diagonal lines indicate extent of the Santa Lucia-Orocopia allochthon; dashes indicate Mesozoic batholithic rocks of the Sierra Nevada that obscure older terrane boundaries; stipple indicates Paleozoic "eugeosynclinal" sequences and Mesozoic plutonic rocks, undifferentiated.

redbeds reflecting a possible rifting event; these beds are overlain by deep-water and sandstone and shale. The Nikolai Greenstone of Wrangellia may be rift related, and the overlying carbonate and clastic rocks suggest later subsidence (Jones et al 1980).

Consuming margins are commonly characterized by volcaniclastic basinal facies of either forearc or backarc settings. These strata lie on igneous basement near the axis of the volcanic arc, but they also lie on oceanic crust in more distal forearc or backarc directions. Disrupted strata of inferred trench and trench-slope environments may lie structurally alongside and/or below the forearc beds. This tripartite division of volcanic arc, coherent forearc, and disrupted trench strata (accretionary or mélange wedge) reflects the popular plate-tectonics model of a subduction setting. Seismic reflection profiling across modern subduction zones has indicated that this model is in many instances too facile. For example, portions of the Middle America Trench have no mélange wedge (von Huene et al 1980), and accretionary prism rocks of the Aleutian margin compose only a small volume, hardly representative of the 60 m.y. of subduction beneath this arc (Scholl & Vallier 1983).

Terranes that are composed all or in part by consuming margin facies include the Hokonui terrane of New Zealand (Figure 3), the Stanley Mountain terrane of California (Figure 4), and the basal portion of Wrangellia; in these instances it is not possible to unequivocally link the submarine fan facies of the forearc to any graywacke sequences of a subduction complex, and all or part of the arc source terrane is also missing. Terranes such as those that compose the Franciscan assemblage of California or the Manley terrane (Figure 1) and other flysch sequences of Alaska [such as the Gravina-Nutzotin assemblage (Berg et al 1972)] are considered by some workers to be arc-related (either forearc or trench deposits). Alternatively, we believe that some of these thick sequences of structurally complicated graywacke represent a variety of deep-marine depositional environments; nonetheless, their postdepositional tectonic histories may be similar in that they all represent the effects of basin closure and terrane amalgamation and accretion. The Molucca Sea is a modern example of this type of setting (Silver & Smith 1983).

Transform margins are typified by wrench tectonics (transpressional and transtensional stresses) that generally result in borderland settings. Lithofacies reflecting nonmarine to deep-marine environments grade rapidly one into the other, both vertically and horizontally. The Upper Cretaceous clastic facies of the Salinia and Tujunga terranes of California reflect a sliding margin history. The "porpoising" (alternate uplift and subsidence in time and space; Crowell & Sylvester 1979) of crustal blocks within wrench-tectonics settings leaves a special sedimentologic imprint

(Howell et al 1980): basins fill rapidly, often with restricted water circulation that promotes accumulation of organic-rich sediments. If the transtensional phase passes into a transpressional one (Harland 1971), strata of the basins are "squeezed" out, and in the process large petroleum traps can be created such as those that developed in Neogene strata along the San Andreas fault system. The transcurrent-faulting effects of wrench tectonics also are evident in the slicing and slivering of terranes into smaller tectonostratigraphic units (dispersion). The collage aspect of much of the western North American Cordillera is in part a result of the northwestward transcurrent dispersion throughout the Cordilleran margin.

Terrane Processes

CONTINENTAL CONSTRUCTION: AMALGAMATION AND ACCRETION Several different processes may affect a terrane during transport from its origin to its place of accretion and consolidation onto a continent (Figure 2). Amalgamation occurs when two terranes collide prior to their accretion onto a continental margin. For example, the Santa Lucia-Orocopia allochthon is composed of the Stanley Mountain, San Simeon, Salinia, and Tujunga terranes (Figure 4); the details of the timing of the amalgamation and accretion of these terranes are discussed below.

Accretion most profoundly affects the tectonic history of a terrane, and in some cases it may severely obscure the evidence of the original geologic history. Thrust faulting appears to play an important role in the accretion process, and many terranes occur as thick stacked thrust packages, but in many cases thrust faults are later reactivated and cut by high-angle faults. Commonly, mélange formation and blueschist metamorphism accompany accretion of blocks in subduction zone settings. For example, in northern California, the metamorphism of the Yolla Bolly terrane of the Franciscan assemblage was possibly associated with a 90 m.y.-old accretion event, and the older blueschist metamorphism of the Franciscan Pickett Peak terrane seemingly reflects an earlier but discrete accretionary episode. Later accretion of seamounts within the Franciscan assemblage resulted in formation of some of the Central belt melange (Blake et al 1982b).

In some cases high-temperature metamorphism and plutonism as well as local anatexis have been documented (Hudson et al 1978, Hudson & Plafker 1982, Monger et al 1982). For example, in the Coast Plutonic Complex of British Columbia, magmatism and metamorphism are thought to have occurred in response to accretion of the Peninsular-Wrangellia-Alexander composite terrane (Monger et al 1982). Ophiolite obduction may occur with the accretion of some terranes. The sedimentary record itself may indicate accretion, including phenomena such as deposition of molasse in the foreland, development of unconformities due to uplift and erosion, changes in sedimentary polarity (the direction in which sediment is shed from a source area), and development of provenancial links (see below) between terranes or between terranes and a continent.

The timing of amalgamation and accretion episodes can be established through the analysis of one or more of the following: (*a*) plutons or batholiths may stitch together two or more terranes, thereby indicating a minimum age of amalgamation [e.g. 60 m.y.-old plutons stitch terranes between the Denali fault and the Border Ranges fault in Alaska (Reed & Lanphere 1969, Hudson 1979), and the 125–90 m.y.-old Peninsular Ranges batholith of southern California stitches the Cortez, Guerrero, and Malibu terranes (Figure 4; Silver et al 1979)]; (*b*) sedimentary basins may overlap two terranes; and (*c*) debris from one terrane may be deposited on another (provenancial linkage). An example of an overlap assemblage is the late Cretaceous Gravina-Nutzotin belt of southern Alaska, which links the Wrangellia and Alexander terranes; an example of provenancial linkage is provided by the Middle to Upper Jurassic strata of the Bowser Basin of British Columbia, which are deposited on the Stikine terrane and contain debris from the Cache Creek terrane (Eisbacher 1974, Monger et al 1982). Unless ties to the continent or continental margin can be demonstrated, the linkages may only indicate intermediate amalgamation episodes rather than accretion.

Postaccretionary consolidation processes also play an important role in the evolution of the terrane collage. This tightening of a loose package of terranes in the North American Cordillera has been interpreted as a major cause of the Laramide orogeny and probably has resulted in deformation in the Cordilleran foreland (Coney 1981).

CONTINENTAL DESTRUCTION AND MODIFICATION: RIFTING AND DISPERSION

Whereas accretion adds to a continental margin, rifting and dispersion attenuate terrane boundaries and erode the margin. Pieces of continental crust may be rifted away from a craton and sent toward a consuming margin. Numerous fragments of this type are found around the margins of Australia, e.g. the Queensland Plateau, the Lord Howe Rise, and the Exmouth Plateau (Figure 1). Dispersion is the process by which previously accreted or amalgamated terranes are faulted into smaller pieces and scattered along the margin. Dispersion may occur by rifting or strike-slip faulting and may operate at the same time as accretionary processes (e.g. during oblique subduction). In western North America, dispersion is presently occurring along right-slip faults such as the San Andreas, Fairweather, Denali, Fraser River, and Tintina fault systems. In Japan, left-slip faults such as the Median tectonic line and, in New Zealand, the right-slip Alpine fault are also slivering and dispersing terranes. Dispersion creates disjunct terranes consisting of stratigraphically correlative but spatially distinct fragments. Wrangellia is a good example of a disjunct terrane; it occurs as several correlative fault-bounded bodies from Oregon through British Columbia to Alaska over a latitudinal spread of nearly 24°. However, based on paleomagnetic data, these bodies originated within less than 7° of each other (Hillhouse & Gromme 1983). Dispersive processes may destroy original accretionary structures and produce wrench-fault basins that cover terrane basement rocks [e.g. the Methow Trough of British Columbia (Figure 1) and numerous basins in the southern California borderland.

Terrane Displacements

Displacement for some terranes has been estimated at several thousand kilometers; for others, the movement is clearly much less. Although much attention has been directed toward "exotic" terranes, terrane *recognition* is not dependent on establishing a minimum amount of relative displacement. The important criterion is that movement be sufficient to completely disrupt original facies relations and thus to render uncertain original genetic relations.

Terrane displacements can be measured in several ways. Traditional geologic methods of measuring fault displacement (e.g. measuring offset of shorelines, dike swarms, fold hinges and other linear elements, or matching offset distinctive rock types or stratigraphic sequences) can determine offsets of generally less than 500 km with relatively good precision. Matching offset biogeographic provinces or climatically controlled lithologies such as redbeds and sabkhas offers less precision but can determine offsets greater than 500 km, and paleomagnetic studies can document latitudinal offsets of greater than 300 km.

Recent plate motion studies have attempted to determine travel paths for terranes whose geologic history and timing of accretion are relatively well known (e.g. Engebretson & Cox 1982, Engebretson 1982). By assuming that a certain terrane rides on one or more plates, such as the Pacific, Farallon, or Kula, during its travel history, it is possible to use one or a combination of plate trajectories to track the terrane from its origin to its accretion at a part of the continent. Reconstructions of oceanic plate motions older than ca. 150 m.y. are uncertain, and are not possible for those older than ca. 180 m.y. owing to the subduction of all pre-early Jurassic oceanic crust. While plate motion trajectories can give a better idea of large amounts of offset, rarely are the geologic history, the timing and location of accretion, or the plate motions sufficiently well known to make such travel histories unequivocal. With more work, plate motion studies may become a feasible way of determining displacement histories for terranes accreted since the

Jurassic, although the complete subduction of some plates makes such reconstructions highly subjective. The plate kinematics must be constrained by the geologic data, e.g. the provenance of strata, within a terrane. For example, Wrangellia preserves an entirely oceanic history in the early Mesozoic, allowing for a trans-Pacific migration. In contrast, upper Mesozoic and Cenozoic sedimentary rocks of Salinia indicate proximity to a continental margin throughout its migratory path.

Terrane analysis involves a combination of as many of the above displacement measurements as possible. If two or more methods are in agreement as to the origin and displacement of a terrane, much more confidence can be placed in paleogeographic reconstructions. An outstanding example of the concordance of geologic, paleontologic, and geophysical data is afforded by Wrangellia. Here, paleomagnetic data indicate equatorial paleolatitudes during the Triassic (Hillhouse 1977), and Triassic carbonate rocks are characteristic of tropical supratidal conditions (Armstrong & MacKevett 1983) and contain high-diversity, endemic Molluscan faunas of low-latitude aspect (Stanley 1982, Tozer 1982, Newton 1983, Silberling & Jones 1983).

Oceanic Plateaus and Terrane Analysis

Oceanic plateaus are anomalously high parts of the sea floor that at present are not parts of continents, active volcanic arcs, or spreading ridges. Plateaus comprise fragments of continents, oceanic islands and seamounts, hotspot tracks, remnant arcs, and other anomalously thick volcanic piles (Figure 1; Ben-Avraham et al 1981, Nur & Ben-Avraham 1982). Examples of continental fragments include the Campbell Plateau and the Lord Howe Rise. Oceanic islands and seamounts, such as the Mid-Pacific Mountains, the Caroline Ridge, and the Galapagos Rise, are abundant throughout the Pacific. Some of these are hotspot tracks, such as the Hawaiian and Emperor seamount chains and the Line Islands Ridge, or remnant arcs, such as the Bowers and Palau-Kyushu ridges. Large plateaus, such as Ontong-Java, Hess Rise, and Shatsky Rise, seem to be characterized by oceanic basalt and sedimentary rocks in their upper portions, but seismic refraction data indicate that they have a continental-like structure at depth (Vallier et al 1981, Nur & Ben-Avraham 1982).

These anomalously thick and low-density bodies are more likely to be obducted than subducted when they encounter a convergent margin. Oceanic plateaus that have collided with fossil and modern subduction zones include the Sea of Okhotsk Plateau, Shirshov Ridge, Nazca Ridge, and Carnegie Ridge (Figure 1). The effects of these collisions include the creation of a volcanic gap and anomalous seismic activity along the subducting margin. In addition, some pieces may be obducted onto the island arc or the continent, as has been proposed for the Ontong-Java Plateau (Coleman & Kroenke 1981) and the Nazca Ridge (Nur & Ben-Avraham 1981).

The presence of similar lithotectonic elements in oceanic plateaus and as onshore accreted terranes suggests that many terranes originate as oceanic plateaus. Nur & Ben-Avraham (1977) have proposed a lost continent of Pacifica as the origin for several terranes of circum-Pacific orogenes, but there is little evidence to support the idea that any of these diverse fragments had a common origin. More likely they began as numerous arcs, seamounts, and continental fragments similar to those present in the southwest Pacific and Indonesian regions today.

AMALGAMATION, ACCRETION, AND DISPERSION OF CIRCUM-PACIFIC TERRANES

The Growth and Shaping of Continents

From the perspective of tectonostratigraphic terrane analysis, the shape of continents results from both terrane accretion and dispersion. Just as the accretion of terranes results in continental growth or outbuilding, the dispersion of terranes, by either rifting or sliding, results in the diminution of continents. The continents of the circum-Pacific can be viewed in this context.

Only a few terranes in the North American Cordillera have well-documented displacement histories, although geologic evidence suggests exotic origins for many. Because the history of terranes in Alaska and British Columbia (Wrangellia, Alexander, Peninsular, Stikine, and Cache Creek) has been extensively discussed and reviewed (e.g. Jones et al 1977, 1981, 1983, Jones & Silberling 1979, Monger et al 1982, Saleeby 1983), this need not be repeated here. Instead, we present below an example of the amalgamation, accretion, and dispersion history of terranes in southern California. We also give some examples from the rest of the circum-Pacific, including Asia, Australia and New Zealand, and South America, although the characteristics of these terranes and their displacement histories are very poorly known; in fact, studies are only now in progress to identify and characterize these terranes (Jones et al 1982, Howell et al 1983, Howell et al, in preparation).

Proterozoic and Phanerozoic of North America

In North America, the Archean massifs that are surrounded by a variety of Proterozoic fold belts (Figure 1) suggest successive terrane accretions between 2 and 1 b.y. B.P. (Condie 1982, Hoffman et al 1982). The combined occurrences of upper Proterozoic tillites along the margin of the present craton and upper Proterozoic rifting sequences outboard of these glacial deposits suggest that North America was the interior part of a much larger continental agglomeration ca. 800 m.y. ago. For most of the Paleozoic, the western Cordilleran and Ouachitan regions remained a passive or trailing margin while the Franklinian and Appalachian margins experienced episodic growth caused by terrane accretion (Williams & Hatcher 1982). By the late Paleozoic, the east and southeast margins of North America were again interior regions of a supercontinent; in the Mesozoic, rifting reshaped these areas. The northern margin was modified in the Paleozoic and in the late Mesozoic by major rifting events. In the Cordilleran region, beginning in the late Paleozoic and still active today, a protracted series of collision events combined with dispersive phenomena to distribute terrane slivers along the western margin of North America.

Southern California

In southern California, allochthonous terranes that were accreted during the Cenozoic are juxtaposed against autochthonous upper Proterozoic and Paleozoic platform strata. A host of pre-Cenozoic dispersive events have been inferred, including left-slip of 600 km along the Mojave-Sonora megashear during the Middle Jurassic (Silver & Anderson 1974). In this region, Proterozoic and Phanerozoic terranes currently outboard of the truncated margin of North America represent a wide variety of different geologic environments.

The Santa Lucia-Orocopia allochthon comprises the San Simeon, Stanley Mountain, Salinia, and Tujunga terranes that amalgamated by the late Cretaceous, prior to their accretion approximately 55 m.y. ago (Figure 4; Vedder et al 1983). Some of the terranes composing Santa Lucia-Orocopia are themselves composite; for example, the Tujunga terrane is composed of at least three Precambrian terranes that represent different crustal levels of basement rocks and a variety of sedimentary environments (Powell 1982). While these rocks could represent different crustal levels of the same block, discrepancies in age and lithofacies relations are sufficient to render the original paleogeographic relations uncertain; hence, they are called separate suspect terranes until genetic linkage can be proven. The Salinia terrane is a middle Cretaceous continental volcanic arc; Ross (1977) has distinguished several subterranes of differing lithologies in the prearc basement. The Stanley Mountain terrane consists of Middle Jurassic ophiolite overlain by pelagic and forearc submarine fan sequences, while the San Simeon terrane is a disrupted terrane composed of Cretaceous chert, graywacke, and greenstone—the well-known Franciscan assemblage.

These terranes were once thought to conform to the typical arc-forearc-subduction complex model (e.g. Dickinson 1970) because they appear to fit into three such belts that stretch along the California continental margin. Much new data argue against this model. Although ophiolite complexes similar in age to the Stanley Mountain terrane occur at the base of the Great Valley sequence (Hopson et al 1981), some of these complexes are sufficiently different in terms of their environments of formation that they are best treated as separate terranes (e.g. Blake et al 1982a). The Franciscan Complex is also composed of numerous discrete terranes that differ in age, metamorphic grade, and tectonic history. Some parts of the Franciscan appear to have originated in the Southern Hemisphere (Alvarez et al 1980), whereas others may be from equatorial or northerly latitudes. The arc-forearc-trench belt is repeated in central and southern California; the eastern belt comprises the Sierra Nevada batholith, the Great Valley sequence, and the Franciscan assemblage, while the western belt is made up of the Salinia, Stanley Mountain, and San Simeon terranes (Figure 4).

There are numerous arguments for and against the role of the San Andreas fault in producing this doubling up of the belts (see Champion et al 1983, Dickinson 1983). For example, strata overlying ophiolite in the Stanley Mountain terrane are extremely similar to coeval strata of the Great Valley sequence (Dickinson 1983). However, paleomagnetic data show that the western belt of terranes originated at far more southerly latitudes than can be attributed to right-slip motion on the San Andreas fault (Champion et al 1980, 1983, McWilliams & Howell 1982). McWilliams & Howell (1982) have documented the displacement and amalgamation history of Stanley Mountain and Salinia; their data suggest that Stanley Mountain was in low northerly or southerly latitudes (14°) in the late Jurassic, then moved to 6° north or south by the early Cretaceous. Salinia and Tujunga were linked by middle to late Cretaceous time, as indicated by pluton stitching and overlap sequences. Stanley Mountain and Salinia were amalgamated by the late Cretaceous; provenancial linkage and overlap assemblages of this age are supported by the paleomagnetic data, which indicate both terranes were at ca. 21°N at this time and then moved together to 25°N through the Paleocene. Displacement for Salinia since Cretaceous time is estimated to be at least 2500 km (Champion et al 1980, 1983). Eocene strata are the oldest to overlap all the terranes of the Santa Lucia-Orocopia allochthon as well as the continental margin, indicating accretion had occurred by that time.

Several terranes of the southern California borderland also show significant northward displacement, but their movement and accretion history is much younger than that of Santa Lucia-Orocopia. Continental strata of the Cortez terrane, Upper Jurassic to Cretaceous island-arc rocks of the Guerrero terrane, and Jurassic metamorphosed arc rocks of the Malibu terrane are stitched together by 125–90 m.y.-old plutons of the

Peninsular Ranges batholith (Figure 4). By at least 90 m.y. ago, Jurassic ophiolite and Cretaceous turbidites of the Nicolas terrane were accreted to the Cortez-Guerrero-Malibu composite and overlapped by forearc basin and submarine fan strata. We are not aware of any stratigraphic sequences reflecting either provenancial linkage involving blueschist metamorphosed rocks of the Catalina terrane or overlap assemblages on this terrane that are older than Miocene. Thus, amalgamation of Catalina to the other terranes could not have occurred prior to this time. The first strata that overlap all the terranes of the borderland and the California margin are upper Miocene. The inference that these terranes were not accreted until late Miocene time is supported by paleomagnetic data, which indicate that Eocene rocks of the Nicolas terrane have been translated 19° northward, and that lower and middle Miocene rocks of the Nicolas and Malibu terranes may have been translated 10–15° northward and rotated as much as 110°, while Miocene rocks of the Stanley Mountain and Tujunga terranes show rotations but no translations (Kamerling & Luyendyk 1979, Champion et al 1981, Hornafius et al 1981).

Asia

Asia is composed of several large cratonal blocks separated by fold belts of various ages (e.g. Figures 1 and 5; Terman 1974, Chinese Academy of Geological Sciences 1975, 1976, 1979, Bally et al 1980). These fold belts formed during major orogenic events, and each is composed of many terranes of oceanic, volcanic arc, and continental affinities. Paleomagnetic data from rocks on several of the cratons indicate that they were widely separated during the Permian. While the Siberian craton was at ca. 57°N, the Omolon terrane (formerly considered part of a Kolyma microplate) lay at 33°N, Sikhote Alin at 34°N, Sino-Korea at 10°N, Yangtze at 2°N, Japan at 5°S (McElhinny et al 1981), and the Malay Peninsula at 15°N (McElhinny et al 1974). Because several of these microcontinents are probably composed of numerous terranes, the Permian Panthalassa Ocean was apparently filled with many isolated fragments, much like the modern southwest Pacific.

Many of the fold belts in Siberia and China are marked by ophiolite belts and blueschist, which provide good evidence for the occurrence and age of fossil subduction zones and ocean closures. Preliminary plate-tectonics reconstructions of China have been proposed by Li et al (1980) and Huang (1978), but much more detailed work is needed to date the ophiolites and blueschist and to discriminate between terranes within the fold belts, so that the characteristics and timing of accretion can be better documented.

On a broad scale, this region has undergone sequentially southward-stepping accretionary events that constructed a collage of terranes against the 1.4 m.y.-old Baikalian rifted margin of southern Siberia. This period of growth began with late Proterozoic accretion of oceanic and arc material in the Baikalides fold belt (Figure 5). Throughout the Paleozoic, accretion occurred in Mongolia and northern China as the Dzungaria-Kasakhstania block approached the Siberian margin. Accretion of this block may have

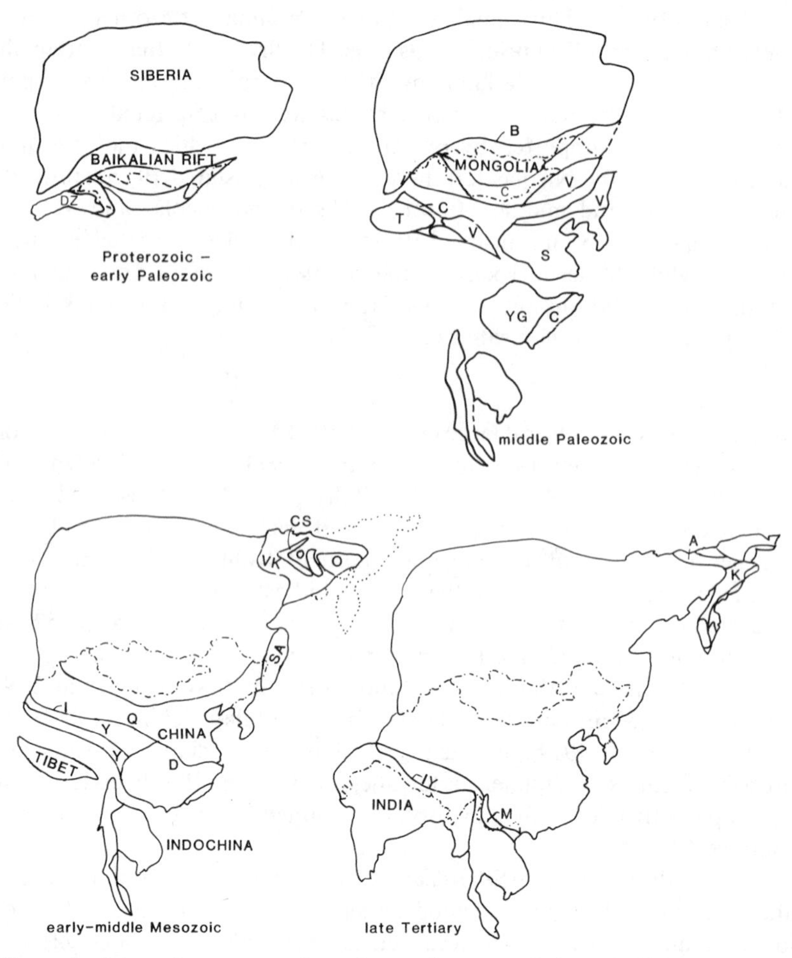

Figure 5 Schematic representation of accretionary evolution of Siberia and Asia. DZ = Dzungaria; C = Caledonian fold belts; B = Baikalian suture; V = Variscan fold belts; T = Tarim; S = Sino-Korea; YG = Yangtze; CS = Cherskiy terrane; VK = Verkhoyansk fold belt; O = Omolon terrane; SA = Sikhote Alin; I = Indosinides sutures; Q-D = Qilian-Dabie Shan suture; Y = Yenshanides fold belts; A = South Anyui; K = Koryak Highlands; M = Mekong fold belt; IY = Indus-Yaluzangbu suture.

produced the Caledonian (early Paleozoic) fold belts of this region. Collision of the Tarim block and the Sino-Korean block in latest Paleozoic was accompanied by accretion of several island arcs, oceanic terranes, and small continental fragments to both the northern and southern margin of these blocks, forming the Variscan fold belts (Figure 5).

During the Triassic, the paleogeography of China was vastly different from that of today. The southern margin of China appears to have been along what is now a major suture marked by several ophiolite belts and blueschist that passes through the Qilian Mountains (North Qilian-Dabie Shan belt, Figure 5). North of the ophiolite belts, Triassic deposits are continental facies; south of the suture are a variety of disparate marine environments (Wang et al 1981) whose boundaries appear to correlate with terrane boundaries. The Triassic to Jurassic Indosinian orogeny appears to have been marked by the collision of the Yangtze block with the continental collage to the north; this may have been coincident with or followed by the collision of Southeast Asia along a major suture marked by ophiolitic and island-arc terranes of the Mekong fold belt. The Mekong fold belt and some of the Indosinian fold belts also record Jurassic to Cretaceous (Yenshanian) accretionary events, which may indicate that collisions of Yangtze with Sino-Korea and of Indochina with the Yangtze block were not complete until Cretaceous time. The Indosinian and Yenshanian fold belts have been considered to fit into a typical backarc-arc-subduction complex model (e.g. Li et al 1980, E. Zhang, written communication, 1982). However, there appears to be more than one island-arc sequence; one or more continental fragments are apparently embedded in the "subduction complex"; and the ages and chemistry of oceanic crustal fragments differ, so that they may or may not represent genetically related backarc basin, arc, or trench basement materials. Thus we suspect that this region is a complex collage of accreted terranes whose paleogeographic relations need to be elucidated. The collision of India with Tibet in the Eocene was the culmination of numerous Jurassic through Cretaceous accretions of arc, ophiolitic, and continental fragments (including Tibet). Continued northward movement of India under Tibet is presently causing strike-slip dispersion of terranes in northern and eastern China.

Northeast Siberia has experienced mostly middle to late Mesozoic accretion of continental fragments, such as the Omolon terrane, and predominantly oceanic material along the eastern and northeastern margins. This region, including the Koryak Highlands and the South Anyui fold belt (Figure 5), contains a complex series of arcs and ophiolitic terranes that amalgamated and accreted during the late Jurassic and Cretaceous (Fujita 1978, Churkin & Trexler 1981, Fujita & Newberry 1982).

Australia and New Zealand

Australia is similar to other continents in having several Archean cratons surrounded by Proterozoic fold belts. Major reshaping of the continent occurred in the late Proterozoic to Cambrian with the development of a rifted margin in what is presently central Australia (Figure 3). Deposits of aulacogen environments (intracratonic basins within the failed arm of a triple junction) resulting from this breakup are still preserved in central and southern Australia. From the early Paleozoic through the early Mesozoic this margin was a consuming margin of Gondwana, and numerous accretionary events occurred. A composite of oceanic, island-arc, and continental fragments accreted during the Paleozoic make up the Lachlan fold belt; late Paleozoic to early Mesozoic island-arc and ophiolite terranes accreted generally during the Triassic compose the New England fold belt (Figure 3). The Paleozoic strata in the Tuhua composite terrane of New Zealand must also have been accreted during the early Mesozoic (Figure 3; E. Scheibner, written communication, 1983). Jurassic and Cretaceous accretionary events are reflected by the additions of the Hokonui, Caples, and Torlesse terranes of New Zealand (Howell 1980).

Australia is different from other circum-Pacific continents in that much of its Mesozoic-Cenozoic history has been dominated by rifting rather than accretionary events. The western margin was rifted in the early Mesozoic during the breakup of Gondwana. In the middle Cretaceous, New Zealand and the Lord Howe Rise rifted away from the eastern margin with the opening of the Tasman Sea, and the southern margin formed as it rifted away from Antarctica. The northwest Australian margin grew during the middle Tertiary as a result of accretion related to the collision of the Banda arc (Figure 3). These rifting events have scattered continental fragments around the margins of Australia and have caused dispersion and deformation of previously accreted arc, oceanic, and marginal basin terranes of New Zealand.

South America

The distribution and nature of tectonostratigraphic terranes in South America is very poorly known. The core of South America comprises the Guyana and Brazilian shields; however, the Brazilian shield appears to be composed of two Archean cratons that are separated by a latest Proterozoic orogenic belt that may contain ophiolitic rocks (Figure 1; Borello 1972, Burke et al 1977, de Almeida 1978). The thick volcanic and plutonic rocks of the Andean arc cover and obscure most of the pre-Jurassic basement of western South America; however, several continental fragments that may have been accreted during the Paleozoic or early Mesozoic have been distinguished (Figure 1). The principal Paleozoic suture appears to be marked in part by the Sierra de la Ventana and in the Precordillera of Argentina (Burke et al 1977). It is currently debated as to whether Precambrian rocks found along the western coast of South America (e.g. the Arequipa massif; Figure 1) are uplifted outliers of the shields or exotic fragments of sialic crust. Paleomagnetic studies are in progress on Devonian strata of the Arequipa massif of Peru to determine whether there has been any relative movement with respect to cratonal South America (M. O. McWilliams, oral communication, 1982).

Because the western margin of South America has been a convergent margin since the Jurassic, we expect to find terranes accreted during the Mesozoic along the Andean margin. In contrast to the central Andes, where only terranes with continental affinity have been identified, well-developed oceanic (including ophiolite and blueschist) sequences are present in the northern and southern sections. These appear to represent a series of accretions of island-arc and oceanic rocks that occurred in the middle Cretaceous and possibly earlier in the southernmost Andes (Dalziel et al 1974) and from the middle Cretaceous to Tertiary in the northern Andes and Caribbean Mountains. These terranes comprise island-arc, oceanic, and continental fragments (Maresch 1974, Case et al 1971, 1983); paleomagnetic studies indicate significant amounts of translation and rotation for several of them (MacDonald & Opdyke 1972, Hargraves & Skerlec 1980, Skerlec & Hargraves 1980, Case et al 1983).

CONCLUSIONS

From studies of the geology and geophysics of the circum-Pacific region, it is apparent that this vast region is a collage of disparate crustal fragments. Throughout Proterozoic and Phanerozoic time, accretion of terranes has resulted in continental growth and rifting, and dispersive processes have operated to modify and attenuate continental margins. These tectonically eroded margins, in addition to oceanic plateaus, may serve as the source areas for crustal blocks that eventually become accreted terranes. Because the present Pacific Ocean is no older than Middle Jurassic, the geologic history of the proto-Pacific Ocean (Panthalassa) and surrounding cratonal regions will only be elucidated through further work to characterize the origins and kinematic histories of allochthonous terranes. Terrane analysis can be used to provide an objective, interdisciplinary approach to interpretation of geological and geophysical data in which genetic relations in time and space must be proven before they are used as the basis for plate-tectonic reconstructions. Within the framework of plate tectonics, this approach enables the refinement of more actualistic models to depict the fragmented nature, structural complexities, and great mobility of active margins.

ACKNOWLEDGMENTS

The data and concepts presented in this review reflect numerous discussions with many colleagues who are experts in various regions of the circum-Pacific. We thank M. C. Blake, Jr., Zvi Ben-Avraham, J. E. Case, Michael Churkin, Jr., P. J. Coney, Kazuya Fujita, J. W. H. Monger, Erwin Scheibner, N. J. Silverling, and Edmund Zhang for sharing their data and ideas. Reviews by Tracy Vallier and Michael Fisher improved the clarity of the manuscript.

Literature Cited

Alvarez, W., Kent, D. V., Premoli-Silva, I., Schweickert, R. A., Larson, R. A. 1980. Franciscan complex limestone deposited at 17° south paleolatitude. *Geol. Soc. Am. Bull.* 91:476–84

Armstrong, A. K., MacKevett, E. M. Jr. 1983. Geologic relations of Kennecott-type copper deposits, Wrangell Mountains, Alaska. Part B. Carbonate sedimentation, sabkha facies, diagenesis, and stratigraphy, lower part Triassic Chitistone Limestone—the ore host rock. *US Geol. Surv. Prof. Pap.* In press

Bally, A. W., Allen, C. R., Geyer, R. B., Hamilton, W. B., Hopson, C. A., et al. 1980. Notes on the geology of Tibet and adjacent areas—report of the American Plate Tectonics Delegation to the People's Republic of China. *US Geol. Surv. Open-File Rep. 80-501.* 100 pp.

Beck, M. E. Jr., Cox, A. 1979. Paleomagnetic evidence for large-scale tectonic rotations and translations along the western edge of North America. In *Cenozoic Paleogeography of the Western United States, Paleogeogr. Symp. 3*, ed. J. M. Armentrout, et al, p. 325. Los Angeles, Calif: Pac. Sect., Soc. Econ. Paleontol. Mineral.

Ben-Avraham, Z., Nur, A., Jones, D. L., Cox, A. 1981. Continental accretion and orogeny: from oceanic plateaus to allochthonous terranes. *Science* 213:47–54

Berg, H. C., Jones, D. L., Richter, D. H. 1972. Gravina-Nutzotin belt—tectonic significance of an upper Mesozoic sedimentary and volcanic sequence in southern and southeastern Alaska. *US Geol. Surv. Prof. Pap. 800-D*, pp. D1–24

Berg, H. C., Jones, D. L., Coney, P. J. 1978. Map showing pre-Cenozoic tectonostratigraphic terranes of southeastern Alaska and adjacent areas. *US Geol. Surv. Open-File Rep. 78-1085, scale 1:1,000,000*

Blake, M. C. Jr., Jones, D. L. 1981. The Franciscan assemblage and related rocks in northern California: a reinterpretation. In *The Geotectonic Development of California, Rubey Vol. 1*, ed. W. G. Ernst, pp. 307–28. Englewood Cliffs, NJ: Prentice-Hall. 706 pp.

Blake, M. C. Jr., Howell, D. G., Jones, D. L. 1982a. Preliminary tectonostratigraphic terrane map of California. *US Geol. Surv. Open-File Report 82-593, scale 1:500,000*

Blake, M. C. Jr., Howell, D. G., Jones, D. L. 1982b. Relation between orogeny, subduction, and changing-plate motion models for the Pacific coast region of the United States. *Eos, Trans. Am. Geophys. Union* 63:911 (Abstr.)

Borello, A. V. 1972. The Precordillera as a type of geosyncline in Argentina. *Int. Geol. Congr., 24th* 3:293–98

Burke, K., Dewey, J. F., Kidd, W. S. F. 1977. World distribution of sutures—the sites of former oceans. *Tectonophysics* 40:69–99

Campa, M. F., Coney, P. J. 1983. Tectonostratigraphic terranes and mineral resource distributions in Mexico. *Can. J. Earth Sci.* In press

Case, J. E., Duran, L. G., Alfonso Lopez, P., Moore, W. R. 1971. Tectonic investigations in western Colombia and eastern Panama. *Geol. Soc. Am. Bull.* 82:2685–2712

Case, J. E., Holcombe, T. L., Martin, R. G. 1983. Map of geologic terranes in the Caribbean region. *Geol. Soc. Am. Mem.* In press

Champion, D. E., Howell, D. G., Gromme, C. S. 1980. Paleomagnetism of the Cretaceous Pigeon Point Formation and inferred northward displacement of 2500 km for the Salinian Block, California. *Eos, Trans. Am. Geophys. Union* 61 : 948 (Abstr.)

Champion, D. E., Howell, D. G., Marshall, M. C. 1981. Paleomagnetic evidence for 3800 km of northwestward translation of San Miguel Island, Southern California Borderland. *Eos, Trans. Am. Geophys. Union* 62 : 855 (Abstr.)

Champion, D. E., Howell, D. G., Gromme, C. S. 1983. Paleomagnetic and geologic data indicating 2500 km of northward displacement for the Salinian, Sur Obispo, Tujunga, and Baldy(?) terranes, California. *J. Geophys. Res.* In press

Chinese Academy of Geological Sciences. 1975. *Geological map of Asia, scale 1 : 5,000,000*

Chinese Academy of Geological Sciences, Compilation Group of the Geological Map of China. 1976. *An Outline of the Geology of China.* Peking. 22 pp.

Chinese Academy of Geological Sciences, Institute of Geology, Section of Structural Geology. 1979. *Tectonic map of China, scale 1 : 4,000,000*

Churkin, M. Jr., Trexler, J. H. Jr. 1981. Continental plates and accreted oceanic terranes in the Arctic. In *The Ocean Basins and Margins, The Arctic Ocean*, ed. A. E. M. Nairn, M. Churkin, Jr., F. G. Stehli, 5 : 120. New York : Plenum

Coleman, P. J., Kroenke, L. W. 1981. Subduction without volcanism in the Solomon Islands arc. *Geo-Mar. Lett.* 1 : 129–34

Condie, K. C. 1982. Plate-tectonics model for Proterozoic continental accretion in the southwestern United States. *Geology* 10 : 37–42

Coney, P. J. 1981. Accretionary tectonics in western North America. *Ariz. Geol. Soc. Dig.* 14 : 23–37

Coney, P. J., Jones, D. L., Monger, J. W. H. 1980. Cordilleran suspect terranes. *Nature* 288 : 329–33

Crowell, J. C., Sylvester, A. G. 1979. Introduction to the San Andreas–Salton Trough juncture. In *Tectonics of the Juncture Between the San Andreas Fault System and the Salton Trough, Southeastern California. A Guidebook*, ed. J. C. Crowell, A. G. Sylvester, pp. 1–13. Dept. Geol. Sci., Univ. Calif., Santa Barbara

Dalziel, I. W. D., De Wit, M. J., Palmer, K. F. 1974. Fossil marginal basin in the southern Andes. *Nature* 250 : 291–94

de Almeida, F. F. M., coordinator general. 1978. Tectonic map of South America. *Geol. Soc. Am. Map Chart Ser. MC-32, scale 1 : 5,000,000*

Dewey, J. F., Bird, J. M. 1970. Mountain belts and the new global tectonics. *J. Geophys. Res.* 75 : 2625–47

Dickinson, W. R. 1970. Relations of andesites, granites, and derivative sandstones to arc-trench tectonics. *Rev. Geophys. Space Phys.* 8 : 813–60

Dickinson, W. R. 1983. Cretaceous sinistral strike slip along Nacimiento fault in coastal California. *Am. Assoc. Pet. Geol. Bull.* 67 : 624–45

Eisbacher, G. H. 1974. Evolution of successor basins in the Canadian Cordillera. In *Modern and Ancient Geosynclinal Sedimentation*, ed. R. H. Dott, R. H. Shaver, pp. 274–91. *Soc. Econ. Paleon. Mineral. Spec. Publ. No. 19*

Engebretson, D. C. 1982. *Relative motions between oceanic and continental plates in the Pacific Basin.* PhD thesis. Stanford Univ., Calif.

Engebretson, D. C., Cox, A. 1982. Plate interactions in the northeast Pacific since the Oxfordian. *Eos, Trans. Am. Geophys. Union* 63 : 911 (Abstr.)

Fujita, K. 1978. Pre-Cenozoic tectonic evolution of northeast Siberia. *J. Geol.* 86 : 159–72

Fujita, K., Newberry, J. T. 1982. Tectonic evolution of northeastern Siberia and adjacent regions. *Tectonophysics* 89 : 337–57

Hargraves, R. B., Skerlec, G. M. 1980. Paleomagnetism of some Cretaceous-Tertiary igneous rocks on Venezuelan offshore islands, Netherlands Antilles, Trinidad and Tobago. *Proc. Caribbean Geol. Conf., 9th, Santo Domingo, Domin. Repub.*, pp. 509–15

Harland, W. B. 1971. Tectonic transpression in Caledonian Spitzbergen. *Geol. Mag.* 108 : 27–42

Hillhouse, J. W. 1977. Paleomagnetism of the Triassic Nikolai greenstone, south-central Alaska. *Can. J. Earth Sci.* 14 : 2578–92

Hillhouse, J. W., Gromme, C. S. 1983. Northward displacement of Wrangellia : paleomagnetic evidence from Alaska, British Columbia, Oregon, and Idaho. *J. Geophys. Res.* In press

Hoffman, P. F., Card, K. D., Davidson, A. 1982. The Precambrian : Canada and Greenland. In *Perspectives in Regional Geological Synthesis : Planning for "The Geology of North America," D-NAG Spec. Publ. 1*, ed. A. R. Palmer, pp. 3–6. Boulder, Colo : Geol. Soc. Am.

Hopson, C. A., Mattinson, J. M., Pessagno, E. A. Jr. 1981. Coast range ophiolite, western California. In *The Geotectonic Development of California, Rubey Vol. 1*, ed. W. G. Ernst, pp. 307–28. Englewood Cliffs, N.J : Prentice-Hall. 706 pp.

Hornafius, J. S., Luyendyk, B. P., Weaver, D. W., Wornardt, W. W., Fuller, M. 1981. Paleomagnetic results from the Monterey Formation in the western Transverse

Ranges, California. *Eos, Trans. Am. Geophys. Union* 62:855 (Abstr.)

Howell, D. G. 1980. Mesozoic accretion of exotic terranes along the New Zealand segment of Gondwanaland. *Geology* 8:487–91

Howell, D. G., Crouch, J. K., Greene, H. G., McCulloch, D. S., Vedder, J. G. 1980. Basin development along the late Mesozoic and Cainozoic California margin: a plate tectonic margin of subduction, oblique subduction and transform tectonics. *Spec. Publ. Int. Assoc. Sedimentol.* 4:43–62

Howell, D. G., Jones, D. L., Schermer, E. R. 1983. Tectonostratigraphic terranes of the frontier circum-Pacific region. *Am. Assoc. Pet. Geol. Bull.* 67:485–86 (Abstr.)

Huang, T. K. 1978. An outline of the tectonic characteristics of China. *Eclogae Geol. Helv.* 71:611–35

Hudson, T. 1979. Mesozoic plutonic belts of southern Alaska. *Geology* 7:230–34

Hudson, T., Plafker, G. 1982. Paleogene metamorphism of an accretionary flysch terrane, eastern Gulf of Alaska. *Geol. Soc. Am. Bull.* 93:1280–90

Hudson, T., Plafker, G., Peterman, Z. E. 1978. Paleogene anatexis along Gulf of Alaska Margin. *Geology* 7:573–77

Irving, E., Yole, R. W. 1972. Paleomagnetism and the kinematic history of mafic and ultramafic rocks in fold mountain belts. *Ottawa, Earth Phys. Branch Publ.* 42:87–95

Irwin, W. P. 1960. Geological reconnaissance of the northern Coast Ranges and Klamath Mountains, California. *Calif. Div. Mines Bull. 179.* 80 pp.

Irwin, W. P. 1972. Terranes of the western Paleozoic and Triassic belt in the southern Klamath Mountains, California. *US Geol. Surv. Prof. Pap. 800-C*, pp. 103–11

Jones, D. L., Silberling, N. J. 1979. Mesozoic stratigraphy—the key to tectonic analysis of southern and central Alaska. *US Geol. Surv. Open-File Rep. 79-1200.* 37 pp.

Jones, D. L., Silberling, N. J., Hillhouse, J. W. 1977. Wrangellia: a displaced terrane in north-western North America. *Can. J. Earth Sci.* 14:2565–77

Jones, D. L., Silberling, N. J., Csejtey, B. Jr., Nelson, W. H., Blome, C. D. 1980. Age and structural significance of ophiolite and adjoining rocks in the upper Chulitna district, south-central Alaska. *US Geol. Surv. Prof. Pap. 1121-A.* 21 pp.

Jones, D. L., Silberling, N. J., Berg, H. C., Plafker, G. 1981. Tectonostratigraphic terrane map of Alaska. *US Geol. Surv. Open-File Rep. 81-792, scale 1 : 2,500,000*

Jones, D. L., Howell, D. G., Schermer, E. R. 1982. Preliminary tectonostratigraphic terrane map of the circum-Pacific region. *Am. Assoc. Pet. Geol. Bull.* 66:972 (Abstr.)

Jones, D. L., Howell, D. G., Coney, P. J., Monger, J. W. H. 1983. Recognition character, and analysis of tectonostratigraphic terranes in western North America. In *Advances in Earth and Planetary Sciences*, ed. M. Hashimoto, S. Uyeda, pp. 21–35. Tokyo: Terra Sci. Publ. Co.

Kamerling, M. J., Luyendyk, B. P. 1979. Tectonic rotation of the Santa Monica Mountains region, western Transverse Ranges, California, suggested by paleomagnetic vectors. *Geol. Soc. Am. Bull.* 90: 331–37

Li, C. (C. Y. Lee), Wang. Q., Zhang, Z., Liu, X. 1980. A preliminary study of plate tectonics of China. *Chin. Acad. Geol. Sci. Bull., Ser. I* 2:11–22

Macdonald, W. D., Opdyke, N. D. 1972. Tectonic rotations suggested by paleomagnetic results from northern Colombia, South America. *J. Geophys. Res.* 77: 5720–30

Maresch, W. V. 1974. Plate-tectonics origin of the Caribbean mountain system of northern South America. *Geol. Soc. Am. Bull.* 85:669–82

McElhinny, M. W., Haile, N. S., Crawford, A. R. 1974. Paleomagnetic evidence shows Malay Peninsula was not a part of Gondwanaland. *Nature* 252:641–45

McElhinny, M. W., Embleton, B. J. J., Ma, X. H., Zhang, Z. K. 1981. Fragmentation of Asia in the Permian. *Nature* 293:212–16

McWilliams, M. O., Howell, D. G. 1982. Exotic terranes of western California. *Nature* 297:215–17

Monger, J. W. H. 1977. Upper Paleozoic rocks of the western Canadian Cordillera and their bearing on Cordilleran evolution. *Can. J. Earth Sci.* 14:1832–59

Monger, J. W. H., Ross, C. A. 1971. Distribution of fusulinaceans in the western Canadian Cordillera. *Can. J. Earth Sci.* 8: 259–78

Monger, J. W. H., Price, R. A., Tempelman-Kluit, D. J. 1982. Tectonic accretion and the origin of the two major metamorphic and plutonic welts in the Canadian Cordillera. *Geology* 10:70–75

Nestel, M. K. 1980. Permian fusulinacean provinces in the Pacific northwest are tectonic juxtapositions of ecologically distinct faunas. *Geol. Soc. Am. Abstr. with Programs* 12:144 (Abstr.)

Newton, C. R. 1983. Paleozoogeographic affinities of Norian (late Triassic) molluscs of the Wrangellian terrane: evidence for an east Pacific origin. *Programs and Abstr., Ann. Meet., Pac. Sect., Am. Assoc. Pet. Geol., 58th, Sacramento, Calif.*

Nichols, K. M., Silberling, N. J. 1979. Early Triassic (Smithian) ammonites of paleoequatorial affinity from the Chulitna terrane, south-central Alaska. *US Geol. Surv. Prof. Pap. 1121-B*, pp. B1–B5

Nur, A., Ben-Avraham, Z. 1977. Lost Pacifica continent. *Nature* 270:41–43

Nur, A., Ben-Avraham, Z. 1981. Volcanic gaps and the consumption of aseismic ridges in South America. In *Nazca Plate, Crustal Formation and Andean Convergence*, ed. L. D. Kulm, et al, pp. 729–40. *Geol. Soc. Am. Mem. No. 154*

Nur, A., Ben-Avraham, Z. 1982. Oceanic plateaus, the fragmentation of continents, and mountain building. *J. Geophys. Res.* 87:3644–61

Packer, D. R., Stone, D. B. 1972. An Alaskan Jurassic paleomagnetic pole and the Alaskan orocline. *Nature Phys. Sci.* 237:25–26

Packer, D. R., Stone, D. B. 1974. Paleomagnetism of Jurassic rocks for southern Alaska, and the tectonic implications. *Can. J. Earth Sci.* 11:976–97

Powell, R. E. 1982. Crystalline basement terranes in the southern eastern Transverse Ranges, California. In *Geologic Excursions in the Transverse Ranges, Southern California*, ed. J. D. Cooper, pp. 109–36. *Geol. Soc. Am. Ann. Meet., 78th, Cordilleran Sect., Guideb.*

Reed, B. L., Lanphere, M. A. 1969. Age and chemistry of Mesozoic and Tertiary plutonic rocks in south-central Alaska. *Geol. Soc. Am. Bull.* 80:23–44

Ross, D. C. 1977. Pre-intrusive metasedimentary rocks of the Salinian block, California—a paleotectonic dilemma. In *Paleozoic Paleogeography of the Western United States, Paleogeogr. Symp. 1*, ed. J. H. Stewart, C. H. Stevens, A. E. Fritsche, pp. 371–80. Los Angeles, Calif: Pac. Sect., Soc. Econ. Paleon. Mineral.

Saleeby, J. B. 1983. Accretionary tectonics of the North American Cordillera. *Ann. Rev. Earth Planet. Sci.* 11:45–73

Scholl, D. W., Vallier, T. L. 1983. Subduction and the rock record of Pacific margins. In *Expanding Earth Symposium, Sydney, 1981*, ed. S. W. Carey, pp. 235–45. Univ. Tasmania, Austral.

Silberling, N. J., Jones, D. L. 1983. Paleontologic evidence for northward displacement of Mesozoic rocks in accreted terranes of the western Cordillera. *Programs with Abstr., Joint Ann. Meet., Geol. Assoc. Can., Mineral. Assoc. Can., Can. Geophys. Union, Victoria, B.C.*, p. A62 (Abstr.)

Silberling, N. J., Jones, D. L., Monger, J. W. H., Blake, M. C. Jr., Howell, D. G., Coney, P. J. 1983. Lithotectonic terranes of the Cordillera of western North America. *US Geol. Surv. Misc. Field Invest. Rep., scale 1:2,500,000.* In press

Silver, E. A., Smith, R. B. 1983. Comparison of terrane accretion in modern Southeast Asia and the Mesozoic North American Cordillera. *Geology* 11:198–202

Silver, L. T., Anderson, T. H. 1974. Possible left-lateral early to middle Mesozoic disruption of the southwestern North American craton margin. *Geol. Soc. Am. Abstr. with Programs* 6:955–56 (Abstr.)

Silver, L. T., Taylor, H. P. Jr., Chappell, B. 1979. Some petrological, geochemical, and geochronological observations of the Peninsular Ranges batholith near the international border of the U.S.A. and Mexico. In *Mesozoic Crystalline Rocks: Peninsular Ranges Batholith and Pegmatites, Point Sal Ophiolite*, ed. P. L. Abbott, V. R. Todd, pp. 83–110. Dept. Geol. Sci., San Diego State Univ., Calif. 286 pp.

Skerlec, G. M., Hargraves, R. B. 1980. Tectonic significance of paleomagnetic data from northern Venezuela. *J. Geophys. Res.* 85:5305–15

Stanley, G. D. Jr. 1982. Triassic carbonate development and reef-building in western North America. *Geol. Rundsch.* 71:1057–75

Terman, M. J., principal compiler. 1974. Tectonic map of China and Mongolia, scale 1:5,000,000. Boulder, Colo: Geol. Soc. Am.

Tozer, E. T. 1982. Marine Triassic faunas of North America: their significance in assessing plate and terrane movements. *Geol. Rundsch.* 71:1077–1104

Vallier, T. L., Rea, D. K., Dean, W. E., Thiede, J., Adelseck, C. G. 1981. The geology of Hess Rise, central north Pacific Ocean. In *Initial Reports of the Deep Sea Drilling Project*, J. Thiede, T. L. Vallier, et al, 62:1031–72. Washington DC: GPO

Vedder, J. G., Howell, D. G., McLean, H. 1983. Stratigraphy, sedimentation, and tectonic accretion of exotic terranes, southern Coast Ranges, California. *Am. Assoc. Pet. Geol. Mem.* 34:471–96

von Heune, R., Aubouin, J., Azema, J., Blackinton, G., Carter, J. A., et al. 1980. Leg 67: the Deep Sea Drilling Project Mid-America trench transect off Guatemala. *Geol. Soc. Am. Bull.* 91:421–32

Wang, Y.-G., Chen, C.-C., He, G.-X., Chen, J.-H. 1981. An outline of the marine Triassic in China. *Int. Union Geol. Sci. Publ. No.* 7. 21 pp.

Williams, H., Hatcher, R. D. Jr. 1982. Suspect terranes and accretionary history of the Appalachian orogen. *Geology* 10:530–36

Ann. Rev. Earth Planet. Sci. 1984. 12: 133–53

KIMBERLITES: COMPLEX MANTLE MELTS

Jill Dill Pasteris

Department of Earth and Planetary Sciences, Washington University, St. Louis, Missouri 63130

INTRODUCTION

Kimberlites are rare ultramafic rocks of mantle origin that are the major primary source of diamonds. They host several types of peridotite (olivine, garnet, orthopyroxene, clinopyroxene) and eclogite (garnet, clinopyroxene) inclusions from the upper mantle. Historically, the inclusions (xenoliths) have been studied more intensely than their hosts and have been used to infer the conditions of temperature, pressure, gas fugacity, and stress in the mantle (e.g. Nixon & Boyd 1973, Dawson 1980, Nixon et al 1981, Eggler 1983). In this review the kimberlites themselves are examined as important indicators of the physical and chemical conditions accompanying melting phenomena in the upper mantle.

OCCURRENCE OF KIMBERLITES

Geographic and Geologic Localization

Kimberlites occur on all the major continents (Figure 1) and range in age from about 1750 m.y. for the Premier Pipe, South Africa, to about 30 m.y. for the Colorado Plateau pipes (see review by Dawson 1980) and about 20 m.y. for the kimberlites and associated rocks in Western Australia (Jaques et al 1982). Kimberlites usually are confined to cratons, which have been tectonically stable for at least 0.5 b.y., and they tend to occur in clusters or provinces, some of the most important of which are in southern Africa, Siberia, West Africa, southwestern United States, and northern Western Australia. The latter locality is an example of where kimberlites are associated with or difficult to distinguish from similar ultramafic rocks that also may bear diamonds (e.g. Prider 1960, Jaques et al 1982).

Kimberlite provinces consistently possess the following geological and

0084–6597/84/0515–0133$02.00

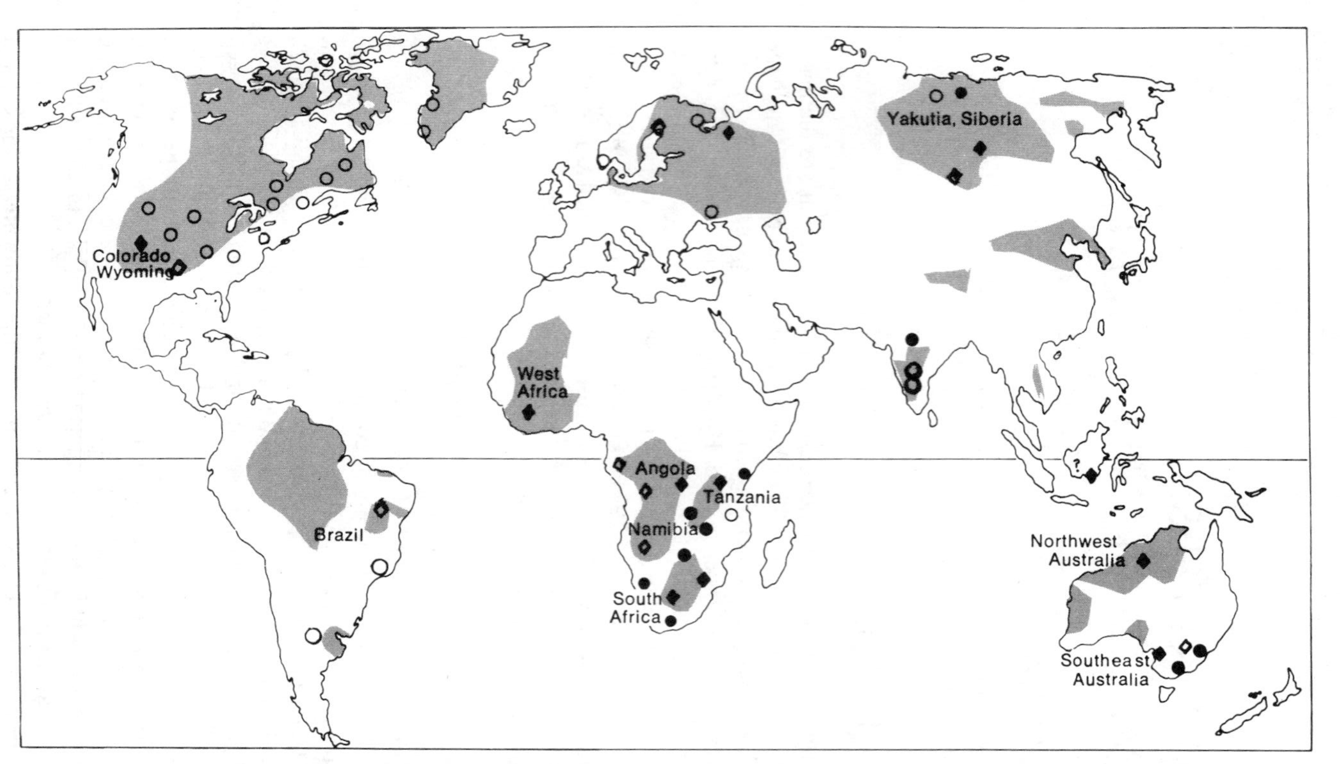

Figure 1 Very generalized representation of kimberlite localities. Shaded regions are cratons $\gtrsim 1500$ m.y. old. Open symbols represent individual kimberlite bodies; closed symbols show provinces. Diamond symbols show localities of diamond-bearing kimberlites. Most data from Ferguson (1980) and Dawson (1980). Figure published with permission of *American Scientist* from Pasteris (1983c), Vol. 71, p. 285.

tectonic features: (*a*) a history of low heat flow followed by high heat flow (Nixon et al 1981), (*b*) earlier extensive volcanic and intrusive activity, such as the Stormberg sequence in South Africa (J. Bristow and J. J. Gurney, personal communication), and (*c*) epeirogenic flexuring directly preceding kimberlite intrusion, as in South Africa and Siberia (Dawson 1980).

Structural analysis of kimberlite terranes through field mapping, geophysical exploration, and satellite imagery (Glover & Groves 1980) has shown that kimberlites frequently are localized by (the intersection of) ancient basement fractures and other crustal lineaments (Dawson 1970). More recent studies suggest that kimberlite intrusions may also be localized by mantle hotspots (Crough et al 1980), by postulated extensions of transform faults (southeastern Australia; Stracke et al 1979), by "prolonged periods of tectonism" associated with continental breakup (West Africa; Haggerty 1982), and by the "intersections of cross-structural lineaments with a rift-parallel zone of weakness" (Parrish & Lavin 1982), as during the opening of the mid-Atlantic (Appalachian kimberlites; Taylor 1982).

Form of Intrusions

Most kimberlites are emplaced by intrusion and occur in diatremes (pipes) or dikes, and rarely in sills. The bodies are small; even the pipes are usually less than 1 km^2 in cross section (Dawson 1980).

Kimberlite diatremes are conical, tapering downward, and frequently become dike-like at depth. Many diatremes apparently result from the abrupt expansion of a confined region of a dike approximately 2–3 km beneath the original land surface. This rapid funneling outward is probably due to the rapid increase in molar volume of water and carbon dioxide at these low pressures and possibly to magmatic reaction with groundwater. Those kimberlites that reach the surface undergo explosive brecciation and fluidization, which shatters wall rocks and rapidly quenches the remaining melt to tuffs (e.g. Hawthorne 1975, Dawson 1980).

Kimberlite diatremes are complex bodies, usually with several petrologically distinct intrusions (Figure 2). In addition, the general character of the kimberlite varies from top to bottom of the pipe owing to the mode of emplacement. Clement (1979) has defined three depth zones in a kimberlite pipe (cf. kimberlite facies of Dawson 1980): crater (≤ 350 m), diatreme (1–2 km below crater), and root zone (≤ 0.5 km below diatreme zone). The hypabyssal-facies kimberlite of the root zone best represents direct precipitation from a melt, whereas the diatreme and crater zones have increasing amounts of brecciated wall rock, kimberlite fragments, and tuffs. The erosional level in a region strongly controls what part of a pipe is accessible. Hence, the interpretation of the mode of emplacement of any kimberlite can be valid only in the context of which pipe level was sampled.

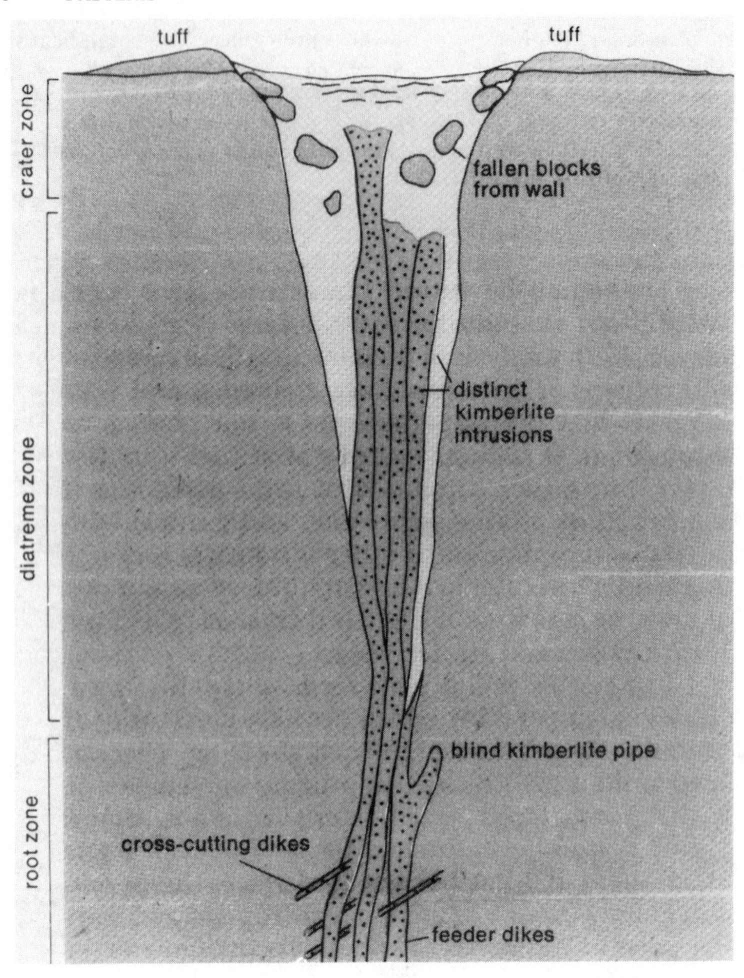

Figure 2 Idealized model of kimberlite pipe or diatreme in vertical section. The root zone lies 3–4 km below the surface. Figure published with permission of *American Scientist* from Pasteris (1983c), Vol. 71, p. 284.

KIMBERLITE PETROGRAPHY

There are many published discussions about the exact definition of a kimberlite (e.g. Clement et al 1977, Mitchell 1979a, Dawson 1980). The complex history of a kimberlite is reflected in its mineralogy, consisting of primary igneous grains that crystallized from the kimberlite melt, grains (xenocrysts) derived from mantle xenoliths, and secondary alteration phases. The following, based on Clement et al (1977), is offered as a working definition: Kimberlite is an ultramafic, potassic, volatile-rich igneous rock, comprised of larger grains (macrocrysts) set in a finer matrix, which gives the rock an inequigranular to porphyritic texture. The macrocrysts are phenocrysts or xenocrysts and are dominated by olivine (the major phase), phlogopite, magnesian ilmenite, magnesian garnet, chromian diopside, and enstatite. These monomineralic grains are set within the matrix, which can exceed 50 vol% of the rock. The matrix contains smaller grains of olivine and commonly phlogopite in a fine-grained, intimate mixture of calcite and serpentine with lesser spinels, ilmenite, perovskite, apatite, and rarely monticellite and/or diopside. Most kimberlites have undergone some late- or postmagmatic hydration with formation of serpentine and/or chlorite, and replacement by calcite. Kimberlites weather easily, so that surface exposures, particularly in South Africa and Brazil, may be comprised of friable, oxide-stained, micaceous "yellow ground."

In hand specimen, kimberlites grade from inequigranular to porphyritic and fragmental in appearance. They frequently contain abundant angular to rounded crustal and mantle xenoliths. The mantle peridotite inclusions show limited evidence of thermal and chemical alteration by the host kimberlite, but crustal xenoliths of shale and limestone often are recrystallized and chemically altered (e.g. Dawson 1980, Clement 1979). Some kimberlites are characterized by autoliths. These are rounded, lapilli-like clusters (frequently several millimeters in diameter) of kimberlite matrix minerals that have nucleated on a small mineral or rock fragment (e.g. Clement 1973, Skinner & Clement 1979). Some kimberlites, especially in dikes, have abundant monomineralic segregations (i.e. intergranular pools) of calcite, phlogopite, serpentine, or devitrifying glass (e.g. Figure 1*b* of Pasteris 1981a). These probably solidified from the last, volatile-enriched portions of the kimberlite melt.

Kimberlites are categorized in several ways. A useful descriptive classification is based on the dominant mineralogy of the rock (see Skinner & Clement 1979), to which may be added its apparent depth level in the original pipe (Clement 1979), such as diatreme-facies phlogopite kimberlite. The abundance and type of xenoliths, the degree and type of alteration, and the texture of the rock (e.g. the presence of flow banding or segregation pools) are additional useful qualifiers.

There are several petrographic controversies about kimberlites, which have significant petrologic implications. The consensus appears to be that kimberlites need not have diamonds, and that not all diamond-bearing porphyritic rocks are kimberlites. However, the question of whether diamonds are xenocrysts or phenocrysts in the kimberlite is still controversial (e.g. Boyd & Finnerty 1980). Although many kimberlites crystallize abundant calcite in their groundmass, carbonatites and kimberlites are distinctly different in their mineralogy, composition, mode of formation, and tectonic setting (Mitchell 1979a). Kimberlites are distinguishable from alnoites, which contain melilite and have higher average Al_2O_3 and CaO contents and lower average MgO and H_2O contents (Dawson 1967, von Eckermann 1967). They also differ from lamproites, which generally contain abundant feldspathoids and/or potassic amphibole and which are poorer in MgO (but richer in SiO_2, Al_2O_3, and Na_2O) than most kimberlites (see, for example, Gupta & Yagi 1980, Jaques et al 1982).

CHARACTERIZATION OF KIMBERLITES

It is difficult to characterize and interpret kimberlites. They contain mantle and crustal xenoliths, phenocrysts and xenocrysts of the same minerals, and abundant secondary phases. The bulk chemical composition of a kimberlite may vary by several weight percent of a major element over a distance of 20 cm (Pasteris, unpublished data). Nevertheless, there are many compositional (major and trace elements) and mineral chemical characteristics shared by almost all kimberlites.

Geochemistry

Kimberlites are geochemically unique among igneous rocks. Dawson (1980) and Wedepohl & Muramatsu (1979) have compiled bulk analyses of many types of fresh, relatively uncontaminated kimberlite, and the latter authors have compared the trace- and major-element abundances for kimberlites with those for garnet peridotite and three types of basalt. Kimberlites are undersaturated, with a silica content $\lesssim 33$ wt%. Compared with other ultramafic rocks, they have high Al_2O_3 and TiO_2 contents, but their total Fe content is about average (~ 9 wt% FeO). Kimberlites have a relatively high alkali content ($\gtrsim 1.5$ wt% $Na_2O + K_2O$) and a K/Na ratio > 3. Their volatile-rich mineralogy provides a variable H_2O and CO_2 content, frequently exceeding 7.5 and 3 wt%, respectively. The large P_2O_5 content (0.5–1.0 wt%) is reflected in the abundance of apatite (Dawson 1967).

The trace-element compositions of kimberlites are even more sensitive than major-element compositions to variations in minor phases, but most kimberlites are strongly enriched in rare earth elements (REE) compared with other ultramafic rocks (Figure 3; Wedepohl & Muramatsu 1979). The major carriers of REE are apatite and especially perovskite (Boctor & Boyd 1979), with contributions from less-abundant monazite, zircon, and garnet. Dawson (1980) and Nixon et al (1981) compared REE data on kimberlites with those on mantle xenoliths, and they found that kimberlites are

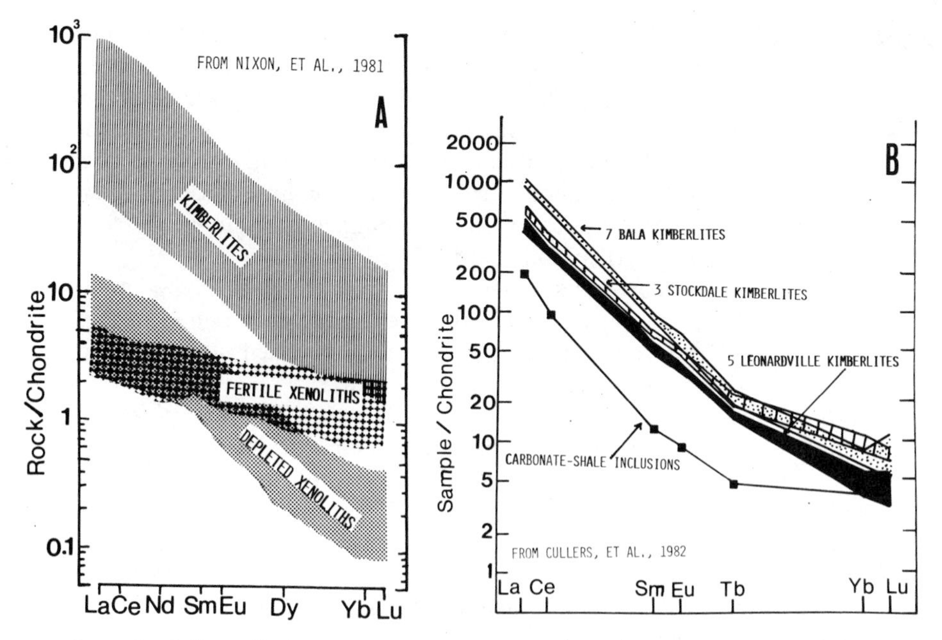

Figure 3 *A*. Generalized REE patterns for 3 types of mantle rocks. Fertile xenoliths may represent the mantle from regions below the depth of kimberlite development. Kimberlites probably form from light-REE-enriched garnet peridotite, whose residue after partial melting resembles the depleted xenoliths. Figure from Nixon et al (1981); reproduced with permission of *Annual Review of Earth and Planetary Sciences*. *B*. Comparison of REE compositions of three groups of kimberlites in Kansas and some carbonate-shale inclusions in one kimberlite. The Bala kimberlites are nonmicaceous types, whereas the Leonardville and Stockdale kimberlites are micaceous. Data from Cullers et al (1982). Copyrighted by *American Mineralogist* and printed with permission.

consistently enriched in total REE by a factor of about 100 over depleted xenoliths (Figure 3*A*) and are strongly enriched in light over heavy REE (La/Lu ratios of 200 to 2000). Furthermore, there appear to be consistent differences in the REE patterns of kimberlites in different provinces or tectonic settings (Nixon et al 1981) and among kimberlites with different mineralogies within the same province (Figure 3*B*; Cullers et al 1982).

Mineralogy

Olivine and phlogopite are the two most abundant primary silicate phases in kimberlites. Olivine compositions range from about Fo_{87} to Fo_{93}, with many phenocrysts between Fo_{91} and Fo_{93} and groundmass olivines usually less than Fo_{91} (Mitchell 1973). Slight iron enrichment from core to rim is common, but the reverse may also occur (Boyd & Clement 1977). Kimberlitic phlogopites have a variable chemistry and origin. Some grains are optically zoned, showing both normal and reverse pleochroism, which is suggestive of variable Fe^{2+}/Fe^{3+} ratios during precipitation (e.g. Smith et al 1978, Boettcher et al 1979, Farmer & Boettcher 1981, Pasteris 1983a). Olivine typically has been replaced by or at least veined and rimmed by serpentine, while phlogopite frequently shows marginal alteration to opaque rims or less commonly to a vermiculite-like phase. Primary kimberlitic diopside is a less common phase but may occur in the matrix of micaceous kimberlites (Dawson et al 1977).

The opaque oxide phases—spinels and ilmenite—have proved to be some of the best petrogenetic indicators in kimberlites because of their abundance, resistance to alteration, highly variable chemistry, and compositional zonation. Kimberlitic spinels frequently are zoned from chromite-rich cores to titanomagnetite-rich rims. The compositions of kimberlitic spinels are not unusual, but their particular compositional zonation distinguishes them from spinels in xenoliths and in other mafic and ultramafic rocks (e.g. Haggerty 1976). Spinel compositions reflect the degree of evolution of their host melts, and the zonation patterns of the spinels can be used to trace the crystallization history of the melt. Different diatremes in a district, as well as individual intrusions within the same diatreme, have compositionally distinct populations of spinels (e.g. Haggerty 1975, Mitchell 1979b, Boctor & Boyd 1982, Agee et al 1982, Pasteris 1983a,b). The relatively small changes in the ratio of $Mg/(Mg + Fe^{2+})$ in the olivines and most spinels in a given kimberlite suggest that the individual melts did not undergo extensive fractionation (e.g. Mitchell & Meyer 1980).

INCLUSIONS: COGNATE OR XENOLITH?

The several major types of mantle inclusions in kimberlites (see summary in Dawson 1980) are discussed here only briefly. The most abundant are peridotites, especially harzburgites and garnet lherzolites. Eclogites are less common in general, but can be locally abundant. Inclusions called megacrysts are essentially monomineralic, coarse-grained (up to tens of centimeters) samples of phases such as garnet, clinopyroxene, orthopyroxene, olivine, phlogopite, and ilmenite. The smaller megacrysts may be disaggregated from polymineralic xenoliths, but the larger grains probably are phenocrysts.

A variety of mafic and ultramafic rocks contain inclusions from the mantle, but according to geobarometric calculations on coexisting silicates in xenoliths (e.g. Boyd 1973, Boyd & Nixon 1973), kimberlites sample the deepest regions (down to about 200 km). Analyses of the modal mineralogy, mineral chemistry, and metamorphic textures of these inclusions have led petrologists to speculate on the structure and composition of the upper mantle (e.g. Nixon et al 1973, Gurney & Harte 1980, Dawson 1980). Implicit in such models is the belief that the xenoliths are representative of the abundance and bulk composition of mantle rocks, i.e. that they are a relatively unbiased and pristine sample (cf. Boettcher et al 1979).

The abundance of papers on xenolith petrology that are presented at "kimberlite conferences" attests to the general, if ill-defined, belief that the kimberlites and their inclusions are related. It appears that some inclusions are accidentally entrained fragments (xenoliths) of the conduit wall through which the kimberlites rose, whereas others are cumulates from the kimberlite melt or from its parent (protokimberlite). The latter are called *cognate* and may include portions of the megacryst suite. The multiphase peridotite inclusions, however, are probably true xenoliths [see Dawson (1980) for a discussion of some of the controversies]. These two types of inclusions provide valuable information on the nature of the kimberlite source rock and on the subsequent development of a mantle melt.

GENERATION AND EVOLUTION OF KIMBERLITE MELTS

Kimberlites are complex rocks composed of xenocrysts, as well as kimberlitic grains. The kimberlite melt may be a hybrid of more than one mantle liquid (e.g. Boyd & Nixon 1973, Mitchell & Meyer 1980, Hunter & Taylor 1983), which may in part account for the many mineralogical types of kimberlite.

The following are some of the considerations necessary in a model of kimberlite development: (*a*) a mantle source that can account for the major- and minor-element geochemistry, (*b*) a melting mechanism for the source rock, (*c*) a mechanism for rapid ascent of the melt (hours to days), (*d*) the evolution of the melt in the upper mantle, (*e*) the melt evolution at shallower depths, (*f*) the time interval involved in the melting and ascent, and (*g*) the relationship of kimberlitic melts to mantle xenoliths and to other mantle-derived rocks.

Kimberlite liquids have been viewed both as products of partial melting (e.g. Dawson 1971) and of fractional crystallization (O'Hara & Yoder 1967). The partial-melting hypothesis is favored because of the mineralogy, bulk composition, and trace-element geochemistry of kimberlites. The liquid could be derived from small amounts of partial melting of garnet peridotite in the upper mantle (e.g. Mitchell & Brunfelt 1975, Cullers et al 1982). Assuming a garnet peridotite source rock, Wedepohl & Muramatsu (1979) used the major- and trace-element geochemistry of kimberlites, nephelinites, alkali olivine basalts, and tholeiitic basalts to infer that these rocks represent progressively greater degrees of partial melting in the mantle. It is also clear that some flux, such as volatiles, must be present in the peridotite in order for it to melt along the assumed continental geothermal gradient (e.g. Wyllie 1979). The mineralogy of kimberlites, theoretical considerations (e.g. Eggler 1976, 1978), and the presence of CO_2-filled fluid inclusions in xenoliths in kimberlites (e.g. Roedder 1965) are evidence for the presence of C-H-O species in these mantle melts.

In simplified form [see Wyllie (1979, 1980) and Eggler (1976, 1978) for details], the partial-melting hypothesis states that at high pressures the volatiles are stabilized in dolomite-phlogopite-bearing peridotite (Figure 4). A density inversion causes the solid peridotite to rise in the form of a teardrop-shaped diapir (Green & Gueguen 1974). Partial melting occurs at the depth where the continental geotherm crosses the solidus for the specific composition of peridotite ($\sim$120–260 km; Wyllie 1980). At decreasing depths, a sequence of breakdown reactions in the peridotite diapir releases CO_2 and H_2O into the increasing volume of melt (Figure 4*A*). The peridotite and its partial melt remain in equilibrium during the rise of the diapir through the mantle, with the changing pressure and composition of coexisting phases controlling how the available volatiles partition between the liquid and vapor phases (e.g. Eggler & Wendlandt 1979). Thus, for approximately the same peridotite source, variations in the degree of partial melting and the depth of melting could produce different liquids, e.g. kimberlitic melts at great depth and carbonatitic, nephelinitic, or basanitic melts at shallower levels (Eggler 1978, Wyllie 1979). A range in the depth of melting could be produced by variations in the volatile (or other element) contents of the peridotite and/or by differences in the local geothermal gradient.

There are several variations on the above model (e.g. Eggler & Wendlandt 1979, Mercier 1979, Wyllie 1979, 1980). However, it is widely accepted that volatiles are essential in explaining the mechanisms of deep melting and shallow-level intrusion (see above), as well as in explaining the trace-element chemistry and silica-undersaturated composition of kimberlites. The latter two features require the presence of a separate CO_2-rich fluid, which would be stable only at $\lesssim$100-km depth. Polymerization of the melt by CO_2 could account for its silica undersaturation. There are apparently both internally generated and externally introduced fluids involved in the liquids.

A "normal" mantle peridotite source rock cannot account for the abundance of certain minor elements in kimberlites, such as K, Ti, F, and S. These apparently must be introduced into the peridotite before it can produce a kimberlite liquid. Several research groups have pointed to the effects of mantle metasomatism, that is, the introduction of mobile elements (particularly potassium) into peridotite. The elements presumably are dissolved in a fluid phase (e.g. Lloyd & Bailey 1975, Harte & Gurney 1975, Boettcher et al 1979, Menzies & Murthy 1980, Bailey 1982). Metasomatism probably also accounts for the veinlets in some xenoliths that contain one or more of the phases amphibole, phlogopite, ilmenite, rutile, and sulfides (Harte & Gurney 1975, Boettcher & O'Neil 1980, Gurney & Harte 1980, Pasteris 1982).

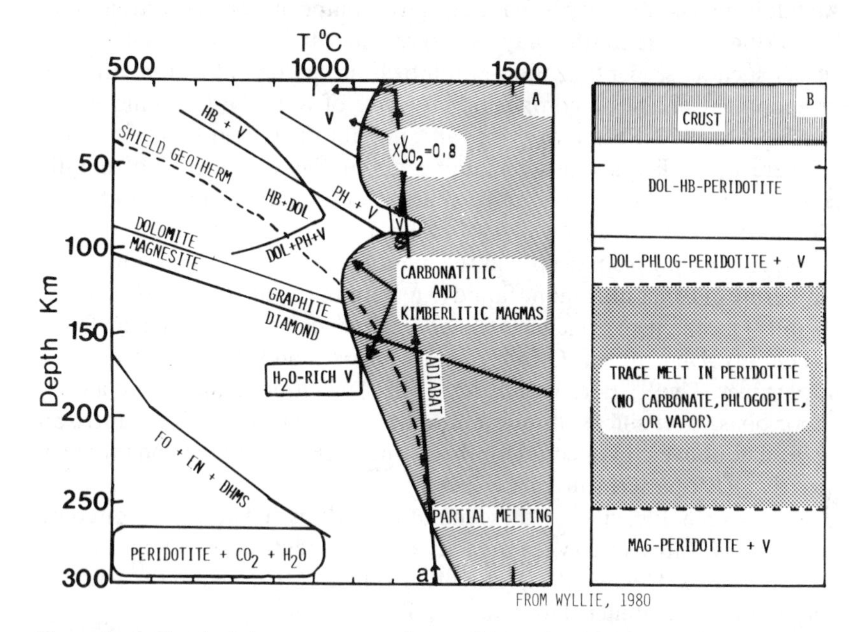

Figure 4 *A*. The shaded area represents the conditions of melting in the system peridotite + CO_2 + H_2O. It is bounded on the left by the volatile-present peridotite solidus for a high CO_2:H_2O ratio. The solid inclined lines show some subsolidus breakdown reactions involving hydrous and carbonate phases and some solid-solid transitions. The shield geotherm represents the postulated thermal gradient under a stable craton. In the depth interval over which the geotherm exceeds the temperature of the solidus (120–260 km), partial melting occurs in the peridotite (see also *B*). In the partial-melt zone, the volatile components are incorporated in solid minerals or dissolved in the trace melt, and kimberlitic and carbonatitic melts form. Also shown is the adiabat, the temperature-depth path that a diapir probably would follow as it rose. At *s*, the peridotite trace liquid associated with the diapir would solidify and release a vapor phase that might fracture and metasomatize the surrounding mantle. It is also possible that reduced C-H-O species are stable at depths >260 km, in which case partial melting may be triggered as the volatiles rise along the adiabat from below *a* and reach the solidus. *B*. Inferred cross section of the crust and upper mantle, assuming the stability of carbonate and hydrous phases. DOL = dolomite, HB = amphibole, PHLOG or PH = phlogopite, V = vapor, MAG = magnesite, FO = forsterite, EN = enstatite, DHMS = dense hydrous magnesian silicates. Adapted from Wyllie (1980), *Journal of Geophysical Research*, Vol. 85, pp. 6905–7. Copyrighted by the American Geophysical Union.

There is undeniable evidence that metasomatism occurred in some mantle xenoliths *before* the kimberlite entrained them (e.g. Gurney & Harte 1980). However, it is difficult to determine over what depth range or time interval metasomatism operates. Hunter & Taylor (1982) showed that the kelyphitic rims commonly found on garnet inclusions in kimberlite contain partial melts that reflect the addition of alkalis and volatiles, and Pasteris (1982) came to similar conclusions for partial melts in pyroxene megacrysts. An additional complexity is that deep-level compositional patterns in both kimberlites and xenoliths may be overprinted by apparently later-stage effects such as serpentinization-graphitization, phlogopitization, and the development of a wide compositional range of K-Fe-Ni-Cu-sulfides (e.g. Clarke et al 1977, Pasteris 1981b, 1982, unpublished data).

There are still questions remaining about the significance of mantle metasomatism. Is it a necessary precursor to the formation of kimberlite (Boettcher et al 1979; cf. Wyllie 1980)? What is the origin of the fluids? [For example, are they from deeper in the mantle or from crystallization of the kimberlite or other ultramafic liquid (e.g. Bergman et al 1981)?] What is the chemical mechanism of element transport (Ryabchikov & Boettcher 1980)?

The involvement of C-O-H fluids has an additional geochemical importance. The REE (especially light-REE) partition strongly into a CO_2 vapor phase over silicate liquids and solid phases (Wendlandt & Harrison 1979). The above may account for the strong REE enrichment and the steep negative REE patterns in kimberlites.

The oxygen fugacity of these mantle fluids is a subject of ongoing controversy. Intrinsic oxygen fugacity measurements on mantle xenoliths (e.g. Arculus & Delano 1981) show fO_2 values at and below the wüstite-magnetite (WM) buffer. However, Fe-Ti oxides in terrestrial basalts (e.g. Haggerty 1976) and CO-CO_2 fluid inclusions in mantle xenoliths (Bergman & Dubessy 1983) suggest much higher fO_2 values, near the quartz-fayalite-magnetite (QFM) buffer. Hammond & Taylor's (1982) recent study on the kinetics of ilmenite-magnetite reequilibration casts suspicion on the use of these phases for geothermometry and oxygen barometry in kimberlites. The consensus at the Third International Kimberlite Conference in September, 1982, was that conditions in much of the upper mantle are more reducing than QFM, and probably near WM (L. Taylor, personal communication). This is supported by Eggler's (1983) calculations on olivine-orthopyroxene-ilmenite megacrysts.

Given the possibility that oxygen fugacities in the mantle may even be too low to stabilize carbonate, Wyllie (1980) has reevaluated the depth and mechanism of the partial melting of peridotite to form kimberlitic liquids (Figure 4). Mixtures of reduced species like CO and CH_4 cannot exist as stable solids at mantle depths and may exist as a separate fluid phase.

Therefore, it may be the influx of a mobile fluid with reduced species that causes melting where the peridotite solidus crosses the geotherm, which in turn could trigger diapiric uprise (cf. Green & Gueguen 1974). The rise and melting continue until the temperature maximum (kink) in the peridotite solidus is reached at about 80-km depth. The melt then crystallizes and produces volatiles that may propagate fractures upward and compositionally alter the surrounding mantle. In fact, Wyllie contends that kimberlites are the cause rather than the products of metasomatism. Subsequent diapirs may produce vapor-undersaturated melts that can reach the surface in a series of intrusions through these conduits (Wyllie 1980).

FURTHER SPECULATIONS AND SUGGESTIONS FOR FUTURE RESEARCH

Numerous studies (see Table 1) have established a framework for interpreting both mantle xenoliths and the multi-intrusion kimberlite bodies that host them. Some of the most productive future kimberlite research may come from an attempt to interpret petrologic studies in terms of Wyllie's (1980) modified diapiric model (in turn based on previous work by Wyllie, Eggler, and coworkers). For instance, the abundances and compositions of opaque oxide phases and, to a lesser extent, phlogopite have been used to distinguish individual intrusions in a kimberlite pipe (e.g. Mitchell & Clarke 1976, Mitchell 1979b, Pasteris 1983a). Mitchell (1979b) has inferred that the distinction between kimberlite and micaceous kimberlite intrusions at a specific locality is imposed in the upper-mantle level. Pasteris (1983a,b) has presented compositional patterns in spinels to suggest that the successive intrusions in a diatreme came from a single evolving reservoir. From apparent correlations between spinel compositions and the diamond content of individual kimberlite intrusives, Pasteris (1983b) has also inferred that the kimberlite reservoir was rising over time.

We have strong evidence of the characteristics of a mantle source rock, C-O-H volatiles, mechanism of rise, and fractionation of the original mantle reservoir and/or the eventual kimberlite melt. One goal is to trace the evolution of an entire system—diapir(s), kimberlite melt(s), fluids. In the same effort we may discover why kimberlites are so localized and how they are related to other ultramafic rocks.

The diapir with its accompanying melt zone is the most remote portion of the system. Marsh (1982) demonstrated that there is little chance that a hot or molten body less than 10 km in diameter will pass through a pristine channelway directly from the upper mantle to the crust. One or more bodies must pass incrementally closer to the surface to prepare the channelway, especially thermally [cf. Wyllie's (1980) multiple diapirs)]. In addition, geochemical preparation, or metasomatism, may be necessary for the development of kimberlites. Fluids could fulfill both of these needs—acting as geochemical "fertilizers" and "fluxes," in addition to transporting heat (Spera 1981) as required in Marsh's model. Mantle xenoliths are the most direct evidence of the diapirs, but it is difficult to assess the degree of thermal, textural, and compositional overprinting on these inclusions during their rise in the diapir and their entrapment in the kimberlite (see

Table 1 Selected compendia on kimberlites and xenoliths

Published Works

Ahrens, L. H., Dawson, J. B., Duncan, A. R., Erlank, A. J., eds. 1975. *Physics and Chemistry of the Earth*, Volume 9. 940 pp. [Papers from 1st International Kimberlite Conference in 1973.]

Bailey, D. K., Tarney, J., Dunham, K. 1980. Issue on "The Evidence for Chemical Heterogeneity in the Earth's Mantle." *Philosophical Transactions of the Royal Society of London Series A*, Volume 297. 357 pp.

Boyd, F. R., Meyer, H. O. A., eds. 1979. *Kimberlites, Diatremes, and Diamonds: Their Petrology and Geochemistry*. Washington DC: American Geophysical Union. 400 pp.

Boyd, F. R., Meyer, H. O. A., eds. 1979. *The Mantle Sample*. Washington, DC: American Geophysical Union. 423 pp. [Papers from 2nd International Kimberlite Conference in 1977.]

Dawson, J. B. 1980. *Kimberlites and Their Xenoliths*. New York: Springer-Verlag. 252 pp.

Frantsesson, E. V. 1970. *The Petrology of Kimberlite*. Canberra: Australian University Press. [Translated by D. A. Brown.]

Glover, J. E., Groves, D. I. 1980. *Kimberlites and Diamonds*. Maylands, Western Australia: Extension Service of the University of Western Australia. 129 pp.

Irving, A. J., Dungan, M. A., eds. 1980. Issue on "Mantle Xenoliths and Their Host Magmas." *American Journal of Science*. Volume 280-A, Parts 1 & 2. 868 pp. (The Jackson volume.)

Journal of Geophysical Research. 1980. Issue honoring George Kennedy. Volume 85, No. B12

Meyer, H. O. A. 1976. The kimberlites of the continental United States: a review. *J. Geol.* 84:337–404

Nixon, P. H., ed. 1973. *Lesotho Kimberlites*. Maseru, Lesotho: Lesotho National Development Corporation. 350 pp.

Sobolev, R. V. 1977. *Deep-Seated Inclusions in Kimberlites and the Problem of the Composition of the Upper Mantle*. Washington DC: American Geophysical Union. 279 pp. [Translated by D. A. Brown.]

Terra-Incognita. 1982. Volume 2, No. 3. [Abstracts for 3rd International Kimberlite Conference in 1982.]

Wagner, P. A. 1971 (reprinted from 1914). *The Diamond Fields of Southern Africa*. Cape Town, South Africa: C. Struik (PTY) Ltd. 355 pp.

Williams, A. F. 1932. *The Genesis of the Diamond*. London: Ernest Benn

Wyllie, P. J., ed. 1967. *Ultramafic and Related Rocks*. New York: John Wiley. 464 pp.

Unpublished Works

Extended abstracts for the 1st and 2nd International Kimberlite Conferences.

Extended abstracts for the 1st and 2nd Kimberlite Symposia in Cambridge, England (sponsored by De Beers Consolidated Mines).

Mercier 1979, Boettcher et al 1979). It is also difficult to assess a xenolith's exact point of origin within a diapir.

It would appear easier to trace the evolution of the kimberlite melt than that of the diapir. However, the kimberlite rock is both a contaminated and altered sample of its parent melt. Megacrysts may provide the necessary link. Hunter & Taylor (1983) and Hunter et al (1983), for instance, have distinguished a Cr-rich and Cr-poor suite of megacryst silicates and ilmenite in a kimberlite from Pennsylvania. They attribute the compositional variations in the megacrysts to the magma mixing that produced the host kimberlite in the upper mantle.

With the admittedly large assumption that megacrysts crystallized from the melts that pond at the thermal maximum ($\sim$80-km depth) on Wyllie's (1980) peridotite solidus, megacrysts record changes in the (proto)-kimberlite melts over time. The individual intrusions in kimberlite pipes may be products of separate diapirs (cf. Wyllie 1980) or sequential segregations from the same rising diapir (cf. Pasteris 1983b). It may be possible to distinguish between these two hypotheses by analyzing the phenocryst and matrix grains in the kimberlite in addition to the megacrysts.

The ideal test for the above working model is to focus on a single well-exposed and well-sampled kimberlite province, such as Kimberley, South Africa (cf. work done by R. H. Mitchell on kimberlites in the more diverse Canadian "province"). The individual intrusions should be distinguished in each pipe. (This has been accomplished in large part by the De Beers Geology Department.) For each intrusion the composition and abundance of megacrysts, xenoliths, and indigenous kimberlite phases (particularly phlogopite, ilmenite, spinels) should be documented. These data would permit the following questions to be answered: If distinct populations are evident, could the individual intrusions represent sequential injections from a fractionating source, particularly a source rising over time? If not, are multiple sources probable, or can any of the rocks represent metasomatism or late alteration of other recognized intrusions? On the regional scale, do the xenoliths from each pipe record the same *P-T* path and interval, and the same mineralogy and textures? Could one diapir have been the source of several diatremes? For instance, do the multi-intrusion pipes within a district show parallel fractionation patterns, or are they very different from each other? Pasteris (unpublished data) has found numerous similarities in the compositions and textures of the opaque phases in several of the Kimberley diatremes, which distinguish them from kimberlites elsewhere. Petrologic differences in the mantle from one region to the other have been inferred from many xenolith studies (see compendia in Table 1). However, we still need to determine the *scales* of lateral and vertical heterogeneity in the mantle that is being sampled by kimberlites. [In practice, the above project would be very difficult because of the need to document the exact location of each sample (see Pasteris 1983b).]

The most elusive components of the entire system are the fluids, despite the fact that their origin is an important problem. One possibility is that large proportions of the fluids are not stored at depth in carbonate phases but rather exist as a reduced free volatile phase. Eggler & Baker (1982) and Wyllie (1980) have considered the effects of mantle fluids so reduced that CO_2 and H_2O are not the dominant species, although they have differing opinions on the likelihood of this possibility. There is some evidence in fluid inclusions from xenoliths of reducing species like CO and CH_4 (e.g. Bergman & Dubessy 1983, Pasteris, unpublished data).

Another interesting possibility is that the volatiles (particularly C-O species) are dissolved in small amounts in very abundant solids such as olivine. Green (1972, 1979) and Green & Gueguen (1983) have documented the exsolution of CO_2 from mantle olivines, whereas Freund and coworkers (Freund 1981, Oberhauser et al 1983) have presented evidence for the dissolution of up to 180 ppm of carbon in olivine (cf. Duba & Shankland 1982). The mobility of carbon, especially in a thermal gradient, is well known (e.g. Friel et al 1977, Duba & Shankland 1982). Therefore, either dissolved carbon or a reduced free volatile phase could provide the mobile volatile flux postulated in Wyllie's (1980) model.

The existence of carbon or other volatiles dissolved in silicates might explain some other geochemical features not specifically addressed in Wyllie's (1980) broad-scale model. For instance, the exact mechanism of partial melting in the mantle is still not fully understood, although it has been studied experimentally in simple systems (e.g. Waff & Bulau 1979). The major difference between the bulk chemistry of fertile and depleted xenoliths (Nixon et al 1981) is the enrichment and depletion, respectively, of these rocks in the more fusible components, such as Fe, Ti, K, and S. What are the large-scale and small-scale mechanisms of extracting these elements from mantle peridotite and partitioning them into kimberlitic and basaltic melts? If such components are concentrated primarily in separate phases such as phlogopite, then melting of individual nonrefractory grains can effectively deplete a mantle source rock. However, how are garnets, pyroxenes, and olivines stripped of their trace contents of fusible elements? Operating on a more local scale, dissolved volatiles could migrate through silicates, acting as scavengers and concentrators of minor elements (see Green 1979, Green & Gueguen 1983, Stosch 1982). Depressurization could localize melting in the inter- and intragranular regions enriched in volatile and fusible components. Partial melts derived this way might be detected primarily in small-scale autometasomatizing effects within xenoliths (Pasteris, unpublished data).

The apparently conflicting evidence on fO_2 levels within specific xenolith and kimberlite suites (Holmes & Arculus 1982) may be an artifact of our interpretation of the analytical techniques. However, it does suggest that the mantle may be heterogeneous with respect to oxygen fugacity. It would be difficult to maintain such gradients, but they might be part of the dynamic process of fluid and diapir movements involved in kimberlite generation. It may be the coincidence of several events that accounts for the mechanism and restricted depth for generation of kimberlites. For instance, a decrease in pressure might exsolve carbon from silicate solids and liquids while a simultaneous increase in oxygen fugacity triggered volatile evolution and consequent melting (cf. Wyllie 1980). The compositions of coexisting oxide phases and the speciation of most fluid inclusions in mantle rocks may only preserve the latest, most oxidized stages of this process. A corollary to this hypothesis is that the heterogeneous distribution of diamonds within specific diatremes may reflect the variations in the fO_2 state of their individual kimberlite intrusions (e.g. Pasteris 1983b).

CONCLUSIONS

What do kimberlites and their inclusions tell us about the mantle? They reveal a heterogeneous upper mantle that is composed predominantly of peridotite. The scale and the dynamics of the compositional and mineralogical heterogeneities are unclear. However, the mantle apparently consists of solid rock with segregations of a crystal-liquid mush (see Gurney & Harte 1980), the localization and proportions of which are partly controlled by the movement of volatiles containing dissolved species. The evidence (isotopic, bulk chemical, trace element, textural, mineralogical, theoretical) seems overwhelming that the movement of fluids is essential to melting and metasomatism in the mantle. Kimberlites, in turn, may be the key to understanding these fluids, as revealed in their volatile-rich groundmass, REE-rich perovskite, redox-sensitive opaque oxides, and fluid inclusions in phenocrysts and xenocrysts. However, the complex history of kimberlites continues to challenge our ability to interpret them.

Acknowledgments

The author has benefited from discussions with many colleagues over the years. She especially appreciates comments on an earlier version of the manuscript by T. Abrajano, S. Bergman, E. Dromgoole, H. Green, R. Mitchell, and L. Taylor. However, the author accepts full responsibility for all interpretations and opinions expressed. Publication of this paper was supported in part by NSF grant #EAR8025255.

Literature Cited

Agee, J. J., Garrison, J. R., Taylor, L. A. 1982. Petrogenesis of oxide minerals in kimberlite, Elliott County, Kentucky. *Am. Mineral.* 67:28–42

Arculus, R. J., Delano, J. W. 1981. Intrinsic oxygen fugacity measurements: techniques and results for spinels from upper mantle peridotites and megacryst assemblages. *Geochim. Cosmochim. Acta* 45: 899–913

Bailey, D. K. 1982. Mantle metasomatism—continuing chemical change within the Earth. *Nature* 296:525–30

Bergman, S. C., Dubessy, J. 1983. CO_2-CO fluids in a composite peridotite nodule: implications for oxygen barometry. *Contrib. Mineral. Petrol.* Submitted for publication

Bergman, S. C., Foland, K. A., Spera, F. J. 1981. On the origin of an amphibole-rich vein in a peridotite inclusion from the Lunar Crater Volcanic Field, Nevada, U.S.A. *Earth Planet. Sci. Lett.* 56:343–61

Boctor, N. Z., Boyd, F. R. 1979. Distribution of rare earth elements in perovskite from kimberlites. *Carnegie Inst. Washington Yearb.* 78:572–74

Boctor, N. Z., Boyd, F. R. 1982. Petrology of kimberlite from the DeBruyn and Martin Mine, Bellsbank, South Africa. *Am. Mineral.* 67:917–25

Boettcher, A. L., O'Neil, J. R. 1980. Stable isotope, chemical, and petrographic studies of high-pressure amphiboles and micas: evidence for metasomatism in the mantle source regions of alkali basalts and kimberlites. *Am. J. Sci.* 280-A: 594–621 (Part 2)

Boettcher, A. L., O'Neil, J. R., Windom, K. E., Stewart, D. C., Wilshire, H. G. 1979. Metasomatism of the upper mantle and the genesis of kimberlites and alkali basalts. In *The Mantle Sample*, ed. F. R. Boyd, H. O. A. Meyer, pp. 173–82. Washington DC: Am. Geophys. Union. 423 pp.

Boyd, F. R. 1973. A pyroxene geotherm. *Geochim. Cosmochim. Acta* 37:2533–46

Boyd, F. R., Clement, C. R. 1977. Compositional zoning of olivines in kimberlite from the De Beers Mine, Kimberley, South Africa. *Carnegie Inst. Washington Yearb.* 76:485–93

Boyd, F. R., Finnerty, A. A. 1980. Conditions of origin of nature diamonds of peridotite affinity. *J. Geophys. Res.* 85:6911–18

Boyd, F. R., Nixon, P. H. 1973. Origin of the ilmenite-silicate nodules in kimberlites from Lesotho and South Africa. See Nixon 1973, pp. 254–68

Clarke, D. B., Pe, G. G., McKay, R. M., Gill, K. R., O'Hara, M. J., Gard, J. A. 1977. A new potassium-iron-nickel sulphide from a nodule in kimberlite. *Earth Planet. Sci. Lett.* 35:421–28

Clement, C. R. 1973. Kimberlites from the Kao pipe, Lesotho. See Nixon 1973, pp. 110–21

Clement, C. R. 1979. The origin and infilling of kimberlite pipes. *Extended Abstr. Kimberlite Symp., 2nd, Cambridge, England, July 1979*

Clement, C. R., Skinner, E. M. W., Scott, B. H. 1977. Kimberlite redefined. *Extended Abstr. Int. Kimberlite Conf., 2nd, Santa Fe, N. Mex.*

Crough, S. T., Morgan, W. J., Hargraves, R. B. 1980. Kimberlites: their relation to mantle hotspots. *Earth Planet. Sci. Lett.* 50:260–74

Cullers, R. L., Mullenax, J., Dimarco, M. J., Nordeng, S. 1982. The trace element content and petrogenesis of kimberlites in Riley County, Kansas, U.S.A. *Am. Mineral.* 67:223–33

Dawson, J. B. 1967. Geochemistry and origin of kimberlite. In *Ultramafic and Related Rocks*, ed. P. J. Wyllie, pp. 269–78. New York: Wiley. 464 pp.

Dawson, J. B. 1970. The structural setting of African kimberlite magmatism. In *African Magmatism and Tectonics*, eds. T. N. Clifford, I. G. Gass, pp. 321–35. Edinburgh: Oliver & Boyd. 461 pp.

Dawson, J. B. 1971. Advances in kimberlite geology. *Earth Sci. Rev.* 7:187–214

Dawson, J. B. 1980. *Kimberlites and Their Xenoliths.* New York: Springer-Verlag. 252 pp.

Dawson, J. B., Smith, J. V., Hervig, R. L. 1977. Late-stage diopside in kimberlite matrix. *Neues Jahrb. Mineral. Mitt.* 1977:529–53

Duba, A. G., Shankland, T. J. 1982. Free carbon and electrical conductivity in the Earth's mantle. *Geophys. Res. Lett.* 9:1271–74

Eggler, D. H. 1976. Does CO_2 cause partial melting in the low-velocity layer of the mantle? *Geology* 4:69–72

Eggler, D. H. 1978. The effect of CO_2 upon partial melting of peridotite in the system Na_2O-CaO-Al_2O_3-MgO-SiO_2-CO_2 to 35 kb, with an analysis of melting in a peridotite-H_2O-CO_2 system. *Am. J. Sci.* 278:305–43

Eggler, D. H. 1983. Upper mantle oxidation state: evidence from olivine-orthopyroxene-ilmenite assemblages. *Geophys. Res. Lett.* 10:365–68

Eggler, D. H., Baker, D. R. 1982. Reduced volatiles in the system C-O-H: implications to mantle melting, fluid formation, and diamond genesis. In *High Pressure Research in Geophysics*, ed. S. Akimoto, M. H. Manghnani, pp. 237–50. Tokyo: Cent. Acad. Publ. 632 pp.

Eggler, D. H., Wendlandt, R. F. 1979. Experimental studies on the relationship between kimberlite magmas and partial melting of peridotite. In *Kimberlites, Diatremes, and Diamonds*, ed. F. R. Boyd, H. O. A. Meyer, pp. 330–38. Washington DC: Am. Geophys. Union. 400 pp.

Farmer, G. L., Boettcher, A. L. 1981. Petrologic and crystal-chemical significance of some deep-seated phlogopites. *Am. Mineral.* 66: 1154–63

Ferguson, J. 1980. Tectonic setting and paleogeotherms of kimberlites with particular emphasis on southeastern Australia. See Glover & Groves 1980, pp. 1–14

Freund, F. 1981. Mechanism of the water and carbon dioxide solubility in oxides and silicates and the role of O^-. *Contrib. Mineral. Petrol.* 76: 474–82

Friel, J. J., Goldstein, J. I., Romig, A. D. 1977. The effect of carbon on phosphate reduction. *Proc. Lunar Sci. Conf., 8th*, pp. 3941–54

Glover, J. E., Groves, D. I., eds. 1980. *Kimberlites and Diamonds.* Maylands, West. Austral: Ext. Serv. Univ. West. Austral. 129 pp.

Green, H. W. 1972. A CO_2 charged asthenosphere. *Nature Phys. Sci.* 238: 2–5

Green, H. W. 1979. Trace elements in the fluid phase of the Earth's mantle. *Nature* 277: 465–67

Green, H. W., Gueguen, Y. 1974. Origin of kimberlite pipes by diapiric upwelling in the upper mantle. *Nature* 249: 617–20

Green, H. W., Gueguen, Y. 1983. Deformation of the mantle and extraction by kimberlite: a case history documented by fluid and solid precipitates in olivine. *Tectonophysics* 92: 71–92

Gupta, A. K., Yagi, K. 1980. *Petrology and Genesis of Leucite-Bearing Rocks.* New York: Springer-Verlag. 252 pp.

Gurney, J. J., Harte, B. 1980. Chemical variations in upper mantle nodules from southern African kimberlites. *Philos. Trans. R. Soc. London Ser. A* 297: 273–93

Haggerty, S. E. 1975. The chemistry and genesis of opaque minerals in kimberlites. *Phys. Chem. Earth* 9: 295–307

Haggerty, S. E. 1976. Opaque mineral oxides in terrestrial igneous rocks. In *Oxide Minerals*, ed. D. Rumble III, pp. Hg101–300. Washington DC: Mineral. Soc. Am.

Haggerty, S. E. 1982. Kimberlites in western Liberia: an overview of the geological setting in a plate tectonic framework. *J. Geophys. Res.* 87: 10811–26

Hammond, P. A., Taylor, L. A. 1982. The ilmenite/titano-magnetite assemblage: kinetics of re-equilibration. *Earth Planet. Sci. Lett.* 61: 143–50

Harte, B., Gurney, J. J. 1975. Ore mineral and phlogopite mineralization within ultramafic nodules from the Matsoku kimberlite pipe, Lesotho. *Carnegie Inst. Washington Yearb.* 74: 528–36

Hawthorne, J. B. 1975. Model of a kimberlite pipe. *Phys. Chem. Earth.* 9: 1–15

Holmes, R. D., Arculus, R. J. 1982. Metal-silicate redox reactions: implications for core-mantle equilibrium and the oxidation state of the upper mantle. *Extended Abstr. Conf. Planet. Volatiles*, pp. 45–46. Houston: Lunar Planet. Inst.

Hunter, R. H., Taylor, L. A. 1982. Instability of garnet from the mantle: glass as evidence of metasomatic melting. *Geology* 10: 617–20

Hunter, R. H., Taylor, L. A. 1983. Magma-mixing in the low velocity zone: kimberlitic megacrysts from Fayette County, Pennsylvania. *Am. Mineral.* 68: In press

Hunter, R. H., Kissling, R. D., Taylor, L. A. 1983. Mid- to late-stage kimberlitic melt evolution: phlogopites and oxides from the Fayette County kimberlite, Pennsylvania. *Am. Mineral.* 68: In press

Jaques, A. L., Lewis, J. D., Gregory, G. P., Ferguson, J., Smith, C. B., et al. 1982. *The ultrapotassic, diamond-bearing kimberlites and lamproites of the West Kimberley region, Western Australia.* Presented at Int. Kimberlite Conf., 3rd, Clermont-Ferrand, France

Lloyd, F. E., Bailey, D. K. 1975. Light element metasomatism of the continental mantle: the evidence and consequences. *Phys. Chem. Earth* 9: 389–416

Marsh, B. D. 1982. On the mechanics of igneous diapirism, stoping, and zone melting. *Am. J. Sci.* 282: 808–55

Menzies, M., Murthy, V. R. 1980. Mantle metasomatism as a precursor to the genesis of alkaline magmas—isotopic evidence. *Am. J. Sci.* 280-A: 622–38 (Part 2)

Mercier, J.-C. C. 1979. Peridotite xenoliths and the dynamics of kimberlite intrusion. In *The Mantle Sample*, ed. F. R. Boyd, H. O. A. Meyer, pp. 197–212. Washington DC: Am. Geophys. Union. 423 pp.

Mitchell, R. H. 1973. Composition of olivine, silica activity and oxygen fugacity in kimberlite. *Lithos* 6: 65–81

Mitchell, R. H. 1979a. The alleged kimberlite-carbonatite relationship: additional contrary mineralogical evidence. *Am. J. Sci.* 279:570–89

Mitchell, R. H. 1979b. Mineralogy of the Tunraq kimberlite, Somerset Island, N.W.T., Canada. In *Kimberlites, Diatremes, and Diamonds*, ed. F. R. Boyd, H. O. A. Meyer, pp. 161–71. Washington DC: Am. Geophys. Union. 400 pp.

Mitchell, R. H., Brunfelt, A. O. 1975. Rare earth element geochemistry of kimberlite. *Phys. Chem. Earth.* 9:671–86

Mitchell, R. H., Clarke, D. B. 1976. Oxide and sulphide mineralogy of the Peuyuk kimberlite, Somerset Island, N.W.T. Canada. *Contrib. Mineral. Petrol.* 56:157–72

Mitchell, R. H., Meyer, H. O. A. 1980. Mineralogy of micaceous kimberlite from the Jos Dyke, Somerset Island, N.W.T. *Can. Mineral.* 18:241–50

Nixon, P. H., ed. 1973. *Lesotho Kimberlites.* Maseru, Lesotho: Lesotho Natl. Develop. Corp. 350 pp.

Nixon, P. H., Boyd, F. R. 1973. Petrogenesis of the granular and sheared ultrabasic nodule suite in kimberlites. See Nixon 1973, pp. 48–56

Nixon, P. H., Boyd, F. R., Bouiller, A. M. 1973. The evidence of kimberlite and its inclusions on the constitution of the outer part of the Earth. See Nixon 1973, pp. 312–18

Nixon, P. H., Rogers, N. W., Gibson, I. L., Grey, A. 1981. Depleted and fertile mantle xenoliths from southern African kimberlites. *Ann. Rev. Earth Planet. Sci.* 9:285–309

Oberhauser, G., Kathrein, H., Demortier, G., Gonska, H., Freund, F. 1983. Carbon in olivine single crystals analyzed by the ^{12}C (d, p) ^{13}C method and by photoelectron spectroscopy. *Geochim. Cosmochim. Acta* 47:1117–30

O'Hara, M. J., Yoder, H. S. 1967. Formation and fractionation of basic magma at high pressure. *Scott. J. Geol.* 3:67–117

Parrish, J. B., Lavin, P. M. 1982. Tectonic model for kimberlite emplacement in the Appalachian Plateau of Pennsylvania. *Geology* 10:344–47

Pasteris, J. D. 1981a. Kimberlites: strange bodies? *Eos, Trans. Am. Geophys. Union* 62:713–16

Pasteris, J. D. 1981b. Occurrence of graphite in serpentinized olivines in kimberlite. *Geology* 9:356–59

Pasteris, J. D. 1982. Evidence of potassium metasomatism in mantle xenoliths. *Eos, Trans. Am. Geophys. Union* 63:462 (Abstr.)

Pasteris, J. D. 1983a. Spinel zonation in the De Beers kimberlite, South Africa—possible role of phlogopite. *Can. Mineral.* 21:41–58

Pasteris, J. D. 1983b. Justification for possible use of indigenous kimberlite minerals in evaluation of diamond potential. *Process Mineral. Spec. Issue, Am. Inst. Min. Metall. Eng.* In press

Pasteris, J. D. 1983c. Kimberlites: a look into the Earth's mantle. *Am. Sci.* 71:282–88

Prider, R. T. 1960. The leucite-lamproites of the Fitzroy Basin, Western Australia. *J. Geol. Soc. Austral.* 6:71–120

Roedder, E. 1965. Liquid CO_2 inclusions in olivine-bearing nodules and phenocrysts from basalts. *Am. Mineral.* 50:1746–82

Ryabchikov, I. D., Boettcher, A. L. 1980. Experimental evidence at high pressure for potassic metasomatism in the mantle of the Earth. *Am. Mineral.* 65:915–19

Skinner, E. M. W., Clement, C. R. 1979. Mineralogical classification of southern African kimberlites. In *Kimberlites, Diatremes, and Diamonds*, ed, F. R. Boyd, H. O. A. Meyer, pp. 129–39. Washington DC: Am. Geophys. Union. 400 pp.

Smith, J. V., Brennesholtz, R., Dawson, J. B. 1978. Chemistry of micas from kimberlites and xenoliths. I. Micaceous kimberlites. *Geochim. Cosmochim. Acta* 42:959–71

Spera, F. J. 1981. Carbon dioxide in igneous petrogenesis. II. Fluid dynamics of mantle metasomatism. *Contrib. Mineral. Petrol.* 77:56–65

Stosch, H.-G. 1982. Rare earth element partitioning between minerals from anhydrous spinel peridotite xenoliths. *Geochim. Cosmochim. Acta* 46:793–811

Stracke, K. J., Ferguson, J., Black, L. P. 1979. Structural setting of kimberlites in south-eastern Australia. In *Kimberlites, Diatremes, and Diamonds*, ed. F. R. Boyd, H. O. A. Meyer, pp. 71–91. Washington DC: Am. Geophys. Union. 400 pp.

Taylor, L. A. 1982. *Kimberlitic magmatism in the eastern United States: relationships to mid-Atlantic tectonism.* Paper presented at Int. Kimberlite Conf., 3rd, Clermont-Ferrand, France

von Eckermann, H. 1967. A comparison of Swedish, African, and Russian kimberlites. In *Ultramafic and Related Rocks*, ed. P. J. Wyllie, pp. 302–12. New York: Wiley. 464 pp.

Waff, H. S., Bulau, J. R. 1979. Equilibrium fluid distribution in an ultramafic partial melt under hydrostatic stress conditions. *J. Geophys. Res.* 84:6109–14

Wedepohl, K. H., Muramatsu, Y. 1979. The chemical composition of kimberlites compared with the average composition of three basaltic magma types. In *Kimberlites, Diatremes, and Diamonds*, ed. F. R. Boyd, H. O. A. Meyer, pp. 300–12.

Washington DC: Am. Geophys. Union. 400 pp.

Wendlandt, R. F., Harrison, W. J. 1979. Rare earth partitioning between immiscible carbonate and silicate liquids and CO_2 vapor: results and implications for the formation of light rare earth-enriched rocks. *Contrib. Mineral. Petrol.* 69:409–19

Wyllie, P. J. 1979. Kimberlite magmas from the system peridotite-CO_2-H_2O. In *Kimberlites, Diatremes, and Diamonds*, ed. F. R. Boyd, H. O. A. Meyer, pp. 319–29. Washington DC: Am. Geophys. Union. 400 pp.

Wyllie, P. J. 1980. The origin of kimberlite. *J. Geophys. Res.* 85:6902–10

Ann. Rev. Earth Planet. Sci. 1984. 12: 155–77

THEORY OF HYDROTHERMAL SYSTEMS

Denis L. Norton

Department of Geosciences, University of Arizona, Tucson, Arizona 85721

INTRODUCTION

Hydrothermal systems are sets of processes that redistribute energy and mass in response to circulating H_2O fluids. These systems are active today in the oceanic and continental crusts, and their fossilized equivalents constitute a substantial portion of the geologic record. Throughout the Earth's history they have effectively stabilized the physical and chemical states of the crust by dispersing perturbations in fluid pressure and density.

Although hydrothermal systems can be generated by many types of perturbations, those related to magma-induced thermal anomalies are the most active class. Vigorous hydrothermal systems typify the roofs of mid-oceanic ridge magma chambers, while less active ones occur along chamber walls and extend to at least the base of the crust (Walther & Orville 1982, Lister 1977, 1980, J. R. Delaney 1980, Gregory & Taylor 1981, Norton & Taylor 1979). The net result is that in the vicinity of these magma chambers, the entire oceanic crust becomes hydrothermally altered and then thermally dehydrated during subduction processes, and is ultimately converted to mineral assemblages stable at high temperatures and pressures (Delany & Helgeson 1978). This process releases H_2O and produces favorable thermal gradients for hydrothermal flow within the subducting lithospheric plate.

Magmas within convergent plate environments transport thermal energy and mass into the continental crust and form batholiths and stocks. The batholiths generate hydrothermal systems equal in magnitude and duration to those associated with the ocean-ridge magma chambers (Taylor 1977, Criss et al 1982). Independent of the source of the parent magmas, as soon as the magma contacts the fluid-rich rocks in the crust, hydrothermal activity begins. This activity precedes magma infiltration and guides it through the crust. Cooled by the circulating fluids, the magma

0084–6597/84/0515–0155$02.00

crystallizes to a pluton whose thermal history is strongly controlled by the hydrothermal system.

Hydrothermal activity is probably a more common phenomenon in the crust than is currently realized because it develops as a natural consequence of any thermal perturbation in fluid-rich rocks. Even though fluid circulation may affect several cubic kilometers of rock, extend to at least the base of the crust, and be active for long periods of time, its presence is generally not recognized unless surface activity or anomalous conditions in shallow drill holes are encountered. Our concept of the abundance of active hydrothermal systems relies heavily on considerations of this type, while in fact many systems do not breach the near-surface environment.

The physics and chemistry of hydrothermal processes form an important basis for reconstructing the history of many geological provinces. These processes, only qualitatively understood from studies of natural systems, can be closely approximated by sets of mathematical equations that are based on the conservation of mass and energy. These idealized mathematical systems portray the processes in a form that facilitates study of the transient character of natural systems and enables us to separate trivial details from the principal controls that affect hydrothermal evolution. Theoretical models constructed in this manner describe the properties and conditions that would likely be encountered if the inaccessible portions of natural systems could be examined. The models use fundamental physical and chemical concepts as a key to reconstructing the geological history of fossilized hydrothermal systems.

THEORY

The theory of hydrothermal systems portrays the transport of thermal, chemical, and mechanical energy by fluid flow and molecular diffusion through a coupled set of partial differential equations. This article reviews the relationships that comprise the foundation for this theory. First, the systematic relationships among the processes that occur in response to thermal perturbations in rocks are examined, and then the effects that the processes have on each other are analyzed. From these considerations a mathematical system is presented that represents the physical and chemical processes in nature, after which the most important equations of state of the rocks and fluids are reviewed. Finally, the main events that occur during the hydrothermal history of a magma are described.

Interactions Between Hydrothermal Processes

The complex patterns of alteration products found in hydrothermal systems result from interactions between physical and chemical processes.

These patterns arise in response to a situation in which the products of the processes affect those processes that formed them (Norton 1979). This feedback effect is common in natural systems. In fact, recognition of this phenomenon has motivated the formulation of mathematical expressions that describe such feedback relationships and the ways in which large groups of processes are interrelated by feedback loops. These studies reveal that when the output from a mathematical expression is input back into the same expression, a numerical pattern is produced that is very sensitive to slight changes in single parameters (Hofstadter 1981).

Alteration patterns observed in fossilized hydrothermal systems vary tremendously from ordered to chaotic. Because this variability appears to be a consequence of the presence of feedback effects, the principal processes are portrayed as time-dependent rates interconnected by feedback loops (Figure 1).

The system is driven by a thermal perturbation that imposes potential

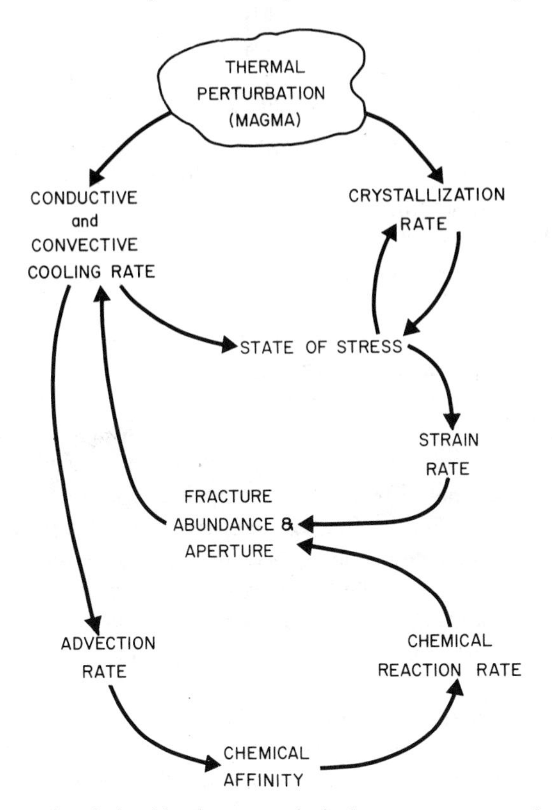

Figure 1 Systematic relationships between principal transport rates and their products. Arrows depict directions of energy, mass flow, and feedback effects of state conditions on rates.

fields for rock deformation, fluid flow, and chemical reactions on the surrounding wall and roof rocks of a magma chamber by the transfer of thermal energy from a magma. Unless the magma itself is permeable to fluid flow, the initial heat-transfer rate through the chamber walls is dominated by conduction. Conductive heat transfer alters both the local state of stress and the principal stress trajectories through expansion of host rock and contraction of magma. Deformation occurs at low differential stresses because of the intrinsically low resistance of rock to stress. The strain rate relieves these stresses and increases rock permeability by increasing either the number of fractures or their apertures and continuities. Because fluid buoyancy forces are always associated with a thermal perturbation in crustal rocks, increased permeability augments the magnitude of fluid flow and consequently increases the convective transfer rate of thermal energy. The overall heat-transfer rate, enhanced by convection, amplifies the stress conditions, increases the strain rate, produces more fractures, and increases the convective rate of heat transfer. Although this condition could cause unlimited growth of cooling and strain rates, chemical alteration products mitigate this cascading feedback effect.

Chemical components are redistributed by fluids flowing across the boundaries between contrasting chemical environments. The advective rate of transport disrupts local chemical equilibrium conditions, increases chemical affinity, and initiates irreversible reactions. Chemical reaction rates that produce alteration assemblages with volumes larger than the original mineral assemblages fill the flow channels by decreasing either the apertures or continuities of the fractures. Fluid-flow rates are decreased by this effect, but the state of stress continues to increase as fluids trapped along the once-continuous flow channel expand in response to the advancing thermal front.

The magma crystallization rate and stress conditions near the chamber are also controlled by the dispersion of energy away from the chamber. Alteration of the magnitude and principal stress trajectories is most pronounced when a volatile-rich phase of the magma is produced during crystallization. Large volume increases caused by the separation of an H_2O-rich phase from the magma lead to increased pressure in the chamber and shift the state conditions of the magma-crystals-fluid mixture away from the solidus toward the magma-crystals stability field. Then, continued heat loss from the magma shifts state conditions back toward the solidus and causes a further increase in pressure. The extent of the magma-induced pressure increase is limited by the strength of the chamber walls and by the increase in confining pressure; when this limit is exceeded, the wall rocks fracture and cause the magma to crystallize locally by adiabatic decompression.

As mutually dependent potential fields for fluid flow, stress, and chemical reaction evolve and produce thermal, mechanical, and chemical alteration effects, alteration minerals and patterns of fracture networks emerge that preserve the local state conditions. For example, the variation from orderly to chaotic fracture patterns noted in the vicinity of ore deposits appears to be caused by feedback. Stockwork fracture (chaos) arises when the feedback effects amplify convective heat transfer and strain rates, but the fractures are not filled rapidly enough with alteration products to slow the heat-transfer rate.

This systematic portrayal of hydrothermal events forms the basis for deriving a system of mathematical equations that describe the coupled rates of change of state properties for idealized geologic conditions.

Mathematical System

The conservation of scalar, vector, or tensor properties λ within a porous medium composed of $\hat{I}$ phases (Figure 2) is derived by associating the volume average of Euler's transport theorem (Slattery 1972, Bear 1972) at a point P with a representative elemental volume (REV) of the system, which is enclosed in a surface S:

$$\frac{\mathrm{d}}{\mathrm{d}t}\int_V \lambda\,\mathrm{d}V = \int_V \left(\frac{\partial \lambda}{\partial t}\right)_{xyz} \mathrm{d}V + \int_S \mathbf{j}_\lambda \cdot \mathrm{d}\mathbf{S}. \tag{1}$$

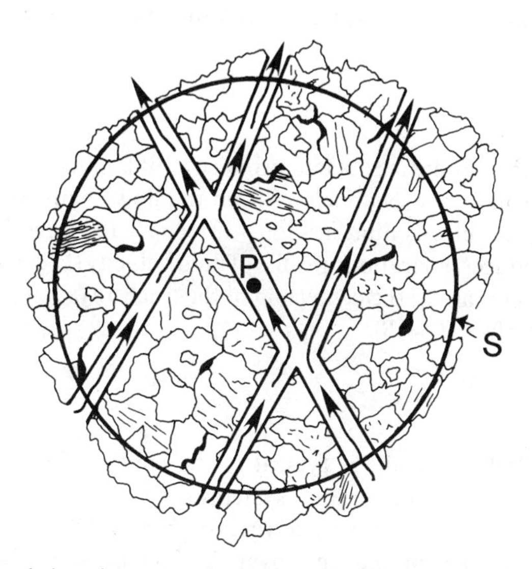

Figure 2 Sectional view of representative elemental volume ΔV of fractured granular rock enclosed in surface S. Arrows in fractures depict average fluid-flow directions. Solid black regions represent intergranular, discontinuous porosity within matrix blocks. P is the mathematical representation of ΔV.

This theorem states that the sum of local variations in λ, $\partial\lambda/\partial t$, and of fluxes in λ, $\mathbf{j}_\lambda$, through the surface **S** represent the total change in λ within V.

Rocks typical of hydrothermal systems are composed of fracture networks and a rock matrix (Norton & Knapp 1977) that is itself reticulated with microscopic cracks and intergranular pores (Simmons & Richter 1976, Sprunt & Nur 1979). All pores, fractures, and cracks are generally filled with fluid, but fluid flow is confined to the fracture networks. The porous matrix is a multiphase assemblage of minerals and aqueous solutions that are locally in chemical equilibrium. The geologic history of this heterogeneous mixture of phases can be deduced by integration of (1) for REVs, in which the material properties and their derivatives are considered continuous and are associated with the mathematical point P. This concept preserves the continuity requirements of differential calculus but requires that the transport properties be treated as interleaved continua, a constraint that is satisfied by obtaining average properties ζ^* of the REV as a function of properties ζ_i of the phases it contains:

$$\zeta^* = \frac{1}{\Delta V}\int \zeta_i \, \mathrm{d}V, \tag{2}$$

where $i = 1, 2, \ldots, \hat{I}$, and the integration is taken over the entire volume of the REV. If the integral is replaced with the sum over all the phases, and if the volume fraction $\phi_i = V_i/\Delta V$ of the ith phase is substituted, the volume-averaged property can be expressed as a function of the volume fraction and ζ_i:

$$\zeta^* = \sum_i \phi_i \zeta_i. \tag{3}$$

Fluxes of energy and mass occur primarily by laminar fluid flow along fracture-controlled flow channels. Within the flow channels, the dynamic fluid viscosity μ_f retards fluid motion $\mathbf{v}$ driven by the force field $\nabla\Phi$. This condition produces a parabolic profile of fluid velocity (Figure 3) that varies from a maximum along the centerline of the fracture ($b = 0$) to zero at the fluid-rock interface ($b = B$):

$$\mathbf{v} = (B^2 - b^2)\frac{\nabla\Phi}{2\mu}. \tag{4}$$

This Poiseuille flow profile is typical of the laminar flow condition that dominates flow regimes in natural systems.

The average velocity within fractures is defined by (2), where $\zeta_i = \mathbf{V}_f$, and the integration is performed only over the flowing fluid. These averaged particle velocities are then averaged over the entire volume of the continua, and that velocity is associated with the point P in Figure 2. This volume-

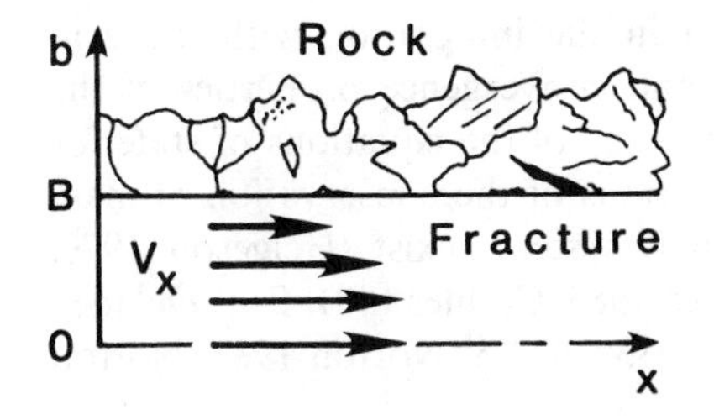

Figure 3 Poiseuille velocity profile for steady-state fluid flow through a single fracture with half-aperture B. Typical apertures in natural systems range from 10 nm to 10 mm.

averaged velocity $\mathbf{V}_f^*$ is consistent with the Darcy fluid velocity (Hubbert 1969):

$$\mathbf{V}_f^* = \mathbf{V}_{\text{Darcy}} = \frac{k}{v_f} \nabla \mathbf{\Phi}, \tag{5}$$

where k is rock permeability, v fluid viscosity, and $\nabla\mathbf{\Phi}$ force per unit volume of fluid.

The advective flux $\mathbf{j}$ caused by fluid flowing through the rock volume is proportional to the Darcy velocity $\mathbf{V}_f^*$ and concentrations $C_{\lambda,f}$ in the fluid phase:

$$\mathbf{j}_{\text{advection},\lambda} = \mathbf{V}_f^* C_{\lambda,f}. \tag{6}$$

Although diffusive flux away from fractures into matrix blocks is a major portion of energy transport, it is only of local importance to mass transport. Diffusive fluxes are caused by molecular interactions along gradients in the volume-averaged energy E^*:

$$\mathbf{j}_{\text{diffusion},E} = \kappa^* \nabla \mathbf{E}^*, \tag{7}$$

where the proportionality constant κ^* is the volume-averaged capacity of the material to transfer energy by molecular-level processes. Fluxes are transformed into rates of change with respect to rock volume by calculating divergence of the flux vector on the surface of the REV. The net rate of change caused by local changes, advection, and molecular interactions forms the basis of a general equation of the conservation of λ. Because (1) represents all elemental volumes of the system, the integral can be replaced by a sum over all phases:

$$\sum_i \left(\frac{\partial \phi_i \lambda_i}{\partial t} \right)_{xyz} + \sum_i \nabla \cdot \mathbf{j}_{\lambda,i} = 0. \tag{8}$$

A set of equations of this form, one for each of the conservative properties, forms the basis for analyzing processes in hydrothermal systems.

The degree to which (8) approximates the actual conditions that prevail in nature depends to a large extent on (*a*) the correspondence of the boundary and initial conditions used in the integration with those in nature, (*b*) the stability, consistency, and convergence properties of the integration techniques, and (*c*) the accuracy of the equations of state for minerals and fluids. Numerous applications of the conservation of mass and energy equations to hydrothermal systems exist (Helgeson 1970, Helgeson et al 1970, Brimhall & Ghiorso 1983, Cathles 1981, P. T. Delaney 1982, Knapp & Norton 1981, Pruess 1983, Bird & Norton 1981, Norton 1982).

EQUATIONS OF STATE

Fluid

Equations of state for fluid phases commonly found in hydrothermal systems are well known (Helgeson & Kirkham 1974a,b, 1976, Helgeson et al 1981). Any uncertainties that do arise are usually a consequence of variations in the composition of natural materials that are not accounted for by the existing data. However, the general behavior of hydrothermal systems can be deduced in spite of minor inadequacies in the data because of both the common occurrence of H_2O-rich fluids and the major effect they have on these environments.

H_2O-rich fluids typically found in the crust dominate hydrothermal transport processes. Although natural fluid compositions are normally not pure H_2O, their behavior is directly dependent on the properties of the phases in the H_2O-SYSTEM (Figure 4). The critical end-point on the liquid-vapor saturation surface causes extreme yet continuous variations in state properties of supercritical fluid. Because the temperature-pressure region where the extrema are found lies midway between normal background conditions in the crust and conditions associated with magmas, supercritical conditions are encountered in each magma-hydrothermal system.

The density of liquid H_2O decreases with increasing temperature from 1.0 g cm^{-3} at low temperatures to 0.01 g cm^{-3} at near-magmatic conditions (Figure 5). At constant pressure, density variations with temperature have a maximum slope in the supercritical region at low pressures (Figure 6). Because all other transport properties of hydrothermal fluids are related to fluid density, the nonlinear variation manifests itself throughout the hydrothermal system.

Fluid buoyancy forces **F** are generated as heat transfer from the magma disturbs ambient surfaces of equal density and pressure so that they no longer coincide:

$$\mathbf{F}_{\text{buoyancy}} = \rho_f \mathbf{g} - \nabla \mathbf{P}. \tag{9}$$

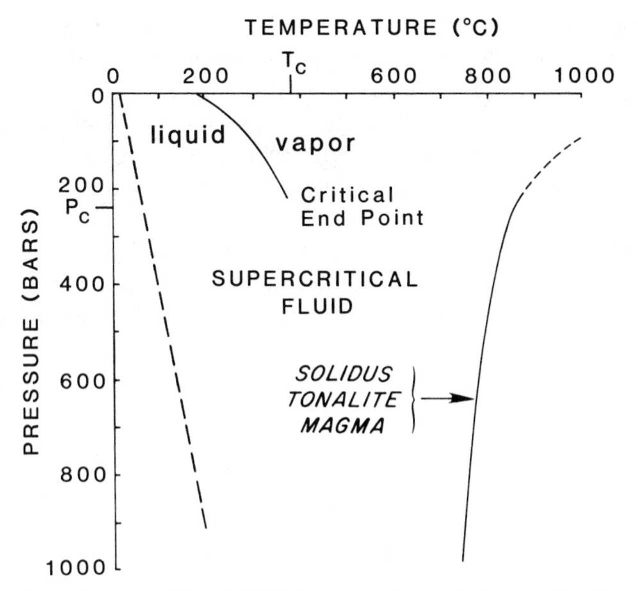

Figure 4 Phase diagram of H_2O-SYSTEM, projected onto *T-P* plane. The diagram shows location of supercritical region with respect to normal geotherm (*dashed line*) and approximate solidus of tonalite composition magma containing ~4 wt% H_2O.

This force per unit volume causes rotational fluid flow in a direction that tends to bring the surfaces of equal density and pressure back into coincidence. The circulation tendency of this field is proportional in magnitude to the gradient in fluid temperature, the isobaric coefficient of thermal expansion α_f (Figure 7), and the mean density ρ_f° of the fluid:

$$\text{curl } \mathbf{F} = \rho_f^\circ \alpha_f \nabla\mathbf{T} \times \mathbf{g}. \tag{10}$$

The fluid mass flux $\mathbf{q}$ (in g s^{-1}), caused by these lateral gradients in density, is inversely proportional to the kinematic viscosity ν_f (Figure 8). Liquid and supercritical fluid viscosity varies from a maximum at low temperatures through a broad minimum at about 225°C, and then increases again. This viscosity minimum lies at about 50°C on the low-temperature side of the maxima in α_f.

The thermal energy content of the fluid phase is commonly expressed in terms of enthalpy H_f (Figure 9). However, convection depends on the gradient of the energy content of the fluid, and the transport equations are conveniently derived in terms of the isobaric heat capacity:

$$\mathbf{j}_{\text{convection,f}} = \mathbf{q}^* H_f = \mathbf{q}^* C_p T^*. \tag{11}$$

The isobaric heat capacity C_p varies from infinity at the critical end-point to values of 1.5 cal g^{-1} $°C^{-1}$ just a few degrees on either side of the end-point (Figure 10). Extreme values of heat capacity coincide with extreme values in α_f. Consequently, the pressure-temperature region of maximum convective heat-transport capacity coincides with that of maximum buoyancy force, and both are bordered on their low-temperature side by the viscosity minima.

Fluid in isolated pores expands against the pore walls as it is heated. In a constant-volume rigid wall pore, this expansion causes the fluid to be compressed by the pressure increase. The isothermal coefficient of com-

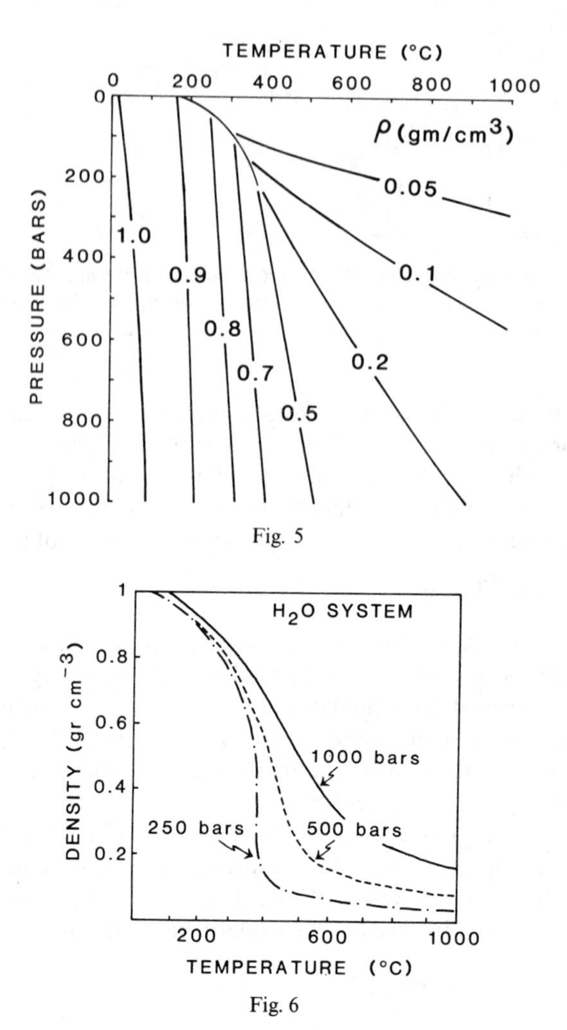

Fig. 5

Fig. 6

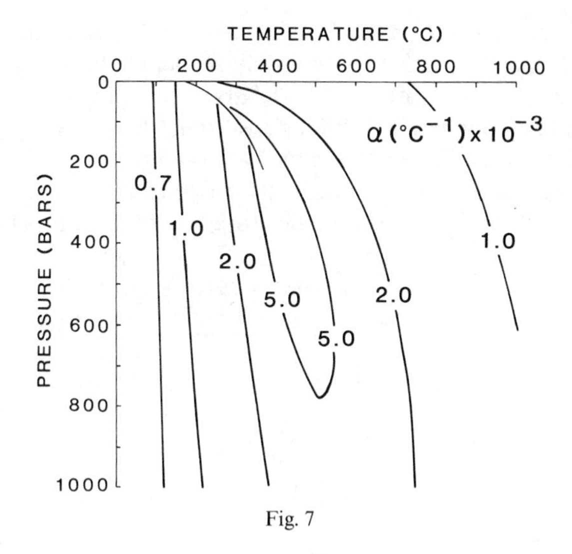

Fig. 7

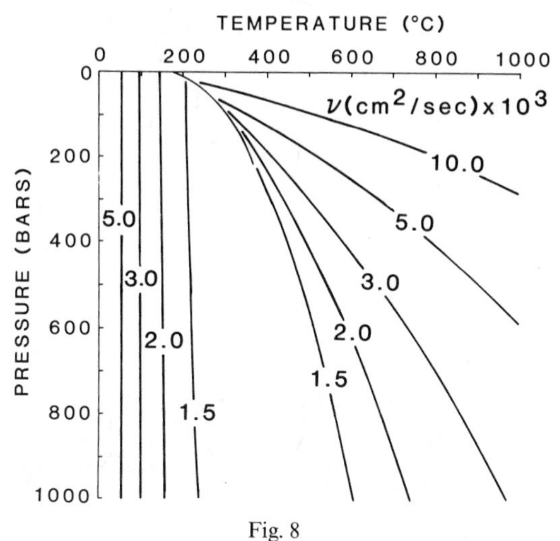

Fig. 8

Figures 5–8 Projections of the H_2O-SYSTEM for density (5 and 6), isobaric coefficient of thermal expansion (7), and kinematic viscosity (8).

pressibility β_f (Figure 11) is large for H_2O-vapor but small for H_2O-liquid and low-temperature supercritical fluid.

The net fluid pressure increase caused by heating a fixed-volume fluid-filled pore is related to the ratio α_f/β_f:

$$dP_f = (\alpha_f/\beta_f)\, dT_f. \tag{12}$$

This ratio ranges from about 1 to 18 bars °C^{-1} over the conditions encountered in hydrothermal systems; consequently, for every degree rise in temperature, increases of from 1 to 18 bars of fluid pressure are generated in isolated pores (Figure 12). Within pores connected to extensive fracture networks, a portion of the pressure increase is dissipated by fluid flow.

The association of aqueous ions and hydrolytic destruction of minerals is strongly dependent on the electrostatic properties of the fluid. The dielectric

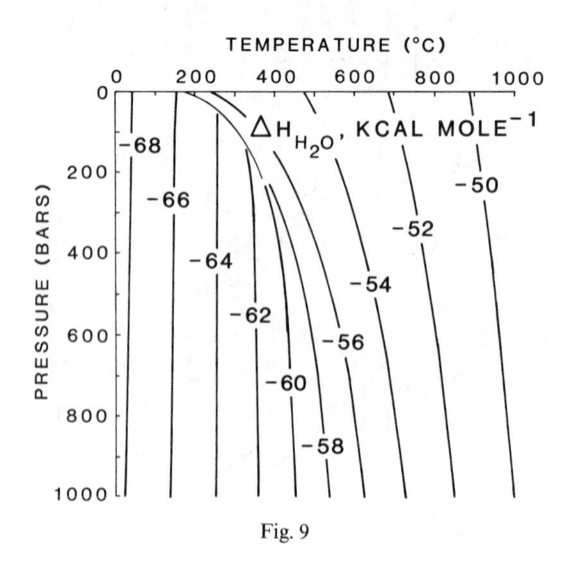

Fig. 9

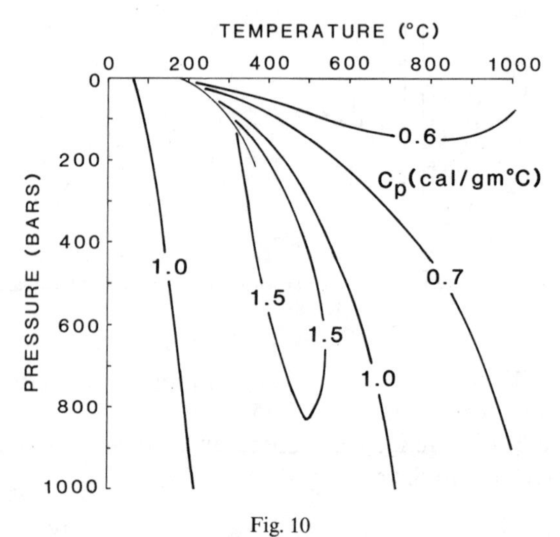

Fig. 10

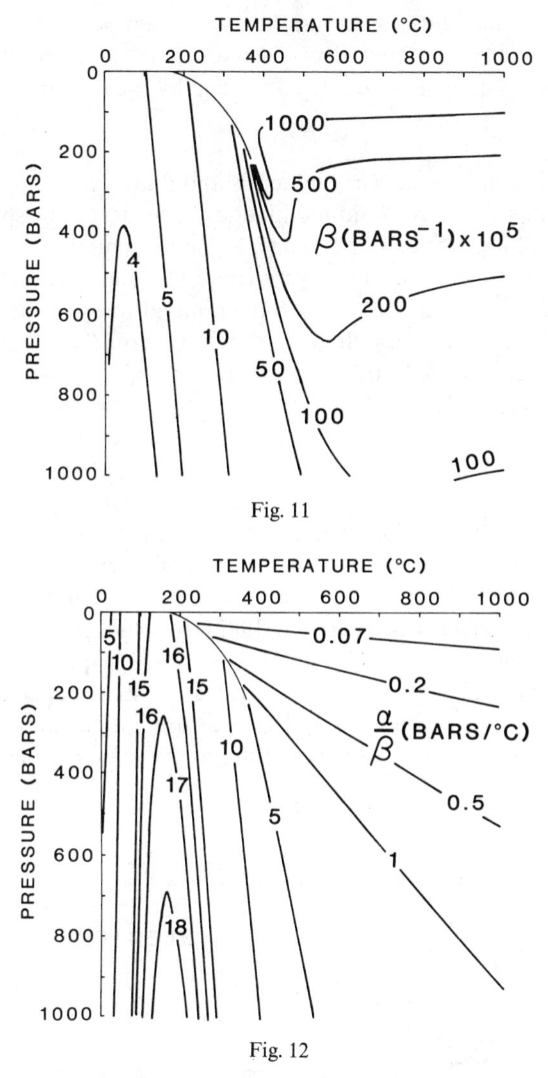

Fig. 11

Fig. 12

Figures 9–12 Pressure-temperature projections of H_2O-SYSTEM for standard molal enthalpy of formation (9), isobaric heat capacity (10), isothermal coefficient of compressibility (11), and the ratio of expansivity to compressibility (12).

constant of liquid-H_2O decreases from a maximum at low temperatures to a minimum in the supercritical region (Figure 13). This variation favors the association of ions at high temperatures and decreases the ionic strength.

The correlation between optimum conditions for anomalous pore-fluid pressures and those for developing fluid-flow fields, together with the coincidence of the fluids' maximum heat-carrying capacity and minimum viscous resistance to flow, demonstrates that state conditions in a hydrothermal system are strongly controlled by the fluid itself (Figure 14).

Minerals

Chemical relationships between minerals and fluids are a function of their standard-state thermodynamic and kinetic properties (Helgeson et al 1978, Helgeson 1979). Although the thermodynamic conditions range from disequilibrium to equilibrium, overall equilibrium is rare because the flowing fluid is in direct contact with the minerals only in the vein and its contiguous aureole. Because fluids in the flow channels and in the matrix blocks communicate only by diffusion, mineral alteration zones tend to develop symmetrically about the flow channels. These zones grow into the matrix blocks in a systematic sequence that advances through continuous irreversible reactions between matrix-block minerals and fluids whose compositions are controlled by fluids in the flow channels. Where the flow channels are less than several meters apart (a common condition in nature), this alteration takes place in a nearly isothermal environment.

Irreversible rates k° for the common rock-forming minerals are nearly instantaneous (Aagaard & Helgeson 1982). Even so, it is the irreversible mineral reaction paths that effectively control the sequence of alteration minerals formed. These standard-state rates are sufficiently large (Figure 15) at elevated temperatures to quickly restore equilibrium between the igneous minerals and the infiltrating fluid as the state conditions change.

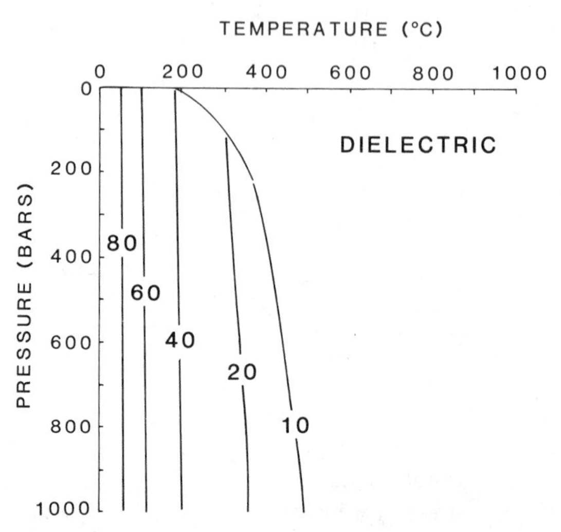

Figure 13 Pressure-temperature projection of H_2O-SYSTEM, for dielectric constant.

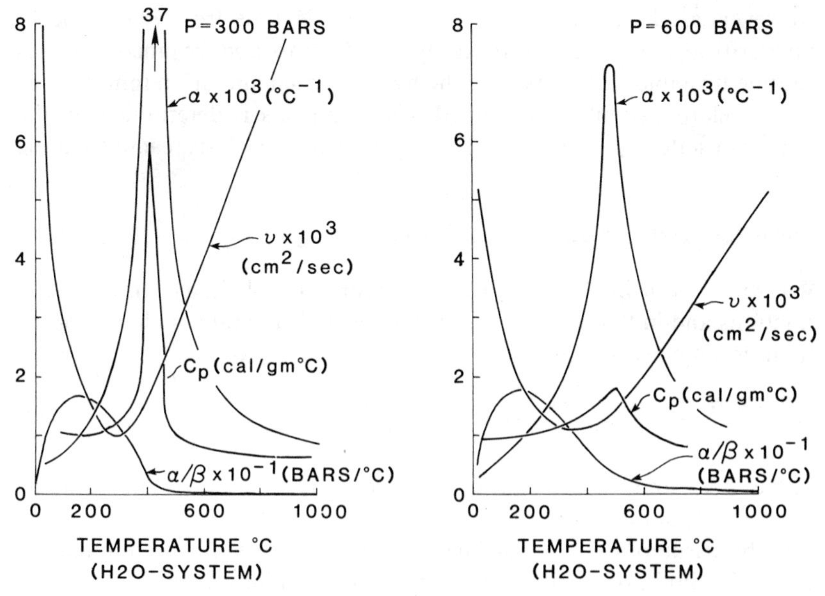

Figure 14 Variation of transport properties in H_2O-SYSTEM at constant pressures of 300 and 600 bars.

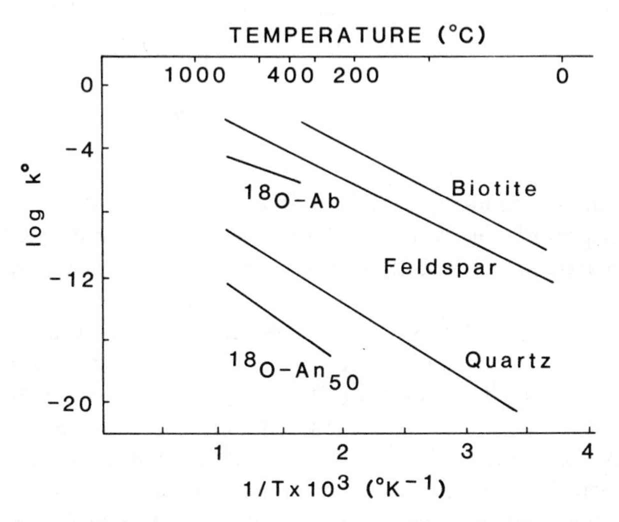

Figure 15 Log of standard-state reaction-rate constant k° as a function of temperature for irreversible dissolution of rock-forming minerals and for "exchange" of ^{18}O with albite and plagioclase An_{50}. For discussion of data and sources, see Norton & Taylor (1979).

However, as the temperature and pressure degrade from magmatic to supercritical, the irreversible reaction rates decrease and the igneous phases become unstable with respect to the hydrothermal alteration minerals.

The relative stability of hydrothermal minerals is determined by the change in state conditions and the properties of hydrolysis reactions of the form

$$\text{Minerals} + \text{H}^+ \rightleftarrows \text{Aqueous Basis Ions.} \tag{13}$$

Variations both in the standard-state properties of the entities in these reactions and in temperature and pressure affect the rate of change of the equilibrium constant for (13):

$$\frac{\mathrm{d}\log K_\mathrm{r}}{\mathrm{d}t} = \frac{\Delta H^\circ_\mathrm{r}}{2.3RT^2}\left(\frac{\mathrm{d}T}{\mathrm{d}t}\right) - \frac{\Delta V^\circ_\mathrm{r}}{2.3RT}\left(\frac{\mathrm{d}P}{\mathrm{d}t}\right), \tag{14}$$

where $\Delta H^\circ_\mathrm{r}$ and $\Delta V^\circ_\mathrm{r}$ are the standard molal enthalpy and volume, respectively, of the reaction.

If the assumption is made that pressure changes are predominantly caused by fluid expansion, (12) can be substituted into (14):

$$\frac{\mathrm{d}\log K_\mathrm{r}}{\mathrm{d}t} = \left(\frac{\Delta H^\circ_\mathrm{r}}{2.3RT^2} - \frac{\Delta V^\circ_\mathrm{r}}{2.3RT}\frac{\alpha_\mathrm{f}}{\beta_\mathrm{f}}\right)\frac{\mathrm{d}T}{\mathrm{d}t}. \tag{15}$$

This equation is consistent with the steady-state pressure conditions required by Darcy flow, e.g. $\mathrm{d}P/\mathrm{d}t = 0$. If the only transient pressure changes considered are those that contribute to fracture, then for the conditions

$$\Delta H^\circ_\mathrm{r} = 0.024\Delta V^\circ_\mathrm{r}\frac{\alpha_\mathrm{f}}{\beta_\mathrm{f}}T, \tag{16}$$

an extremum in the $\log K_\mathrm{r} = f(\text{time})$ occurs and steady-state equilibrium conditions prevail. The principal changes in the mineral content of rock must then arise predominantly from changes caused by advection of ions [see Equation (6)].

For all minerals, the terms $\Delta H^\circ_\mathrm{r}/RT^2$ and $\Delta V^\circ_\mathrm{r}/RT$ in (14) range from near zero at low temperatures to extreme negative values at temperatures and pressures in the supercritical region, then increase toward zero at elevated temperatures (Figure 16). The general magnitude of these terms is similar for all minerals; however, larger negative heats and volumes of reaction are more typical of phyllosilicates than of tectosilicates. Consequently, the micas are stable when in contact with solutions that have lower ratios of basis ion to hydrogen ion activities than required by feldspars.

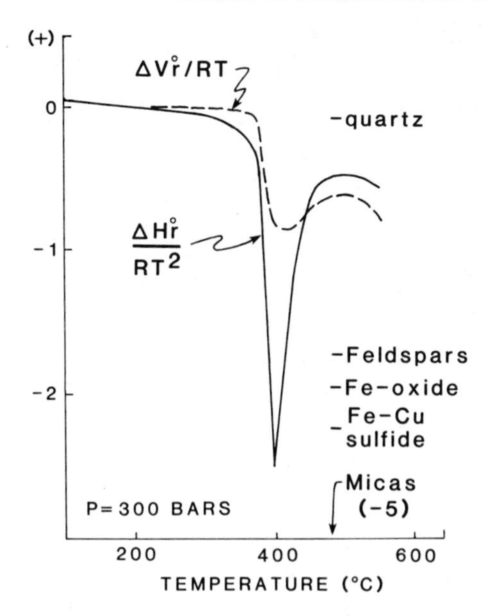

Figure 16 Schematic variations of parameters in Equation (14) as a function of temperature. Relative positions of minima for several mineral groups are represented by labels.

Rock

Permeability, one of the most important rock properties, represents viscous interactions between fracture surfaces and fluid as it flows through interconnected fracture networks (Figure 1). Although permeability is a critical rock property, it is poorly understood (Brace 1980). Numerous geometric models of fracture networks have been derived and used to describe hydrothermal fluid flow (Norton & Knapp 1977, Madden 1983), yet the inescapable fact remains that permeability of active hydrothermal systems is not directly measurable, nor can it be closely approximated by idealized fracture geometries.

Mineral-filled fractures, or veins, are the fossil record of fracture networks through which hydrothermal fluids once flowed; their geometry, concentration, and composition record variations in chemical, mechanical, and thermal state conditions that occurred during the thermal event. Although veins appear continuous over distances of several meters, they are in fact composed of many less-continuous pores interspersed along the failure surface between rock asperities that hold the pores open against confining stresses P_c (Figure 17). In plan view, vein walls are irregular surfaces with elliptical outlines, and in cross section their widths vary from a

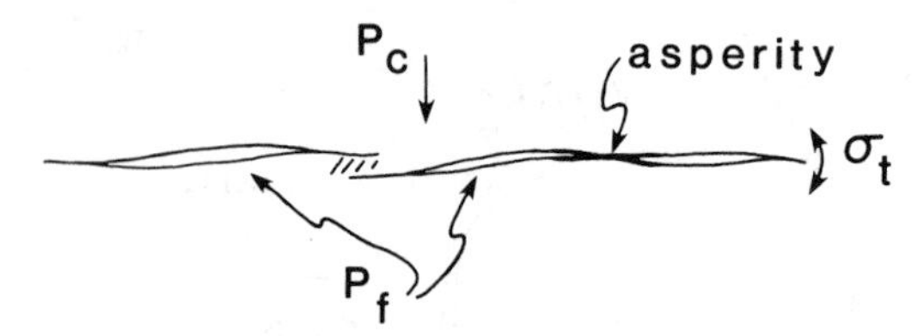

Figure 17 Schematic sectional view of macroscopic fracture geometry and stress conditions around fluid-filled pores. Fracture is composed of discontinuous cracks interconnected by microcracks and held open against confining stress P_c by asperities and a tangential stress σ_t that is generated by fluid pressure P_f.

maximum value near their centroids to zero along their perimeters. This form is similar to that produced by deflection by uniform pressure against a confined elastic plate (Johnson 1970). The morphology of veins portrays what was once the locus of relatively discontinuous yet interconnected pores, asperities, and zones of microcracks linking adjacent fractures and forming a continuous fluid-flow path.

The fracture-yield strength of matrix blocks transected by fractures and riddled with grain-boundary microcracks is negligible. Even the strength of the bond between mineral grains is minimal; consequently, fluid pressures P_f in these cracks can increase only to the confining pressure. These conditions are consistent with the provisional failure law that relates local stresses to volumetric changes in terms of effective pressure P_e:

$$P_e = P_c - P_f + \tau, \tag{17}$$

where the rock strength τ is considered zero, and the confining pressure P_c is either the overburden load $\rho_f gz$ in a homogeneous stress field or the trace of the stress tensor in an anisotropic field.

Fluid pressure increases when the temperature in a pore isolated from the principal flow channels is increased [Equation (12)]. When the rate of pore-fluid pressure increase is greater than the rate of increase in confining pressure, the effective pressure shifts toward negative values. At fixed positions in the system, P_e is a simple function of heating rate and α/β. When $P_e < 0$, the fluid cavity fails and a fracture propagates parallel to the direction of the maximum stress trajectory.

The geometry of a typical fluid-filled pore provides a large mechanical leverage to fluid pressure because in sectional view the ratio of the major axis a to the minor axis b is large ($\sim 10^3$ to 10^4). Therefore, the tangential stress at the crack tip is

$$\sigma_t = P_f(1 + 2a/b). \tag{18}$$

Extremely small increases in temperature cause pores of large a/b to propagate.

THE HYDROTHERMAL HISTORY

The extent to which hydrothermal systems contribute to the geological history of a region can be approximated by the theoretical relationships discussed above. Because the processes just described are operative in all hydrothermal systems, consider now the physical and chemical conditions likely to prevail during the history of a magmatic thermal perturbation.

Throughout large regions of the crust, the dispersion of energy and mass away from magma chambers causes thermal, chemical, and mechanical alteration of rocks, and initiates hydrothermal processes during the early stages of increased heat flow that precede magmatic activity. They accompany the flow of magma into the crust and then envelop the chamber as it inflates with magma. At this stage of chamber growth, hydrothermal processes are limited to contacts between the magma and host rocks; however, this sheath of activity quickly expands into the wall and roof rocks of the chamber, as well as into the crystalline portions of the magma, as soon as these regions achieve sufficient permeability ($>10^{-14}$ cm^2). The rate at which this zone of activity expands is commensurate with the relative permeabilities of the host and igneous rocks. During these early stages the magma is impermeable, so the activity expands preferentially into the host rocks. Then, as the magma crystallizes and fractures, fluids flow through the pluton in proportion to the extent of fracturing. Where conduction is the dominant process of thermal energy dispersion from the magma, the convecting hydrothermal fluid around the magma has a minor effect on its crystallization rate. The exception to this condition occurs in cases where hydrothermal fluids can flow through the magma; however, this appears to be an uncommon condition in nature.

As magmas rich in volatiles cool, fractionate into more volatile-rich phases, and partially crystallize, they form distinct subchambers. Within these subchambers magma pressure steadily increases with further cooling and exsolution of H_2O from the magma. Although this pressure increase initially shifts the system away from the solidus, it eventually exceeds both confining pressure and rock strength, thereby causing the subchamber walls to fracture. As magma fills these fractures they propagate further, but this expansion in turn reduces local pressure, which promotes solidification of the magma. These sudden pressure decreases cause exsolution of volatiles from the residual magma, thereby propagating fractures and increasing crystallization rates.

Host rocks are stressed by a combination of magma pressure and differential thermal expansion of their minerals and fluids. Magma pressure decreases to inconsequential values at distances away from the magma greater than the radius of the magma, and at times greater than one third the total thermal life of the system. While magma is present in the chamber, the contact region is fractured by the magma-induced stresses and by the expansion of pore fluids. Pore-fluid expansion continues to be effective long after the magma has crystallized and ceases only when the zero effective pressure front has reached its maximum extent. This expansion causes the rock matrix to fracture, and the system returns to an equilibrium stress condition.

Fractures propagate in response to the fluid expansion within a single pore and interconnect other pores, which may contain fluids of different densities. These thermally and compositionally related density gradients cause fluids to convect through the growing fracture network. This network forms a broad region around the pluton that expands along with the advancing zero-effective pressure front. Within this region the orientations of an individual fracture are controlled at early stages by the magma pressure, and subsequently by the thermal field. Only during the waning stages of activity do the regional stress orientations control the fracture orientations (Roberts 1970, Koide & Bhattacharji 1975, Knapp & Norton 1981).

Continuous shifts in fracture orientation are accompanied by variations in the composition of fluids that flow through the fractures. Even though the temperatures and pressures in regions of upwelling fluid coincide closely with the supercritical region (Figure 18), the conditions along the

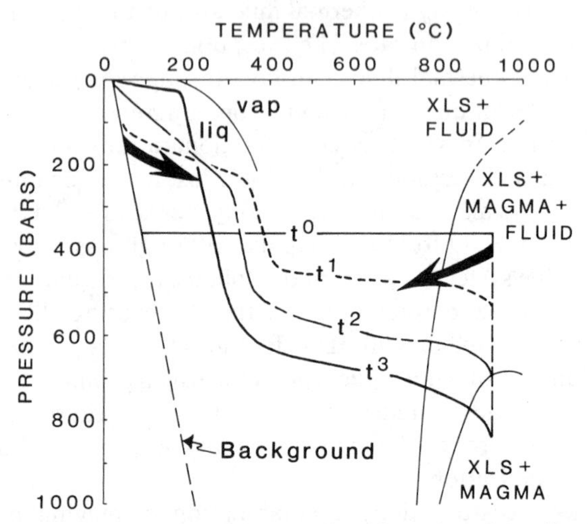

Figure 18 Pressure-temperature conditions predicted for region of upwelling hydrothermal fluid over a sequence of relative times t^0 to t^3. Sketch is based on an idealized magma-hydrothermal system that contains pure H_2O and a 4 wt% tonalite magma (see Norton 1982). Heavy arrows depict general trend of *T-P* conditions for host-rock (*upper left*) and pluton (*lower center*) environments.

flow paths that supply fluid to this upflow zone cover a much wider range. Also, these fluids are derived from increasingly more distant source regions, and as the source regions expand both laterally and into the roof of the pluton, a broad spectrum of chemical environments is sampled (Norton 1979). The net result is that an extreme diversity of fluid compositions is introduced into the local system.

Peak development of fluid-flow rates coincides with the formation of new fractures as rapidly as the old ones become tightly filled veins. During this stage, maximum metasomatic changes occur because large fluid velocities coincide with steep gradients in the activities of aqueous ions [Equation (6)]. Metasomatism is predominant along the flow channels, though only for short distances into the matrix blocks. Pervasive metasomatic alteration of matrix blocks occurs only where the flow channels are closely spaced.

The magma completely crystallizes and the pluton fractures during this main stage of hydrothermal activity. Rapid temperature decreases occur in the pluton, and an instability develops between the minerals and aqueous phase, which were initially in equilibrium at magmatic conditions. This shift favors equilibrium between the fluid phase and minerals that are stable at hydrothermal temperatures and pressures. The particular alteration assemblage that forms is determined by the activities of aqueous ions in the infiltrating aqueous phase and the reactant magmatic igneous assemblage. But the advection rate of aqueous components, which is proportional to these ion activities, depends indirectly on rock permeability. Therefore, the stable alteration assemblage is ultimately determined by the pluton's permeability. This conclusion appears to explain some of the extreme variations in mineral alteration seen in plutons.

Overall thermal decline is associated with both the collapse of flow channels from decreasing fluid pressure and the infilling by relatively large-volume alteration minerals, such as clays, that are stable at low temperatures. The last vestige of activity occurs in the upper portions of the chamber roof. This final stage is one of minor activity at depth, but at the Earth's surface it leaves a false impression near the surface of high subsurface temperatures.

SUMMARY

The theoretical hydrothermal system portrayed depicts conditions as they might appear in natural systems. Each of the processes and controls discussed is an obvious consequence of the theory, but few have ever been directly observed. This theory can be improved through imaginative field studies of fossilized systems, extensions of state properties to more commonly encountered natural materials, and improved formulations of the mathematical system.

ACKNOWLEDGMENTS

The contents of this article were influenced by many students and colleagues who have studied and contributed to this topic over the years. I particularly wish to thank R. N. Villas, R. B. Knapp, R. Capuano, B. Moskowitz, J. Knight, J. Johnson, J. R. Delaney, T. Gerlach, D. K. Bird, H. P. Taylor, and H. C. Helgeson for contributing to these thoughts. I am grateful to Emily C. Creigh for arranging the words and text and to Ann Troutner for drafting the figures. A special thanks goes to James Johnson, Rik Lantz, Daniel Barker, and J. A. Woodward for their thorough and timely reviews. Financial support for these studies has been provided by the National Science Foundation's Geochemistry Program and the Department of Energy's Office of Basic Science.

Literature Cited

Aagaard, P., Helgeson, H. C. 1982. Thermodynamic and kinetic constraints on reaction rates among minerals and aqueous solutions. I. Theoretical considerations. *Am. J. Sci.* 282:237–85

Bear, J. 1972. *Dynamics of Fluids in Porous Media.* New York: Elsevier. 764 pp.

Bird, D. K., Norton, D. L. 1981. Theoretical prediction of phase relations among aqueous solutions and minerals: Salton Sea geothermal system. *Geochim. Cosmochim. Acta* 45:1479–93

Brace, W. F. 1980. Permeability of crystalline and argillaceous rocks: status and problems. *Int. J. Rock Mech. Min. Sci. Geomech. Abstr.* 17:241–51

Brimhall, G. H., Ghiorso, M. S. 1983. Origin and ore-forming consequences of the advanced argillic alteration process in hypogene environments by magmatic gas contamination of meteoric fluids. *Econ. Geol.* 78:73–90

Cathles, L. M. 1981. Fluid flow and genesis of hydrothermal ore deposits. *Econ. Geol.* 75:424–57

Criss, R. E., Lanphere, M. A., Taylor, H. P. Jr. 1982. Effects of regional uplift, deformation, and meteoric-hydrothermal metamorphism on K-Ar ages of biotites in the southern half of the Idaho batholith. *J. Geophys. Res.* 87:7029–46

Delaney, J. R. 1980. High-temperature sulfide-bearing hydrothermal systems on the Mid-Atlantic Ridge at 23.6 north. *Geol. Soc. Am. Abstr. with Programs* 12:411

Delaney, P. T. 1982. Rapid intrusion of magma into wet rock: groundwater flow due to pore pressure increases. *J. Geophys. Res.* 87:7739–56

Delany, J. M., Helgeson, H. C. 1978. Calculation of the thermodynamic consequences of dehydration in subducting oceanic crust to 100 kb and >800°C. *Am. J. Sci.* 278:638–86

Gregory, R. T., Taylor, H. P. Jr. 1981. An oxygen isotope profile in a section of Cretaceous oceanic crust, Samail Ophiolite, Oman: evidence for ^{18}O buffering of the oceans by deep (>5 km) seawater-hydrothermal circulation at mid-ocean ridges. *J. Geophys. Res.* 86:2737–55

Helgeson, H. C. 1970. A chemical and thermodynamic model of ore deposition in hydrothermal systems. *Mineral. Soc. Am. Spec. Publ.* 3:155–86

Helgeson, H. C. 1979. Mass transfer among minerals and hydrothermal solutions. In *Geochemistry of Hydrothermal Ore Deposits*, ed. H. L. Barnes, pp. 568–610. New York: Wiley. 798 pp. 2nd ed.

Helgeson, H. C., Brown, T. H., Nigrini, A., Jones, T. A. 1970. Calculation of mass transfer in geochemical processes involving aqueous solutions. *Geochim. Cosmochim. Acta* 34:569–92

Helgeson, H. C., Kirkham, D. H. 1974a. Theoretical prediction of the thermodynamic behavior of aqueous electrolytes at high pressures and temperatures. I. Summary of the thermodynamic/electrostatic properties of the solvent. *Am. J. Sci.* 274:1089–1198

Helgeson, H. C., Kirkham, D. H. 1974b. Theoretical prediction of the thermodynamic behavior of aqueous electrolytes at high pressures and temperatures. II. Debye-Huckel parameters for activity coefficients and relative partial molal properties. *Am. J. Sci.* 274:1199–1261

Helgeson, H. C., Kirkham, D. H. 1976.

Theoretical prediction of the thermodynamic behavior of aqueous electrolytes at high pressures and temperatures. III. Equation of state for aqueous species at infinite dilution. *Am. J. Sci.* 276:97–240

Helgeson, H. C., Delany, J. M., Nesbitt, H. W., Bird, D. K. 1978. Summary and critique of the thermodynamic properties of rock-forming minerals. *Am. J. Sci.* 278-A:1–220

Helgeson, H. C., Kirkham, D. H., Flowers, G. C. 1981. Theoretical prediction of the thermodynamic behavior of aqueous electrolytes at high pressures and temperatures. IV. Calculation of activity coefficients, osmotic coefficients, and apparent molal and standard and relative partial molal properties to 600°C and 5 kb. *Am. J. Sci.* 281:1249–1536

Hofstadter, D. R. 1981. Metamagical themas. *Sci. Am.* 245:22–43

Hubbert, M. K. 1969. *The Theory of Ground-Water Motion and Related Papers*. New York: Hafner. 310 pp.

Johnson, A. M. 1970. *Physical Processes in Geology*. San Francisco: Freeman. 577 pp.

Knapp, R. B., Norton, D. L. 1981. Preliminary numerical analysis of processes related to magma crystallization and stress evolution in cooling pluton environments. *Am. J. Sci.* 281:35–68

Koide, H., Bhattacharji, S. 1975. Formation of fractures around magmatic intrusions and their role in ore localization. *Econ. Geol.* 70:781–99

Lister, C. R. B. 1977. Qualitative models of spreading-center processes, including hydrothermal penetration. *Tectonophysics* 37:203–18

Lister, C. R. B. 1980. Heat flow and hydrothermal circulation. *Ann. Rev. Earth Planet. Sci.* 8:95–117

Madden, T. R. 1983. Microcrack connectivity in rocks: a renormalization group approach to the critical phenomena of conduction and failure in crystalline rocks. *J. Geophys. Res.* 88:585–92

Norton, D. L. 1979. Transport phenomena in hydrothermal systems: the redistribution of chemical components around cooling magmas. *Bull. Mineral.* 102:471–86

Norton, D. L. 1982. Fluid and heat transport phenomena typical of copper-bearing pluton environments, southeastern Arizona. In *Advances in Geology of the Porphyry Copper Deposits, Southwestern North America*, ed. S. R. Titley, pp. 59–72. Tucson: Univ. Ariz. Press. 560 pp.

Norton, D. L., Knapp, R. B. 1977. Transport phenomena in hydrothermal systems: nature of porosity. *Am. J. Sci.* 277:913–36

Norton, D. L., Taylor, H. P. Jr. 1979. Quantitative simulation of the hydrothermal systems of crystallizing magmas on the basis of transport theory and oxygen isotope data: an analysis of the Skaergaard intrusion. *J. Petrol.* 20:421–86

Pruess, K. 1983. Heat transfer in fractured geothermal reservoirs with boiling. *Water Resour. Res.* 19:201–8

Roberts, J. L. 1970. The intrusion of magma into brittle rocks. In *Mechanism of Igneous Intrusion*, ed. G. Newall, N. Rast, pp. 287–338. London: Gallery. 362 pp.

Simmons, G., Richter, D. 1976. Microcracks in rocks. In *The Physics and Chemistry of Minerals and Rocks*, ed. R. G. J. Strens. New York: Wiley. 695 pp.

Slattery, J. C. 1972. *Momentum, Energy, and Mass Transfer in Continua*. New York: McGraw-Hill. 679 pp.

Sprunt, E. S., Nur, A. 1979. Microcracking and healing in granites: new evidence from cathodoluminescence. *Science* 205:495–97

Taylor, H. P. Jr. 1977. Water/rock interactions and the origin of H_2O in granitic batholiths. *J. Geophys. Res.* 133:509–58

Walther, J. V., Orville, P. M. 1982. Volatile production and transport in regional metamorphism. *Contrib. Mineral. Petrol.* 79:252–57

Ann. Rev. Earth Planet. Sci. 1984. 12:179–204

SEDIMENTATION IN LARGE LAKES

Thomas C. Johnson

Limnology Program and Department of Geology,
University of Minnesota, Duluth, Minnesota 55816[1]

INTRODUCTION

Sedimentary deposits in large lakes have been studied less than those of any other major environment on Earth. Much has been learned about the thickness, structure, and composition of sediments in the oceans, rivers, and subaerial environments; and our understanding of climatic and tectonic processes and variability has improved considerably within the past two decades from detailed stratigraphic studies of, for example, deep-sea sediments, continental ice sheets, and deposits in small lakes. The history preserved in sediments of modern large lakes, however, has been hardly tapped, and our understanding of the physical, chemical, and biological processes associated with sedimentation in these lakes is in its infancy.

The large lakes of the world are precious resources that provide their nearby inhabitants with abundant fresh water, food, economical transportation, and recreation. They also provide additional benefits to the geologist: unique paleoclimatic records, and useful basins in which to test models of aquatic sedimentary processes.

This paper summarizes the global distribution of large lakes and focuses upon their geological attributes.

GLOBAL DISTRIBUTION OF LARGE LAKES

The distribution, origin, and evolution of lake basins have been examined in some detail during the past century by Davis (1882, 1887), Penck (1894), Russel (1895), and Murray (1910), among others. Hutchinson (1957)

[1] Present address: Duke University Marine Laboratory, Beaufort, North Carolina 28516.

0084–6597/84/0515–0179$02.00

summarized these and other papers and developed a genetic classification scheme for 76 types of lake basins.

Rather than employ the detailed classification of Hutchinson (1957), the 25 largest lakes on Earth are recognized here as being simply of either tectonic or glacial origin (Table 1). The large lakes of glacial origin are all located north of 40°N latitude along arcs of maximum glacial scour between Precambrian shield rocks to the north and Paleozoic or younger sedimentary strata to the south (Figure 1; White 1972, Sugden 1978). The lakes of tectonic origin are mostly located along major rift zones; these include Lake Baikal and all of the great lakes in the East African Rift Valley except Lake Victoria. The rift zones, however, do not mark *major* plate boundaries, as illustrated, for example, in Dewey (1972). The East African Rift Valley is within the African plate, and the Baikal Rift is within the Eurasian plate. Some large lakes are found along major plate boundaries of convergence. These include Lakes Nicaragua and Titicaca, as well as the

Table 1 The 25 largest lakes (by volume) in the world, with approximate locations and origins (from Herdendorf 1982)

Name	Latitude	Longitude	Origin
1. Caspian	42°00′N	50°00′E	Tectonic
2. Baikal	54°00′N	109°00′E	Tectonic
3. Tanganyika	6°00′S	29°30′E	Tectonic
4. Superior	47°30′N	88°00′W	Glacial
5. Malawi	12°00′S	34°30′E	Tectonic
6. Michigan	44°00′N	87°00′W	Glacial
7. Huron	45°00′N	81°15′W	Glacial
8. Victoria	01°00′S	33°00′E	Tectonic
9. Great Bear	66°00′N	121°00′W	Glacial
10. Great Slave	62°40′N	114°00′W	Glacial
11. Issykkul	42°30′N	77°15′E	Tectonic
12. Ontario	43°30′N	77°45′W	Glacial
13. Aral	45°00′N	60°00′E	Tectonic
14. Ladoga	61°00′N	31°30′E	Glacial
15. Titicaca	15°45′S	69°30′W	Tectonic
16. Reindeer	57°15′N	102°15′W	Glacial
17. Helmand	31°00′N	61°15′E	Tectonic
18. Erie	42°15′N	81°15′W	Glacial
19. Hovsgal	51°00′N	100°30′E	Tectonic
20. Winnipeg	52°30′N	97°45′W	Glacial
21. Kivu	02°00′S	29°15′E	Tectonic
22. Nipigon	49°45′N	88°30′W	Glacial
23. Melville	53°45′N	59°30′W	Glacial
24. Onega	61°30′N	35°45′E	Glacial
25. Maracaibo	09°45′N	71°30′W	Tectonic

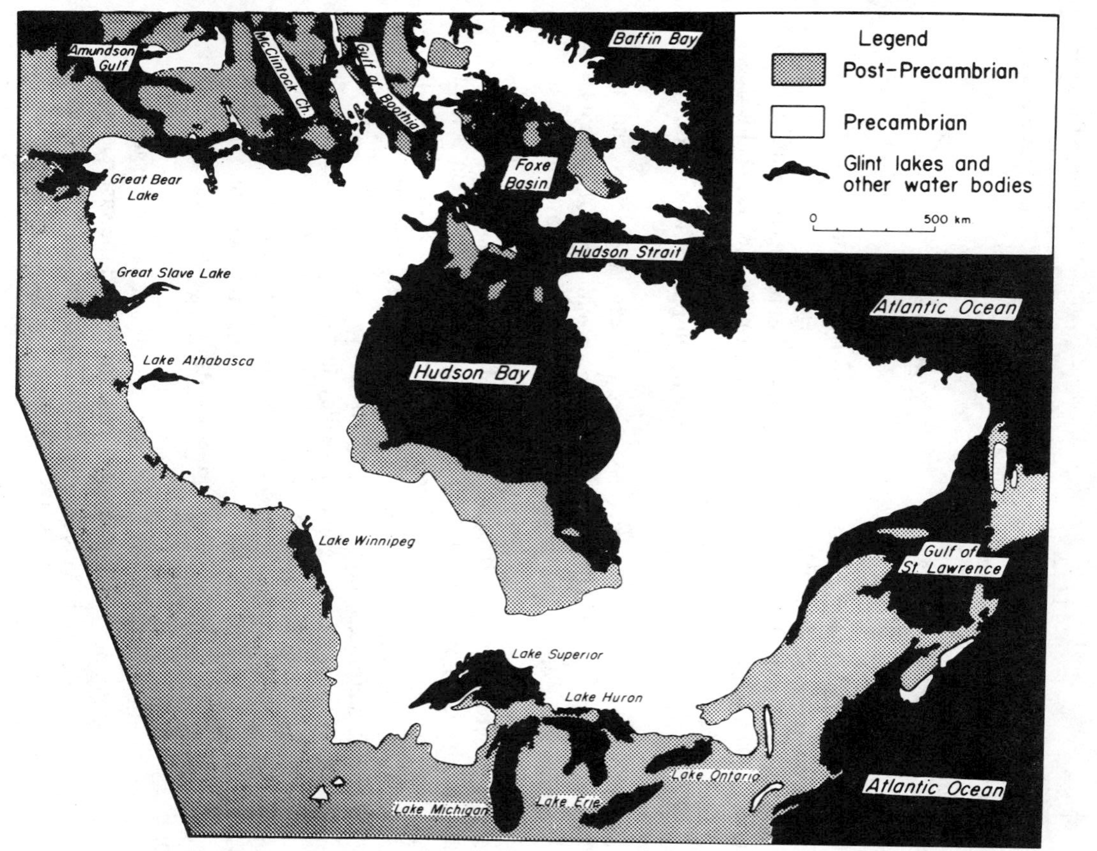

Figure 1 Locations of large glacial lakes in North America, showing their relation to Precambrian and younger rocks. Reproduced with permission from White (1972).

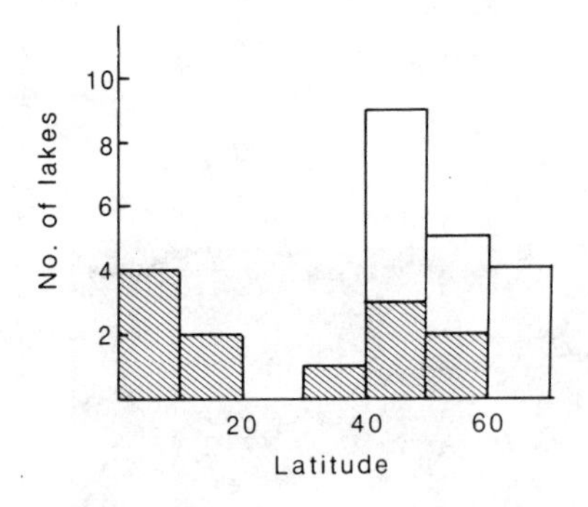

Figure 2 Latitudinal distribution of the 25 largest lakes on Earth. The shaded areas represent tectonic lakes, nonshaded areas represent glacial lakes.

Caspian Sea. Lake Maracaibo is on a transform boundary between the Caribbean and South American plates.

The latitudinal distribution of the 25 largest lakes reveals the importance of climatic conditions for lake formation. Only one of the lakes is located between 20° and 40° latitude (Figure 2), where the subtropical high-pressure zone creates arid conditions that inhibit the development of large bodies of water.

There are several important differences between lakes of glacial and tectonic origin. The glacial lakes are all younger than about 10,000 yr. By contrast, some tectonically formed lakes, such as Baikal, Tanganyika, and Malawi, are several million years old. Tectonic lakes tend to have thick sediment sequences [for example, over 3 km in Lake Tanganyika (Rosendahl & Livingstone 1983)], whereas glacial lakes, such as Lake Superior, have at most a few hundred meters of glacial and lacustrine sediment overlying bedrock (e.g. Wold et al 1982). The deepest lakes on Earth are all tectonic, although some glacial lakes, such as Great Slave and Superior, are deeper than some rift-valley lakes, such as Turkana. The relatively great depths of Lake Superior and Great Slave Lake may be due in part, however, to their locations on ancient rift zones (Hinze & Wold 1982, Hoffman et al 1974) and their young ages (i.e. thin sediment fill). Rift-valley lakes tend to have high length-to-width ratios and relatively small drainage basins.

Throughout most of Earth history there were no ice ages, so large lakes of tectonic origin undoubtedly prevailed and contributed the vast majority of lacustrine deposits to the geological record. Their abundance certainly has varied through time, with the peaks occurring during periods of continental rifting in nonarid regions.

A UNIQUE PALEOCLIMATIC RECORD

The sedimentary record of some of the large tectonic lakes is potentially the most detailed and valuable record available for the past few million years.

Sedimentation rates in large lakes are two orders of magnitude faster than in the deep sea, so there is good potential for improved resolution of paleoclimatic fluctuations beyond what is available from the deep-sea record.

Accurate age dating and assessment of sedimentation rates in lacustrine deposits are possible for the past 100 yr using lead-210 (e.g. Bruland et al 1975, Robbins & Edgington 1975, Evans et al 1981), but this isotope is not suitable for dating older deposits. Pollen analysis has been marginally successful in dating sediment in the Laurentian Great Lakes (e.g. Kemp et al 1974, 1977, 1978), but again only in sediments younger than 200 yr and relying solely upon the depth of the *ambrosia* rise that signals early European settlement. Pollen is poorly preserved in older deposits in these lakes because of the well-oxygenated bottom waters and low total organic carbon (TOC) content (typically 3–4%) in the bottom sediments. The "half life" for TOC degradation in Lake Superior is about 95 yr (Johnson et al 1982).

Radiocarbon dating is difficult because there usually is not enough shell material present in the sediment to date, so TOC must be analyzed. TOC contains a complex mixture of organic compounds derived from terrestrial, littoral, planktonic, and bacterial sources (Barnes & Barnes 1978). Kemp & Johnston (1979) reported that plankton is the primary source of TOC in Lakes Ontario, Erie, and Huron, and Johnson et al (1982) concluded the same for Lake Superior. Rea et al (1980) and Meyers et al (1980), on the other hand, suggested that much of the TOC in sediments of Lakes Michigan and Huron is allocthonous. Radiocarbon dating of TOC that contains an unknown percentage of old, detrital carbon yields errors of more than 1000 yr in many Holocene sediment dates (Rea et al 1981). C-14 dating can occasionally provide reliable dates, however, as in a core from Lake Victoria, East Africa (Kendall 1969).

The most promising dating technique for Holocene sediment in the Laurentian Great Lakes is the paleomagnetic method. Secular variations in magnetic inclination and declination preserved in lake-floor sediment have been correlated among cores in Lakes Superior, Huron, and Erie (Creer et al 1976, Mothersill 1979, 1981, Creer & Tucholka 1982, Johnson & Fields 1983). The technique is difficult to use in cores that contain hiatuses, and uncertainties still exist in the assignment of absolute ages to Holocene paleomagnetic events. It is still the best technique presently available, however, for dating sediments younger than 10,000 yr.

Sediments that contain ash layers, as is the case in many tectonic lakes, can be dated by tephrochronology. Degens et al (1973) used ash layers and other structures to correlate stratigraphy in cores from Lake Kivu, but the ash layers of Holocene age could be used only for relative age dating. Tuffs in older lacustrine, alluvial fan, and fluvial deposits of Plio-Pleistocene age exposed near Lake Turkana have been dated by K-Ar and $^{40}Ar/^{39}Ar$ total degassing geochronological methods (e.g. Fitch & Miller 1976). These techniques usually cannot provide reliable dates on tuffs that are younger than about 500,000 yr. They also require a large sample size, which limits their application in sediments recovered by cores. Another technique—fission-track dating—may become useful for dating volcanogenic sediment in lakes, but it has not yet gained widespread use in lacustrine studies.

Sedimentation rates published for several large lakes are summarized in Figure 3. There is surprisingly little difference between the sedimentation rates observed in tectonic versus glacial lakes, even though the relief and hence, the sediment supply rate per unit area of drainage basin, should be higher around tectonic lakes. Two factors mitigate the relief effect, however. (*a*) The rift-valley lakes typically have small drainage basins relative to their

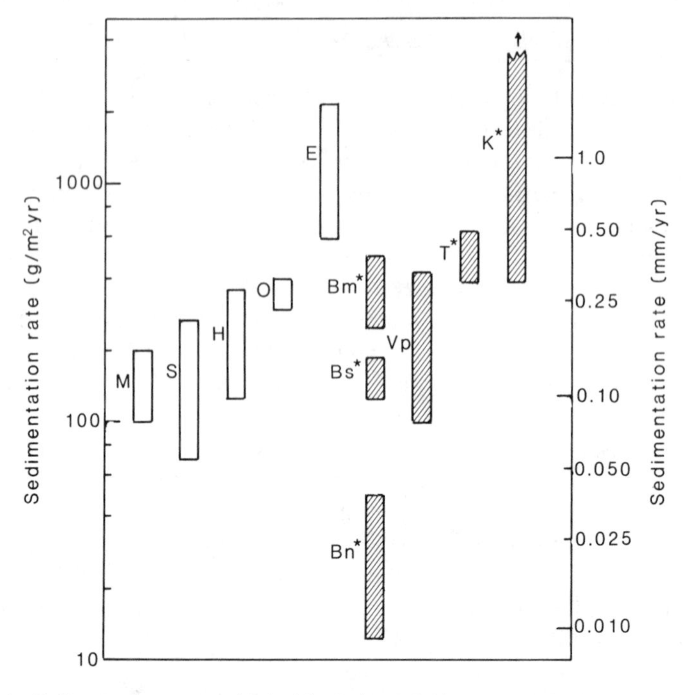

Figure 3 Sedimentation rates published for Lakes Michigan (M), Superior (S), Huron (H), Ontario (O), Erie (E), Baikal [northern basin (Bn), middle basin (Bm), southern basin (Bs)], Victoria [core P22 (Vp)], Tanganyika (T), and Kivu (K). Sedimentation rates are expressed in mass influx (*left scale*) unless designated by a star, which notes a linear sedimentation rate (*right scale*). The two ordinate scales are aligned assuming a sediment porosity of 85%. Open rectangles: glacial lakes; shaded rectangles: tectonic lakes. Sedimentation rates are mean values for the glacial lakes and Baikal, and all available values for the other three lakes. Source publications are cited in the text.

size (although this is not always true), and (*b*) unconsolidated glacial deposits that are easy to erode abound in the drainage basins of glacial lakes. These factors alone cannot explain the slight difference in sedimentation rates, however. The range of sedimentation rates encountered in offshore basins of Lake Superior is 70–250 g m^{-2} yr^{-1} (Kemp et al 1978, Evans et al 1981), which nearly equals the range of 250–500 g m^{-2} yr^{-1} in the central basin (where sedimentation rates are fastest) of Lake Baikal (Mizandrontsev 1982). Yet the relief in the Baikal drainage basin greatly exceeds that of the Superior drainage basin, and the ratio of drainage basin area to lake surface area is about 18 : 1 for Lake Baikal and only about 1.5 : 1 for Lake Superior. One explanation for their similar sedimentation rates might be that about 45% of the floor of Lake Superior is swept by currents generated by surface waves (Johnson 1980a), so sedimentation is focused into the deeper basins. This is also the case in Lake Huron (Thomas et al 1973) and the other Laurentian Great Lakes. Lakes that are deeper and have smaller surface areas than Superior and Huron probably are not subjected to such intense sediment focusing.

Sedimentation rates of 100–500 g m^{-2} yr^{-1} translate to linear sedimentation rates of 2–5 m $(1000\ yr)^{-1}$. This allows time resolution of fluctuations in sediment input that is on the order of decades. Sediments recovered from drilling in Clear Lake, California, for example, contain a record 180 m long that is estimated to be 130,000 yr old (Sims et al 1983). This indicates a sedimentation rate of about 1.4 cm per decade. A pollen curve for the site suggests the presence of several warm and cold periods during the past 130,000 yr, some of which are correlated with the deep-sea oxygen isotope stages 1–6 (Adam 1983). A 200-m-long core recovered from Lake Biwa, Japan, covers a time span of 565,000 yr (Yamamoto 1976); the sedimentation rate is slower than in Clear Lake, yet still allows far greater temporal resolution of climatic fluctuation than is available in marine cores.

The lacustrine sedimentary record is not as blurred by bioturbation as the marine record is. The Laurentian Great Lakes have a young benthic population dominated by amphipods, oligochaetes, and nematodes, most of which are shorter than 5 mm and restricted to the top 4 cm of sediment (e.g. Heuschele 1982, Robbins 1982). Analyses of Pb-210 and Cs-137 profiles in Lakes Michigan, Huron, Erie, and Superior generally show the biologically mixed layer to be shallower than 5 cm (Robbins & Edgington 1975, Robbins et al 1978, Robbins 1980, Evans et al 1981). Rift-valley lakes in tropical regions are anoxic below 100–200 m (Beadle 1981), so sediments are not mixed biologically in the deep basins, and laminated sediments may prevail [e.g. Lake Tanganyika (Degens et al 1971), Lake Kivu (Degens et al 1973, Stoffers & Hecky 1978), Lake Malawi (G. Müller, personal communication)]. Laminated sediments are also common in Lake Turkana, East

Africa, even though the lake is oxygenated throughout its water column (Yuretich 1979).

Large lakes can respond to changes in climate in many ways. Variations in rainfall can influence the input of terrigenous sediment, and in the case of many tropical African lakes, these variations can affect lake level by several meters (Figure 4) and thereby convert a closed-basin lake (with no river outflow) to an open-basin lake, or vice versa (e.g. Kendall 1969, Hecky & Degens 1973, Stoffers & Hecky 1978, Hastenrath & Kutzbach 1983). This may then have an effect on the overturn of the water column, the occurrence of anoxia in the bottom waters, and the production and input of biogenic sediments to the lake floor. Aeration of bottom waters also could affect the kinds of authigenic minerals forming in the sediments, as could fluctuations in lake-water alkalinity [demonstrated in lakes of the Eastern Rift Valley by Cerling (1979)]. Hecky & Degens (1973) and Hecky (1978) reported evidence for lake-level fluctuations of 400 m in Lake Kivu and suggested fluctuations of 600 m for Lake Tanganyika. Capart (1949) interpreted echo soundings from Lake Tanganyika to show low stands of lake level at −550 m and −850 m. Such fluctuations strongly affected water chemistry and structure in Lake Kivu, but had much less limnological impact on Lake Tanganyika because of its great volume (Livingstone 1975, Livingstone & Van der Hammen 1978).

Wind intensity and direction influence the size of surface waves and the depth to which wave scour affects sedimentation. Storm waves prevent sediment accumulation in offshore Lake Superior wherever the lake floor is shallower than 100 m (Johnson 1980a). The grain-size distribution,

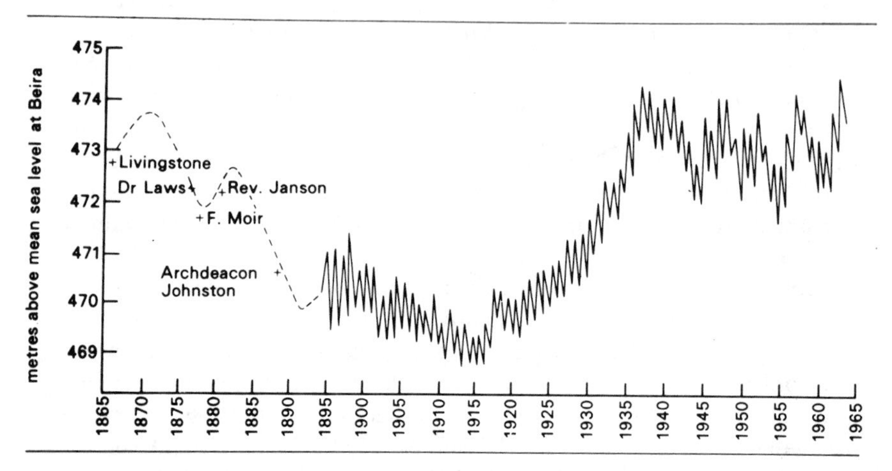

Figure 4 Lake Malawi water levels 1865–1963 (after Pike & Rimmington 1965). Reproduced with permission from Beadle (1981).

consequently, is strongly depth dependent (e.g. Thomas et al 1972, 1973, 1976, Sly et al 1983), and since sediment composition varies with grain size (e.g. Mothersill 1976, Cahill 1981), this, too, is affected by surface waves. Quantitative models of the impact of waves upon sedimentation in lakes have been attempted by Håkanson (1977, 1982), Johnson (1980b), and Lick (1982).

The effect of waves upon sediment entrainment is seasonal in the Laurentian Great Lakes. Sediments transported to the lakes by rivers in the winter and spring are deposited nearshore in shallow water. Waves and currents at these times of the year create high-turbidity events and transport the recently deposited sediment into deeper water. During most of the rest of the year, fluvial discharge is low, the winter and spring supply of sediment has been removed, and wave stresses that caused high turbidity during the spring no longer have as great an effect on sediment resuspension (Chambers & Eadie 1981, Lick 1982).

Storms passing over large lakes set up standing waves both on the lake surface and along the pycnocline of a stratified lake. If a lake is sufficiently large, the Coriolis effect causes the waves to rotate about one or more amphidromic points and take the form of Kelvin or Poincaré waves (Mortimer 1974). Strong bottom currents can be generated in association with these waves that fluctuate at near-inertial frequency and are focused by bathymetric features that cause localized erosion even in water depths of several hundred meters (Dell 1976, Johnson et al 1980; Figure 5). Strong deep-water currents generated by passing storms are more likely to occur in relatively shallow lakes with large surface areas than in deeper lakes with smaller surface areas. For example, side-scan SONAR records from western Lake Superior reveal pronounced lineations over much of the basin floor that probably resulted from bottom currents (Flood & Johnson 1983), whereas no such features are seen in the few side-scan SONAR records available from Lake Tanganyika (Johnson et al 1983b).

By monitoring the response of large, shallow lakes to passing storms, one could obtain a paleowind record with better time resolution than could be obtained from oceanic sediments, and with more sensitivity than could be expected from sediments of small lakes that are protected from the wind by surrounding vegetation. Grain-size analyses of the silt fraction of sediments accumulating in a region of Lake Superior that is influenced by bottom currents show, for example, that wind intensity or the frequency of storms in the Great Lakes region of North America were greater in the early Holocene than at present (Halfman & Johnson 1983; Figure 6). This is the first paleowind record reported for a large lake.

Seasonal winds over permanently stratified lakes, such as many of those in East Africa, can promote seasonal upwelling of nutrient-rich deep water

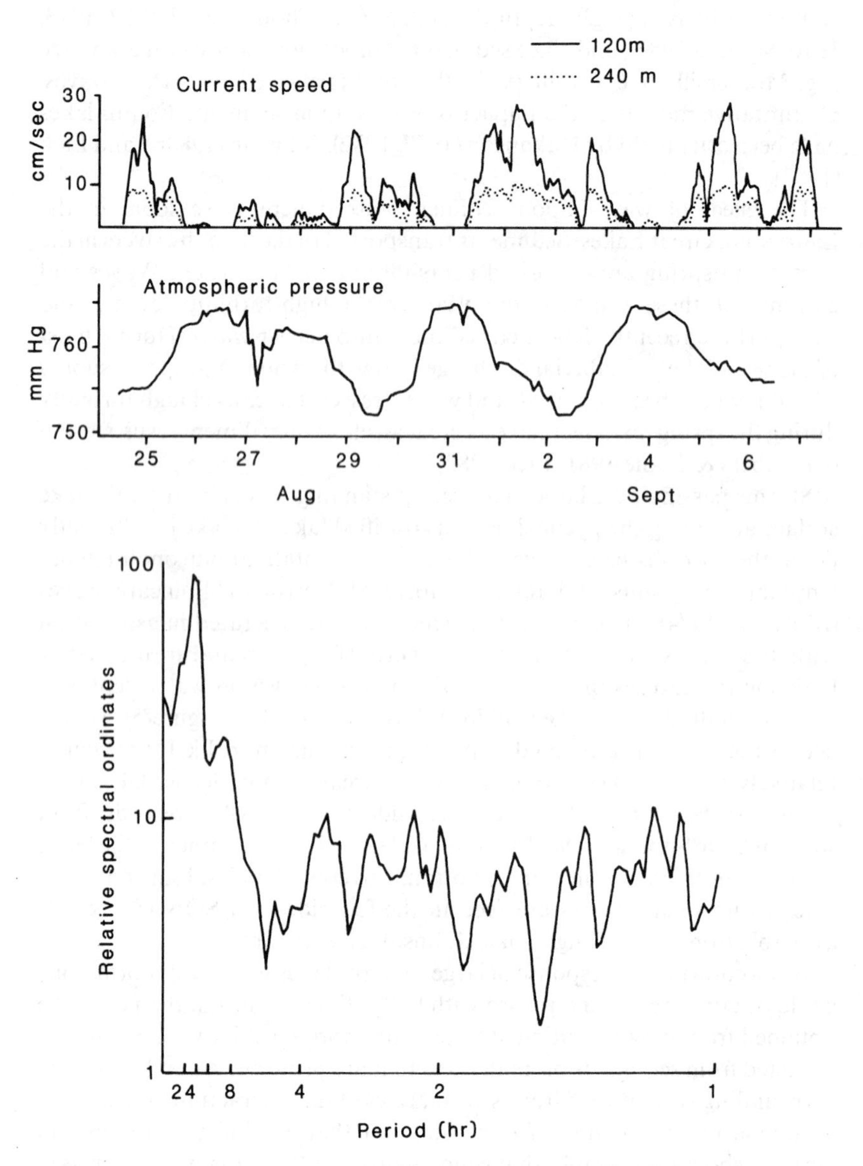

Figure 5 (*top*) Deep-water currents measured at 120 and 240-m depths in Lake Superior and atmospheric pressure measured at Grand Marais, Michigan, in a two-week period during the summer of 1978; (*bottom*) autocorrelation of the east-west current component at 120 m, showing the strong inertial period of about 16 hr (redrawn from Carlson 1982).

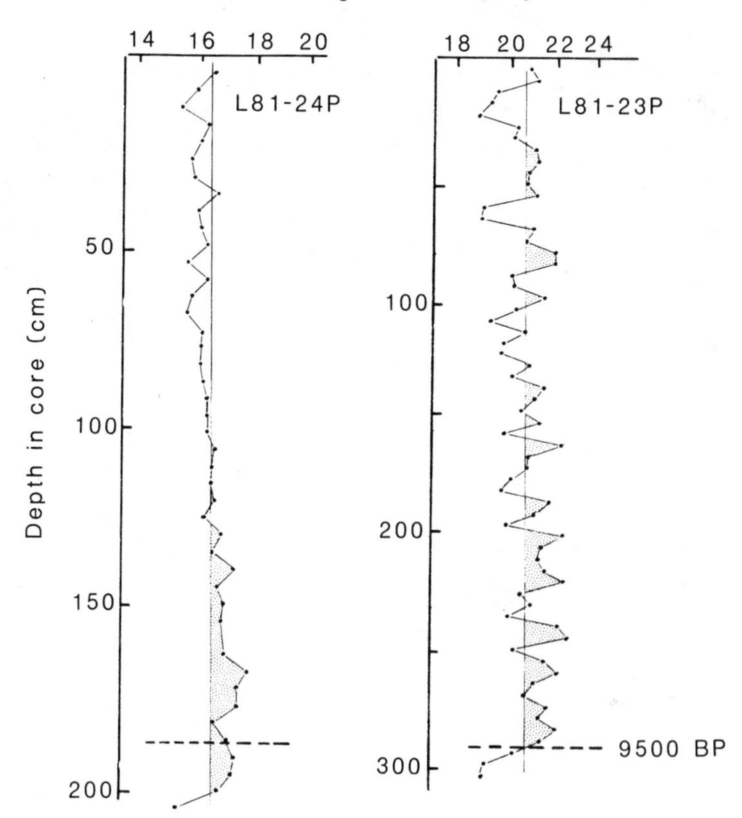

Figure 6 Median grain size of the silt fraction vs depth in Holocene sediment from Lake Superior, revealing evidence for stronger deep-water currents in the early than late Holocene. The dashed horizontal lines near the base of each core indicate the glacial-postglacial boundary at 9500 yr BP (redrawn from Halfman & Johnson 1983).

(Figure 7) and, consequently, algal blooms. This should cause seasonal variation in biogenic components to the sediments that may be discernable in annual laminations, similar to that described for Lake Zürich deposits by Kelts & Hsü (1978).

One aspect of sedimentation in large lakes that must be considered in paleolimnological investigations is the regional variability in surface sediment composition within a lake. This effect has been known since at least Caspari (1910), yet paleoclimatic interpretations still are based occasionally upon single-core analyses that do not fully consider how representative the core site is of the sampled basin. The distribution of diatoms in Lake Superior sediments, for example, is extremely patchy and is

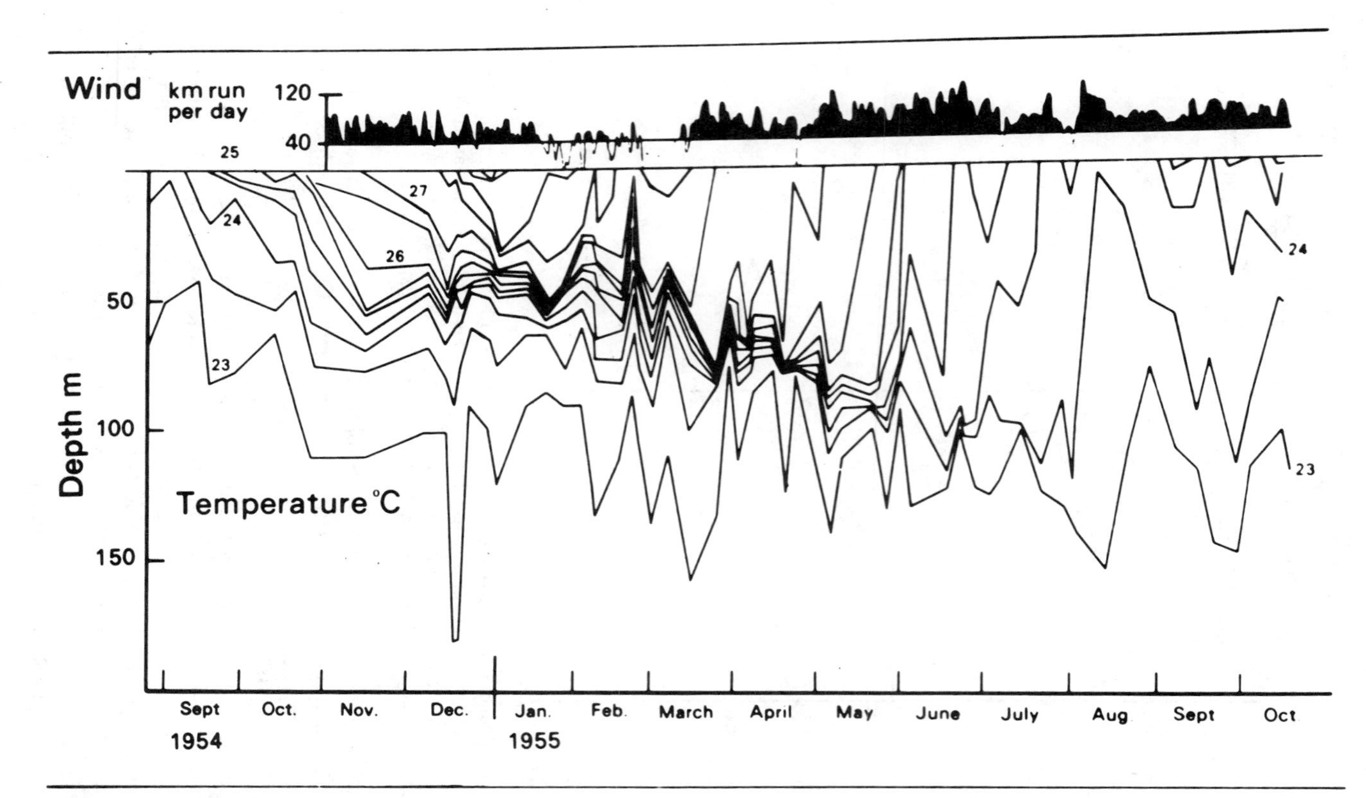

Figure 7 Seasonal isotherms (to depths of 200 m) and wind speed at Lake Malawi, Nkata Bay (after Jackson et al 1963). Reproduced with permission from Beadle (1981).

related as much to physical as to biological processes (Thayer et al 1983). Surface sediment mineralogy and core stratigraphy vary considerably in Lake Tanganyika (Degens et al 1971, Stoffers & Hecky 1978), Lake Kivu (Degens et al 1973, Stoffers & Hecky 1978), Lake Turkana (Yuretich 1979), Lake Malawi (G. Müller, personal communication) and Lake Victoria (Mothersill 1976). The same is true for the Laurentian Great Lakes and undoubtedly for other large lakes as well. This heterogeneous mosaic of surface sediments and stratigraphy requires that the sedimentary processes within a lake basin be examined in detail before any paleolimnological interpretations of one or more cores are made.

LARGE LAKES AS MODEL DEPOSITIONAL BASINS

The largest lakes on Earth are sufficiently large to respond more like ocean basins than like small lakes to the driving physical forces, and yet are sufficiently small to be measured adequately and economically for the various parameters affecting sedimentation within them. Thus large lakes are recognized as ideal model basins in which to study sedimentary processes, and have been since at least the time of Louis Ferdinand Comte de Marsili in the early 1700s [Pfannenstiel (1970) in Kelts (1978)].

The morphology of subaqueous canyons and their formation and/or maintenance by turbidity currents were first studied in alpine lakes by Forel (1885) and Heim (1876, 1888). Turbidity currents and turbidites have been studied extensively in Lake Mead, Nevada (Bell 1942, Gould 1951), Walensee, Switzerland (Lambert et al 1976), Lake Zürich (Kersey & Hsü 1976), and Lake Baikal (Karabanov 1982).

Perhaps the best-studied sublacustrine fan is that formed by iron ore tailings disposed continuously in Lake Superior between 1955 and 1980. Daily discharge averaged about 40,000 metric tons and formed a 20 km^2 fanlike feature that is incised with leveed fan valleys (Normark & Dickson 1976b). The growth pattern of the fan was determined from seismic reflection profiles and sediment core analyses, and was found to support the model for submarine-fan growth developed by Normark (1970, 1974, Normark & Dickson 1976b). Continuous measurement of bottom currents for 30 weeks over the sublacustrine fan recorded 25 episodes of turbidity currents. The currents were observed to pass through a hydraulic jump and achieved velocities consistent with speeds calculated by equations developed for submarine turbidity currents (Normark & Dickson 1976a).

Contourites, or interbedded sands and muds deposited under relatively fast clear-water currents, are deposited in water depths of about 250 m in southern Lake Superior (Johnson et al 1980). The lake-floor morphology, acoustic stratigraphy, and sediment stratigraphy within the region of the contourites were recently studied in detail by Flood & Johnson (1983), Halfman & Johnson (1983), and Johnson et al (1983a). An attractive aspect of these studies was that since biological activity was expected to be minimal, implying that physical processes would dominate the formation of sedimentary structures, it was anticipated that the sediments could be studied under relatively simple conditions. It was found, however, that unique regional factors in addition to bottom currents strongly affect sedimentation in this area, including complex ring-shaped depressions of glacial origin(?) on the lake floor (Flood & Johnson 1983) and an active community of sculpins that, through their nest making, form dish laminations that are the dominant sedimentary structures found in the sediments (Johnson et al 1983a). Nevertheless, the field of contourites in Lake Superior is easily accessible and well suited for continued analysis of the bottom currents that form these deposits.

Contourites have not been studied in other lakes but have been reported to exist in Lakes Malawi and Tanganyika (Rosendahl & Livingstone 1983; Figure 8), and deep-water currents form erosional channels in Lake Kivu (Wong & Von Herzen 1974). The currents forming contourites in the rift-valley lakes may be driven by a density differential rather than by wind, although this remains to be investigated.

Numerous other physical processes of sedimentation can be studied under relatively well-measured conditions in lakes; these include slope instability and slumping, delta development, and the effects of earthquakes and faulting upon subaqueous deposition. Many of these processes are most common in rift-valley lakes in remote regions that have not yet been studied in detail.

A general model of rift-valley lake sedimentation in a tropical setting was developed by LeFournier (1980; Figure 9) and related to premarine deposits of the West African continental margin. The model includes three major phases of sedimentation. In the first phase, thick sequences of alluvial conglomerates and arkoses are deposited in asymmetric subbasins bounded on one side by a major normal fault and on the other by a monoclinal flexure. These coarse detrital deposits interfinger with turbidites that contain freshwater ostracodes in the shaly intervals, signifying a lacustrine environment. The second stage of deposition is characterized by organic-rich marly muds that are deposited in an anoxic deep-water environment. These black sapropels may be laminated and locally diatomaceous, and the organic carbon content may exceed 10 wt%. In the third stage, green shales and fine sands are deposited, with sands becoming more frequent at the top. The series is covered by interbedded red shales and cross-bedded fluvial sands. This is a sequence of prograding deltaic sands and muds overlain by fluvial deposits, and it marks the final filling of the lake depression (LeFournier 1980).

LeFournier's (1980) model of deposition in a rift-valley lake is dependent upon climatic conditions that promote anoxia in the deep basins. The deep lakes of the East African Rift Valley are subjected to only minor seasonal temperature fluctuations (Griffiths 1972), and sediment-water reactions generate a slight salinity increase with depth that inhibits overturn (Beadle 1981). Lake Baikal, by contrast, experiences bi-annual overturn because of the large seasonal variation in air temperature, and it therefore has well-oxygenated deep water. The sequence of deposition in Lake Baikal should be the same as that described by LeFournier (1980) except that the organic-rich, laminated sapropel of the tropical lakes would be replaced by a nonlaminated, less organic mud.

A general stratigraphic model for a large glacial lake is considerably different than the rift-lake model. Deep drilling in Lake Superior in 1961 and 1962 (Zumberge & Gast 1961), coupled with coring and seismic reflection profiling in many of the Laurentian Great Lakes (e.g. Farrand

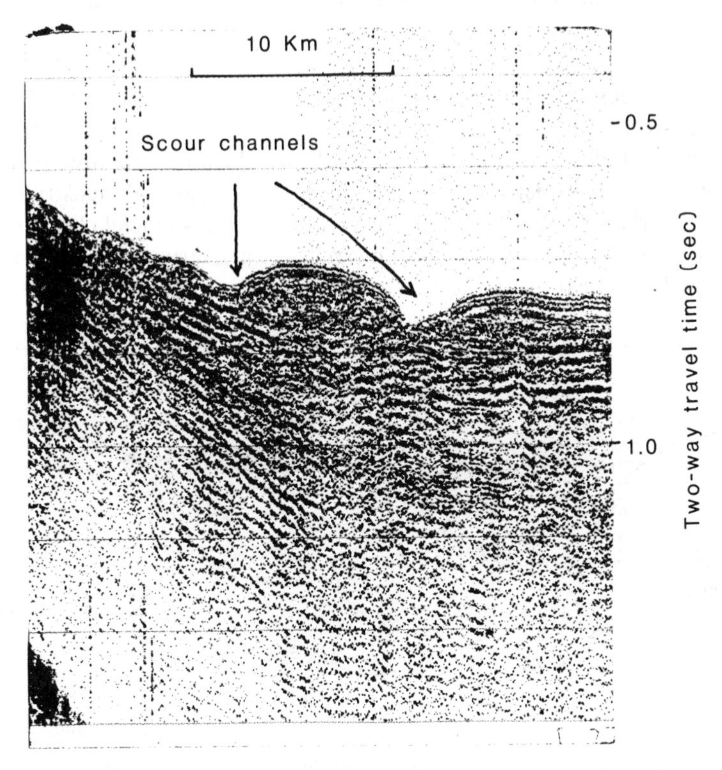

Figure 8 Seismic reflection profile from northern Lake Malawi, showing channels scoured by bottom currents and an interposed mound of possible contourites. Water depth is about 650 m. Profile obtained by B. R. Rosendahl.

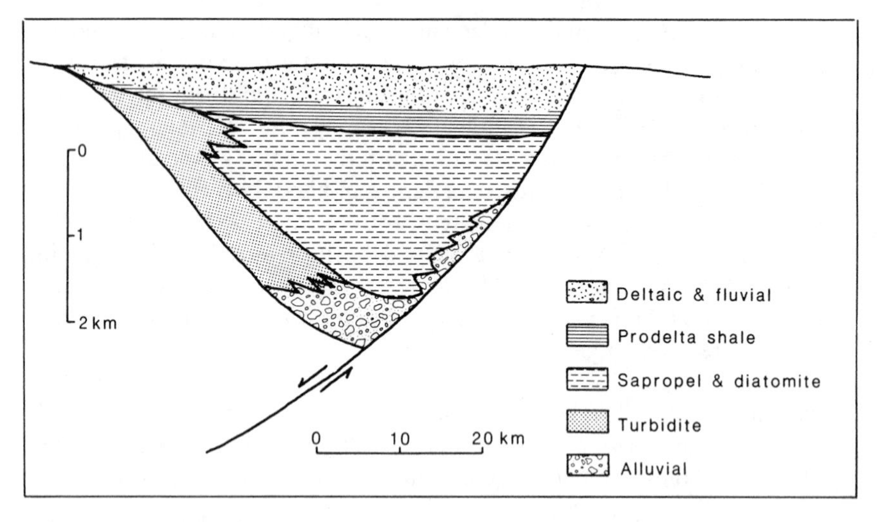

Figure 9 Schematic diagram of rift-valley lake stratigraphy, based upon model of LeFournier (1980).

1969, Lineback et al 1971, 1974, 1979, Dell 1972, Wickham et al 1978, Landmesser et al 1982), provides the data for such a model. Glacial deposits, including till and outwash sands, overlie an ice-scoured bedrock topography. These deposits are overlain by varved glacial-lacustrine clay that is overlain by homogeneous, brown or gray, postglacial clay. In time the lake basins will fill in, with organic-rich gyttja overlying the deep-water clays and with peat overlying the shallow-lacustrine gyttja (Figure 10). The

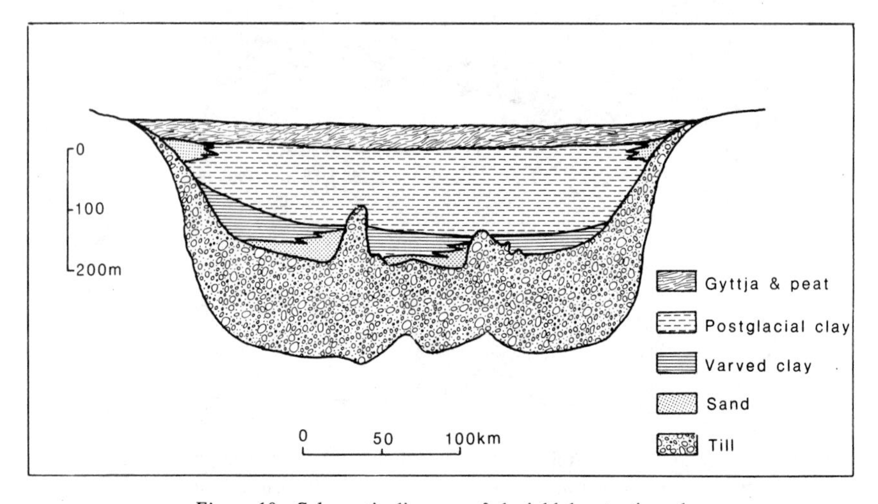

Figure 10 Schematic diagram of glacial lake stratigraphy.

thickness of glacial lake deposits in general cannot be expected to exceed a few hundred meters, whereas the thickness of tectonic lake deposits may be several kilometers.

Large lakes provide convenient model basins for examination of chemical as well as physical sedimentary processes. The most dramatic chemical events occur in saline lakes. Good summaries of saline brine evolution and evaporite mineral formation are provided by Eugster & Hardie (1978) and Eugster (1980), while carbonate deposition in lakes is nicely summarized by Kelts & Hsü (1978). Neither topic is reviewed here.

The question of silicate mineral authigenesis in freshwater lakes is an interesting one that is not resolved because of the difficulty of distinguishing between authigenic and detrital clay-sized silicates; nonetheless, it is a question of fundamental importance to our understanding of sediment-mineral interaction. Jones & Bowser (1978) cite several papers that document the detrital sources of silicate minerals in lake sediments, and they correctly state the lack of definitive evidence for silicate authigenesis (except for nontronite formation in Lake Malawi; Müller & Forstner 1973).

Kramer (1967) and Sutherland (1970) employed mineral-water equilibria concepts and calculations to suggest that trends in Great Lakes water composition were consistent with equilibrium with respect to kaolinite and Ca-smectite in the underlying sediment, which would imply that these minerals are authigenic.

Geochemical mass balances can provide valuable insight into the role of sediments in controlling lake-water composition. Mass-balance calculations are worth attempting in large lakes where water residence times exceed several decades, yet they have rarely been tried, even though the measurement of inputs and outputs can be made more precisely in a large lake than in the sea.

A geochemical mass balance for silica in Lake Superior by Johnson & Eisenreich (1979) is summarized in Figure 11. It reveals that the major inputs of silica to the water column are advection from rivers and diffusion from sediment pore waters. The annual biological uptake of silica for diatom frustule production is estimated to exceed the annual input from external sources by about one order of magnitude, and this dictates that about 90% or more of the diatom frustules formed dissolve and do not become preserved in the bottom sediments. Perhaps the most interesting result of the silica budget investigation is the realization that diatoms do not constitute the major sedimentary sink for silica and that therefore silicate authigenesis must be an important process (Johnson & Eisenreich 1979). The authigenic silicates have not been identified unequivocally, but possible candidates are X-ray-amorphous ferro-aluminosilicates (Nriagu 1978) and smectite (Johnson & Eisenreich 1979).

Schelske (1983) suggests that silica concentrations in Lake Superior may not be in a steady state, and that silicate authigenesis therefore may not be as important as Johnson & Eisenreich (1979) conclude. There are not enough historical data available to determine if steady-state conditions exist, but there is no a priori reason to believe they do not. So the question of silicate mineral authigenesis in freshwater lakes remains unanswered; it is perhaps the most interesting geochemical problem to be resolved in large lakes today.

Authigenic formation of iron and manganese oxides and iron phosphates is commonplace in sediments of several large lakes (e.g. Rossmann & Callender 1969, Müller & Forstner 1973, Robbins & Callender 1975, Dell 1973). Many of these oxides can be formed by early diagenesis in sediments that have strong gradients in interstitial-water pH and redox potential. In Lake Malawi, however, the occurrence of nontronite and limonite in sediment layers (sometimes as ooids) is attributed to leaching of basin sediment by geothermal spring water, followed by precipitation in oxic shallow water (Müller & Forstner 1973). This process is suggested by Müller & Forstner as a new model for the genesis of iron formations.

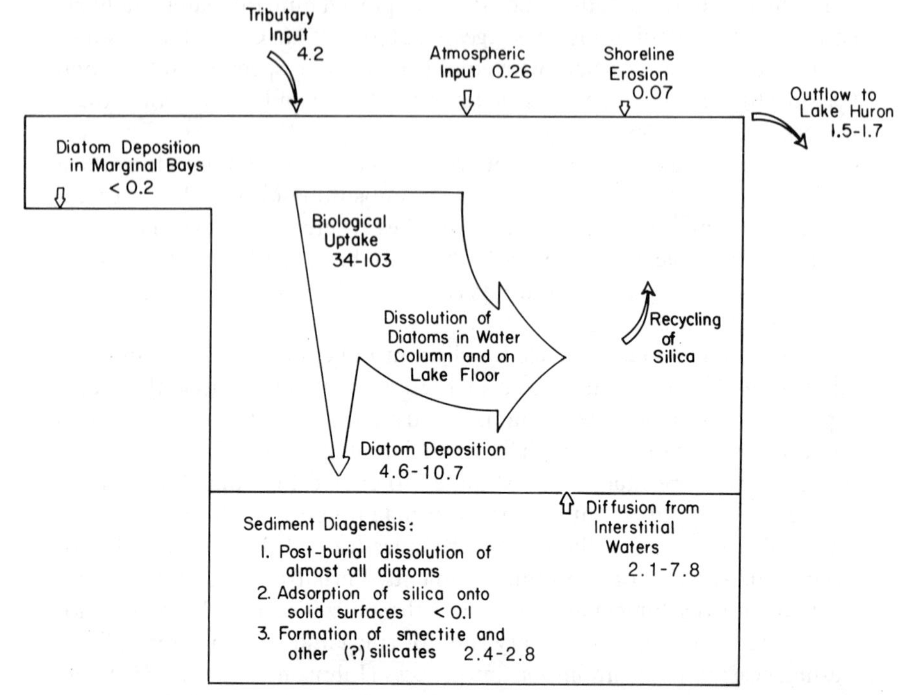

Figure 11 Silica budget for Lake Superior. All units of silica flux are in units of 10^8 kg SiO_2 yr^{-1}. Figure modified from Johnson & Eisenreich (1979).

Models of biological mixing of sediments were first developed for the oceans (e.g. Berger & Heath 1968, Guinasso & Schink 1975) and have been applied and refined for the Laurentian Great Lakes by the examination of such radionuclides as Cs-137, Pb-210, and isotopes of plutonium (Robbins et al 1977, 1979, Krezoski et al 1978, Robbins 1982). A quantitative steady-state mixing model is found to describe adequately the profiles of radioactivity and heavy metals in the sediments; the biologically homogenized zone varies usually from 1 to 5 cm in thickness and is consistent with the observed vertical distribution of oligochaete worms and amphipods (Robbins 1982). The depth of bioturbation has been mapped in great detail in southern Lake Huron (Robbins 1982). The mixing depth is greatest in the deepest areas of the basins and correlates with the annual influx of organic carbon. Studies such as this promise to provide valuable new insights into the biological impact upon sedimentation, which will have applications to marine as well as lacustrine depositional environments.

The fates of various anthropogenic contaminants are being investigated and modeled in the Laurentian Great Lakes. Most contaminants adsorb onto particles once they are introduced into a lake, and then are transported, deposited, perhaps resuspended and redeposited, and finally buried with the natural sediment. The anthropogenic substances usually are not inert as they travel with the sediment, however, for they come into contact with filter and detritus feeders, bacteria, and other biota. Biochemical reactions occur that may strip the contaminants from the sediment, remobilize them in solution, or incorporate them into the food chain.

Concentrations of PCBs and organochlorine pesticides have been measured in over 1200 samples of surface sediment from Lakes Erie, Ontario, Huron, Michigan, and Superior to establish baseline levels and to identify sources (Frank et al 1977, 1979a,b, 1980, 1981). Concentrations are higher in depositional than in nondepositional zones, and are also higher in Lakes Erie and Ontario than in the upper Great Lakes.

Historical trends of PCB deposition are being determined from down-core analysis of PCBs and Pb-210 (Eisenreich & Johnson 1983). Box cores frequently have lower PCB concentrations in the upper 0–0.25 cm of sediment than in the 0.25–0.5 cm segment. This phenomenon is due either to a decreased influx of PCBs in recent years or to a sampling problem (Eisenreich et al 1980, Eisenreich & Johnson 1983). PCBs were first introduced to Lake Superior sediments in the 1940s, and their appearance in older sediment is attributed to bioturbation.

It is difficult to reconcile the vertical distributions of PCBs and Pb-210 in the same sediment cores. There is little or no correlation between the thickness of the biologically homogenized zone as determined by Pb-210 analysis and the maximum depth of measurable PCBs. In some cores the layer of relatively uniform PCB concentration is thicker than that of uniform Pb-210 activity, and in other cores the opposite is the case (Eisenreich et al 1980). Differences in these profiles can be attributed to differences in input rates and chemical behavior of PCBs and Pb-210 in lake-floor sediments. Both PCBs and Pb-210 are derived primarily from atmospheric sources (Bruland et al 1975, Eisenreich et al 1981), but the Pb-210 input rate has remained relatively constant while the PCB input rate has varied considerably during the past few decades. Pb-210 decays with a half life of about 22 years, whereas PCBs exhibit greater stability. PCBs may be extracted from the sediment by benthic organisms to a greater degree than Pb-210 is, and the chemical remobilization and diffusion processes certainly are different for the two species.

Quantitative models of the behavior of PCBs and other contaminants in Great Lakes sediments are being developed, but they depend upon new information concerning historical loadings (e.g. Eisenreich & Johnson 1983), residence time in the water column (e.g. Bierman & Swain 1982, Eisenreich & Johnson 1983), bioturbation (e.g. Robbins 1982), and degradation processes (e.g. Ballschmiter et al 1980). These models are crucial to our ability to predict better the degree and duration of environmental degradation as new contaminants appear or are recognized within the world's great lakes.

CONCLUSIONS

The large lakes on Earth are of glacial or tectonic origin. The glacial lakes are concentrated along zones of maximum glacial scour between Precambrian shield rocks to the north and Paleozoic and younger strata to the south. Tectonic lakes are formed primarily in continental rift zones, but they also are found along convergent and transform plate margins. The large rift-valley lakes are much older than glacial lakes and contain thicker sediments within their basins.

Physical sedimentary processes form a complex distribution of sediment texture and composition on the floors of large lakes. Sediment focusing by waves and currents creates lag deposits of coarse-grained sediment in high-energy environments, and chaff deposits of fine-grained, organic-rich sediment in protected basins. Sediment mineralogy, chemical composition, and fossil content are all affected by the focusing process, even in water depths of several hundred meters. Slumping and turbidity currents are locally important in areas of high sediment input and tectonic activity, and regions of high primary productivity are affected by winds, currents, and availability of nutrients.

Chemical sedimentary processes affect both lake-water composition and sediment mineralogy to varying degrees. Large lakes have water residence times of decades to centuries, so chemical reactions between sediments and water have more impact upon their composition than is the case in small lakes. Closed-basin lakes can become quite saline and precipitate a host of carbonate, zeolite, and evaporite minerals. Open-basin lakes have fresher water but still provide adequate conditions for authigenesis of various oxides, phosphates, silicates, and other mineral groups. Anoxic conditions are common in several deep rift-valley lakes in East Africa, and these promote formation of various sulfides within the deep anoxic waters, and oxides and silicates where the chemocline intersects the lake floor.

Sediment compositional response to physical and chemical processes results in the deposition of a unique historical record in the basins of large lakes. Sedimentation rates are high, and bioturbation is frequently absent or minimal, so the record can be resolved to time spans of decades or less. Variability in wind intensity can strongly affect sediment texture in high-energy environments and biogenic content in regions of upwelling. Rainfall variation changes terrigenous input to a lake and in some cases strongly affects lake-water salinity and authigenic mineral formation. Tectonic activity influences the abundance of ash layers, mass wasting, and at times even the hydrological regime of a large lake. Many of these processes affect small lakes as well, but large lakes are more likely to reflect large-scale, regional processes rather than just local phenomena, and in the case of tectonic lakes, they are more apt to contain a long, continuous sediment record spanning millions of years.

Comprehensive models can be developed for sedimentation in large lakes because the variables involved can be measured with relative ease compared with oceanic measurement. This has been demonstrated most successfully for the study of turbidity currents and their depositional products in North American and alpine lakes. Models are also being developed for sediment entrainment by waves, bioturbation, and other physical processes; and geochemical mass balances for silica and various microcontaminants are being refined. The results of these advances will have applications not only to lakes but to the oceans as well.

Acknowledgments

I thank S. J. Eisenreich, R. E. Hecky, J. LeFournier, D. A. Livingstone, B. R. Rosendahl, and H. J. Schrader for thoughtful discussions during the past year that have influenced my ideas on large lakes. Funding for my research on Lake Superior during the past six years has been provided by the Environmental Protection Agency, the National Science Foundation, and the Minnesota Sea Grant Program. My introduction to field work on the East African lakes has been funded by Project Probe, under the direction of B. R. Rosendahl at Duke University.

Literature Cited

Adam, D. P. 1983. *Pollen zonation and informal climatic units for Clear Lake, California, cores CL-73-4 and CL-73-7.* Presented at Ann. Meet. Cordilleran Sect. Geol. Soc. Am., 79th, Salt Lake City

Ballschmiter, K., Zell, M., Neu, H. J. 1978. PCBs: will they never degrade? *Chemosphere* 1:173

Barnes, M. A., Barnes, W. C. 1978. Organic compounds in lake sediments. See Lerman 1978, pp. 127–52

Beadle, L. C. 1981. *The Inland Waters of Tropical Africa.* London/New York: Longman. 475 pp. 2nd ed.

Bell, H. S. 1942. Density currents as agents for transporting sediment. *J. Geol.* 50: 512–47

Berger, W. H., Heath, G. R. 1968. Vertical mixing in pelagic sediments. *J. Mar. Res.* 26:134–43

Bierman, V. J. Jr., Swain, W. R. 1982. Mass balance modeling of DDT dynamics in Lakes Michigan and Superior. *Environ. Sci. Technol.* 16:572–79

Bruland, K. W., Koide, M., Bowser, C., Goldberg, E. D. 1975. Lead-210 and pollen geochronologies in Lake Superior sediments. *Quat. Res.* 5:89–98

Cahill, R. A. 1981. *Geochemistry of Recent Lake Michigan Sediments.* Champaign: Ill. Inst. Nat. Res., State Geol. Surv. Div. Circ. 517. 94 pp.

Capart, A. 1949. Sondages et carte bathymetrique. In *Exploration Hydrobiologique du Lac Tanganyika (1946–1947)*, 2:1–16. Brussels: Inst. R. Sci. Nat. Belg.

Carlson, T. W. 1982. *Deep-water currents and their effect on sedimentation in Lake Superior.* PhD thesis. Univ. Minn., Minneapolis. 174 pp.

Caspari, W. A. 1910. The deposits of the Scottish fresh-water lochs. In *Bathymetrical Survey of the Scottish Fresh-Water Lochs*, ed. J. Murray, L. Pullar, 1:261–74. Edinburgh: Challenger Off.

Cerling, T. E. 1979. Paleochemistry of Plio-Pleistocene Lake Turkana, Kenya. *Palaeogeogr. Palaeoclimatol. Palaeoecol.* 27:247–85

Chambers, R. L., Eadie, B. J. 1981. Nepheloid and suspended particulate matter in southeastern Lake Michigan. *Sedimentology* 28:439–47

Creer, K. M., Tucholka, P. 1982. Construction of type curves of geomagnetic secular variation for dating lake sediments from east central North America. *Can. J. Earth Sci.* 19:1106–15

Creer, K. M., Anderson, T. W., Lewis, C. F. M. 1976. Late Quaternary geomagnetic stratigraphy recorded in Lake Erie sediments. *Earth Planet. Sci. Lett.* 31:37–47

Davis, W. M. 1882. On the classification of lake basins. *Proc. Boston Soc. Nat. Hist.* 21:315–81

Davis, W. M. 1887. On the classification of lake basins. *Science* 10:142

Degens, E. T., von Herzen, R. P., Wong, H. K. 1971. Lake Tanganyika: water chemistry, sediments, geological structure. *Naturwissenschaften* 58:229–41

Degens, E. T., von Herzen, R. P., Wong, H. K., Deuser, W. G., Jannasch, H. W. 1973. Lake Kivu: structure, chemistry and biology of an East African rift lake. *Geol. Rundsch.* 62:245–77

Dell, C. I. 1972. The origin and characteristics of Lake Superior sediments. *Proc. Great Lakes Res. Conf., 15th*, pp. 361–70

Dell, C. I. 1973. Vivianite—an authigenic phosphate mineral in Great Lake sediments. *Proc. Conf. Great Lakes Res., 16th*, pp. 1027–28

Dell, C. I. 1976. Sediment distribution and bottom topography of southeastern Lake Superior. *J. Great Lakes Res.* 2:164–76

Dewey, J. F. 1972. Plate tectonics. *Sci. Am.* 226:34–45

Eisenreich, S. J., Johnson, T. C. 1983. PCBs in the Great Lakes: sources, sinks, burdens. In *PCBs: Human and Environmental Hazards*, ed. F. J. D'Itri, M. A. Kamian, pp. 49–76. Boston: Butterworth

Eisenreich, S. J., Hollod, G. J., Johnson, T. C., Evans, J. E. 1980. PCB and other microcontaminant-sediment interactions in Lake Superior. In *Contaminants and Sediments*, ed. R. A. Baker, 1:67–94. Ann Arbor, Mich: Ann Arbor Sci. Publ.

Eisenreich, S. J., Looney, B. B., Thornton, J. D. 1981. Airborne organic contaminants in the Great Lakes ecosystem. *Environ. Sci. Technol.* 15:30–38

Eugster, H. P. 1980. Geochemistry of evaporitic lacustrine deposits. *Ann. Rev. Earth Planet. Sci.* 8:35–63

Eugster, H., Hardie, L. A. 1978. Saline lakes. See Lerman 1978, pp. 237–94

Evans, J. E., Johnson, T. C., Alexander, E. C. Jr., Lively, R. S., Eisenreich, S. J. 1981. Sedimentation rates and depositional processes in Lake Superior from Pb-210 geochronology. *J. Great Lakes Res.* 7:299–310

Farrand, W. R. 1969. The Quaternary history of Lake Superior. *Proc. Conf. Great Lakes Res. 12th*, pp. 181–97

Fitch, F. J., Miller, J. A. 1976. Conventional potassium-argon and argon-40/argon-39 dating of volcanic rocks from East Rudolf. In *Earliest Man and Environments in the Lake Rudolf Basin*, ed. Y. Coppens, F. C. Howell, G. L. Isaac, R. E. F. Leakey, 1:123–47. Univ. Chicago Press. 615 pp.

Flood, R. D., Johnson, T. C. 1983. Side-scan SONAR targets in Lake Superior—evidence for current transport of bottom sediments. *Sedimentology*. In press

Forel, F. A. 1885. Le ravin sous-lacoustre des fleuves glaciaires. *C. R. Acad. Sci. Paris* 101:725–28

Frank, R., Holdrinet, M., Braun, H. E., Thomas, R. L., Kemp, A. L. W., Jaquet, J. M. 1977. Organochlorine insecticides and PCBs in sediment of Lake St. Clair (1970 and 1974) and Lake Erie (1971). *Sci. Total Environ.* 8:205–27

Frank, R., Thomas, R. L., Holdrinet, M., Kemp, A. L. W., Braun, H. E. 1979a. Organochlorine insecticides and PCB in surficial sediments (1968) and sediment cores (1976) from Lake Ontario. *J. Great Lakes Res.* 5:18–27

Frank, R., Thomas, R. L., Holdrinet, M., Kemp, A. L. W., Dawson, R. 1979b. Organochlorine insecticides and PCB in the sediments of Lake Huron (1969) and Georgian Bay and North Channel (1973). *Sci. Total Environ.* 13:101–17

Frank, R., Holdrinet, M., Braun, R. L., Rasper, J., Dawson, R. 1980. Organochlorine insecticides and PCB in surficial sediments of Lake Superior (1973). *J. Great Lakes Res.* 6:113–20

Frank, R., Holdrinet, M., Gross, D. L., Davies, T. T. 1981. Organochlorine insecticides and PCB in surficial sediments of Lake Michigan (1975). *J. Great Lakes Res.* 7:42–50

Galazy, G. I., Belova, V. A., Loot, B. F., Zolotarev, A. G., eds. 1982. *Particularities of Formation of Terrigenic Sediments in the Baikal Depression, Int. Union Quat. Res. Congr., 11th, Moscow*. 35 pp.

Gould, H. R. 1951. Some quantitative aspects of Lake Mead turbidity currents. *Spec. Publ. Soc. Econ. Paleontol. Mineral.* 2:34–52

Griffiths, J. F. 1972. *Climates of Africa*. Amsterdam/London/New York: Elsevier. 604 pp.

Guinasso, N. L., Schink, D. R. 1975. Quantitative estimates of biological mixing rates in abyssal sediments. *J. Geophys. Res.* 80:3032–43

Håkanson, L. 1977. The influence of wind, fetch, and water depth on the distribution of sediments in Lake Vanern, Sweden. *Can. J. Earth Sci.* 14:397–412

Håkanson, L. 1982. Bottom dynamics in lakes. *Hydrobiologia* 91:9–22

Halfman, J. D., Johnson, T. C. 1983. The sediment texture of contourites in Lake Superior. *Geol. Soc. London Spec. Publ.* In press

Hastenrath, S., Kutzbach, J. E. 1983. Paleoclimatic estimates from water and energy budgets of East African lakes. *Quat. Res.* 19:141–53

Hecky, R. E. 1978. The Kivu-Tangangyika basin: the last 14,000 years. *Pol. Arch. Hydrobiol.* 25:159–65

Hecky, R. E., Degens, E. T. 1973. Late Pleistocene-Holocene chemical stratigraphy and paleolimnology of the Rift Valley lakes of central Africa. *Woods Hole Oceanogr. Inst. Tech. Rep. WHOI 73-28*. Unpublished manuscript.

Heim, A. 1876. Bericht und Expertengutachten uber die im Februar und September 1875 in Horgen vorgekommen Rutschungen. *Bericht der Expertenkommission*. Zurich: Hofer und Burgen. 22 pp.

Heim, A. 1888. Die Catastrophe von Zug 5. Juli 1887. *Gutachten der Experten*. Zurich: Hofer und Burgen. 57 pp.

Herdendorf, C. E. 1982. Large lakes of the world. *J. Great Lakes Res.* 8:379–412

Heuschele, A. 1982. Vertical distribution of profundal benthos in Lake Superior sediments. *J. Great Lakes Res.* 8:603–13

Hinze, W. J., Wold, R. J. 1982. Lake Superior geology and tectonics—overview and major unsolved problems. In *Geology and Tectonics of the Lake Superior Basin*, ed. R. J. Wold, W. J. Hinze, *Geol. Soc. Am. Mem.* 156:273–80

Hoffman, P., Dewey, J. F., Burke, K. 1974. Aulacogens and their genetic relation to geosynclines, with a Proterozoic example from Great Slave Lake, Canada. *Soc. Econ. Paleontol. Mineral. Spec. Publ.* 19:38–55

Hutchinson, G. E. 1957. *A Treatise on Limnology: Geography, Physics and Chemistry*, Vol. 1. New York/London/Sydney/Tokyo: Wiley. 540 pp.

Jackson, P. B. N., Iles, T. D., Harding, D., Fryer, G. 1963. *Report on a Survey of Northern Lake Nyasa by the Joint Fisheries Research Organization, 1953–1955*. Zomba: Malawi Gov. Printer

Johnson, T. C. 1980a. Late-glacial and postglacial sedimentation in Lake

Superior based on seismic-reflection profiles. *Quat. Res.* 13:380–91

Johnson, T. C. 1980b. Sediment redistribution by waves in lakes, reservoirs and embayments. *Proc. Symp. Surf. Water Impoundments*, ed. H. Stefan, pp. 1307–17. New York: Am. Soc. Civil Eng.

Johnson, T. C., Eisenreich, S. J. 1979. Silica in Lake Superior: mass balance considerations and a model for dynamic response to eutrophication. *Geochim. Cosmochim. Acta* 43:77–91

Johnson, T. C., Fields, J. 1983. Paleomagnetic dating of postglacial sediment, offshore Lake Superior. *Chem. Geol.* In press

Johnson, T. C., Carlson, T. W., Evans, J. E. 1980. Contourites in Lake Superior. *Geology* 8:437–41

Johnson, T. C., Evans, J. E., Eisenreich, S. J. 1982. Total organic carbon in Lake Superior sediments: comparisons with hemipelagic and pelagic marine environments. *Limnol. Oceanogr.* 27:481–91

Johnson, T. C., Halfman, J. D., Busch, W. H., Flood, R. D. 1983a. Effects of bottom currents and fish on sedimentation in a deep-water, lacustrine environment. *Geol. Soc. Amer. Bull.* Submitted for publication

Johnson, T. C., Rosendahl, B. R., Halfman, J. D. 1983b. *Comparison of side-scan SONAR records from Lake Superior and Lake Tanganyika.* Presented at Ann. Meet. Int. Assoc. Great Lakes Res., 26th, Oswego, N.Y.

Jones, B. F., Bowser, C. J. 1978. The mineralogy and related chemistry of lake sediments. See Lerman 1978, pp. 179–236

Karabanov, E. B. 1982. The role of suspension flows in formation of bottom sediments in Lake Baikal. See Galazy et al 1982, pp. 8–12

Kelts, K. 1978. *Geological and sedimentary evolution of Lakes Zurich and Zug, Switzerland.* PhD thesis. Eidg. Tech. Hochsch., Zurich. 250 pp.

Kelts, K., Hsü, K. J. 1978. Freshwater carbonate sedimentation. See Lerman 1978, pp. 295–324

Kemp, A. L. W., Johnston, L. M. 1979. Diagenesis of organic matter in the sediments of Lakes Ontario, Erie and Huron. *J. Great Lakes Res.* 5:1–10

Kemp, A. L. W., Anderson, T. W., Thomas, R. L., Mudrochova, A. 1974. Sedimentation rates and recent sediment history of Lakes Ontario, Erie and Huron. *J. Sediment Petrol.* 44:207–18

Kemp, A. L. W., MacInnes, G. A., Harper, N. S. 1977. Sedimentation rates and a revised sediment budget for Lake Erie. *J. Great Lakes Res.* 3:221–33

Kemp, A. L. W., Dell, C. I., Harper, N. S. 1978. Sedimentation rates and a sediment budget for Lake Superior. *J. Great Lakes Res.* 4:276–87

Kendall, R. L. 1969. An ecological history of the Lake Victoria basin. *Ecol. Monogr.* 39:121–76

Kersey, D. G., Hsü, K. J. 1976. Energy relations of density-current flows: an experimental investigation. *Sedimentology* 23:761–89

Kramer, J. R. 1968. Mineral-water equilibria in silicate weathering. *Int. Geol. Congr., 23rd*, 6:149–60

Krezoski, J. K., Mozley, S. C., Robbins, J. A. 1978. Influence of benthic macroinvertebrates on mixing of profundal sediments in southeastern Lake Huron. *Limnol. Oceanogr.* 23:1011–16

Lambert, A., Kelts, K., Marshall, N. 1976. Measurements of density underflows from Walensee, Switzerland. *Sedimentology* 23:87–105

Landmesser, C. W., Johnson, T. C., Wold, R. J. 1982. Seismic reflection study of recessional moraines beneath Lake Superior and their relationship to regional deglaciation. *Quat. Res.* 17:173–90

LeFournier, J. 1980. Dépôts de preouverture de l'Atlantique Sud. Comparaison avec la sedimentation actuelle dans la branche occidentale des Rifts Est-Africains. In *Recherches Geologiques en Afrique*, 5:127–30

Lerman, A., ed. 1978. *Lakes: Chemistry, Geology, Physics.* New York/Heidelberg/Berlin: Springer-Verlag. 363 pp.

Lick, W. 1982. Entrainment, deposition and transport of fine-grained sediments in lakes. *Hydrobiologia* 91:31–40

Lineback, J. A., Gross, D. L., Meyer, R. P., Unger, W. L. 1971. High-resolution seismic profiles and gravity cores in southern Lake Michigan. *Environ. Geol. Notes, Ill. State Geol. Surv. 47.* 41 pp.

Lineback, J. A., Gross, D. L., Meyer, R. P. 1974. Glacial tills under Lake Michigan. *Environ. Geol. Notes, Ill. Geol. Surv. 69.* 48 pp.

Lineback, J. A., Gross, D. L., Dell, C. I. 1979. Glacial and postglacial sediments in Lakes Superior and Michigan. *Geol. Soc. Am. Bull.* 90:781–91

Livingstone, D. A. 1975. Late Quaternary climatic change in Africa. *Ann. Rev. Ecol. Syst.* 6:249–80

Livingstone, D. A., Van der Hammen, T. 1978. Palaeogeography and palaeoclimatology. *Nat. Resour. Res.* 14:61–90

Meyers, P. A., Bourbonniere, R. A., Takeuchi, N. 1980. Hydrocarbons and fatty acids in two cores of Lake Huron sediments. *Geochim. Cosmochim. Acta* 44:1215–21

Mizandrontsev, I. B. 1982. The rates of

sedimentation in the Baikal basin. See Galazy et al 1982, pp. 12–16

Mortimer, C. H. 1974. Lake hydrodynamics. *Mitt. Int. Ver. Theor. Angew. Limnol.* 20:124–97

Mothersill, J. S. 1976. The mineralogy and geochemistry of the sediments of northwestern Lake Victoria. *Sedimentology* 23:553–65

Mothersill, J. S. 1979. The paleomagnetic record of the late Quaternary sediments of Thunder Bay. *Can. J. Earth Sci.* 16:1016–23

Mothersill, J. S. 1981. Late Quaternary paleomagnetic record of the Goderich basin, Lake Huron. *Can. J. Earth Sci.* 18:448–56

Müller, G., Forstner, U. 1973. Recent iron ore formation in Lake Malawi, Africa. *Miner. Deposita* 8:278–90

Murray, J. 1910. The characteristics of lakes in general, and their distribution over the surface of the globe. In *Bathymetric Survey of the Scottish Fresh-Water Lochs*, ed. J. Murray, L. Pullar, 1:514–658. Edinburgh: Challenger Off. 658 pp.

Normark, W. R. 1970. Growth patterns of deep-sea fans. *Bull. Am. Assoc. Pet. Geol.* 54:2170–95

Normark, W. R. 1974. Submarine canyons and fan valleys: factors affecting growth patterns of deep-sea fans. *Soc. Econ. Paleontol. Mineral. Spec. Publ.* 19:56–68

Normark, W. R., Dickson, F. H. 1976a. Manmade turbidity currents in Lake Superior. *Sedimentology* 23:815–31

Normark, W. R., Dickson, F. H. 1976b. Sublacustrine fan morphology in Lake Superior. *Bull. Am. Assoc. Pet. Geol.* 60:1021–36

Nriagu, J. O. 1978. Dissolved silica in pore waters of lakes Ontario, Erie, and Superior sediments. *Limnol. Oceanogr.* 23:53–67

Penck, A. 1894. *Morphologie der Erdoberflache*, Vols. 1, 2. Stuttgart: Engelhorn. 471 pp., 696 pp.

Pfannenstiel, M. 1970. Das Meer in der Geschichte der Geologie. *Geol. Rundsch.* 60:3–72

Pike, J. G., Rimmington, G. T. 1965. *Malawi: A Geographical Study*. Oxford Univ. Press. 229 pp.

Rea, D. K., Bourbonniere, R. A., Meyers, P. A. 1980. Southern Lake Michigan sediments: changes in accumulation rate, mineralogy and organic content. *J. Great Lakes Res.* 6:321–30

Rea, D. K., Owen, R. M., Meyers, P. A. 1981. Sedimentary processes in the Great Lakes. *Rev. Geophys. Space Phys.* 19:635–48

Robbins, J. A. 1980. Sediments of southern Lake Huron: elemental composition and accumulation rates. *Rep. 600/3-80-080*, US Environ. Prot. Agency, Duluth, Minn. 307 pp.

Robbins, J. A. 1982. Stratigraphic and dynamic effects of sediment reworking by Great Lakes zoobenthos. *Hydrobiologia* 92:611–22

Robbins, J. A., Callender, E. 1975. Diagenesis of manganese in Lake Michigan sediments. *Am. J. Sci.* 275:512–33

Robbins, J. A., Edgington, D. N. 1975. Determination of recent sedimentation rates in Lake Michigan using Pb-210 and Cs-137. *Geochim. Cosmochim. Acta* 39:285–304

Robbins, J. A., Krezoski, J. R., Mozley, S. C. 1977. Radioactivity in sediments of the Great Lakes: post-depositional redistribution by deposit-feeding organisms. *Earth Planet. Sci. Lett.* 36:325–33

Robbins, J. A., Edgington, D. A., Kemp, A. L. W. 1978. Comparative lead-210, cesium-137 and pollen and geochronologies of sediments from Lakes Ontario and Erie. *Quat. Res.* 10:256–78

Robbins, J. A., McCall, P. L., Fisher, J. B., Krezoski, J. R. 1979. Effect of deposit feeders on migration of cesium-137 in lake sediments. *Earth Planet. Sci. Lett.* 42:277–87

Rosendahl, B. R., Livingstone, D. A. 1983. Rift lakes of East Africa: new seismic data and implications for future research. *Episodes* 1983:14–19

Rossmann, R., Callender, E. 1969. Geochemistry of Lake Michigan manganese nodules. *Proc. Conf. Great Lakes Res., 12th*, pp. 306–16

Russel, I. C. 1895. *Lakes of North America; A Reading Lesson for Students of Geography and Geology*. Boston/London: Ginn. 125 pp.

Schelske, C. L. 1983. Geochemical silica mass balances in Lake Michigan and Lake Superior. Unpublished manuscript

Sims, J. D., Rymer, M. J., Perkins, J. A. 1983. *Late Quaternary stratigraphy and paleolimnology, Clear Lake, California*. Presented at Ann. Meet. Cordilleran Sect. Geol. Soc. Am., 79th, Salt Lake City

Sly, P. G., Thomas, R. L., Pelletier, B. R. 1983. Interpretation of moment measures derived from water-lain sediments. *Sedimentology* 30:219–34

Stoffers, P., Hecky, R. E. 1978. Late Pleistocene-Holocene evolution of the Kivu-Tanganyika basin. *Spec. Publ. Int. Assoc. Sedimentol.* 2:43–55

Sugden, D. E. 1978. Glacial erosion by the Laurentide ice sheet. *J. Glaciol.* 20:367–91

Sutherland, J. C. 1970. Silicate mineral stability and mineral equilibria in the Great Lakes. *Environ. Sci. Technol.* 4:826–33

Thayer, V. L., Johnson, T. C., Schrader, H. J. 1983. The distribution of diatoms in Lake Superior sediments. *J. Great Lakes Res.* 9:497–507

Thomas, R. L., Kemp, A. L. W., Lewis, C. F. M. 1972. Distribution, composition and characteristics of the surficial sediments of Lake Ontario. *J. Sediment. Petrol.* 42:66–84

Thomas, R. L., Kemp, A. L. W., Lewis, C. F. M. 1973. The surficial sediments of Lake Huron. *Can. J. Earth Sci.* 10:226–71

Thomas, R. L., Jaquet, J. M., Kemp, A. L. W., Lewis, C. F. M. 1976. Surficial sediments of Lake Erie. *J. Fish. Res. Board Can.* 33:385–403

White, W. A. 1972. Deep erosion by continental ice sheets. *Geol. Soc. Am. Bull.* 83:1037–56

Wickham, J. T., Gross, D. L., Lineback, J. A., Thomas, R. L. 1978. Late Quaternary sediments of Lake Michigan. *Environ. Geol. Notes, Ill. State Geol. Surv. 84.* 26 pp.

Wold, R. J., Hutchinson, D. R., Johnson, T. C. 1982. Topography and surficial structure of Lake Superior bedrock as based on seismic reflection profiles. In *Geology and Tectonics of the Lake Superior Basin*, ed. R. J. Wold, W. J. Hinze, *Geol. Soc. Am. Mem.* 156:257–72

Wong, H. K., Von Herzen, R. P. 1974. A geophysical study of Lake Kivu, East Africa. *Geophys. J. R. Astron. Soc.* 37:371–89

Yamamoto, A. 1976. Paleoprecipitational change estimated from the grain size variations in the 200 m-long core from Lake Biwa. In *Paleolimnology of Lake Biwa and the Japanese Pleistocene*, ed. S. Horie, 4:179–203

Yuretich, R. F. 1979. Modern sediments and sedimentary processes in Lake Rudolf (Lake Turkana) Eastern Rift Valley, Kenya. *Sedimentology* 26:313–31

Zumberge, J. H., Gast, P. 1961. Geological investigations in Lake Superior. *Geotimes* 6:10–13

Ann. Rev. Earth Planet. Sci. 1984. 12: 205–43

PRE-QUATERNARY SEA-LEVEL CHANGES

Anthony Hallam

Department of Geological Sciences, University of Birmingham, Birmingham B15 2TT, England

INTRODUCTION

While the paramount importance of global sea-level changes has been widely accepted by Quaternary geologists throughout this century, it is only within the last decade that such changes in the rest of the Phanerozoic have received much attention. There are two principal reasons for this arousal of interest. First, increased exploration of the stratigraphic record across the Earth, both on land and under the sea, and significant improvement in biostratigraphic correlation have drawn attention to what appear to have been globally synchronous events. Second, oceanographic research has demonstrated the importance of tectonically controlled changes in the cubic capacity of the ocean basins, notably variations in ridge volume, which provide a ready mechanism for sea-level changes even at nonglacial times.

The adjective *eustatic* was first proposed for global changes of sea level by Suess (1906) in his great treatise *The Face of the Earth.* (The corresponding noun is *eustasy*, not eustacy as it is commonly misspelt; the unconvinced reader need only think of ecstasy.) Eustatic sea-level changes can be inferred, as Suess recognized, by (*a*) plotting the temporal spread of marine sediments over the continents and (*b*) estimating the depositional depth variations in marine stratal sequences, which can be correlated across and between continents. To these methods can be added seismic stratigraphy, in which biostratigraphically age-determined seismic sequences on the continental margins are analyzed in terms of cycles of coastal onlap and offlap.

When the results of these methods are broadly in accord, confidence in a eustatic interpretation is increased, but the range of confidence can vary considerably. It is widely accepted that there was an exceptionally high sea

0084–6597/84/0515–0205$02.00

level in the late Cretaceous and an equally marked low one at the Paleozoic-Mesozoic boundary, but smaller-scale changes and short-term oscillations have proved more controversial. This is because of the complicating effects of regional tectonics and the limits of biostratigraphic resolution in the all-important matter of intra- or intercontinental correlations.

If global "signals" can be satisfactorily disentangled from local or regional tectonic "noise," eustatic changes can be identified and accepted because well-understood processes are known to produce such variations. The alternative suggestion that different continents have moved up and down in concert (Sloss & Speed 1974) is not only less economical but demands recourse to otherwise unsupported speculations about the thermal and tectonic behavior of both the continents and the upper mantle.

In this review, eustasy is first treated in terms of shorter-phase oscillations and then in terms of changes through the whole Phanerozoic, involving both major oscillations and longer-term secular trends. (The Precambrian must obviously be excluded because of the inability to obtain sufficiently refined correlation and the uncertainty as to which strata are marine in origin.) Correlation of sea-level variations with changes in sedimentary facies, isotope ratios, and faunas, and the underlying causes of these variations, are then briefly discussed.

SEA-LEVEL OSCILLATIONS

Facies Analysis

The most obvious and longest-practiced method of inferring eustatic changes is to analyze the facies of continental sequences in terms of alternating marine and nonmarine strata in more marginal marine situations, and shallower- and deeper-water deposits in more offshore, fully marine situations. If shallower-water marine phases of sedimentation can be correlated over considerable distances with regressive episodes, and deeper-water marine phases with transgressive episodes in the marginal environments, then purely local controlling factors can be ruled out, and events of at least regional significance are indicated. Similarly, the correlation of shallowing events from terrigenous siliciclastic to calcareous facies within marine deposits can be used to exclude an interpretation involving merely increased influx of sands and muds from some adjacent land area, leading perhaps to a regression by sediment progradation.

Figure 1 illustrates this point. The sandier units on the left side of the diagram, with, for instance, trough cross-bedding and oscillation ripples, exhibit facies characteristics suggesting shallower-water deposition than the intervening more argillaceous units. They also correlate biostratigraphically with the erosional events and shallow-water facies of the purely carbonate sequence on the right (such as bored and truncated hardgrounds, intraformational conglomerates, dolomites, benthos-rich, stromatolitic, and birdseye limestones), better than, say, with intervening deeper-water micrites with pelagic fauna. The former area could signify an environmental setting marginal to a delta, the latter an isolated carbonate bank, such as the Bahamas, or a marginal marine platform free from terrigenous siliciclastic influx, such as the Yucatan or the southern side of the Persian Gulf. Note that the intervening area, a deeper-water basin characterized by a relatively high subsidence and sedimentation rate, contains a uniform argillaceous facies. Siliciclastic muds may be deposited over a considerable depth range, and regional or eustatic sea-level changes may find no expression in the sedimentation. Furthermore, subsidence in a basin will counteract sea-level fall and may conceal it if the rates of vertical movement match each other.

The degree of confidence in eustatic interpretation clearly depends on (*a*) the reliability of bathymetric estimates, (*b*) the precision of biostratigraphic correlation, and (*c*) the extent of such correlation, preferably from continent to continent and at least across large areas of continent. Whereas there is in many if not most cases a high measure of consensus about the *relative* depositional depths of particular marine facies, based on an array of sedimentological and paleoecological evidence, *absolute* determinations are at best only very approximate (Hallam 1981a). The quality of correlation obviously varies with the available fossils, with the most precise age determinations probably being given by ammonites and planktonic foraminifera. Unfortunately, the deposits of the shallower or more marginal

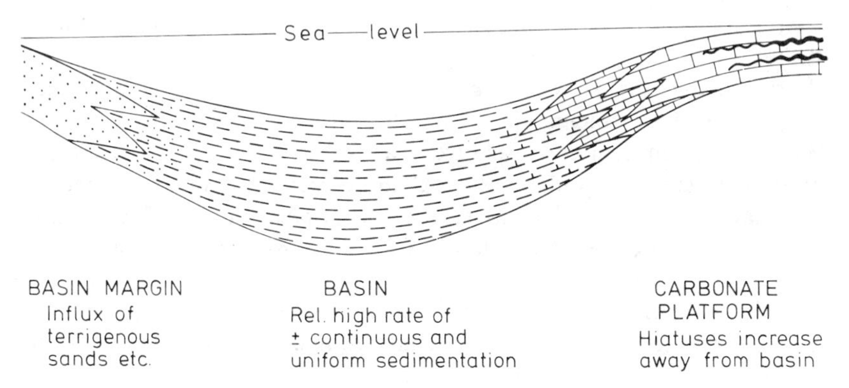

Figure 1 Illustration of how sea-level fluctuations affect different sedimentary regimes. Two episodes of shallowing give rise to spreads of sand into the left-hand basin margin, and corresponding spreads of shallower and deeper neritic limestone facies into marls and shales of the right-hand basin margin. These episodes of sea-level fall are represented by hiatuses in the carbonate platform sequence on the right and lack any representation in terms of facies changes in the basin center.

marine environments, which are best for depth estimates, are frequently the least likely to contain good biostratigraphic indices. Allowance must also be made for at least a limited amount of diachronism in following transgressions laterally because continents are unlikely ever to have been as flat and smooth as a billiard table.

Examples of facies analysis leading to eustatic interpretations, covering a wide stratigraphic range, are given by Brenchley & Newall (1980), Lenz (1982), Ramsbottom (1979), Hallam (1978, 1981b,c), Hancock & Kauffman (1979) and Olsson et al (1980).

Seismic Stratigraphy

Seismic stratigraphy, as developed in recent years by the major oil companies, is essentially a geological approach to the interpretation of subsurface data produced by seismic reflection profiling, and has been widely used in the interpretation of continental margins. Its application to the study of eustasy has been pioneered by Vail et al (1977).

The primary seismic reflections evidently tend to follow bedding surfaces and unconformities with velocity-density contrasts. One might naturally assume that such reflections would normally bound major lithological units, even when these cut across stratification surfaces, but Vail et al, on the basis of careful studies well controlled by borehole logging, insist that in fact they follow chronostratigraphic boundaries that parallel such surfaces. This is a crucial assumption in all that follows, but the supporting evidence has to be taken largely on trust because little of it has been published.

In what might conveniently be called the Vail technique, major stratigraphic units consisting of a relatively conformable succession of strata, with upper and lower boundaries defined by unconformities, are classified as *depositional sequences*. The age of the boundary unconformities is determined by tracing them laterally into conformable successions. Within a given region, a relative sea-level rise is inferred from the progressive landward onlap of littoral and/or coastal nonmarine deposits in a marine sequence, which is termed *coastal onlap*, and a relative fall is inferred from the downward shift of coastal onlap. Regional sea-level curves are drawn on the basis of this type of analysis, from which chronostratigraphic correlation charts are constructed. If cycles of relative rise and fall can be correlated in several widely separated regions across the world, the underlying control is held to be eustatic.

The great strength of seismic stratigraphy is that vast and otherwise inaccessible terrains can be rapidly analyzed in terms of shifting packages of sediment; given adequate borehole coverage for good facies and biostratigraphic control, there is no reason why valid and important results should not be obtained. One should take full account of the possible drawbacks, however. In practice it is doubtful if the chronostratigraphic precision (which in the pre-Cenozoic is based mainly on palynology) usually matches that achievable in studying sections exposed on land, and information on pre-Mesozoic rocks is scanty. Likewise, facies details are not normally available, and the fundamental assumption that impedance contrasts invariably follow chronostratigraphic surfaces is open to question. Most important, details of the evidence supporting the eustatic claims of the Exxon group (Vail et al 1977) are not published, and hence their claims cannot be checked directly. One wonders if teams from other oil companies would interpret the same data in a similar way, or how useful the Vail curves have proved in prediction within the industry.

Until the supporting evidence becomes publicly available, probably the best one can hope for is to match the Vail eustatic curves against those produced using other techniques. Conformity in results should increase confidence in interpretation.

Distinction of Eustatic from Local and Regional Tectonic Events

It is obviously of great importance to be able to disentangle the effects of local and regional tectonics when evaluating the role of eustasy in a given case. Thus a change in a sequence from shallower- to deeper-water facies could be the result either of a rise in sea level or an increase in the rate of subsidence uncompensated by an increased sedimentation rate. In extreme cases the influence of local tectonics is obvious, as evidenced, for instance, by sharp angular unconformities, fault-bounded basins with stratigraphically thick but laterally restricted scarp-front conglomerates, and abrupt lateral variations in stratal facies and thickness. Over extensive areas, however, such telltale features are absent (most notably in cratonic regimes), and facies changes tend to be more subtle and widespread. These cases require careful and detailed facies analysis utilizing the widest possible array of sedimentological, paleoecological, and stratigraphic evidence (e.g. Corbin 1980, Loughman 1982, Phelps 1982).

As regards *local* changes, involving areas on the order of thousands to tens of thousands of square kilometers, the technique proposed by Hallam & Sellwood (1976) may prove helpful. Most cratonic regimes can readily be divided into a series of basins and swells, respectively characterized by greater and lesser thicknesses of stratal sequences of a given age, which correspond to higher and lower mean sedimentation rates. For a given regime, the mean thicknesses of successive stratigraphic intervals are determined, as is the percentage deviation from the mean for each time interval for the various basins and swells. Areas subjected to minimal local tectonic disturbance should record a more or less constant basin or swell tendency over the time period in question, whereas areas where sedimentation has been strongly affected might be expected to show frequent deviations, both positive and negative, from the mean.

An example from the Jurassic of England is presented in Figure 2. In this case the dominant tectonic influence is taphrogenic, with the rate of subsidence being controlled primarily by vertical movements of the horst-and-graben type, either within the Hercynian basement or the Jurassic itself. The best candidates for sequences recording eustatic signals in terms of shallowing and deepening events are clearly those with little evidence of local tectonic disturbance, such as North Creake in Figure 2.

It is also desirable to distinguish *regional* events of subsidence or uplift, involving areas on the order of hundreds of thousands or more square kilometers. Thus, Officer & Drake (1982) point out a 140-m relative change in elevation northeastward along the east coast of North America, over a

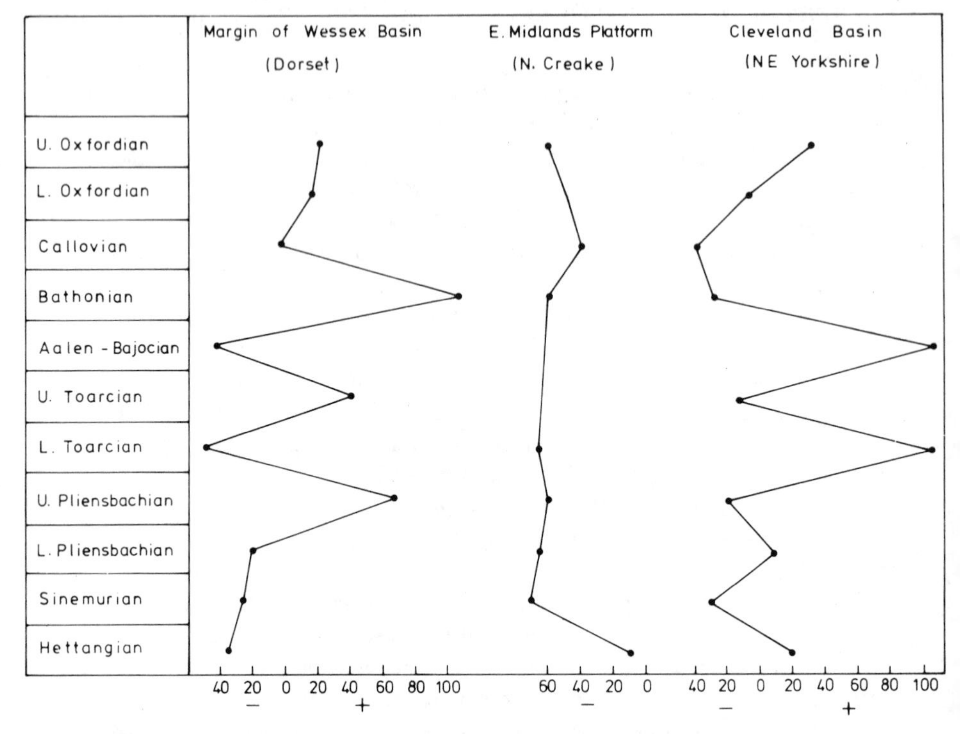

Figure 2 Graphical plots illustrating fluctuations in the rate of subsidence in different English localities throughout the Jurassic. Numbers at bottom are percentage deviation from mean stage thickness of Anglo-Welsh sections. Note the striking lack of correlation between Dorset and Yorkshire (based on data presented in Hallam & Sellwood 1976).

distance of 3000 km. This has apparently taken place over the last 18,000 years, giving a rate of elevation of Florida with respect to the Scotia Shelf of 0.8 cm yr^{-1}. Bally (1981) relates widespread unconformities in the stratigraphic record to major plate reorganizations that are not necessarily orogenic events, and he questions the eustatic interpretation of Vail et al (1977).

Watts (1982), following McKenzie (1978), has developed a model to account for the subsidence of passive continental margins, in which thermal contraction follows initial rifting and extension, and sediment loading occurs with the flexure of a progressively more rigid basement subsequent to the rifting phase. There is initially a significant onlap of sediments onto the basement due to the abrupt transition from fault-controlled Airy-type subsidence to flexure-controlled subsidence, followed by a slower rate of progressive onlap due to an increase in flexural rigidity with age. This pattern of onlap closely matches that attributed by Vail to the rise of sea level, and Watts claims a correlation between the beginning of Vail's supercycles and the age of the rift-drift transition at the continental margins formed by the breakup of Pangea. Watts concedes that a few transgressions are too widespread, however, to be accounted for solely by this regional subsidence.

One effective, though imprecise, way of disentangling regional epeirogenic effects from eustatic ones is to construct a general eustatic curve by plotting the areal distribution of marine sediments and making best estimates of shoreline positions for successive time intervals on a global scale; these are then matched against the pattern of transgression-regression and shallowing-deepening episodes in a given region. A marked disparity signifies regional epeirogenic involvement. Thus, the regressive character of the Bajocian (Middle Jurassic) of the North Sea is anomalous in global terms and indicates significant regional tectonism involving both uplift and subsidence; this interpretation is supported by the evidence of contemporary faulting and volcanicity (Hallam 1978).

Bond (1976) proposes a method for assessing the role of regional subsidence. If we assume that the sediment supply keeps the depositional interface at depositional base level, the maximum thickness of sediment is given by

$$2.4t = h + 3.4t - 3.4h,$$

where h is rise of sea level, t is the thickness, and 2.4 and 3.4 g cm^{-3} are the mean sediment density and mantle density, respectively. Any excess sediment thickness must be attributed to subsidence.

Bond concludes from his analysis that the late Cretaceous sea-level rise that flooded nearly half of North America, implying a rise of over 300 m, was insufficient to account for the thickness of nearly half of the Upper Cretaceous transgressive deposits of the Gulf and Atlantic coastal plains and the Western Interior. A corresponding amount of subsidence must therefore be invoked in addition.

The immense thickness of deposits in the western part of the Western Interior Cretaceous seaway province has long indicated a high rate of subsidence and, hence, a foredeep adjacent to the rising Laramide mountains. Hancock & Kauffman (1979) show that the increasing proportion of regressive facies from Santonian times onward is anomalous in global terms and relates to the Laramide uplift to the west. Australia is likewise anomalous in that the sea retreated from most of the continent following a widespread Aptian-Albian transgression (Brown et al 1968), whereas in the rest of the world the sea-level maximum was not reached until Campanian times. Although this does not necessarily indicate uplift of the whole of Australia, because the sea could have reached the interior through one or more confined areas or straits, epeirogenic movements on the regional rather than the local scale seem to be implied.

The Pliensbachian and Toarcian stages of the Lower Jurassic provide an especially instructive case history in how regional tectonics may be disentangled from eustatic factors in accounting for geographic and temporal variations in marine sedimentation (Hallam 1978, 1981b; Figure 3). In the British area terrigenous sands and silts were introduced for the first time in the Jurassic into what had been a predominantly argillaceous sequence approximately at the Lower-Upper Pliensbachian boundary; and the top of the Pliensbachian is marked everywhere either by sandstones or oolitic ironstones that give clear evidence of having been deposited in shallower water than the more argillaceous or locally calcareous strata above and below. Relatively shallow-water sandstones or limestones also mark the top of the stage in western France (Normandy) and the western parts of the Iberian peninsula. The British deposits are overlain in most areas by a Lower Toarcian unit of laminated organic-rich shales, which is widespread in northwest Europe and is universally accepted as a relatively deep-water deposit.

Therefore, in terms of a purely eustatic model, the top of the Pliensbachian (Spinatum Zone) would naturally be taken as marking the lowest stand of sea level. There is, however, an important regional unconformity directly *below* the Spinatum Zone, recorded almost everywhere in Great Britain (Phelps 1982), that does not accord with this simple picture.

Studies elsewhere serve to resolve this problem. Southeastward into eastern France and southern Germany the arenaceous deposits disappear, replaced by calcareous and argillaceous deposits. These indicate a predominantly westerly land source, but the Lower Toarcian organic-rich shale is still present (schistes cartons in France, Posidonienschiefer in Germany). Facies analysis suggests a general pattern of marine deepening through the later Pliensbachian continuing to an early Toarcian maximum, followed by shallowing. Such a pattern corresponds more closely to the world picture as revealed by the deposits of western North and South America and eastern Asia, the only other regions outside Europe where adequate stratigraphic information is available (Hallam 1981c). The

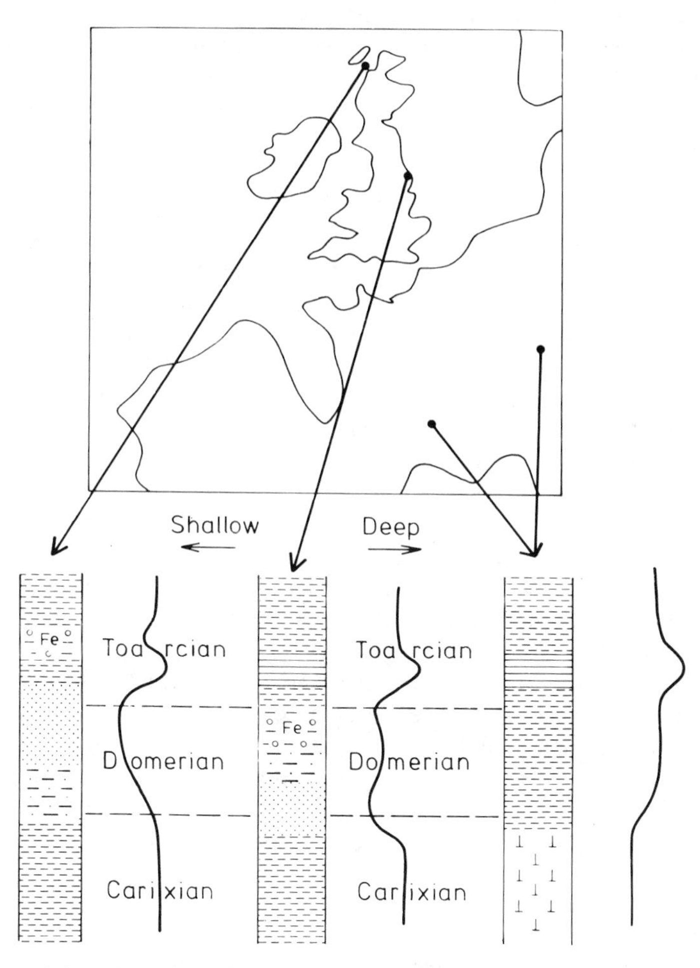

Figure 3 Schematic sections of Lower Jurassic in different parts of western Europe, with accompanying inferred bathymetric curves. Carixian = Lower Pliensbachian, Domerian = Upper Pliensbachian.

deposits on the western margins of Europe therefore appear to reflect regional uplift to the west, which can plausibly be related to taphrogenic horst-and-graben phenomena anticipating the later opening of the Atlantic (Hallam & Sellwood 1976). It is important to note, however, that the eustatic signal is still recorded in the British area, with the shallowest-water deposits of the youngest Pliensbachian recording a transgressive episode following widespread uplift and erosion.

The Lower Jurassic of northwest Europe provides another illuminating example of the interaction of regional tectonic and eustatic factors. It is the only region outside the Pacific margins with marine Hettangian, and global analysis suggests a Hettangian rise of sea level following an end-Triassic fall (Hallam 1981b,c). The sea apparently flooded into northwest Europe as a result of both this rise and the widespread crustal attenuation due to extension (Hallam & Sellwood 1976).

Rates and Amount of Sea-Level Change

While it is clearly important to determine as precisely as possible the rates of oscillatory sea-level rise and fall, we are still sadly lacking in reliable quantitative estimates because of various methodological difficulties. Rises are easier to document than falls, provided the biostratigraphic correlation is good. Marine transgressions and the correlative deepening phases in fully marine sequences can often be precisely dated, but the regressive facies commonly associated with eustatic falls may lack stratigraphically useful fossils. Additionally, erosion following regression may lead to the removal of part of the sedimentary sequence. There is furthermore the problem that coarsening-upward siliciclastic sequences can result either from sea-level falls or from the filling in of a marine basin by sediments at a time of stillstand. In extreme cases, this may even occur during times of sea-level rise if the rate of shallowing due to sediment influx exceeds the rate of rise (Vail et al 1977).

There are a number of ways of circumventing this last difficulty. If shallowing-upward siliciclastic sequences can be correlated with shallowing-upward carbonate sequences elsewhere, then a good case for sea-level fall on at least a regional scale can be established (Hallam 1978).

Another approach is indicated by the work of Thiede (1981). The occurrence of neritic fossils in pelagic sediments of the Deep-Sea Drilling Project (DSDP) cores from the Pacific documents the repeated injection of skeletal debris derived from shallow-water provinces into adjacent deep-sea basins through the late Mesozoic and Cenozoic, during short-lived intervals when little or no benthic sediment was so transported. Thiede demonstrates that such episodes correlate in almost all cases with low stands of sea level in the scheme of Vail et al (1977). Short-phase eustatic falls, leading to transient land connections, are also the most plausible explanation for the pronounced similarities of North American and European late Jurassic terrestrial vertebrate faunas, because paleogeographic data point to shallow marine separation of the two continents at this time (Prothero & Estes 1980).

In the facies analysis method of studying eustasy, approximate order-of-magnitude estimates of the extent of sea-level oscillations can be made by assigning likely depths of deposition to the deepest- and shallowest-water strata of given marine cycles and then by obtaining best estimates of age from biozonal and radiometric data. By assuming depth ranges of a few tens of meters, which seem plausible for the deposits in question, Hallam (1978) and Ramsbottom (1979) both obtained figures of up to a few centimeters per 1000 yr of Jurassic and Carboniferous age. The data of Olsson et al (1980) indicate a major short-phase sea-level fall at a comparable rate of ~ 3 cm per 1000 yr during the early Oligocene.

A more direct method of estimation, proposed by Vail et al (1977), is to measure the vertical and horizontal components of coastal onlap from seismic stratigraphic data from the continental margins. However, Watts (1982) points out that they did not correct for flexural effects that vary as a function of time and position. Furthermore the Vail techniques for estimating sea-level rise and fall are quite different from each other.

There is as yet no accord about the relative rates of sea-level rise and fall and the proportion of time occupied by stillstand, although there is of course no reason why these should not vary significantly in particular instances. On the basis of facies analysis, Hallam (1978) argued that the predominant mode for the northwest European Jurassic was probably a rapid rise and fall interrupted by a longer phase of stillstand, while Ramsbottom (1979) suggested that the mode for the Carboniferous of the same region was a gradual or rapid rise, followed by brief stillstand and a very rapid fall (Figure 4).

Vail et al (1977) proposed a general Phanerozoic mode of gradual or moderately rapid rise decelerating with time, minimal stillstand, and extremely rapid, "geologically instantaneous" fall. This was based on a direct correspondence between relative changes of coastal onlap with relative changes of sea level. It is now admitted (Vail & Todd 1981) that this mode is erroneous because it takes no account of the facies differences between coastal plain (marine) and alluvial plain (continental) facies, and in fact the landward boundary that should be plotted is the upper limit of the coastal plain. Where an alluvial plain is present, the relative sea-level boundary corresponds with the facies change between the coastal and alluvial plain. Unfortunately, seismic stratigraphic techniques do not often permit this identification. Charts of relative changes of coastal onlap

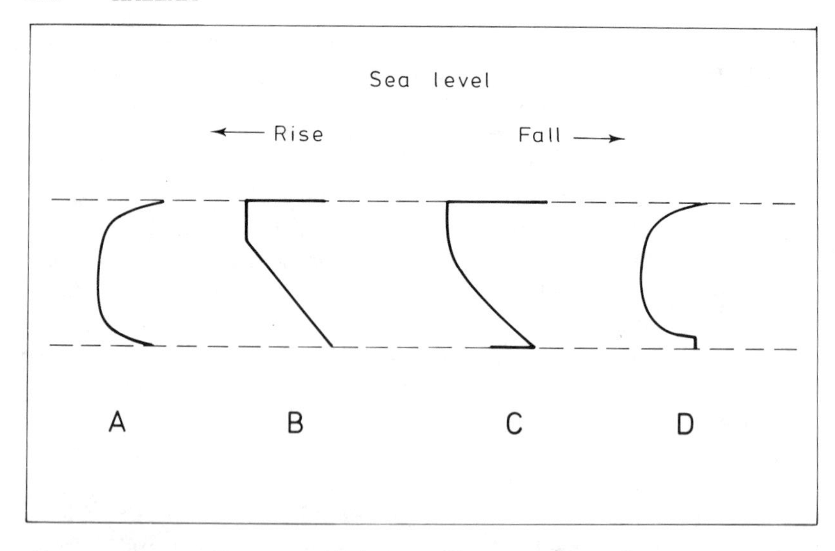

Figure 4 Diagrams illustrating the character of sea-level rises and falls through time in eustatic cycles. *A*. The most probable mode for Jurassic cycles, based on European sections (Hallam 1978). *B*. The Carboniferous mode, based on British sections (Ramsbottom 1979). *C*. The mode of second- and third-order cycles, based on seismic stratigraphy (Vail et al 1977). *D*. Revised mode for Jurassic cycles, based on seismic stratigraphy (Vail & Todd 1981).

typically show abrupt shifts from widespread to restricted, whereas charts of relative changes of sea level commonly show more-gradual shifts. Vail & Todd's amended oscillatory eustatic curves for the Jurassic, based on North Sea data, correspond quite closely to those inferred independently from my facies analysis of marine sections exposed on land (Figure 4).

CHANGES THROUGH THE PHANEROZOIC

The only comprehensive Phanerozoic eustatic curve in existence is that of Vail et al (1977, Part 4, Figure 1). This shows both first- and second-order cycles. The first-order cycles are defined by low sea-level stands at the beginning of the Cambrian and in the Permian-Triassic and Quaternary, and high stands in the Ordovician-Silurian and late Cretaceous, with the late Cretaceous sea level attaining the highest value of any in the Phanerozoic. Superimposed on this gross cyclicity are a series of second-order cycles with a characteristic sawtooth pattern: gradual, decelerating rises followed by sudden falls. Smaller third-order cycles with a similar shape are indicated in other more detailed diagrams.

The Vail Phanerozoic curve is only partly based on continental margin seismic stratigraphy; the Paleozoic part is based on the North American cratonic sequences of Sloss (1963). Thus its quality and reliability are variable through time, even accepting the criteria and assumptions of the Vail team. It has been pointed out above that the sawtooth shape of the second- and third-order cycles is no longer accepted by Vail, and the lack of third-order cycles in the Cretaceous part of the detailed Jurassic-Tertiary curve (Vail et al 1977, Part 4, Figure 2) is strictly a manifestation of Exxon management policy in releasing data. Furthermore, although data from widely separated regions across the world are considered, there is a strong weighting in favor of the Atlantic margins of Europe and North America and the Gulf Coast region.

The broad trends of the curve can be tested by the independent technique of estimating the proportion of continents covered by the sea at successive time intervals. This method is thought to give reliable first-order approximations, but claims to high precision should be viewed with reservation. Because of subsequent erosion, the feather edge of marine sedimentary units does not necessarily correspond closely with the original limits of deposition, and as Cogley (1981) has pointed out, most paleogeographic analyses have ignored the 16% of continents covered by present-day shelf seas because of inadequate data. If only present land areas are considered, then unless ancient and modern shorelines coincide everywhere, estimates of the extent of ancient epicontinental seas must exceed the extent of modern seas.

Areal plots of the degree of continental inundation, based on the paleogeographic maps of Strakhov (1948) and Termier & Termier (1952), show a significant departure from the smoothed Vail curve, insofar as the extent of marine cover of the continents is considerably greater in the early to mid-Paleozoic than in the late Mesozoic (Egyed 1956, Holmes 1965, Hallam 1977a). On the other hand, Wise (1974) used Schuchert's (1955) paleogeographic maps of North America to argue for a condition of essentially constant continental freeboard throughout the Phanerozoic, interrupted by short-term oscillations of sea level, 80% of which remained within about 60 m of a normal freeboard level ~20 m above the present level. The maps were drawn several decades ago and do not take into account extensive modern work that demonstrates a much greater spread of sea over North America in Paleozoic times. My areal plot of Schuchert's maps produces a curve closely resembling the smoothed first-order cycle curve of Vail et al, but by taking into account the new data, the revised curve strongly resembles my plots of the global Strakhov and Termier & Termier data (as well as those of the Soviet Union) based on a comprehensive series of paleogeographic maps (Hallam 1977a).

These results suggest that Phanerozoic sea level was at a maximum not in the late Cretaceous, as proposed by Vail et al (1977), but in the late

Ordovician, when nearly two thirds of the continents were inundated. In addition, the late Devonian and early Carboniferous seas were also slightly more extensive than those of the late Cretaceous, which covered about one third of the present continental area. Because the probability of losing the stratigraphic record through subsequent deep burial, metamorphism, and erosion must increase with time, paleogeographic maps for older periods are likely to underestimate the former extent of marine cover. Moreover, the proportion of carbonates to terrigenous siliciclastics on the cratons is significantly higher in the Paleozoic than in the Mesozoic and Cenozoic. Since erosion rates apparently increase exponentially with elevation (Garrels & McKenzie 1971), this implies sediment sources that were both more areally restricted and topographically subdued (Hallam 1977a).

The hypsometric curve of continental elevations can be used to estimate the sea-level change that will flood varying portions of the continents, on the assumption that the ancient hypsometric curves were similar to the present one (Forney 1975). Forney's method was developed by Bond (1978), who used the scatter of data for different continental elevations to distinguish between sea-level changes and the vertical motions of large continental surfaces. Because of post-Cretaceous continental uplift, use of the present hypsometric curve in conjunction with data on marine spread for earlier times is likely to overestimate the amount of sea-level rise required by a variable amount.

The reliability of the hypsometric method can be evaluated against other methods for the late Cretaceous high stand, which has received much attention in the last few years. Plots of the data derived from Strakhov's global maps indicate that a maximum of ~40% of the continental area was flooded, indicating by the hypsometric method a rise of ~350 m; the corresponding Termier & Termier figure is ~220 m (Hallam 1977a). Strakhov's figure compares much more closely than Termier & Termier's to my (42%) and Bond's (1976) figure (45%) for North America, and for a variety of reasons Strakhov's data are here considered to give a more accurate, though less precise, picture of Phanerozoic sea-level change in general.

Sleep (1976) observed that the inferred late Cretaceous coastline in Minnesota changed little in elevation, which implies only a slight amount of local tectonic disturbance. If epeirogenic "noise" is minimal, the height of sea level may be estimated directly; thus, Sleep arrived at a figure of 375 m for late Turonian–early Coniacian times, which compares very well with the Strakhov-derived figure and with Bond's (1976) figure of 390 m for the highest Cretaceous stand. Bond (1978) subsequently arrived at a lower global estimate of 150–200 m as a result of taking into account variations in continental hypsometries, while Watts & Steckler (1979) proposed an even lower figure of ~100 m as a result of their study of the subsidence history of the continental margin of eastern North America.

By using the entirely independent method of measuring the changing volumes through time of the oceanic ridge system, Pitman (1978) obtained a figure of 350 m for 85 m.y. B.P., which is in close agreement with the earlier figures cited above.

Hancock & Kauffman (1979) recognized from their study of the late Cretaceous deposits of western Europe and the United States Western Interior a late Campanian–mid Maastrichtian maximum of ~650 m. Their method involves first determining the present height of sea level for a given stage in tectonically undisturbed areas, then adding the thickness of Upper Cretaceous marine sediments up to that stage and the assumed depth of sea at the time of deposition. Since their figure is considerably higher than the others, it seems possible that Hancock & Kauffman might have (*a*) overestimated the depth of deposition or (*b*) underestimated the amount of subsequent epeirogenic uplift of certain areas such as the Western Interior. On the other hand, the value of Watts & Steckler (1979), based only on a regional study, seems too low; hence the figure of ~350 m, which falls almost halfway between the extreme values, is here accepted as being the most reasonable, having been arrived at, more or less, by three independent methods.

The general smoothed trend of the eustatic curve presented in Figure 5*A* has been determined on the basis of the extent of marine epicontinental cover as estimated from paleogeographic maps, with the sea-level values estimated by the method of Forney (1975), using his hypsometric curve. For the Ordovician maximum, the Strakhov-derived global figure is ~50% marine cover (~450 m above present), whereas the corresponding figures derived from the more detailed USSR and North American data are ~60% and ~600 m (Hallam 1977a). While there is an admitted danger of overestimating the ancient sea level because of subsequent epeirogenic uplift, the Strakhov maps, which are several decades old, are thought to underestimate the areal spread of sea in the early Paleozoic to a significantly greater extent than for later periods (for the reasons outlined above), and the higher figure of 600 m is here accepted as being of the right order of magnitude.

As regards the second highest sea-level stand, in the late Devonian and early Carboniferous, the Strakhov and North American figures correspond closely (~40%, 380 m; ~43%, 400 m, respectively), although the results for the USSR are somewhat higher (Hallam 1977a).

Apart possibly from the earliest Cambrian, the lowest stand was at the Permian-Triassic boundary, for which time Forney (1975) estimates a value ~40 m below the present. This is supported by the analysis of Cogley

(1981), who draws attention to the existence of continental Permo-Triassic deposits beneath large areas of present shelf seas.

Superimposed on the general trend are a series of oscillations corresponding to Vail's second-order cycles and to successive marine transgressions and regressions over continental margins and interiors. In many

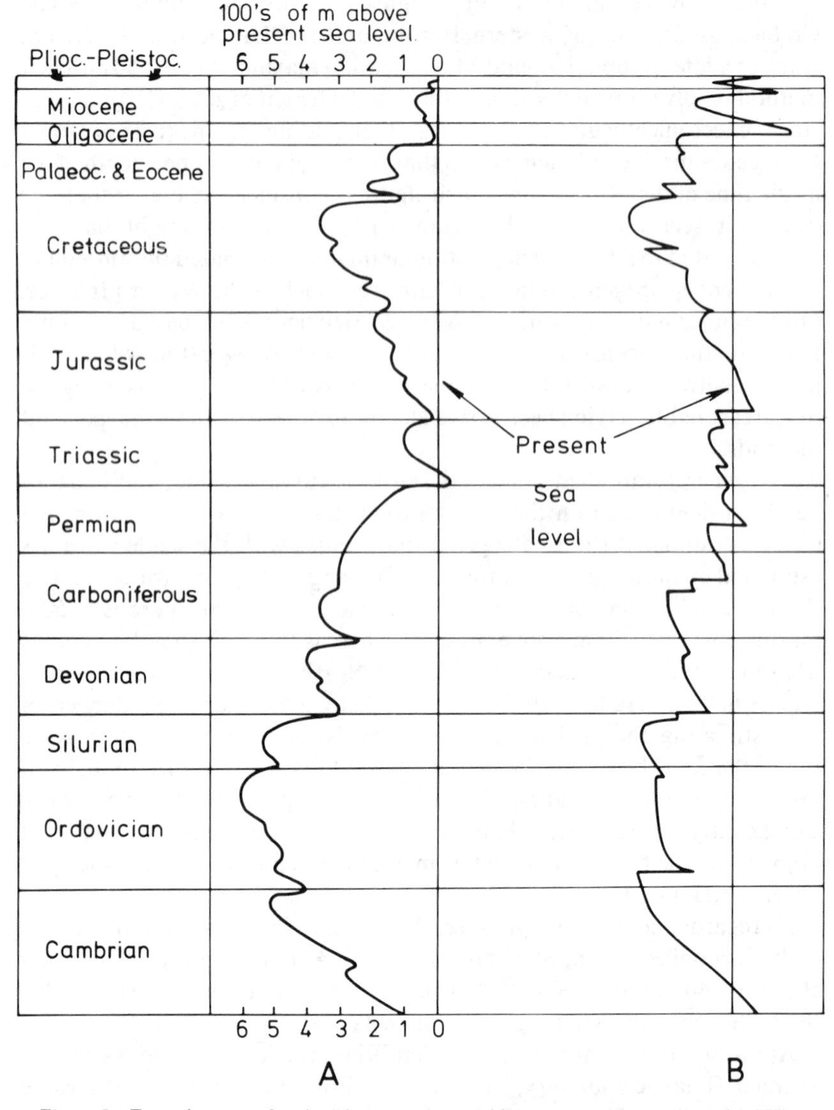

Figure 5 Eustatic curves for the Phanerozoic. *A*. This paper. *B*. After Vail et al (1977).

cases, data are presently inadequate to determine the extent of such sea-level changes, so that there is here a considerable element of "guesstimation" in producing the curve. The evidence used in producing the curve is based not only on reports addressed specifically to eustasy that deal with more than one continent, but also on general stratigraphic accounts of cratonic regions across the world: in Eurasia (Ager 1980, Nalivkin 1973, Lee 1939), Africa (Furon 1963), Australia (Brown et al 1968), and North America (Levin 1978). Only the more important events are noted.

Paleozoic

The Vail curve (Figure 5*B*) shows a progressive rise of sea level through the Cambrian from a low stand slightly lower than today at the Precambrian-Cambrian boundary. The sea-level rise is supported by stratigraphic evidence of transgressive deposits on cratons across the world (Matthews & Cowie 1979). The recognition in recent years of Tommotian deposits in a number of places suggests that perhaps the earliest Cambrian seas were more extensive than indicated in the Vail curve. Oscillations within the Cambrian have not yet been clearly established; the likeliest candidate is a regression followed by a transgression across the early–mid-Cambrian boundary, the evidence for which is especially well marked in Europe. The Vail curve indicates that the Paleozoic seas reached their maximum extent in the late Cambrian (or the beginning of the Ordovician). This is not accepted here; instead, the late Ordovician seas are considered to have been slightly more extensive.

Ordovician and Silurian eustasy is discussed by Leggett et al (1981), Lenz (1982), and Fortey (1984); where there is disagreement, the results of Fortey are favored because his evidence spans the largest number of regions across the world. There is general support for Vail of a significant fall of sea level at the end of the Cambrian, followed by an early Ordovician (Tremadoc) rise. Minor sea-level falls also mark the Tremadoc-Arenig and Arenig-Llanvirn boundaries. While the early part of the mid-Ordovician (Llanvirn) was a time of relatively low Ordovician sea level, there was a notable rise later on, in the Llandeilo, which is marked by an especially wide spread of sea in North America. The biggest Ordovician transgression however, was in the Caradoc, when the seas reached their greatest extent. (In eastern North America this phenomenon is obscured by the effects of the Taconic orogeny.) There is general agreement that sea level dropped sharply at the end of the period (Ashgill) and was followed by a rapid rise at the start of the Silurian (Llandovery). In the Vail curve only a very minor fall is indicated, shortly *before* the end of the Ordovician. From their facies analysis in Norway, Brenchley & Newall (1980) suggest that the end-Ordovician sea-level fall was 50–100 m.

The Silurian appears to correspond with one major eustatic cycle, the Llandovery transgression continuing to a maximum at about the Llandovery-Wenlock boundary, and then followed by a progressive fall to the end of the period, although this latter trend might have been interrupted by a minor Ludlow rise. The indication in the Vail curve that sea level fell at the end of the Silurian to its then lowest stand since the early Cambrian is well supported by cratonic stratigraphy across the world, but the sharpness of the fall is greatly exaggerated.

The lowest sea-level stand might actually have been in the late Gedinnian, shortly after the beginning of the Devonian, because there are a number of areas in Europe and North America where marine conditions may have continued from the Silurian (House 1975a). House's (1975a,b) analysis of Devonian facies and faunas concentrates on Europe and North America but also takes into account evidence from other continents. He infers eustatic control for major early mid-Devonian (Eifelian) and early late Devonian (Frasnian) transgressions, leading to progressive onlap and a Frasnian maximum spread of sea. A number of minor transgressive phases of less certain origin are recorded from the Siegenian to the early Famennian, and major eustatically induced regressions in the early (Gedinnian and Emsian) and late (Famennian) Devonian. There is broad agreement with the Vail curve insofar as progressive (but stepwise) sea-level rise was followed near the end of the period by a more rapid sea-level fall.

Sea level began to rise again at or slightly before the Devonian-Carboniferous boundary (House 1975b) and reached a Carboniferous maximum in the late Dinantian (Visean), which marks the last really extensive global spread of neritic carbonates. Areal plots of the spread of marine sediments, either on the global or continental scale, show a significant reduction from the Lower to the Upper Carboniferous (Hallam 1977a), supporting a sea-level fall in the Vail curve from the Mississippian to the Pennsylvanian. That this was a sudden and considerable fall, the fourth biggest in the Paleozoic according to the Vail curve, is based on evidence of significant regression in the North American midcontinent. According to Ramsbottom (1981), this is more likely to be the result of epeirogenic uplift on a regional scale than a sea-level fall. His own analysis of British Carboniferous deposits, using the Bond (1976) technique for eliminating the effects of regional tectonism, suggest a ~300-m rise of sea level through the Dinantian and Namurian, followed by a period of stillstand in the Westphalian. He notes, however, evidence of widespread nonsequences at the Lower-Upper Carboniferous boundary in several continents. Here, a more modest sea-level lowering from the Lower to the Upper Carboniferous than that proposed by Vail is accepted. It is likely that the regressive nonmarine facies widely developed in the late

Carboniferous is associated with the onset of a major phase of the Hercynian Orogeny in several continents. That these deposits are frequently as widespread as the underlying marine deposits argues against a significant sea-level drop, because they must have been laid down close to base level.

There has been no modern analysis of oscillatory eustatic changes across the Carboniferous-Permian boundary, but Schopf's (1974) analysis of the areal spread of marine sediments indicates a continuing regression through the Permian, accelerating toward the end of the period. Forney's (1975) analysis, using Schopf's data and the hypsometric method, suggests a late Permian sea-level fall of between 125 and 225 m. While early Permian epicontinental seas were evidently more extensive than late Permian ones (which were probably even less extensive than today), there seems to be little support for the Vail proposal of a sudden and marked mid-Permian sea-level fall, followed by a slight rise to the end of the period. Sea level must have been at its lowest at the very end of the period; and even in those few areas where there is an apparently continuous marine transition to the Triassic, as in the Caucasus, the Salt Range of Pakistan, East Greenland, and South China, argument persists about whether or not nonsequences occur at the boundary.

Mesozoic

Like the Silurian, the Triassic exhibits a simple first-order cycle of transgression followed by regression that intercontinental analysis suggests was under eustatic control (Schopf 1974, Hallam 1981b). The sea began to spread in the early Triassic and withdrew from East Greenland and the West Australian coastal zone at the end of this time, not to return until the Jurassic. There was a rapid transgression in the early mid-Triassic (Anisian), with the seas spreading again after a minor interruption to a maximum in Ladinian-Carnian times, when about 17% of the continental area was inundated. Thereafter, sea-level lowering is indicated by shallowing of facies within the Tethyan zone and passage to nonmarine facies elsewhere.

According to Vail et al (1977), sea level reached its Triassic maximum height in the Norian; this is presumably based on the fact that the spread of both marine and continental sedimentary deposits reached a maximum at that time. This is certainly true of the Atlantic margin regions of Europe and North America. There are, however, two alternative explanations for the apparent paradox in these areas of onlap of more regressive facies; both are thought to be more plausible explanations than Vail's because there is no indication from either facies or stratigraphy that shallow epicontinental seas were choked by heavy terrigenous sediment influx. One explanation is that the spread of sediments increased through Triassic times as Hercynian mountains (probably horst blocks) were progressively worn down by erosion and intervening depressions filled up, leading to sediment "spill-over" onto the adjacent peneplains (Hallam 1981b). The second alternative involves flexural onlap on structural highs in regions of stretched lithosphere (Dewey 1982, Watts 1982).

There was a relatively sharp sea-level fall at the end of the Triassic, immediately preceded over a wide area by a transgression that might have been eustatically controlled (Hallam 1981b). Thus, across the Triassic-Jurassic boundary the only epicontinental sea outside the Pacific margins was in northwest Europe, and its presence there was, as already noted, at least partly related to subsidence of continental crust as a consequence of tectonic stretching. There is clear evidence, furthermore, of a regression followed by a hiatus at the system boundary. Thereafter, the *leitmotif* of Jurassic eustasy was one of interrupted sea-level rise to an Oxfordian-Kimmeridgian maximum, when epicontinental seas covered a quarter of the present continental area (Hallam 1978). The revised Jurassic eustatic curve of Vail & Todd (1981), based on a detailed analysis of North Sea data, shows broad overall agreement with mine, but there are important differences of detail. The most glaring is the sudden eustatic fall seen in the Vail & Todd curve at the end of the Hettangian, which was so large that sea level was not restored to its Norian and Hettangian height until mid-Jurassic (Bajocian) times. This is quite unacceptable on grounds of distribution and character of marine early Jurassic facies on the continents (Hallam 1981c, Loughman 1982). The early Sinemurian in fact was a time of significant eustatic rise, not fall, and the sea reached areas never covered in Hettangian or Norian times.

Whereas the different eustatic curves agree about important transgressive pulses in the Toarcian, Bajocian, and Callovian and about regression in the Aalenian, the Kimmeridgian, rather than the Oxfordian, is accepted by Vail & Todd (1981) as the time of the most significant late Jurassic sea-level rise. However, this relates at least partly to a major regional event in northwest Europe leading to the deposition of the Kimmeridge Clay. Global data suggest a sea-level fall near the end of the Jurassic (Hallam 1978), but according to Vail & Todd sea level did not fall significantly until the end of the Berriasian (early Cretaceous), an event corresponding in Europe to the rapid change from calcareous (Purbeck) to siliciclastic (Wealden) facies. This is evidently the result of disregarding upward facies changes indicating regression in favor of determining the time of downward shift of coastal onlap. Vail & Todd's curve shows a succession of short-phase eustatic oscillations for Kimmeridgian-Tithonian times, but this is more likely to relate to North Sea tectonics.

The Tithonian part of the Jurassic eustatic curve is, together with the Bathonian, the least well established because of difficulties of biostratigraphic correlation, but there was probably a transgressive pulse early in the stage. This is clearly suggested by data from the Andes, where marine Tithonian succeeds regressive evaporites and red beds of Upper Oxfordian and Kimmeridgian age. While the earlier regression, markedly counter to the world trend, clearly relates to Andean tectonics, it is worth noting that the Sinemurian, late Pliensbachian–early Toarcian, Bajocian, and Callovian pulses of sea-level rise are all clearly recorded by transgressive events in the southern Andes. Thus, the early Tithonian transgression may also have a eustatic component, especially as it appears that it correlates at least approximately with a relatively deep-water organic-rich shale unit in northwest Europe and western Siberia (Upper Kimmeridgian in the English sense, Lower Tithonian in the French sense).

Using the hypsometric method, a sea-level rise through the Jurassic of slightly under 200 m is indicated. The results of three different methods of estimating the amount of eustatic rise through the early Jurassic are presented here. The hypsometric figure is ~110 m (Hallam 1981c), while the Vail & Todd figure, based on seismic stratigraphy, is ~75 m. The Bond (1976) technique can be applied to the English Lower Jurassic best in an area of stable platform comparatively resistant to subsidence, using data from the North Creake borehole mentioned earlier; the ammonite evidence indicates that no significant hiatuses are present (Hallam & Sellwood 1976). The value arrived at is ~70 m, but since the sequence is mainly argillaceous, some increase should be allowed for compactional effects. The level of agreement between the three methods is encouraging.

The Vail curve for the Cretaceous shows a general rise to a Campanian-early Maastrichtian maximum (followed by a sharp and considerable end-Maastrichtian fall) from a low early Valanginian stand after a sudden end-Berriasian fall. A minor mid-Aptian and major mid-Cenomanian sudden fall interrupt the general trend.

In the early Cretaceous, the important early Hauterivian rise claimed by Vail is supported by evidence of a major transgression at this time (Cooper 1977). Cooper also argues for a notable eustatically controlled regression within the Aptian. For the younger Cretaceous there is a broad agreement between the eustatic curves of Vail et al (1977) and Hancock & Kauffman (1979). Hancock & Kauffman conclude that sea level rose from an early Albian stand close to that of today, to a late Campanian–early Maastrichtian peak of ~650 m above present; this was followed by a rapid fall considerably thereafter in the late Maastrichtian. The only notable interruption was a sharp fall in the late Turonian (not the mid-Cenomanian as maintained by Vail). Five eustatically controlled transgressive peaks are recognized: late Albian, early Turonian, Coniacian, mid-Santonian, and late Campanian–early Maastrichtian.

Tertiary

Within the Tertiary, the most important change concerns the Oligocene. According to Vail et al (1977), there was a huge sea-level fall of nearly 400 m, by far the biggest in the Phanerozoic, that took place suddenly in the late Oligocene in a mere geological instant. This appears to be based on data from four regions: the North Sea, northwest Africa, the San Joaquin Basin of California, and the Gippsland Basin of Australia (Vail et al 1977, Part 4, Figure 5). Of these, only the North Sea and northwest Africa show a pronounced drop, and the Gippsland Basin hardly records the change at all. Yet the Vail curve is heavily weighted, without explanation, in favor of the North Sea.

Although there is strong evidence for a major Oligocene regression, the Vail estimates of the timing, extent, and speed of the eustatic fall are open to question. Thus, Olsson et al (1980), on the basis of a facies and biostratigraphic study of the United States Atlantic coastal plain, argue cogently for a moderately rapid fall of perhaps as much as 150 m in the *early* Oligocene to account for a widespread gap between youngest Eocene and early late Oligocene. This timing of the major fall and subsequent rise is supported by evidence from the Gulf Coast (Murray 1961), western Europe (Pomerol 1973), Australia (Carter 1978), and southern Africa (Siesser & Dingle 1981).

The *leitmotif* of Tertiary eustasy is, of course, one of more or less progressive fall through the era, with Paleocene seas being more extensive than those of the Eocene, and Oligocene seas less extensive still. Because, however, of the dramatic early Oligocene fall, Miocene seas were *more* extensive than the Oligocene, reaching their maximum toward the middle of the period, or just after (Serravalian), suggesting a sea-level peak at this time. Among the more important oscillatory changes, it is generally agreed that there was a rapid end-Miocene fall of perhaps as much as 100 m (Adams et al 1977, Loutit & Keigwin 1982), followed by an early Pliocene rise. It is here suggested that the end-Paleocene regression was more important than the mid-Paleocene one proposed by Vail et al. It is uncertain whether the Eocene transgressive cycle reached its eustatic peak early or late in the period.

Smaller-scale eustatic cycles than those portrayed in Figure 5, such as Vail's third-order Mesozoic and Cenozoic cycles or those reported by Cooper (1977), Hallam (1978), and Ramsbottom (1979) among others, are naturally more difficult to establish because a smaller eustatic change is more likely to be swamped by local or regional tectonic processes. Each individual case must be scrutinized carefully, using the most refined biostratigraphic data available and the largest possible number of regions, together with an analysis directed at eliminating the effect of noneustatic events.

CORRELATION WITH SEDIMENTARY FACIES AND ISOTOPIC CHANGES

The correlation between spreads of regressive siliciclastic, carbonate, and evaporite facies and sea-level low stands is too obvious to pursue further, but some other, less obvious relationships warrant brief attention.

A strong association between the stratigraphic distribution of finely laminated organic-rich shales and marine transgressive episodes has been recognized throughout the Phanerozoic (Hallam & Bradshaw 1979, Jenkyns 1980, Leggett et al 1981). Notable examples include widespread deposits in the Caradoc, the Llandovery, the Devonian-Carboniferous boundary, the early Sinemurian and Toarcian, the Callovian, the early Tithonian, and the Aptian-Albian and Cenomanian-Turonian boundaries. The last two have been related to oceanic "anoxic events" (Jenkyns 1980), but the lack of DSDP data for times prior to the Callovian rules out the confirmation of this for older deposits. The most important reason for the association is probably that the early stages of transgression over the continents are characterized by broad stretches of poorly oxygenated shallow water (with restricted circulation with the open ocean) that provide a short transit for organic matter from productive surface water to the bottom sediments. Consequently, there is less oxidation and greater retention of organic matter (Hallam 1981a, Arthur & Jenkyns 1981).

Whereas phosphorite genesis correlates in general with times of high sea level and warm climate, it correlates more specifically with marine transgressions, whereby a number of shallow pericontinental and epicontinental sites are available for phosphate fixation. In other words, such shallow seas tend to sequester nutrients and in consequence are highly fertile. The resultant high plankton productivity will in turn lead to reduced oxidation in this particular setting and therefore to the greater initial retention of organic phosphorus. On the other hand, oceanic anoxic events correlate with high sea-level stands (Jenkyns 1980) but do not generally correlate well with major phosphorogenic episodes. This may be because abundant phosphorus is fixed in the deep ocean by organic carbon and is therefore relatively unavailable in the epicontinental seas (Arthur & Jenkyns 1981).

Scholle & Arthur (1980) found that the Aptian-Albian and Cenomanian-Turonian anoxic events correlated with episodes of heavy $\delta^{13}C$ carbonate carbon. This was tentatively attributed to the increased preservation by burial of light $\delta^{13}C$ organic carbon during anoxic episodes, so that the dissolved carbon remaining in the oceanic reservoir became progressively heavier isotopically. An alternative explanation is proposed by Loutit & Keigwin (1982), whose isotopic study of carbonate carbon records an abrupt shift coinciding with the previously reported end-Miocene eustatic fall. They argue that the shelves have acted as carbon sinks during regressions, when erosion supplies previously deposited organic matter, depleted in ^{13}C, to the open ocean. Thus oceanic $\delta^{13}C$ values can be decreased by introducing isotopically light organic matter during sea-level low stands.

As regards regressive episodes, Hallam & Bradshaw (1979) noted for the western European Jurassic a common association with oolitic ironstones, which frequently cap shallowing-upward marine sequences. Van Houten & Karasek (1981) report a similar relationship for the Devonian of Libya. On occasions, Jurassic ironstones may actually occur at the base of a transgressive sequence, but the facies relationships always indicate that they are among the shallowest-water, most "regressive" deposits. In general, the formation of oolitic ironstones is related to episodes of relatively high sea level, mild climate, and tectonic stability accompanied by reduced influx of siliciclastic detrital sediment. The facies associations of the ironstones indicate repeated shallowing-upward sequences, abrupt waning of sediment supply, and rapid renewed sea-level rise; the ironstones tend to occur at the sharpest lithofacies discordance (Van Houten & Bhattacharyya 1982).

Another highly intriguing relationship, in this case between sea level and the $^{87}Sr/^{86}Sr$ ratio of seawater, has been pointed out by Spooner (1976). The ratio of seawater strontium (0.7091) is lower than the ratio of dissolved strontium delivered to the oceans by continental runoff ($\sim$0.716). This difference can be accounted for by the exchange with the isotopically lighter Sr of oceanic crust during the process of hydrothermal convection within spreading ridges. Spooner demonstrated a good correlation between estimates of the increase in land area in the last 70 m.y., related to fall of sea level, and the strontium isotope ratio. This implies that there is covariation between the increase of seawater $^{87}Sr/^{86}Sr$ and the increase in the continental runoff flux.

Spooner assumed that land area and continental runoff are directly proportional, but consideration of the Phanerozoic as a whole, based on the marine carbonate analyses of Veizer & Compston (1974), suggests a more complicated picture. Their data indicate variation in the sea-water ratio from a maximum in the early Cambrian (0.7093) that is comparable to the present-day value (0.7091) to a minimum in the late Jurassic (Oxfordian; 0.7069).

In Figure 6*B* the broad trend of Phanerozoic change in the strontium isotope ratio is plotted adjacent to the eustatic curve of Figure 5*A*. Whereas there is good general agreement between the trend and falling sea level since the Cretaceous, as Spooner pointed out, the minimum is in the late Jurassic, not the late Cretaceous, as might have been predicted from Spooner's hypothesis. Similarly, the Permo-Triassic interval fails to show the

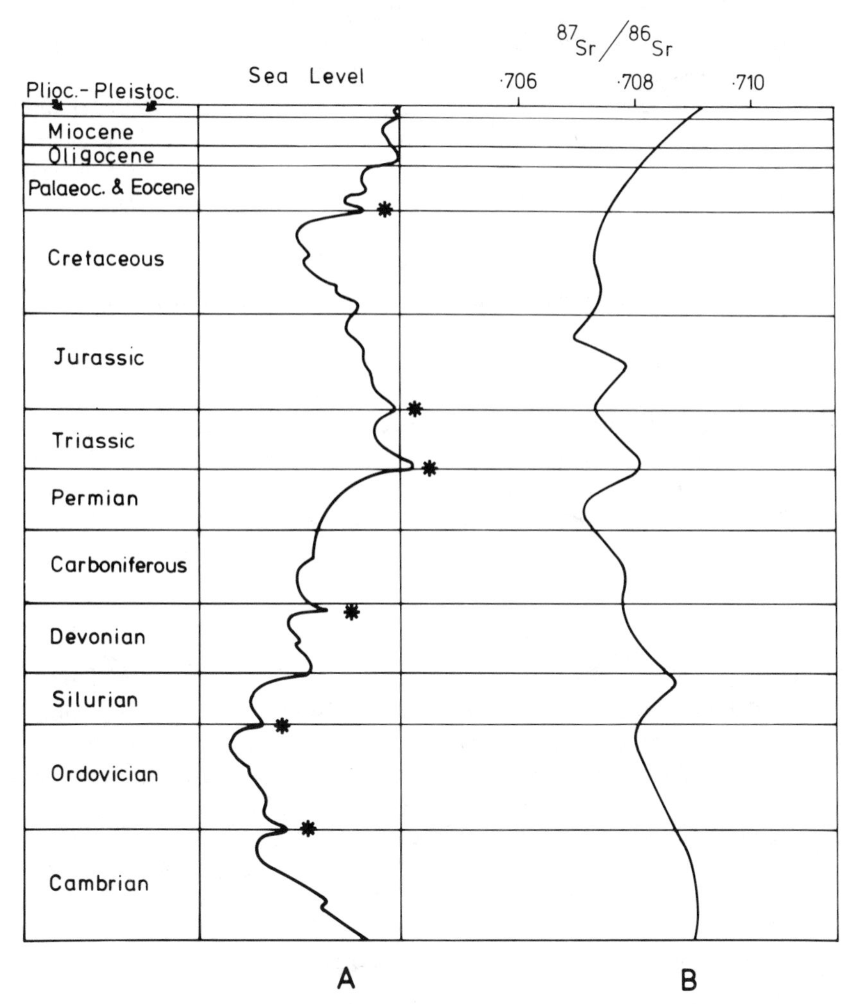

Figure 6 *A*. The Phanerozoic eustatic curve of Figure 5*A*, with asterisks signifying the mass-extinction phases of Newell (1967). *B*. Curve representing the changing $^{87}Sr/^{86}Sr$ ratio through the Phanerozoic, based on Veizer & Compston (1974, Figure 5). In the Veizer & Compston diagram the curve is represented by a band signifying a modal tendency of a wide scatter of data points.

conspicuous peak that would be expected from a simple runoff–land area correlation, with the curve from the Carboniferous rising, not falling. If, however, climatic variations are considered, some form of runoff hypothesis can be salvaged. Thus the Permo-Triassic was a time of widespread aridity after the humidity of the Carboniferous (Frakes 1979), so that runoff would have been correspondingly reduced. Furthermore, there is good evidence that runoff increased from the Jurassic to the Cretaceous. Thus, contrary to Frakes' (1979) claim that a trend of increasing global aridity through the Jurassic continued into the Cretaceous, Middle and Upper Jurassic evaporite-bearing red beds in the United States Western Interior are succeeded by Cretaceous coal measures, while in large areas of western and southern Europe, carbonate- and evaporite-bearing "Purbeck" facies are succeeded by coarse siliciclastic paralic and fluvio-deltaic "Wealden" facies. A similar pattern of change in the Middle East persuaded Murris (1980) of a Jurassic to Cretaceous change to a more humid climate. More generally, a global diminution in evaporite deposition (Meyerhoff 1970, Gordon 1975) is consistent with a change from a more arid Jurassic to a more humid Cretaceous climate.

In earlier times, the late Silurian–early Devonian isotope peak corresponds well with the conspicuous sea-level low in the eustatic curve and thus with a simple runoff–land area relationship, as does the fall through the Devonian and the fall from the Cambrian to a late Ordovician trough. The general level of the early Paleozoic curve is anomalously high compared with the rest of the Phanerozoic, since the eustatic curve indicates a minimum of exposed land. It should, however, be borne in mind that continental runoff must have been higher before the Devonian because of the absence or sparsity of vegetation cover. (This is an evolutionary phenomenon, of course, and not a function of the degree of aridity.)

Thus, while a good correlation may exist between the amount of continental runoff and the strontium isotope ratio of seawater, this amount is a function of at least three variables: area of exposed land, climate, and extent of vegetation cover. Further isotope data are required to confirm the reliability of the Veizer & Compston curve. If the effect of sea-level change can be partitioned out, there is promise of an independent monitor both of climatic change and of the evolution of terrestrial vegetation in the Paleozoic.

CORRELATION WITH BIOGEOGRAPHY AND ORGANIC EVOLUTION

Compared with the open ocean, epicontinental seas are much more likely to have been affected by changes in the physical environment. Even a quite modest change of sea level could have had significant environmental consequences, not only in the extension of marine habitats but in variations of temperature, salinity, and oxygen content within large bodies of shallow water with somewhat restricted interchange with ocean water (Hallam 1981a).

In terms of biogeography one might reasonably predict that organisms occurring in the shallower-water deposits of, for instance, cratonic interiors should exhibit a greater tendency to endemism than deeper-water, more offshore organisms because of their greater tendency to isolation at times of regression. That this is true for groups as different, and as widely separated in time, as Ordovician trilobites (Fortey 1975, 1984) and Jurassic bivalves (Hallam 1977b) suggests the existence of a phenomenon of more general importance.

Furthermore, it follows that times when epicontinental seas were widespread should correlate with times when the incidence of cosmopolitan faunas was relatively high; conversely, provincial marine differentiation of neritic organisms should be high at times of regression. This prediction is borne out for Jurassic mollusks (Hallam 1977b), and, more generally, for the Paleozoic. Thus there was a relatively high incidence of endemism in the early Devonian, late Carboniferous, and late Permian, and a low incidence in the late Ordovician, mid-Silurian, and late Devonian to early Carboniferous (Ziegler et al 1981). The degree of endemism does not, however, invariably fall with sea level, or at least with the reduced extent of sea; much depends on the particular geographic situation. Thus the creation of a new seaway in the North American Western Interior in late Cretaceous times led to the evolution of endemic ammonites (Kennedy & Cobban 1976) and bivalves (Kauffman 1973).

The grosser patterns of evolution can be considered in terms of extinctions and radiations. Since episodes of mass extinction and radiation tend to coincide for a diversity of groups of different biology and habitat, this argues against the Darwinian view that emphasizes the paramount importance of biotic competition, and in favor of first-order control by the physical environment (Hallam 1983b).

Newell (1967) was the first to relate explicitly episodes of mass extinction among diverse animal families to eustatic falls of sea level. Of the major extinction phases he recognized (at or close to the end of the Cambrian, Ordovician, Devonian, Permian, Triassic, and Cretaceous), all but the first have recently been confirmed by the statistical analysis of Raup & Sepkoski (1982). Details of the groups concerned are given in Newell (1967) and Hallam (1981a).

Newell's extinction phases are indicated in Figure 6*A*, which confirms his conclusion about the correlation with sea-level falls. The detailed analysis by Schopf (1974) and Simberloff (1974) demonstrates that the fall in marine invertebrate diversity through the Permian correlates well with the inferred fall of sea level and accords well with ecological predictions based on the so-called "species-area relationship." As the area of epicontinental sea habitat declines, so the extinction rate increases; hence, the fall in diversity. There are some statistical grounds for believing that the species-area relationship may hold for the Phanerozoic as a whole (Sepkoski 1976, Flessa & Sepkoski 1978).

Doubt remains in particular cases about whether the extinction rate increased directly because of regressions, or whether it was because of the widespread anoxic bottom conditions that characteristically mark the beginning of the succeeding transgression (e.g. Hallam 1981b). The latter phenomenon was clearly more significant in causing the mass extinction of marine invertebrates in Europe during the early Toarcian (Hallam 1977b). Many species endemic to Europe disappeared, and the seas were subsequently repopulated by immigrants from the proto-Pacific. This example shows that physical events on a regional scale may be significant promoters of extinction where endemism is involved, although the overall controlling event was in fact eustatic. Whether, more generally, the immediate cause of extinction was regression leading to local emergence and more variable water temperatures and salinities in the shallower seas that remained, or instead was widespread anoxia, the end result would clearly have been increased environmental stress and drastic reduction in habitable area. As might be anticipated, the invertebrates most vulnerable to extinction, apart from ammonites and planktonic forams, tended to be those adapted to shallower-water reefal or perireefal habitats (Hallam 1981a).

The full ecological implications of sea-level fall have yet to be worked out. A brief consideration of the well-documented Pleistocene sea-level oscillations caused by glaciation and deglaciation suggests that these events do not by and large correlate with episodes of pronounced extinction or speciation. Indeed, the characteristic response of both terrestrial and marine organisms to the pronounced climatic changes of the Pleistocene has been to migrate to ecological refuges, in effect to track their environment. It might therefore be wondered why organisms did not respond in a similar way to the much slower changes of sea level in the lengthy periods of climatic equability.

At least two possibilities suggest themselves. Perhaps the Pleistocene sea-level falls, though dramatically rapid in geological terms, were too short-lived to have the kind of environmental impact required to cause extinction. Or perhaps the increasingly unstable environments of the late Cenozoic associated with climatic deterioration caused a selection for eurytopic organisms well adapted to withstand environmental instability. In contrast, the comparative stability of, for instance, Mesozoic environments might have allowed the establishment of complex ecosystems characterized by comparatively stenotopic organisms, which would have been vulnerable to even modest environmental vicissitudes.

Another point to be noted from Figure 6*A* is that not all major sea-level falls correlate with significant mass extinction events at the family level, although this does not of course rule out extinctions at generic and species levels. This is perhaps most clear in the case of the late Silurian–early Devonian eustatic low stand. Perhaps the most critical event is a relatively sudden fall, as appears to have happened at the end of the Ordovician, Permian, Cretaceous, and probably Triassic, that is too rapid for many or most organisms to adapt to.

It is especially intriguing to observe that the end-Permian, end-Triassic, and end-Cretaceous mass extinctions of marine groups coincide with mass extinctions of large terrestrial reptiles (Bakker 1977, Tucker & Benton 1982, Cooper 1982). Another such correlation is in the late Eocene, a time of mass extinction of archaic land mammals (Colbert 1955) and marine microfauna (Benson 1975, Corliss 1979). The controlling event in this case was probably a pronounced climatic decline associated with Antarctic glaciation, which had as one by-product a marked sea-level fall (see below), but such a significant glacially induced event can be ruled out for the other three correlated extinction events. A climatic factor may, however, be involved, because marine regressions will serve to increase the seasonal extremes of continental climate. In particular, Pangea at the end of the Paleozoic must have experienced a climate of extreme continentality, not only because of its coherence (Valentine & Moores 1972) but also because of the high albedo of extensive low-latitude deserts (Barron et al 1980).

Attention has recently been focused on an extraterrestrial catastrophe to account for the end-Cretaceous extinction of the dinosaurs and other organisms, with either asteroid or cometary impact giving rise, respectively, to inhibition of photosynthesis by ejected dust (Alvarez et al 1980) or oceanic poisoning of microplankton leading to increased CO_2 production and hence atmospheric temperature rise (Hsü et al 1982). The end-Cretaceous event was by no means catastrophic to many marine and continental groups, however, least of all apparently to tropical land plants, the group that should have been most affected by photosynthesis inhibition (Hancock 1967, Bakker 1977, Clemens et al 1981, Birkelund & Hakansson 1982). The end-Permian and end-Triassic reptile extinctions indicate that the dinosaur extinction was not unique, and in general the end-Cretaceous extinction event affected the world's fauna much less than the end-Permian one, when it is estimated that perhaps as many as 96% of marine species died out (Raup 1979).

The extinction of calcareous plankton at the end of the Cretaceous is really a greater enigma than the contemporary or near-contemporary extinction of the dinosaurs in terms of plausible terrestrial events. In particular it is difficult to see how such pelagic groups as globigerinid forams should have been affected by epicontinental sea regression. Nevertheless, Hart (1980) has proposed a factually well-supported model that relates their explosive diversification in the late Cretaceous to the increasing depth of epicontinental seas. The obvious implication, surprising as it may seem, is that the *decreasing* depth of such seas might have had a deleterious influence.

Not only is there in general a good correlation between eustatically induced regressions and increased extinction rates, but the converse is also true: major transgressions frequently correlate with mass radiations of marine organisms, presumably as a consequence of improvements in habitat area and quality. Thus, both the Cambrian and Ordovician invertebrate radiations have been related to contemporary sea-level rise (Brasier 1979, House 1967). Among many other examples, the successive Triassic and Jurassic eustatic rises correlate with two successive phases of ammonite radiation. Most spectacular of all is the explosive diversification of a large number of groups in the mid-Cretaceous (such as coccolithophores, foraminifera, diatoms, dinoflagellates, deep-sea ichnofauna, veneroid bivalves, neogastropods, crabs, and teleost fish) near the start of the biggest rise of sea level since the mid-Paleozoic (Hallam 1983b).

Smaller-scale evolutionary effects should not be neglected. Fortey (1984) has observed that the generic longevity of trilobites that lived in deeper-water, extracratonic habitats is greater than shallower-water, more inshore forms. This is presumably a consequence of the lower predictability and higher stress of the shallower-water habitats, promoting higher extinction and hence higher speciation rates at times of sea-level change on either the regional or global scale. A similar pattern has long been recognized among Jurassic ammonites. The deeper-water suborder Phylloceratina had higher generic longevity and are hence much less useful stratigraphically than the shallower-water Ammonitina. Even within the Ammonitina, there may be depth-related differences. Thus, Phelps (1982) found that within the Pliensbachian family Liparoceratidae, the species longevity of *Liparoceras*, inferred on facies grounds to have lived in a deeper-water habitat, is greater than contemporary shallower-water *Androgynoceras*. I have recognized a similar pattern among Sinemurian Arietitidae. The large genera of this family, such as *Arietites* and *Coroniceras*, are common only in shallower-water facies and had only brief time spans, whereas the small *Arnioceras* is abundant only in deeper-water shale facies and had an appreciably greater range in time.

There is increasing evidence that for many species, population increases ("radiations"), migrations, and extinctions can be related to transgressions and regressions on a regional or global scale in a way parallel to that for mass radiations and extinctions (e.g. Hallam 1983a). This suggests that events such as those at the end of the Paleozoic and Mesozoic may be merely the end members of a whole spectrum of terrestrially induced physical events that affected the biosphere.

CAUSES OF SEA-LEVEL CHANGE

Of the various possible causes of eustatic changes, only two are of any significance: melting and freezing of polar ice caps, and changes in the volume of the ocean basins (Donovan & Jones 1979). These have been, respectively, termed glacio- and tectonoeustasy (Fairbridge 1961).

With regard to glacioeustasy, the complete melting of all present land ice should cause a sea-level rise of 40 to 50 m (Pitman 1978). Sea-level depression at the time of the maximum volume of Pleistocene ice sheets is less certain, but Donovan & Jones (1979) make an approximate estimate of 100 m. For the late Tertiary, oxygen isotope data have suggested to Shackleton & Kennett (1975) that the Antarctic ice cap was first established about mid-Miocene times. The end-Miocene eustatic fall and subsequent early Pliocene rise can therefore plausibly be attributed to glaciation and deglaciation phenomena, respectively (Adams et al 1977).

There is abundant evidence from terrestrial plants, marine macrobenthos, and microplankton and oxygen isotopes of a dramatic fall in global temperatures across the Eocene-Oligocene boundary. Together with evidence of increased current scouring on the deep ocean floor, the marine data indicate the formation of the layer of cold, deep water known as the *psychrosphere* and an increased rate of oceanic circulation (Hallam 1981a). Sedimentary evidence from the peri-Antarctic sea floor, such as dropstones, has been held to indicate the initiation of sea (but not land) ice at this time (Kennett 1977). The evidence cited here shows that the most dramatic sea-level fall in the Tertiary took place in the late Eocene to early Oligocene, implying rather strongly that substantial glacial buildup had occurred on the Antarctic continent, even though the oxygen isotope data have been held not to support this. In fact, a shift to heavier isotopes can indicate *either* colder water *or* raised salinity due to the extraction of isotopically lighter water to produce polar ice, *or* both.

Further back in time there is no evidence of polar ice caps, and abundant evidence of global climatic equability, until the mid-Permian (Frakes 1979); consequently, any sea-level changes must have been tectonoeustatic in origin. Even for the time when the Gondwana ice sheets were in existence, purely glacioeustatic interpretations pose problems. Thus, on a small scale, the frequently assumed minor eustatic fluctuations produced by waxing and waning of ice sheets during alternating glacial and interglacial episodes carry implications for sedimentation in, for instance, the English Upper Carboniferous Coal Measures, deposited in a paralic setting close to sea level. In such a setting, eustatic lows should be signified by horizons of deep erosional channeling, but such features are rare (A. P. Heward, personal communication). On a larger scale, the waning and ultimate disappearance of the Gondwana ice sheets in the mid- to late Permian coincided with a *fall*, not a rise, of sea level, implying that glacioeustatic effects were swamped by tectonoeustatic effects.

It has been widely assumed that the end-Ordovician sea-level fall and the early Silurian rise were glacioeustatic in origin, with both related to the growth and decay of the Saharan ice sheet (Frakes 1979), but no other convincing case has been made for glacioeustasy in the Paleozoic; thus, tectonoeustasy must have been the process of overriding importance.

Turning then to tectonoeustasy, attention has been concentrated on the late Cretaceous rise and subsequent fall. Hays & Pitman (1973) correlated the rise with an episode of accelerated seafloor spreading, which would have caused a significant expansion in volume of the oceanic ridge system. This interpretation is based on the age-depth relationship of mid-oceanic ridges, which approximately follows a time-dependent exponential cooling curve (Pitman 1978). However, a more recent analysis of Cretaceous magnetic anomalies and a revised chronology significantly reduce the rate of acceleration from the early to the late Cretaceous demanded by the quantitative analysis of Hays & Pitman (1973) and Pitman (1978) (Hallam et al 1984).

Two other processes have been proposed to account for these events. The mid- to late Cretaceous was a time of significant increase in the length of the ocean ridge system, with the Atlantic Rift extending north and south and the Gondwana continents disintegrating. This must have caused a rise of sea level even without any acceleration in the spreading rate. Post-Cretaceous fall could, to some extent, relate to the progressive consumption through subduction beneath Asia and the Americas of ridges separating the Kula from the Farallon, and the Farallon from the Phoenix plate, together with the Pacific-Kula ridge (Hallam 1977a, 1980).

Schlanger et al (1981) argue that the late Cretaceous sea-level rise was mainly the result of mid-plate volcanism producing swells on the Pacific and Farallon plates, which began to subside about 70 m.y. ago, thereby causing sea-level fall.

All these tectonoeustatic processes, which have a common cause in variations of heat flow from the mantle, could have operated to variable extents throughout the Phanerozoic, and a condition of stable sea level for geologically significant periods of time should in consequence be considered an exceptional circumstance. It is sometimes possible to propose a plausible link between plate tectonic events and eustasy. Thus, Anderton (1982) explicitly relates the Cambrian sea-level rise to the opening of the Iapetus Ocean, associated with the creation of a spreading ridge. Valentine & Moores (1972) argued that the late Paleozoic sea-level fall was the result of the suturing of several continents to produce the supercontinent of Pangea, which caused the cessation of spreading, the collapse of ocean ridges, and the deformation and underthrusting during continent-continent collisions leading to emergence.

The Mesozoic rise of sea level corresponds broadly to the subsequent disintegration of Pangea and the concomitant growth of new spreading ridges, but new DSDP evidence (Sheridan 1983) supports the contention made on other grounds that the oldest, central sector of the Atlantic did not commence opening until late mid-Jurassic times (Hallam 1980). Thus the mid-Triassic and early Jurassic sea-level rises are unaccounted for by the Pangea disintegration.

According to Pitman (1978), the maximum rate of sea-level change producible by the growth and decay of ocean ridges is ~ 1 cm/10^3 yr, about three orders of magnitude slower than Pleistocene glacioeustatic change. This rate is slightly slower than the very approximate figures for short-term eustatic oscillations of Hallam (1978) and Ramsbottom (1979) cited earlier, and much slower than that for eustatic falls derivable from the data of Vail et al (1977). As noted earlier, the Vail "geologically instantaneous" eustatic falls appear to be an artifact of the method of analysis, but there remains a problem concerning the rates of sea-level oscillations, nevertheless.

Pitman's (1978) solution, based on a quantitative analysis of the interaction of sea-level change with sedimentation and subsidence on passive continental margins, is to demonstrate that transgressive and regressive events may not be simply indicative of eustatic rise and fall but of changes in the rate of sea-level change. Thus a decrease in the rate of sea-level rise and an increase in the rate of sea-level fall may result in regressions, while an increase in the rate of sea-level rise and a decrease in the rate of sea-level fall may produce transgressions.

By checking, however, continental margin stratigraphy with correlative events in the deep sea (Thiede 1981) or cratonic interiors, it is sometimes possible to eliminate such "hinge" phenomena from consideration and to indicate genuine sea-level oscillations. If the inferred rates of sea-level change appear too high, it could be that (*a*) the facies depth range in inferred eustatic sedimentary cycles has been overestimated, (*b*) the time interval has been underestimated, and/or (*c*) undetected regional tectonic factors have intervened. Only detailed facies analysis in a refined biostratigraphic framework will convincingly resolve such questions.

There remains the problem of how to account for the long-term change. When the considerable changes related apparently to the successive formation and disintegration of Pangea are allowed for, there appears a secular trend toward fall in sea level through the course of the Phanerozoic. To some extent this can be accounted for by successive phases of orogeny ("Caledonian," "Hercynian," etc), resulting in marginal continental accretion along subduction zones leading to increased mean continental thickness. This cannot, however, be the whole story because shield areas, unaffected by Phanerozoic orogeny, also exhibit evidence of the more or less secular regression (Hallam 1977a).

Since a loss of ocean water is extremely unlikely, as is extensive underplating of continental crust away from subduction zones, the most plausible explanation involves a gradual reduction through time of heat flow from the mantle (Turcotte & Burke 1978, Brown 1984). As the oceanic lithosphere has cooled, thermal expansion has diminished and hence the ocean basins have deepened by a modest amount through the course of the Phanerozoic, causing seawater to be drained off the continents.

Such an interpretation carries with it, of course, the implication that Precambrian seas were generally more extensive than even those of the Paleozoic. The late Precambrian regression, giving rise to a hiatus sometimes known as the *Lipalian interval*, and subsequent Cambrian transgression are more likely to have a tectonoeustatic origin rather than a glacioeustatic one (Matthews & Cowie 1979). The regression is perhaps analogous to that which took place at the end of the Paleozoic.

SUMMARY

Eustasy can be studied using a variety of methods, including areal plots of the changing temporal distribution of marine deposits, facies analysis of stratigraphic sequences, and seismic stratigraphy, allied with the best available means of biostratigraphic correlation. The results of these various methods are then compared for use in eliminating the complicating effects of local and regional tectonics in the interpretation of sea-level oscillations. The determination of the rate and amount of sea-level change is also discussed.

Use is made of areal plots, in conjunction with hypsometric data and a variety of stratigraphic sequence evidence, to produce a eustatic curve for the pre-Quaternary Phanerozoic. Notwithstanding its necessarily tentative and provisional nature, this curve is considered to be a more accurate representation of Phanerozoic eustasy than that of Vail et al (1977). There is an overall trend of declining sea level from a late Ordovician high stand ~600 m above the present, on which several major and a larger number of minor oscillations are imposed. The most important oscillations are (*a*) a late Paleozoic fall to an end-Permian low stand and (*b*) a late Mesozoic rise culminating in the late Cretaceous, when sea level was ~350 m higher than today. A more accurate eustatic curve will only be produced by a collaborative effort involving detailed biostratigraphic correlation, stage-by-stage facies analysis, paleogeographic map production, seismic stratigraphy, and regional tectonic analysis.

A strong correlation exists between marine transgressions and the distribution of organic-rich shales and phosphorites, as does a somewhat weaker correlation between regressions and oolitic ironstones. Correlations with $^{13}C/^{12}C$ and $^{87}Sr/^{86}Sr$ are also noted. As regards the latter, there appears to be covariation between the increase in the $^{87}Sr/^{86}Sr$ ratio in seawater and the amount of continental runoff. Data for the whole Phanerozoic suggest that runoff is a function of the continental area, the degree of aridity, and the evolutionary development of a mantle of terrestrial vegetation in the early to mid-Paleozoic.

There is also a good correlation between eustatic changes and events in the biosphere. Times of low sea level correlate with episodes of increased endemicity and extinction of marine organisms, while times of high sea level are associated with episodes of increased pandemicity and radiations. The celebrated end-Cretaceous mass extinction is not unique but simply one of the most spectacular end-members of a whole series of extinction events that appear to relate to global or regional regressions of corresponding magnitude. This makes it questionable that extraterrestrially induced catastrophes need be invoked.

Changes in the volume of the ocean basins (tectonoeustasy) are generally more important than glaciation and deglaciation in controlling sea-level variations prior to the Quaternary, but the early Oligocene and late Ordovician regressions are probably due to the growth of polar ice caps. The gross features of the Phanerozoic eustatic curve are primarily the result of plate tectonic processes involving continental collision and disintegration (with concomitant changes in ocean ridge volume) and a secular withdrawal of epicontinental seas due to ocean-basin deepening related to the slow cooling of the lithosphere. Smaller-scale oscillatory changes are more controversial and demand more intensive study using a wide array of stratigraphic, sedimentological, and paleoecological methods.

Bearing in mind one of the morals of the continental drift controversy, and the hostility of an earlier generation of geophysicists to Wegener's proposals for the controlling processes, we should seek to establish eustatic oscillations firmly on empirical grounds and make the results as quantitative as possible, rather than dismiss them as unlikely because no appropriate controlling force can be conceived. At least three tectonoeustatic processes can be proposed, but too little is known of their relative importance in the past to permit categorical statements at the present time. On the other hand, the effects of regional as opposed to local epeirogenic movements have not always been adequately appreciated by students of eustasy; these need much more research.

According to Brown (1984), "Throughout the Earth's history, internal heat production has provided the dominant constraint on lithosphere evolution." It is becoming increasingly evident that it has also exercised a major effect on biosphere evolution, thereby helping to vindicate the claim made many years ago by Chamberlin (1909) that significant organic changes through time have ultimately been under the control of diastrophism.

Literature Cited

Adams, C. G., Benson, R. H., Kidd, R. B., Ryan, W. B. F., Wright, R. C. 1977. The Messinian salinity crisis and evidence of late Miocene eustatic changes in the world ocean. *Nature* 269: 383–86

Ager, D. V. 1980. *The Geology of Europe.* London: McGraw-Hill. 535 pp.

Alvarez, L. W., Alvarez, W., Asaro, F., Michel, H. V. 1980. Extraterrestrial cause for the Cretaceous-Tertiary extinction: experiment and theory. *Science* 208: 1095–1108

Anderton, R. 1982. Dalradian deposition and the late Precambrian-Cambrian history of the N. Atlantic region: a review of the early history of the Iapetus Ocean. *J. Geol. Soc. London* 139: 423–34

Arthur, M. A., Jenkyns, H. C. 1981. Phosphorites and paleoceanography. *Oceanologica Acta. Proc. Int. Geol. Congr., 26th, Geol. Oceans Symp., Paris, 1980*, pp. 83–96

Bakker, R. T. 1977. Tetrapod mass extinctions—a model of the regulation of speciation rates and immigration by cycles of topographic diversity. In *Patterns of Evolution as Illustrated by the Fossil Record*, ed. A. Hallam, pp. 439–68. Amsterdam: Elsevier

Bally, A. W. 1981. Basins and subsidence. *Am. Geophys. Union Geodyn. Ser.* 1: 5–20

Barron, E. J., Sloan, J. L., Harrison, C. G. A. 1980. Potential significance of land-sea distribution and surface albedo variations as a climatic factor: 180 m.y. to the present. *Palaeogeogr. Palaeoclimatol. Palaeoecol.* 30: 17–40

Benson, R. A. 1975. The origin of the psychrosphere as recorded in changes in deep-sea ostracode assemblages. *Lethaia* 8: 69–83

Birkelund, T., Hakansson, E. 1982. The terminal Cretaceous extinction in Boreal shelf seas—a multicausal event. *Geol. Soc. Am. Spec. Pap. No. 190*, pp. 373–84

Bond, G. 1976. Evidence for continental subsidence in North America during the late Cretaceous global submergence. *Geology* 4: 557–60

Bond, G. 1978. Speculations on real sea-level changes, and vertical motions of continents at selected times in the Cretaceous and Tertiary periods. *Geology* 6: 247–50

Brasier, M. D. 1979. The Cambrian radiation event. In *The Origin of Major Invertebrate Groups*, ed. M. R. House, pp. 103–59. London/New York: Academic

Brenchley, P. J., Newall, G. 1980. A facies analysis of Upper Ordovician regressive sequences in the Oslo region, Norway—a record of glacioeustatic changes. *Palaeogeogr. Palaeoclimatol. Palaeoecol.* 31: 1–38

Brown, D. A., Campbell, K. S. W., Crook, K. A. W. 1968. *The Geological Evolution of Australia and New Zealand.* Oxford: Pergamon. 409 pp.

Brown, G. C. 1984. Processes and problems in the continental lithosphere: geological history and physical implications. In *Geochronology and the Geological Record*, ed. N. J. Snelling. Geol. Soc. London. In press

Carter, A. N. 1978. Contrasts between oceanic and continental "unconformities" in the Oligocene of the Australian region. *Nature* 274: 152–53

Chamberlin, T. C. 1909. Diastrophism as the ultimate basis of correlation. *J. Geol.* 17: 689–93
Clemens, W. A., Archibald, J. D., Hickey, L. J. 1981. Out with a whimper not a bang. *Paleobiology* 7:293–98
Cogley, J. C. 1981. Late Phanerozoic extent of dry land. *Nature* 291:56–58
Colbert, E. H. 1955. *Evolution of the Vertebrates.* New York: Wiley. 479 pp.
Cooper, M. R. 1977. Eustacy during the Cretaceous: its implications and importance. *Palaeogeogr. Palaeoclimatol. Palaeoecol.* 21:165–208
Cooper, M. R. 1982. A mid-Permian to earliest Jurassic tetrapod biostratigraphy and its significance. *Arnoldia Zimbabwe* 9:77–104
Corbin, S. G. 1980. *A facies analysis of the Lower-Middle Jurassic boundary beds of north-west Europe.* PhD thesis. Univ. Birmingham, Engl.
Corliss, B. H. 1979. Response of deep-sea benthonic Foraminifera to development of the psychrosphere near the Eocene/Oligocene boundary. *Nature* 282:63–65
Dewey, J. F. 1982. Plate tectonics and the evolution of the British Isles. *J. Geol. Soc. London* 139:371–412
Donovan, D. T., Jones, E. J. W. 1979. Causes of world-wide changes in sea level. *J. Geol. Soc. London* 136:187–92
Egyed, L. 1956. The change of the Earth's dimensions determined from paleogeographical data. *Geofis. Pura Appl.* 33:42–48
Fairbridge, R. W. 1961. Eustatic changes of sea level. In *Physics and Chemistry of the Earth,* ed. L. H. Ahrens, pp. 99–185. Oxford: Pergamon
Flessa, K. W., Sepkoski, J. J. 1978. On the relationship between Phanerozoic diversity and changes in habitable area. *Paleobiology* 4:359–66
Forney, G. G. 1975. Permo-Triassic sea level change. *J. Geol.* 83:773–79
Fortey, R. A. 1975. Early Ordovician trilobite communities. *Fossils and Strata* 4: 339–60
Fortey, R. A. 1984. Global earlier Ordovician transgressions and regressions and their biological implications. In *Aspects of the Ordovician System,* ed. D. L. Bruton, pp. 37–50. Oslo: Universitetsforlaget
Frakes, L. A. 1979. *Climates Through Geologic Time.* Amsterdam: Elsevier. 310 pp.
Furon, R. 1963. *The Geology of Africa.* Edinburgh: Oliver & Boyd. 377 pp.
Garrels, R. M., McKenzie, F. T. 1971. *Evolution of Sedimentary Rocks.* New York: Norton. 397 pp.
Gordon, W. A. 1975. Distribution by latitude of Phanerozoic evaporite deposits. *J. Geol.* 53:671–84
Hallam, A. 1977a. Secular changes in marine inundation of USSR and North America through the Phanerozoic. *Nature* 269: 769–72
Hallam, A. 1977b. Jurassic bivalve biogeography. *Paleobiology* 3:58–73
Hallam, A. 1978. Eustatic cycles in the Jurassic. *Palaeogeogr. Palaeoclimatol. Palaeoecol.* 23:1–32
Hallam, A. 1980. A reassessment of the fit of Pangaea components and the time of their initial breakup. In *The Continental Crust and its Mineral Deposits,* ed. D. W. Strangway, pp. 375–87. *Geol. Assoc. Can. Spec. Pap. 20*
Hallam, A. 1981a. *Facies Interpretation and the Stratigraphic Record.* San Francisco/Oxford: Freeman. 291 pp.
Hallam, A. 1981b. The end-Triassic bivalve extinction event. *Palaeogeogr. Palaeoclimatol. Palaeoecol.* 35:1–44
Hallam, A. 1981c. A revised sea-level curve for the early Jurassic. *J. Geol. Soc. London* 138:735–43
Hallam, A. 1983a. Patterns of speciation in Jurassic *Gryphaea. Paleobiology* 8:354–66
Hallam, A. 1983b. Plate tectonics and evolution. In *Evolution from Molecules to Men,* ed. D. S. Bendall, pp. 367–86. Cambridge Univ. Press.
Hallam, A., Bradshaw, M. J. 1979. Bituminous shales and oolitic ironstones as indicators of transgressions and regressions. *J. Geol. Soc. London* 136:157–64
Hallam, A., Sellwood, B. W. 1976. Middle Mesozoic sedimentation in relation to tectonics in the British area. *J. Geol.* 84:302–21
Hallam, A., Hancock, J. M., LaBrecque, J. L., Lowrie, W., Channell, J. E. T. 1984. Jurassic and Cretaceous geochronology and Jurassic to Palaeogene magnetostratigraphy. In *Geochronology and the Geological Record,* ed. N. J. Snelling. Geol. Soc. London. In press
Hancock, J. M. 1967. Some Cretaceous-Tertiary marine faunal changes. In *The Fossil Record,* ed. W. B. Harland, et al, pp. 91–103. Geol. Soc. London
Hancock, J. M., Kauffman, E. G. 1979. The great transgressions of the late Cretaceous. *J. Geol. Soc. London* 136:175–86
Hart, M. B. 1980. A water depth model for the evolution of the planktonic Foraminiferida. *Nature* 286:252–54
Hays, J. D., Pitman, W. C. 1973. Lithospheric plate motion, sea level changes and climatic and ecological consequences. *Nature* 246:16–22

Holmes, A. 1965. *Principles of Physical Geology*. London: Nelson. 1288 pp. 2nd ed.

House, M. R. 1967. Fluctuations in the evolution of Palaeozoic invertebrates. In *The Fossil Record*, ed. W. B. Harland, et al, pp. 41–54. Geol. Soc. London

House, M. R. 1975a. Facies and time in Devonian tropical areas. *Proc. Yorks. Geol. Soc.* 40: 233–87

House, M. R. 1975b. Faunas and time in the marine Devonian. *Proc. Yorks. Geol. Soc.* 40: 459–90

Hsü, K. J., He, Q., McKenzie, J. A., Weissert, H., Perchnie, K., et al. 1982. Mass mortality and its environmental and evolutionary consequences. *Science* 216: 249–56

Jenkyns, H. C. 1980. Cretaceous anoxic events: from continents to oceans. *J. Geol. Soc. London* 137: 171–88

Kauffman, E. G. 1973. Cretaceous Bivalvia. In *Atlas of Palaeobiogeography*, ed. A. Hallam, pp. 353–83. Amsterdam: Elsevier

Kennedy, W. J., Cobban, W. A. 1976. Aspects of ammonite biology, biostratigraphy and biogeography. *Spec. Pap. Palaeontol.* 17: 1–94

Kennett, J. P. 1977. Cenozoic evolution of Antarctic glaciation, the circum-Antarctic ocean, and their impact of global paleoceanography. *J. Geophys. Res.* 82: 3843–60

Lee, J. S. 1939. *The Geology of China*. London: Murby. 528 pp.

Leggett, J. K., McKerrow, W. S., Cocks, L. R. M., Rickards, R. B. 1981. Periodicity in the early Palaeozoic marine realm. *J. Geol. Soc. London* 138: 167–76

Lenz, A. 1982. Ordovician to Devonian sea-level changes in western and northern Canada. *Can. J. Earth Sci.* 19: 1919–32

Levin, H. L. 1978. *The Earth Through Time*. Philadelphia: Saunders. 530 pp.

Loughman, D. L. 1982. *A facies analysis of the Triassic-Jurassic boundary beds of the world, with special reference to north-west Europe and the Americas*. PhD thesis. Univ. Birmingham, Engl.

Loutit, T. S., Keigwin, L. D. 1982. Stable isotope evidence for latest Miocene sea-level fall in the Mediterranean region. *Nature* 300: 163–66

Matthews, S. C., Cowie, J. W. 1979. Early Cambrian transgression. *J. Geol. Soc. London* 136: 133–35

McKenzie, D. P. 1978. Some remarks on the development of sedimentary basins. *Earth Planet. Sci. Lett.* 40: 25–32

Meyerhoff, A. A. 1970. Continental drift. II. High-latitude evaporite deposits and geologic history of Arctic and North Atlantic Oceans. *J. Geol.* 78: 406–44

Murray, G. E. 1961. *Geology of the Atlantic and Gulf Coastal Provinces of North America*. New York: Harper. 692 pp.

Murris, R. J. 1980. Middle East: stratigraphic evolution and oil habitat. *Am. Assoc. Pet. Geol. Bull.* 64: 597–618

Nalivkin, D. V. 1973. *Geology of the USSR*. Edinburgh: Oliver & Boyd. 855 pp.

Newell, N. D. 1967. Revolutions in the history of life. *Geol. Soc. Am. Spec. Pap. No. 89*, pp. 63–91

Officer, C. B., Drake, C. L. 1982. Epeirogenic plate movements. *J. Geol.* 90: 139–53

Olsson, R. K., Miller, K. G., Ungrady, R. E. 1980. Late Oligocene transgression of middle Atlantic coastal plain. *Geology* 8: 549–54

Phelps, M. C. 1982. *A facies and faunal analysis of the Carixian-Domerian boundary beds in north-west Europe*. PhD thesis. Univ. Birmingham, Engl.

Pitman, W. C. 1978. Relationship between eustacy and stratigraphic sequences of passive margins. *Geol. Soc. Am. Bull.* 89: 1389–1403

Pomerol, C. 1973. *Stratigraphie et Paléogéographie. Ere Cenozoïque (Tertiaire et Quaternaire)*. Paris: Doin. 269 pp.

Prothero, D. R., Estes, R. 1980. Late Jurassic lizards from Como Bluff, Wyoming and their palaeobiogeographic significance. *Nature* 286: 484–86

Ramsbottom, W. H. C. 1979. Rates of transgression and regression in the Carboniferous of NW Europe. *J. Geol. Soc. London* 136: 147–53

Ramsbottom, W. H. C. 1981. Eustacy, sea level and local tectonism, with examples from the British Carboniferous. *Proc. Yorks. Geol. Soc.* 43: 473–82

Raup, D. M. 1979. Size of the Permo-Triassic bottleneck and its evolutionary implications. *Science* 206: 217–18

Raup, D. M., Sepkoski, J. J. 1982. Mass extinctions in the marine fossil record. *Science* 215: 1501–3

Schlanger, S. O., Jenkyns, H. C., Premoli-Silva, I. 1981. Volcanism and vertical tectonics in Pacific Basin related to global Cretaceous transgressions. *Earth Planet. Sci. Lett.* 52: 435–49

Scholle, P. A., Arthur, M. A. 1980. Carbon isotope fluctuations in Cretaceous pelagic limestones: potential stratigraphic and petroleum exploration tool. *Am. Assoc. Pet. Geol. Bull.* 64: 67–87

Schopf, T. J. M. 1974. Permo-Triassic extinctions: relation to sea-floor spreading. *J. Geol.* 82: 129–43

Schuchert, C. 1955. *Atlas of Paleogeographic Maps of North America*. New York: Wiley. 155 pp.

Sepkoski, J. J. 1976. Species diversity in the Phanerozoic: species-area effects. *Paleobiology* 2:298–303

Shackleton, N. J., Kennett, J. P. 1975. Paleotemperature history of the Cenozoic and the initiation of Antarctic glaciation: oxygen and carbon isotope analyses in DSDP sites 277, 279 and 281. In *Initial Reports of the Deep Sea Drilling Project*, J. P. Kennett, R. E. Houtz, et al, 29:743–55. Washington DC: GPO

Sheridan, R. E. 1983. Phenomena of pulsation tectonics related to early rifting of the eastern North American continental margin. *Tectonophysics* 94:169–85

Siesser, W. G., Dingle, R. V. 1981. Tertiary sea-level movements around Southern Africa. *J. Geol.* 89:83–96

Simberloff, D. 1974. Permo-Triassic extinctions: effects on an area on biotic equilibrium. *J. Geol.* 82:267–74

Sleep, N. H. 1976. Platform subsidence mechanisms and "eustatic" sea level changes. *Tectonophysics* 36:45–56

Sloss, L. L. 1963. Sequences in the cratonic interior of North America. *Geol. Soc. Am. Bull.* 74:93–113

Sloss, L. L., Speed, R. C. 1974. Relationship of cratonic and continental margin tectonic episodes. In *Tectonics and Sedimentation*, ed. W. R. Dickinson, pp. 89–119. *Soc. Econ. Paleont. Mineral. Spec. Publ. No. 22*

Spooner, E. T. C. 1976. The strontium isotopic composition of sea water, and seawater-oceanic crust interaction. *Earth Planet. Sci. Lett.* 31:167–74

Strakhov, N. M. 1948. *Outlines of Historical Geology.* Moscow: GPO. 756 pp. (In Russian)

Suess, E. 1906. *The Face of the Earth*, Vol. 2. Oxford: Clarendon. 759 pp.

Termier, H., Termier, G. 1952. *Histoire Géologique de la Biosphère*. Paris: Masson. 550 pp.

Thiede, J. 1981. Reworked neritic fossils in Upper Mesozoic and Cenozoic Central Pacific deep-sea sediments monitor sea-level changes. *Science* 211:1422–24

Tucker, M. E., Benton, M. J. 1982. Triassic environments, climates and reptile evolution. *Palaeogeogr. Palaeoclimatol. Palaeoecol.* 40:361–79

Turcotte, D. L., Burke, K. 1978. Global sea-level changes and the thermal structure of the earth. *Earth Planet. Sci. Lett.* 41:341–46

Vail, P. R., Todd, R. G. 1981. Northern North Sea Jurassic unconformities, chronostratigraphy and sea-level changes from seismic stratigraphy. In *Petroleum Geology of the Continental Shelf of North-West Europe*, pp. 216–35. London: Heyden

Vail, P. R., Mitchum, R. M., Todd, R. G., Widmier, J. M., et al. 1977. Seismic stratigraphy and global changes of sea level. In *Stratigraphic Interpretation of Seismic Data*, ed. C. E. Payton, pp. 49–212. *Am. Assoc. Pet. Geol. Mem. No. 26*

Valentine, J. W., Moores, E. M. 1972. Global tectonics and the fossil record. *J. Geol.* 80:167–84

Van Houten, F. B., Bhattacharyya, D. P. 1982. Phanerozoic oolitic ironstones—geologic record and facies model. *Ann. Rev. Earth Planet. Sci.* 10:441–57

Van Houten, F. B., Karasek, R. 1981. Sedimentologic framework of the late Devonian oolitic iron formation, Shatti Valley, west-central Libya. *J. Sediment. Petrol.* 51:415–27

Veizer, J., Compston, W. 1974. $^{87}Sr/^{86}Sr$ composition of seawater during the Phanerozoic. *Geochim. Cosmochim. Acta* 38:1461–84

Watts, A. B. 1982. Tectonic subsidence, flexure and global changes of sea level. *Nature* 297:469–74

Watts, A. B., Steckler, M. S. 1979. Subsidence and eustacy at the continental margins of eastern North America. In *Deep Drilling Results in the Atlantic Ocean: Continental Margins and Paleoenvironment*, ed. M. Talwani, W. Hay, W. B. F. Ryan, pp. 218–34. *Am. Geophys. Union Maurice Ewing Ser. 3*

Wise, D. U. 1974. Continental margins, freeboard and the volumes of continents and oceans through time. In *The Geology of Continental Margins*, ed. K. Burke, C. L. Drake, pp. 45–58. Berlin/New York: Springer

Ziegler, A. M., Bambach, R. K., Parrish, J. T., Barrett, S. F., Gierlowski, E. H., Parker, W. C., Raymond, A., Sepkoski, J. J. 1981. Paleozoic biogeography and climatology. In *Paleobotany, Paleoecology and Evolution*, ed. K. J. Niklas, pp. 231–66. New York: Praeger

Ann. Rev. Earth Planet. Sci. 1984. 12: 245–92

RATES OF EVOLUTION AND THE NOTION OF "LIVING FOSSILS"

Thomas J. M. Schopf

Department of the Geophysical Sciences and Committee on Evolutionary Biology, University of Chicago, Chicago, Illinois 60637

INTRODUCTION

In one sense, any science is only as sound as the knowledge of its state variables. For paleontology, several of those variables are tied up in the topic of rates of evolution. So uncertain is the mean duration of species that when problems in paleontology are subjected to computer simulation, the values that are used extend over four orders of magnitude (10^3 to 10^7 yr; Raup 1981). It should come as no surprise that conclusions over this 10^4 range support vastly different world views about the extent of determinism versus stochasticity in major evolutionary events. On the one hand, this lack of precision means that virtually any answer is possible. On the other hand, it means that the traditional paleontologic assertion in favor of specific deterministic stories must be tempered by a high degree of uncertainty.

The chief body of material to be considered, which was not known to those who formulated the currently prevailing paleontological views on rates of evolution [especially Simpson (1944, 1949, 1953, and largely reaffirmed even in 1983)] lies in information provided on the molecular biology of genomic change. The genome is simply a convenient term for the genetic constitution of an individual or species. Genomic compatibility lies at the basis of reproductive compatibility, and thus genomic change introduces the notion of geonomically regulated reproductive *in*compatibility.

I consider in this article two interrelated aspects of speciation theory. The first is an attempt to integrate the main results from molecular biology that

0084–6597/84/0515–0245$02.00

bear on the topic of rates of evolution. Armed with this perspective, especially as applied to rates of speciation, I then consider the second topic, the now more than 40-year-old view that one can obtain reliable data on rates of species evolution by looking at "living fossils." These taxa, it turns out, are not ancient species, but species with a few prominent primitive traits. Thus, I conclude that "living fossils" are not a problem in speciation theory but rather a problem for developmental biology. More generally, the whole "bradytelic" (extremely slowly evolving) population of species appears to be a mirage resulting from the belief that a literal reading of the fossil record was the best way to proceed in determining rates of evolution. Most probably there is but a single population of speciation rates, and biases of various sorts are responsible for what appears to be slowly evolving species.

AN EVOLUTIONARY HIERARCHY

Rates of evolution may be dissected into many parts. In the present article, the parts considered are the genome, individual, deme (= local population), species, community and biosphere. Figure 1 shows an inferred relationship among these units, or, as I have called them, interacting gears. I have purposely used the metaphor of the gear to emphasize the interconnection between levels in the hierarchy. Also shown in Figure 1 is what changes within a gear (i.e. DNA within a genome, morphology of an individual, etc), and the different modes of explanation for the motor driving each gear. These modes of explanation are (*a*) deterministic explanations, which focus on positive selection, (*b*) limit explanations, which focus on mechanistic constraints, and (*c*) stochastic explanations, which suggest that any of several possible outcomes could in fact occur.

Figure 1 An evolutionary hierarchy. The figure shows the gears (levels in the hierarchy); interactions between the gears; what it is that changes within a gear; three different emphases in the "motors" that may drive each gear [(A) positive deterministic selection, (B) effects of mechanistic constraints, and (C) matters of chance, so as to indicate that among the possible patterns of each gear, the one that obtains in nature was not necessarily preselected]; the turnover time for each gear; and the gear mechanic who customarily studies each level.

Selfish DNA is sensu Doolittle (1982); individual selection sensu Darwin (1859); group selection sensu Wynne-Edwards (1962); species selection sensu Gould (1982) or Stanley (1979); community selection sensu Odum (1959, 1969), Dunbar (1960), and Margalef (1968); and Gaia sensu Lovelock (1979). In each of these emphases on determinism, the gear is treated as though it were an "essence" or an "individual." With regard to gear interactions, the level of success in explaining the interaction is poor for development, good (kin selection) for social groupings, good (clines) for migration, good(?) for trophic relations, and fair for geologic and climatologic control of environments. I especially thank Antoni Hoffman for comments on an earlier draft of this figure (April 1982).

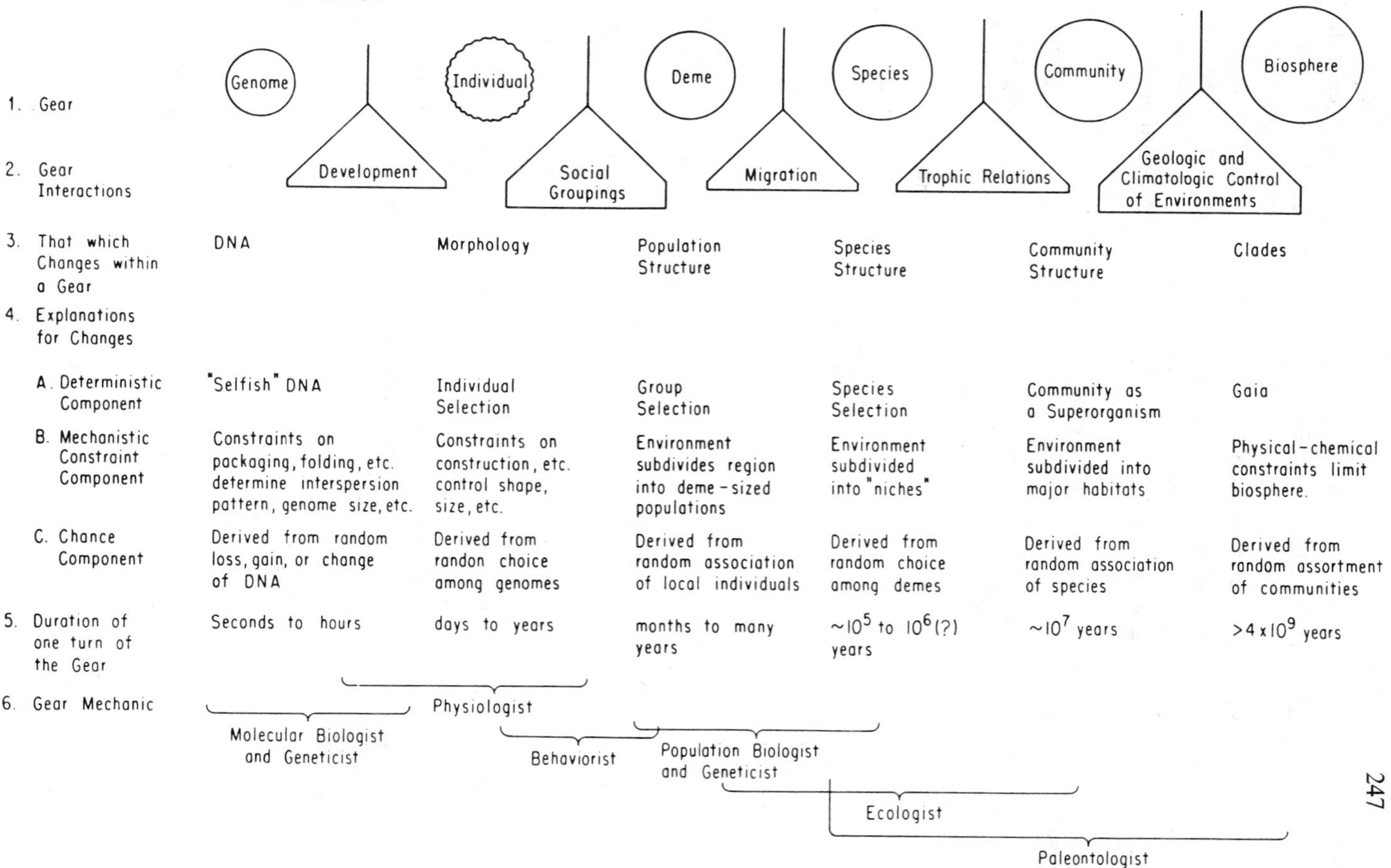
AN EVOLUTIONARY HIERARCHY
1. Gear
2. Gear Interactions
3. That which Changes within a Gear
4. Explanations for Changes
A. Deterministic Component
B. Mechanistic Constraint Component
C. Chance Component
5. Duration of one turn of the Gear
6. Gear Mechanic
Genome
Development
Individual
Social Groupings
Deme
Migration
Species
Trophic Relations
Community
Geologic and Climatologic Control of Environments
Biosphere
DNA
Morphology
Population Structure
Species Structure
Community Structure
Clades
"Selfish" DNA
Individual Selection
Group Selection
Species Selection
Community as a Superorganism
Gaia
Constraints on packaging, folding, etc. determine interspersion pattern, genome size, etc.
Constraints on construction, etc. control shape, size, etc.
Environment subdivides region into deme-sized populations
Environment subdivided into "niches"
Environment subdivided into major habitats
Physical-chemical constraints limit biosphere.
Derived from random loss, gain, or change of DNA
Derived from randon choice among genomes
Derived from random association of local individuals
Derived from random choice among demes
Derived from random association of species
Derived from random assortment of communities
Seconds to hours
days to years
months to many years
~10^5 to 10^6(?) years
~10^7 years
>4 x 10^9 years
Molecular Biologist and Geneticist
Physiologist
Behaviorist
Population Biologist and Geneticist
Ecologist
Paleontologist

I will treat "rates of evolution" by considering the rate of change of each gear (i.e. level) in the evolutionary hierarchy of Figure 1. But before turning to that, there is the independent question of how many truly different rates there may be. That is, if it is possible to "explain" changes at one level in the hierarchy by recourse to a lower level in the hierarchy, then the rate of change at the lower level will preset the rate of change at the higher level. Thus in assessing rates of evolution, one needs to know the extent to which different levels are reducible to their next lower level.

This question of the reality and independence of different levels was addressed by Eldredge (1982b, p. 339), who stated, ". . . most of the estimators of evolutionary rates devised over the past 40 years have actually been geared to estimating rates of transformation [between levels]." Thus the relationship of morphologic rates to speciation rates has been, as he says, a question of whether rates of phenotypic (or genotypic) transformation control rates of species formation (a gradualist view), and not whether these factors are reversed, with the rates of species formation governing phenotypic (and genotypic) rates (a punctuationalist view). For the evolutionary hierarchy shown in Figure 1, the question is whether each level should be treated as composed of "individuals" of that level, individuals that are divorced in principle from other levels; or whether instead each level is to be thought of as a class whose composition consists of the members of the next lower level (as, say, species within a community).

I believe that despite Eldredge's excellently reasoned argument, there is still room for the transformationist view, for the following reason. Fundamentally, each level is a composite accumulation of members of the next lower level. By this view it is in the sheer addition of demes that a species may come to have "emergent" properties, such as its geographic range or temporal persistence. And it would be through the cumulative addition of species that a community would obtain distinct "emergent" properties, such as "complexity," and so forth. Thus to a reasonable degree, the properties of a level are understandable in terms of studying the interactions of the units accumulated at that level. In this sense, the levels would be understood by a "transformationist" view.

It is in the same way that the gas laws in chemistry are a scale-dependent phenomenon. That is, the gas laws are a function of the existence of millions of molecules, and they cannot be explained by recourse to the properties of an individual molecule. By this view, each level in the hierarchy is a class of composites of the next lower level. Only in this one restricted (but significant) sense do different levels obtain properties that, once obtained, allow one to talk and reason as though the new level was an "individual." The "individual" properties are those that scale, and scale alone, brings to a category. Thus the answer to the question of whether different levels can be treated as individuals or as simply an accumulation of the traits of the next lower level, and thus as a class, may be either yes or no, depending upon the context.

In the present article, and as a general principle, special explanations are not sought at a given level unless properties at that level cannot be accounted for by processes acting at a lower level. This is what is known as "pragmatic reductionism" (A. Hoffman, in preparation, personal communication, 26 May 1983). Thus, for example, only if transformationist mechanistic explanations fail to account for the origin of a given level will an explanation be sought at the level itself.

The view that any given level achieves its individuality through an accumulation of members of the next lower level is a different view than one or another version of species selection as the trait of an "individual" (Eldredge 1982b and elsewhere, Gould 1982, Stanley 1979, Vrba 1980). In the present viewpoint, one would seek a rate of speciation at the species level per se only if species could be shown to have a process that results from, and is caused by, being a separate unit in the hierarchy, such as might occur in a scale-dependent reaction. One would accept "species selection" if and only if the level of the species had emergent properties directly related to the process of speciation that were not accounted for by reference to what happens at the level of demes, or of individuals or of genomes (i.e. of some lower level). As is discussed below, the main challenge of molecular biology to evolutionary biology on the speciation problem is that whole levels can be leapfrogged.

Units of Evolutionary Change

I focus on the six units of evolutionary change illustrated in Figure 1. Other emphases are of course possible. One could focus on rates of protein evolution, of chromosome evolution, or of various morphometric characters such as tooth size or leaf shape. Proteins, chromosomes, and teeth and leaves, however, are parts of larger wholes, and it is the larger wholes that are emphasized in this article.

1. GENOME The genome can be considered to consist of two types of DNA, each of which may vary in content and/or amount from individual to individual and, to a greater extent, from deme to deme. Type one DNA is genic DNA (i.e. it codes for specific proteins, and through the expression of these genes to the potential for most of the morphologic and physiologic change). Type two DNA is nongenic DNA (i.e. it does not code for specific structural products and consists chiefly of various classes of repetitive DNA and their messenger RNAs). Over evolutionary time, type two DNA can expand or contract in size to a much greater extent than type one DNA, and

in doing so it can change genome size at measurable rates. In particular, the buildup of families of repetitive DNA (including transposable elements) is the major quantitative difference in genomes over time scales of 10^3 to 10^5 yr (Britten 1982), i.e. on time scales of interest in the origin of species. Individual members of several families of type two DNA have the capacity to move from one place on a chromosome to another place on the same or a different chromosome (Shapiro 1983a). In that process, this nongenic DNA influences the expression of genetic information (especially the regulation of protein synthesis and the control of patterns of differentiation). Thus, although type two DNA is "nongenic," it can have a strong effect on gene expression. Depending upon the species, 25 to 95% of the DNA of many eukaryotic organisms is type one, and 75 to 5% is type two.

Genome sizes vary in prokaryotes from approximately 10^{-2} to 10^{-3} picograms (pg), and in eukaryotes from approximately 10^{-2} to 5×10^2 pg (Cavalier-Smith 1982). As the genome size in eukaryotes becomes larger, the proportion of the genome that is type two DNA increases. Nuclear DNA has both type one and type two DNA, but mitochondrial and chloroplast DNA is chiefly type one.

The unit for measuring genome change is mutation in DNA (Figure 1). Mutation is used here in the broad sense of the creation of new hereditary types (earlier broad usage remarked on by Schopf 1981). Mutation in its various parts is simply a set of biochemical processes, like modes of digestion or locomotion, each with its own enzymatic machinery. For eukaryotes, a reasonable equilibrium value of misincorporation of a base is of the order 1 in every 10^8 base pair replications. [Neel (1983) summarized the following data: Estimates of mutation observable by electrophoretic methods are of the order 2.2×10^{-6} per locus per generation. If (*a*) this method identifies approximately one third of the charge-change mutations, and if (*b*) loss-of-activity mutants are at least twice as common as electrophoretic mutations, and if (*c*) there are 500 to 1000 bases per polypeptide, then this yields a per base per generation rate of 1.4–2.6×10^{-8}. Total per locus rates appear to be on the order of 10^{-5} per generation. Neel cited a variety of data consistent with these values. Loeb & Kunkel (1982) cited a prokaryotic range of average frequency of base pair substitution of 10^{-7} to 10^{-11} misincorporations per base pair replicated.]

The misincorporation of bases is under enzymatic, i.e. genomic, control. Owing to mutation in DNA polymerase, the rate of incorrect base substitution in DNA replication can be as frequent as 1 in every 10^4 bases, or as infrequent as 1 in every 10^{10} bases [relative to a background of 1 in 10^8 bases in bacteriophage T4, where the process has been most thoroughly studied; see the summaries in Ripley (1983) and Ripley & Shoemaker

(1983)]. In the bacterium *Escherichia coli*, the *mut*D gene (= *dna*Q) increases the rate of incorrect base substitution by a factor of 10^2 to 10^4 above normal values (Echols et al 1983). Both the T4 and the *E. coli* mutations represent errors in the proofreading function of DNA polymerase (specifically in the 3′ to 5′ exonuclease proofreading activity). In eukaryotes, the DNA polymerase δ also has a 3′ to 5′ proofreading function (Hübscher 1983, p. 7), and it seems only a matter of time before mutations in this polymerase are also documented. Mutation rate may also be influenced by specific base sequences [palindromes leading to hairpins in DNA secondary structure (Ripley & Glickman 1983)]. This sequence-directed mutagenesis is under enzymatic control to the extent that specific DNA structures are enzymatically controlled. A third enzymatically mediated reaction, that of postreplication mismatch repair, is also significantly involved in correcting spontaneous errors of DNA replication (Glickman 1982).

Genetic control of the rate of mutation is not limited to the substitution of bases. The movement of transposable elements from one chromosome position to another often results in mutations that alter gene expression. For experimental convenience, such elements in eukaryotes have been most thoroughly studied in yeast in their Ty elements (Williamson 1983, Roeder & Fink 1983), but they are also documented in protozoans (both trypanosomes and ciliates), nematodes (Liao et al 1983), insects (*Drosophila*), mammals, and plants [Rubin (1983); other papers in Shapiro (1983a) and Whitney & Lamoreux (1982)]. In yeast, the *Spm* locus acts as a repressor of transposition (Roeder et al 1980, Roeder & Fink 1983, G. R. Fink, personal communication, April 1983). When *Spm* is blocked, Ty elements move at a rate that is 10^2 to 10^3 times higher than normal. Other repressors are known to act on the rate of transposition in maize and in *Drosophila* (Modolell et al 1983). More explicitly, the whole phenomenon of hybrid dysgenesis in *Drosophila* is controlled by repression of transposable elements (summary in Bregliano & Kidwell 1983). In other words, the rate of transposition is under genomic (enzymatic) control.

It may seem a long way from yeast and *Drosophila* to clams, brachiopods, trilobites, and therapsids; and it may be. But the only reason that yeast and insects are used is convenience. There is one single message from the sum total of data on the biochemical control of rate of mutation: *Mutation was too important to leave to chance.* The rates of both point mutations and of regulatory mutations are under genomic enzymatic control, and they are subject to selection and other evolutionary processes. This general point has been made over the past four decades (Sturtevant 1937, Buzzati-Traverso 1955, Kimura 1967, and others), but most evolutionists have downplayed mutation relative to other factors. With current molecular knowledge of the role of transposition, the fundamental importance of mutation in evolution needs to be reevaluated.

The rate of change of mitochondrial DNA is approximately 10 times more rapid than for nuclear DNA (Brown et al 1979, 1982). Accordingly, there are now described a wide variety of both DNA-segments and of proteins with considerably different rates of change. The amount of change is proportional to time over some significant fraction of the life history of these molecules. The result in many cases appears to be a variety of useful but imprecise stochastically ticking clocks that measure time over durations of a few millions of years to hundreds of millions of years. Of recent paleontological interest is the extrapolation of the haemoglobin clock that indicated "that the initial radiation of the animal phyla occurred at least 900–1000 million years ago" (Runnegar 1982). [However, if the haemoglobin clock has varied significantly in rate over geologic time (Goodman 1981, Czelusniak et al 1982), then Runnegar's conclusion is in question.] The significant fact is that molecular clock data are starting to feed back into the solution of problems of particular paleontological interest, other than in hominoid evolution. A compendium of paleontological dates for times of divergence of taxa of interest to molecular biologists in plotting molecular clock data is being prepared (J. J. Chiment & T. J. M. Schopf).

The lowest rate of base substitution of nuclear DNA applies to the so-called replacement sites, which are those bases in the DNA sequence that code for amino acids and lead to functional products. Even here there is variation not only from protein to protein, but within a protein from "conserved" to "variable" regions. Rates of change on the order of 0.1 percent per million years may be representative, with values an order of magnitude lower recorded for sites in transfer RNA molecules (Brown et al 1982).

The highest rate of base substitution of nuclear DNA occurs in the so-called silent sites. These are the bases in the DNA sequence that do not code for amino acids (i.e. pseudogenes, introns of genes, and bases that do not influence the amino-acid codons). This "free-floating" rate may be constrained chiefly by the various factors limiting the rate of mutation itself. Silent sites may change at a rate on the order of 1.0 percent per million years, or 1 in 10^8 sites per year (Brown et al 1982). Miyata et al (1982) cite a value of 6×10^{-9} substitutions per site per year for silent substitutions, with a rate of change in pseudogenes of approximately half of this. [Both Brown et al (1982) and Miyata et al (1982) discuss earlier literature on this topic.]

Within local populations, the significant fact is that the equilibrium rate of mutation is high enough so that mutant individuals are always present, a point made 30 years ago for bacteria by Demerec (1950). If the normal, equilibrium base substitution rate is 1 in 10^8 bases in *E. coli*, and since the genome size of *E. coli* is 4.5×10^6 base pairs (bp) (i.e. of the order of 10^7 bases), then 1 in 10 of any of the *E. coli* colonies will have a base substitution. There are on the order of 10^{10} viable bacteria (including *E. coli*) in the human gut. If the *E. coli* rate is typical of these prokaryotes, then every 60 minutes of the bacterial cell cycle 10 percent of those gut inhabitants, i.e. 10^9 bacteria, would produce daughter cells with a base substitution. This is an enormous investment in the probability of being able to take advantage of some change in the habitat. Or consider the case of yeast Ty 1 transposable elements. If the equilibrium *rate* of transposition is 1 in 10^7 to 10^9 loci (V. M. Williamson, personal communication, April 1983), and if the yeast genome size is of the order of 5×10^3 loci, then 1 in 5×10^3 to 10^5 cells are undergoing transposition every few hours. Consider a grape with on the order of 10^6 to 10^8 yeast cells on its surface. On that grape there are on the order of 5 to 5×10^3 cells undergoing transposition at each cell division. Every vineyard probably has hundreds of thousands of grapes. Thus one can see that transposition can be a significant factor in the natural history of yeast as it and its nutritional substrate continuously change.

Presumably, the majority of these mutations are homozygous lethal, but persist perfectly well in a recessive state, as in *Drosophila*. Woodruff & Thompson (1980, p. 132) summarized data which shows that "from 10% to 60% of the autosomes of wild lines of different *Drosophila* species contains at least one recessive lethal mutation." Earlier, Dobzhansky (1955) had made the same point in discussing the "ubiquity of deleterious recessives in *Drosophila* populations." But what is lethal one day may be viable the next, as the local ecology is always changing. Thus, in a population sense, mutation is simply another type of adaptational process by which individuals and local populations maintain flexibility. This is the view expressed by Lewontin (1957) in his idea of polymorphism as a "homeostatic device" at the species level. And it is the view described from other evidence by Wallace (1977, 1978) when he concluded that "a genetic load, rather than being a burden on a population, is essential for the persistence of the population through time in a variable environment" (Wallace 1978, p. 9).

In type one DNA, mutations seemingly are not readily accepted into the genome, and the following generalizations seem to apply (Lewontin 1974, Ramshaw et al 1979, Coyne 1982): For this genomic DNA, 50 to 65% of the genes appear to be monomorphic (with no or very, very few variants, even given the high-resolution electrophoretic methods). Among polymorphic loci, it appears that only a very few common segregating alleles (two or three) exist in most cases, with a large number of very rare alleles. Thus monomorphic and these polymorphic loci appear to be under very strict selection. The number of loci that tolerate a large number of moderately common alleles may be a very small percentage (of order 5%) of type one DNA, and a yet smaller percentage (of order <1%) of the genome DNA. From a speciation point of view, allelic variation in type one DNA appears *not* to be the *focus* of evolutionary change, but like good marker genes, these alleles record the process of ongoing evolution. In type two DNA, mutations appear to be accepted much more readily because most type two DNA is not translated into a protein product, nor may a significant fraction of it even be transcribed. Selection is much looser than in type one DNA.

Once mutations occur, they may be brought to high frequency within a deme either by external or internal factors. Externally, if mass mortality occurs for any reason, previously rare alleles fortuitously carried by the survivors may experience a rapid increase and achieve a wide distribution. Internally, any of the various mechanisms of sequence homogenization can bring a particular DNA sequence to uniformity throughout a deme. A "gene conversion" mechanism, or "sequence correction" mechanism (often called *concerted evolution*), has the potential to bring a single sequence to dominance (reviewed by Weiner & Denison 1983, Nagylaki 1983). It is presumably in this fashion that major changes occur in the copy number of both multigene families and type two DNA.

In considering a total rate of mutation for the genome, it appears that the most important component will be type two DNA because this part both changes most rapidly and is present in a significant percentage. Paradoxically, this nongenic DNA has clear mutational effects on genic DNA. Genomic processes that lead to the expansion or contraction of a family of nongenic, transposable, repetitive DNA would be expected to have significant phenotypic effects. The rate of mutation is a biological variable.

2. INDIVIDUAL The scale of measurement of individual change should be in units of morphology (Figure 1), but such units are poorly defined. The greater the number of morphological features that might show change, the greater the change that is observed. In fact, rates of morphologic change and durations of "species" are inversely related, with "complex" forms of animals appearing to evolve about a factor of 10 faster than "simple" forms (Schopf et al 1975). Thus the paleontological rule that "only complicated animals evolve" is chiefly a commentary about the perceived complexity of organisms rather than about the extent of evolution per se. This same general principle is seen among plants. Stebbins (1950, p. 515) wrote, "It is a well-known fact that the commonly preserved fossils, particularly the leaf impressions of the flowering plants, are the least diagnostic of all plant

parts." Partly for this reason, then, "evolutionary conservatism and stability are much easier to demonstrate by means of fossil evidence than is rapid progress or the differentiation of the modern families and orders" (p. 517–18).

At present we lack a precise comparative morphologic metric for creatures as unlike as, say, jellyfish, starfish, and bony fish. In limited cases, comparative metrics are possible, as demonstrated by Cherry et al (1982) for the vertebrate groups mammals, lizards, and amphibians; the authors "aimed at measuring homologous traits." Eldredge (1982b, p. 341) wrote, "Surely, by comparing taxonomic rates between sister taxa (which are, by definition, of the same age and almost certain to be equally 'morphologically complex'—cf. Vrba's antelopes) we can sidestep whatever problems differential morphological complexity may pose." But sister groups are not always clearly determined, nor do they necessarily have comparable fossil records. This approach also will not aid the comparison of groups (such as mammals versus clams) that are widely separated on the phylogenetic tree and for which homologous traits are difficult to discern.

In paleontological materials, we also lack information on geographic variation (and hence within-species morphologic variation). This variation has been extremely difficult to document because one cannot tell whether or not deposits a few kilometers from each other are of the exact same geologic age (Cisne et al 1982). A difference of only a thousand years is trivial paleontologically, but it is significant if one is seeking to ascribe a pattern to geographic variation versus migration. Accordingly, uncertainty over stratigraphic correlation prevents us from obtaining reliable data on within-taxon morphologic variation.

Even if such data on geographic variability were widely available, the relevance of this information for depicting evolutionary scenarios through time would be in question. The reason for this resides in genotype-environment interactions, and what is known as the norm of reaction of a genotype in different environments. That is, the characters displayed by a specific genotype are different in different environments (Lewontin 1957, Van Valen 1969, p. 205). This viewpoint has been expressed for at least 25 years, but until recently has been subject to little experimental verification. To quote the major experimental study on this topic, evidence is presented "that without knowing the norms of reaction, the present distribution of environments, the present distribution of genotypes, and without then specifying which environments and which genotypes are to be fixed or eliminated, it is impossible to predict whether the total variation would be increased, decreased, or remain unchanged by environmental or genetic changes, or what the outcome of natural selection would be" (Gupta & Lewontin 1982, p. 948). A similar point is made by Templeton & Johnston

(1982), who write with regard to r- and K-selection, "Thus, an understanding of the genetic basis of the straits under selection is absolutely critical for predicting the types of life histories that will evolve under certain ecological conditions. The classic r- and K-framework is simply inadequate for this task." The significance of this for paleontology is that since the required knowledge is impossible to obtain retrospectively, then the attribution of "responsibility" for evolutionary patterns to environment or to genotype must necessarily be questionable. It is the interaction of particular genotypes with particular environments that yields particular results; retrospectively, these data are unobtainable.

The rate of evolution of morphology might conceivably be understood from a knowledge of the molecular basis for the maintenance and change of form. (This is dealt with more completely in the discussion on living fossils.) Little of particular use to this problem is known from molecular developmental biology. Indeed, the Group Report of the Dahlem Conference on Development and Evolution concluded, "We do not know the mechanisms by which gene activity affects the development of an individual animal, therefore, we cannot come to useful specific conclusions regarding genomic correlates of evolutionary change at the morphological level" (Dawid et al 1982, p. 20). In summary, knowledge of the rate of evolution of individuals based on morphologic change has a long way to go.

3. DEME Individuals are organized into local populations or demes (Figure 1). On an evolutionary time scale, what changes can either be the population structure within a deme, the demise or continuation of the deme as a group, or the distribution of deme sizes (Wright 1982a; also see Wilson 1973). Population structure involves birth, fecundity and death schedules, and social structure—in short, all of the factors that control population size and the organization of individuals within the local population.

Both ecologic factors (via migration and size limitations) and biologic factors (via social structure, etc) may influence the rate of change of deme size and the distribution of deme sizes, depending upon the species. For whatever reason, populations that are subdivided into small demes appear to be most prone to generating reproductively isolated descendants. With regard to mammals, social structure has been greatly emphasized in the maintenance of small demes and as a contributory factor in speciation (Bush et al 1977, Bush 1981, Templeton 1983). Among marine invertebrates, sessile and nearly sessile forms that brood larvae (and hence yield larvae that do not feed, but that settle near adults) have been discussed in the maintenance of local demes and in the tendency for such local demes to evolve toward reproductive isolation (Gooch 1975, Schopf 1977a; molluscan data summarized by Hansen 1982; many data on small demes summarized by Bush 1975; paleontologic implications emphasized by Jablonski & Lutz 1983, p. 50).

Opposing deme isolation is, of course, gene flow (Endler 1977; see below). From an evolutionary perspective, it would be of great help to have field data on the rate of establishment and the rate of mortality of local demes. To establish such information requires a comprehensive overview of the distribution of a species over a considerable period of time.

The major new focus in deme evolution is D. S. Wilson's book (1980) covering patch selection, which Van Valen (1980) considers "a major conceptual advance, almost revolutionary." The argument is simply that variation among local demes, together with differential survival and differential productivity, will result automatically in changes in phenotype and genotype frequencies. Although most authors (Van Valen included) stress the role of competition among demes, this is of course not necessary because any (stochastic) process leading to differential survival will have the same evolutionary result. Deme survival and extinction may in fact be much more significant than individual survival and extinction in leading to the evolution of new species.

4. SPECIES Probably every evolutionary biologist who has ever lived has had ideas on the origin of species and the rate of species evolution. Books, symposia, articles, and seminars all attest to the continuing universal interest in these topics. This interest persists in part because most biologists believe that there are multiple modes of speciation, often related to the biology of the particular group one works on (and hence multiple expert answers should be possible). The pluralist attitude toward species formation is encouraged by the summary literature. Authors from at least Dobzhansky (1937) onward (e.g. Mayr 1963, Grant 1971, etc) have included multiple (pre- and postmating) isolating mechanisms, or a table listing multiple modes of speciation (Bush 1975, Templeton 1984), or both. Finally the pluralist frame of reference is encouraged by the difficulty of knowing whether a species, genus, or family of (say) salamanders is the same as the nominal equivalents of (say) birds. As Van Valen (1973b) wrote in an excellent paper on the topic, "It is commonly believed that the question [of equivalence of taxonomic categories] cannot be answered."

Among these pluralist views, the so-called biological species concept promulgated by Dobzhansky (1935, 1937), and followed by a large number of subsequent authors, has received the most attention. By this definition, a species is composed of individuals and demes, which together form a naturally occurring interbreeding unit. Such units were recognized customarily by similarity in morphology, but especially since 1970, morphologic evidence of genetic relatedness has been shown often to be not as sensitive as karyotypic or electrophoretic indicators of a shared genome and an absence of gene flow (and it is a shared genome and the lack of gene flow that this concept emphasizes).

The ultimate test of genomic sharing is often taken to be whether or not viable hybrids form. This test is justified by recalling that speciation is a process that may take considerable time, and the extent of hybrid formation is used as the assay for how far along two presumptive species are in that process. As Lewontin (1981, p. 47) wrote, "The essence of [Dobzhansky's species] concept is that a *continuous process* of genetic differentiation between populations finally results in a stage of total genetic isolation. Speciation is understood as a *stage* in evolutionary divergence, but as a critical stage, because after this point the units are genetically independent" (italics his). Because two populations can be found at any stage of species formation, there is an enormous range in the extent of gene flow (and in the extent to which gene flow is expressed in morphologic traits). As Templeton (1981) noted, "many species differences contribute little or nothing to reproductive isolation." This disparity between the extent of morphologic differentiation and the existence of reproductive isolation also contributes to a pluralist view of species formation. In discussions of rates of evolution involving fossils, the main (and usually only) concern is "hard part" morphology: the species is purely a morphologic unit. Thus, data on rates of species evolution derived from paleontology are limited to data on morphologic change.

Paleontologic rates In a steady-state world, rates of speciation are usually given as the inverse of the length of species duration. If there occur today on the order of 10^7 species, and if the mean duration were 10^7 yr (Raup 1978), then at steady state the rate is 1 species origination and 1 species extinction per year. Alternatively, if species durations are on the order of 10^6 or 10^5 yr (see below), then at *steady state* the rates of extinction and speciation are on the order of 10 to 100 per year. Can one determine rates of species extinction or production to better than three orders of magnitude?

There have been approximately 100 yr of world-wide searching for species (less in the enormous tropical rain forests and in the vast expanses of the deep ocean, and more in highly populated areas). During this time span, at a production rate of 1 species/yr, there would have been 100 originations and extinctions somewhere in the oceans or on land. At a higher rate of 100 species/yr, there would have been approximately 10,000 originations and extinctions spread over the Earth's surface. Would a change as extensive as this be readily apparent to us?

In the continental United States, perhaps 100 plant species have become extinct out of a pool of 20,000 species, and in Hawaii perhaps 225 have become extinct out of a pool of 2200 species (MacBryde & Altevogt 1977). At maximum, a number approaching 10% of the flora has become extinct in 100 yr, or about 0.1% of the flora per year. Moreover, each plant extinction has a multiplier effect, so that on the order of 10 to 30 animal species disappear with it [P. H. Ravin in Ehrlich (1980)]. Since there are on the order of 10 to 30 times more animal species than plant species, the percentage change of the biota as a whole does not differ much from the estimate based on plant species alone. This highest present rate of change of 0.1% per year is attributed to the effect of human intervention.

Based on the ancient world, it may be more difficult to determine true rates of origination and extinction than it is in the modern world. The great majority of organisms that have become part of the described fossil record is that small subset that originally existed in considerable abundance and that were long lived. I have suggested elsewhere (Schopf 1982), that a variety of biases conspire to make it appear as though even those species that do occur in the fossil record have their geologic duration extended by a factor of 10 to 100 times the true duration. Thus, I only summarize the results here. As shown in Table 1, that study took into account four biases: the bias of morphologic complexity; the bias toward collecting common long-lived taxa whereas most taxa are rare (and short-lived); the bias of reporting data

Table 1 Summary of four prominent biases[a] that influence inferred species durations (after Schopf 1982)

Bias	Effect on increasing inferred species durations (estimated order of magnitude)
1. Absence of preserved morphological complexity makes organisms appear to evolve slowly.	Factor of 10
2. Common taxa are fewer, but longer-lived than the more numerous but more rarely collected shorter-lived taxa. Most estimates of durations based on common taxa.	Factor of 2 to 10
3. Duration of taxa are based on the duration of standard geologic stages, thus presenting a minimum duration.	Factor of 2 or more
4. The polytypic (lumper's) species concept prevails; shorter-lived taxa (e.g. sibling species) are pushed into longer-lived taxa.	Factor of 2 to 10

[a] The biases are multiplicative (i.e. correct for morphologic complexity, and one still must correct for preferential use of common taxa; correct for these, and one still must correct for minimum duration set by stage length; correct for these and one still must correct for sibling species and for the lumper's philosophy). The net effect of these biases is that mean durations of species as commonly recorded may be overestimates of true species durations by a factor of 10 to 100.

to a cumulative stratigraphic unit (the stage); and the bias of lumping variation in the polytypic species concept, thus pushing species that are only slightly different morphologically into broad, longer-lived taxa. The conclusion from that study is that whereas species durations are inferred from a literal reading of the fossil record to be on the order of 10^7 yr duration, the true figure may be on the order of 2×10^5 yr.

Actual values on estimates of the duration of species-level taxa over geologic time (which incorporate various degrees of bias) encompass in 1983 the same approximate range that was given 30 years earlier. I cite here some more recent numbers, but it should be admonished that virtually all of these values need to be modified to take into account the time interval over which the observation is made [Gingerich (1983), and explained below (p. 270)]. Simpson (1952) estimated mean species duration as a range of 0.5 to 5 m.y. Specialists in 1977 dealing with five major groups (bryozoans, bivalves, ammonites, graptolites, and mammals) cite mean values of 0.5 to 15 m.y. (summarized in Schopf 1977b, p. 559). Other values for species-level groups come from work on the Upper Cretaceous Western Interior fauna (chiefly data from ammonites and bivalves), with many values in the 0.5 to 3.0 m.y. range (Kauffman 1977); Polish Miocene bivalves (93 species), with a mean duration of 19.5 m.y. (Hoffman & Szubzda-Studencka 1982); and birds. [Vuilleumier (1984) estimates that 10% of the species that arose in the late Pleistocene are extinct, 32% of those from the middle Pleistocene, and 72% of those from the lower Pleistocene. (The Pleistocene is approximately 2 m.y. long.) Furthermore, at the generic level, among birds, "Eocene-Miocene faunas are composed entirely of extinct taxa."] The desert plant genus *Oxystylis* evolved during the past 15,000 yr (Iltis 1957). Stanley [1976, 1979, (Chapter 9)] utilized a Lyellian method for estimating mean species durations by noting how far back into time one must go before 50% of the Recent fauna had not yet evolved, and then doubling this value to get the inferred species duration. He obtained values from approximately 1 to 25 m.y. for 15 major groups of organisms, although some revision seems necessary (Wilson 1983). Raup (1978) introduced a method for back-estimating mean species duration based on the values of species turnover that would be consistent with known values of generic survivorship of marine invertebrates. This back calculation yielded a mean species duration of 11 m.y.

Aside from the question of actual values of particular rates is that of variability in rates. Current views are highly influenced by empirical plots (Van Valen 1973a) and simulated patterns of evolution (Raup et al 1973) that resulted in Van Valen's law: within an ecologically homogeneous taxonomic group, extinctions occur at a stochastically constant rate (Raup 1975).

Simpson (1983), Wei & Kennett (1983), and others objected to this "law" for two reasons. First, there is the "mass of data showing greatly different extinction rates in different taxa, or in the same taxon at different times" (Simpson 1983, p. 135). Different rates in different taxa was also an initial observation of Van Valen's; this is why he limited his observations to a given adaptive zone. These differences, I would claim, in any event may be more apparent than real. Several taxon-specific biases exist that even-out extinction rates for different major taxa as discussed above (p. 259). It is fair to say that all would agree that very different taxonomic groups are observed to have very different *empirical* rates of evolution. A separate question is the extent to which there exist taxon-dependent biases (in particular, the ease with which evolutionary change can be observed in different groups—see Table 1). This question probably will not be answered to general satisfaction in the near future, but will continuously reside in the background until a much deeper understanding can be brought to bear either on the topic of the origin of species or on the recognition of sibling species in the fossil record.

Simpson's second complaint, that the same taxon has markedly different rates at different times, has, of course, also long been noted. On a per capita per million years basis, rates of origination or extinction will vary quite a bit. During the initial phases of a radiation, taxa will be relatively short-lived with lots of descendants, and during a "mass extinction," most taxa will die off. This is quite consistent with Van Valen's law *so long as there is not preferential extinction for taxa of a given age*. Whether or not mass extinctions are *also* statistically equivalent among different major taxa has not been adequately tested, but it appears to be possible provided that biogeographic factors are taken into account. At the time of the mass extinctions, some major taxa disappear and others do not. The difficulty with making this an observation consistent with a deterministic explanation is that taxa are not randomly distributed around the Earth. Extinctions may preferentially hit one area more than another—what I call the Dresden effect. Those living in Dresden prior to the fire bombings were not preferentially less fit, or less well adapted in any common sense of the word, than those living, say, in Greifswald. Similarly, in the case of the dinosaurs, it is possible, or even probable, that the last dozen families were all represented in one general biogeographic province (Schopf 1983). Hence, any factor that seriously perturbed that province (such as the lowering of sea level leading to the major climatic change that apparently significantly reduced habitable area) does away with dinosaur habitats, but organisms in other biogeographic provinces may have been unaffected. It does not seem fair to ask the question whether dinosaurs (or trilobites, or other major groups) were or were not more prone to extinction than contemporary groups unless one makes allowances for different biogeographic distributions and changes in habitable area. To paraphrase Van Valen (1982), it is necessary to pay close attention not merely to the process of statistical sampling, but also to the nature of the organism distribution itself. Arguments for selectivity in extinction must take into account the biogeographic factor before determinism can be invoked. In summary of this section, traditional values of species duration have not changed during the past 40 years, but a much more acute sensitivity to biases both inherent in the fossil record and from group to group have cast considerable doubt on these previously inferred values. Accordingly, durations of biological species may be much shorter than traditional paleontological values, and species may be statistically equivalent in many situations.

A unifying mode of speciation? To date "there is no single universal marker of speciation" (Templeton 1982). The question is, does this necessarily mean that there are multiple modes of speciation, or, instead, does it mean that at present we lack sufficient information and that in due course we should expect to find a "universal marker"? Data from molecular biology may provide an independent way to obtain data on the rate of species origination. For example, if speciation were a function of genomic processes that, as a byproduct, resulted in reproductive isolation of local populations, then one should measure the amount and rate of genomic change and combine these results with any additional differentiation caused by geographic isolation. Any small change in the genome of a single individual would, it is hypothesized, be able to be maintained within the local deme to which that individual belonged, but demes in different parts of the geographic range might become genetically differentiated.

During the past 15 years, a considerable change has occurred in ideas on the development of genetic differentiation. In previous times, the focus would have been on genes as they were then known, i.e. the structural genes whose frequencies have become recorded so thoroughly in the myriad of electrophoretic studies. At present, the emphasis has shifted to the nongenic portion of the genome (i.e. type two DNA). This shift began as early as 1970, when it was discovered that different closely related species have different satellite DNAs and different sequences of middle repetitive DNA, much of it associated with what had been seen in chromosomal staining and referred to as heterochromatin [literature reviewed in Schopf (1981, pp. 149–56)]. A further aspect of genomically originating variation includes chromosomal mutations associated with type two DNA. Concerning these changes, Cullum & Saedler (1981, p. 146) wrote, "It would be possible to get 'accidental speciation' driven by a chromosome rearrangement that has no phenotypic effect." Similar conclusions regarding frequently occurring "accidental speciation," as opposed to precise "adaptive" speciation, have been reached by several molecular biologists over the past decade. [Note the multitudinous chromosomal data emphasized by White (1978) and other data reviewed in Schopf (1981); see also Dover 1978.] The differences among these authors in the particular mechanism that they emphasize are far outweighed by the general similarity in a reliance on genomic changes per se as the driving force behind reproductive isolation.

Recently, it has been observed that various hybrid dysgenesis systems may be involved in the development of reproductive isolation between populations carrying different transposable elements (Bingham et al 1982). Accordingly, hybrid dysgenesis may be a biochemically driven mechanism for the origin of species, with its own equilibrium rate (Thompson & Woodruff 1978, Kidwell 1983, Bregliano & Kidwell 1983, Rose & Doolittle 1983).

Although we are not yet in a position to accept or reject a broadly based biochemical origin of species, four arguments of varying strength suggest consideration of a unifying, single primary cause for the development of reproductive isolation.

1. There are currently estimated to be on the order of 10^6 to 10^7 eukaryotic entities called species. These species are assigned to approximately 80 different phylum-level subdivisions of the animal, plant, fungal, and protoctist kingdoms (Margulis & Schwartz 1982, Bold et al 1980). Virtually all of these phylum-level divisions have had an independent history for at least the past 6×10^8 yr, and the kingdom-level divisions for $\sim 10^9$ yr. If species durations are on the order of 10^6 yr, there have been 6×10^2 independent turnovers through the various lineages in each of these phyla. Over geologic time since the start of the Cambrian (6×10^8 yr ago), there may have been on the order of 10^8 to 10^9 independent species that have at one time or another been on the surface of the Earth. If modern species of different phyla are the same sort of entity (see points 2 and 3 below), and if the 80 phylum-level lineages to which the current crop of 10^7 species belong have been separated for $\simeq 6 \times 10^8$ yr, then this suggests the occurrence of a common mode of speciation in the earliest eukaryotic lineages (perhaps as related to meiosis and sexuality), and that this mechanism has been responsible for successive rounds of descendants in successive lineages. White (1978, p. 347) remarked that "much of the overemphasis on geographic isolation as a factor in speciation, which has characterized the literature in recent decades, is due to a failure to appreciate the sheer numbers of species on limited areas of the earth's surface." Thus, the first argument for a unified view of speciation is the enormous number of entities called species that have been in phylogenetically distinct lineages for $\simeq 6 \times 10^8$ yr. The problem of the origin of species

is *not* the problem of the origin of *Drosophila melanogaster*, or of any particular other species. The problem is how to account for 10^9 equivalent entities.

2. One can walk into any natural history or paleontologic museum in the world and discuss basic problems in systematics with the curator of any group of animals, and usually even with curators of plants or protoctists. (I do not know about fungi.) Problems encountered by every systematist include difficult genera, sibling species, patterns of radiation, the nature of higher categories, etc. The discussion could, for example, range over corals, rodents, and palms—taxa that are characterized by vastly different extents and modes of dispersal, behavior, or morphology. It appears that the common language in systematics grows out of common problems and bespeaks the same basic underlying unit—the species. This, in turn, implies that species are independent of taxon-specific features within a kingdom, and probably among kingdoms. Thus, the second argument for a unified view of speciation is the taxon independence of species.

3. One can obtain species in virtually every environment on the surface of the Earth. Regions with the highest diversities include such enormously disparate places as the continental slope and continental rise (i.e. deep sea), tropical rain forests, and coral reefs. These habitats have little in common in terms of the time scale of environmental change or of their physical, chemical, and biological characteristics. Yet the taxonomic problems associated with each area, and for each group in each area, are much the same (Schopf 1984). The well-known latitudinal diversity gradient of increasing diversity toward the tropics may be no more than another expression of the species-area curve, with the larger tropical faunal provinces therefore having a larger number of species (Schopf et al 1978). This view of independence of the process of speciation vis-a-vis environment is *not* to negate the obvious fact that every species is adapted to the particular habitat where you find it. Rather, every species is specialized for its particular habitat to roughly the same degree, and the overall resultant of different diversities in different habitats is set by equilibrium matters in species/area relationships (see Schopf 1979, Figure 2).

This discovery of the independence of environmental type and the extent of speciation is the result of observations made during the past 25 years. At the well-publicized Princeton meeting of 1946 (Jepsen et al 1949), Stebbins (1949, p. 241) could write, "Rates of evolution are determined primarily by the relation between the evolving population and its environment, and secondly by forces inherent in the population itself." At present, one cannot be so certain, and in fact the reverse may be true (reviewed in Schopf 1984): internal factors may outweigh environmental considerations.

I would only add that no one calls a halt to such processes as speciation and says, come back next week and start up where you left off. (Yet scientists, from bacteriologists to paleontologists, still write as follows: "Thus if the environment remains constant, so will a species, although perhaps it will gradually become more nearly perfect as the years go by." But the environment is always changing, and with it the metaphor of a static adaptive peak. Probably a more accurate mental image is that of a set of intersecting wave crests that is in the process of change even as a particle is brought to some local height.)

Since the 1946 Princeton meeting, it has become abundantly clear that the mutation rate is biologically set (reviewed above), and that relatively stable habitats (as in the deep sea) have as rich a species composition as do much less stable habitats (such as coral reefs or tropical rain forests). Moreover, the time scale over which geologically rapid change occurs (say $10°C/10^6$ yr, or only 10^{-5}°C/yr) is so slight that if steady acclimation (and no speciation) were meant to be the rule of life, the biochemical and physiological flexibility exists to provide it. (One reviewer wrote, "It's cheating to say that $10°C/10^6$ years is really the same as 10^{-5}°C/year. Climate changes in steps, not in gradual series!" But what are these steps? Any "step" must of necessity be transitional, though probably erratic in rate. Perhaps this will be called punctuated equilibrium versus climatic gradualism.) Thus, the third argument for considering the existence of a unifying primary cause for the development of reproductive isolation is the *broad-scale* environmental independence of the mechanism of speciation.

[The strongest line of argument *against* giving primacy of influence to genomic factors in speciation is the occurrence of fertile hybrids between "good species" that are seemingly separated on *nongenomic* grounds. Maynard Smith (1982, p. 382) wrote that "there are numerous cases in which two related species of animals are isolated by behavioral differences in courtship but, if interspecific mating is brought about in captivity, the resulting hybrids are perfectly fertile. This commonly happens, for example, in birds, grasshoppers and *Drosophila*." With regard to plants, Stebbins (1950) cited cases involving the plain tree *Platanus* and *Catalpa*. The quality and extent of this evidence need to be examined in detail (i.e. whether it extends through F_2 and F_3 generations without any loss of reproductive potential and whether the survival of the hybrids is equivalent to that of pure stocks). Hybrid zones are typically 10 to 100 km (Endler 1977, p. 156, Moore 1977). In order to evaluate the contribution (if any) of changes in nuclear DNA to modes of speciation, it may be crucially important to have a clear understanding at the molecular level of several examples of hybrid fertility between closely related "species." (This topic is of course tied up with that of what constitutes a species, which is discussed at the beginning of this section. The question is, which genomic patterns cause speciation, and which are the effects of speciation?)]

4. White (1969, 1978, p. 324) concluded that "over 90 percent (and

perhaps over 98 percent) of all speciation events are accompanied by karyotypic changes, and that in the majority of these cases the structural chromosomal rearrangements have played a primary role in initiating divergence." Because karyotypic changes are especially mediated by type two DNA (moderately and highly repetitive DNA; see Schopf 1981), the factors that control these genomic and chromosomal changes would seem to be the underlying mechanistic cause that contributes to genetic differentiation.

What seems clear is that there are mechanisms for generating sequence differences in local populations, and that closely related species differ noticeably in some part of the nongenic DNA, even when there is extraordinary similarity in the genic parts of the genome (Schopf 1981, p. 152). What does *not* seem clear is how differences in nongenic DNA are mechanistically related in a simple way to incipient reproductive isolation, i.e. in decreasing gene flow. The most common mechanistic causal connection has been by inferring problems in meiosis. That is, two individuals may mate and be fertile, but subsequent gametes suffer considerable loss of viability owing to difficulties in pairing at meiosis. However, the molecular basis for pairing is still not understood (Maguire 1983), although somatic pairing in polytene chromosomes decreases with decreasing similarity of the DNA in chromosomal bands (Riede & Renz 1983). As several authors have pointed out, any difficulty in pairing is not an automatic consequence of different amounts of nongenic DNA on different chromosomes (data summarized in Schopf 1981), and is probably sequence specific. This remains a mystery.

The various counterexamples of hybrids between "good species" can be reconciled with a single genomically regulated mechanism of speciation. In order to do this, one might suppose that transposons or other representatives of other DNA repetitive families of type two DNA were inserted in genes that influence behavior. In natural populations, transposition evidently occurs at a very high rate (Montgomery & Langley 1983), but many transposition events may enter nongenic DNA where no discernible phenotypic effect may be evident on structural genes, although there may be an effect on development. One place that transposition placement has a clear phenotypic effect is on pigmentation, as in maize, various flowers, *Drosophila*, and mammals. Pigmentation is often associated with behavior patterns in both animals and plants (through coevolution with animals). [The classical explanation for "Indian corn" (with its transposon-rich genome) is that its pigmentation led it to have been selected.]

In evolutionary time, presumably the critical feature is for selection to enhance viability in individuals with insertions. It is at this stage that models of adaptive divergence (Templeton 1982) may apply. Speciation may be very rapid or very slow, depending in part on the population structure of demes (see above) as well as on the traits that are subject to selection, thereby accounting for the pluralism usually attributed to modes of speciation. The hypothesis of a unifying mechanism of speciation is that underlying the diminution of gene flow (and the development of reproductive isolation) are genomic changes that result in reproductive incompatibilities, either directly [as in the excision of heterochromatin in different stages of embryogenesis in different species of the copepod *Cyclops* (Beermann 1977)] or indirectly (as in adaptive divergence acting upon genomically caused changes in traits). Current known foci of insertional activity include nutritional substrates for yeast, and morphologic developmental patterns in *Drosophila* (Bender et al 1983). The idea is that insertional sequences may lead to behavioral differences that over time may lead to reproductive isolation between demes, but that in the short term do not otherwise impair the ability to have fertile offspring.

Two immediate objections are made to this view of a unifying theme:

1. Taxa, even at the kingdom level, are not equivalent to each other, and so a search for a single mode of speciation is ridiculous. My response to this is that there may or may not be clear differences among the kingdoms of fungi, protoctists, plants, and animals in preferred mode of speciation. They last shared a common ancestor so long ago that some aspect of meiosis (for example) might be just different enough that all descendants of one stock may differ in a critical step from all descendants of another stock. Nevertheless, the basic molecular machinery appears to be very similar in all these groups. If fungi, protoctists, plants, and animals are somewhat different in mode of speciation, the reason cannot be habitat (they occur in the same general places and must respond to the same physiological constraints, such as temperature, osmotic balance, and so forth); a basic difference in mode of speciation must therefore be because of a difference in history. Within a kingdom, differences in deme structure (and so forth) may modify the rate rather than the basic process.

2. Hybridization occurs between good species that are separated (apparently) by nongenomic factors (such as behavior). My response to this objection is that a difference in behavior (or other "nongenomic" factor) must have a genomic basis. Furthermore, although hybrids may form under forced conditions, this does not negate a genomic basis for aspects that keep species apart in nature. Thus hybridization (as a proxy for monitoring the potential for gene flow) may be one step removed from the "true" factor leading to reproductive isolation.

Each of the various varieties of speciation connected with genomic change has a strong "accidental" component. This may be why, as Shapiro (1983b) put it, "In many cases, the differences between species appear to have more to do with genome organization (chromosome structure, ploidy, distribution and structure of repetitive elements, sex determination) than with adaptation to different ecological niches. . . ." The image is one of production of reproductive isolation in local demes (often peripheral because they have the highest probability of being isolated), with species adaptation initially entering into the equation only to the extent that all incipient species may be equally well adapted to where they occur. The further expansion of some demes rather than others may be owing to the vagaries of local possibilities for migration. The question of which local demes survive may be as much a matter of good luck as of good genes, especially if very little of the initial differentiation involves the genic portion of the DNA. Since most species are rare, it would appear that there is a good chance that local demes of species, which are rare to begin with, may be physically isolated from other demes for a length of time sufficient to build up nongenic DNA barriers to reproductive isolation. The time course of this event, however, is unknown, and it is a critical gap in knowledge of the process of speciation. A strong emphasis on small population size and on local demes in the incipient stages of achieving reproductive isolation has especially been stressed by Wright (1982b).

In the ideas on speciation presented above, the rate-limiting step may be one of three events: either (*a*) the rate of genomic change leading to incipient reproductive isolation, (*b*) the rate of colonization of a region and subdivision of the species into local demes, or (*c*) the rate of coalescence of local demes and thus the breakdown of geographic isolation. In general, the time course of DNA change must be less than the time course of faunal and floral distribution—otherwise, patterns of geographic distribution, such as faunal and floral provinces, would not exist.

In summary, the view supported is that there is both genic selection and genotypic selection [following Wright (1982a) and over the previous 50 years], but with selection significantly weaker (and the role of chance stronger) as one moves away from strictly biochemical factors at the gene level (type one DNA) to more flexible alternative pathways for survival at the individual level, and even more so at the deme level (Figure 2). The role of selection and the role of chance may be on a sliding scale, dependent upon where in the hierarchy one is focusing attention. In general, speciation is seen as having a very strong accidental component, with many alternative pathways yielding "equally fit" descendants. The duration of species (and thus the rate of species origination in a world with finite resources) is not agreed upon owing to the difficulty of discerning biological species in the fossil record. It appears that the true duration of biological species may have a mean value of $\sim 10^5$ yr.

5. COMMUNITY Rates of community evolution depend upon one's notion of what is a community. In a definition where community membership is strictly specified, then the community "evolves" whenever the member species or genera change. In a definition where community membership is loosely specified (as, say, in "level-bottom" marine community or "reef" community), then a community type may be stated to persist for a very long time, even for hundreds of millions of years (graphically expressed in Schopf 1972, p. 19). Aarssen & Turkington (1983) indicate five levels of community change and three ways the term "community evolution" has been used.

Once again at the community level is the question of whether the community should be considered as an "individual" (Wilson 1976), or whether it is merely a collection of species and therefore should be considered as a "class." The view of the community as an individual, indeed in the older literature as a superorganism, stresses youth, maturity, old age, homeostasis, etc (see caption to Figure 1). Still not settled are the questions of whether or not there are "rules of assembly" of communities, or the extent to which interspecific competition plays a significant part in shaping species assemblages [all sides represented in Strong et al (1984)], and whether coevolution is significant in community or evolution [Futuyma & Slatkin (1983, p. 400) think not]. Until these questions are clearly resolved, i.e. until the reality of communities as classes or as individuals is clear, it may make little sense to talk about "community evolution" in any paleontological context if one uses the word "community" with a meaning similar to its use in living floras and faunas. A thorough recent analysis of community evolution in a paleontological context concluded that much of community paleoecology is an epiphenomenal science (Hoffman 1979).

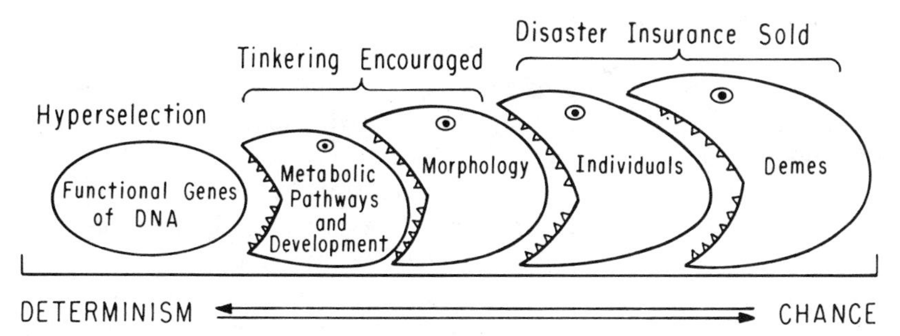

Figure 2 Diagram indicating gradient in the role of determinism versus chance. Functional genes appear to be highly constrained by selection; metabolic pathways and development and morphology offer multiple ways to yield the same functional product—F. Jacob's (1977, 1983) "tinkering"; and survival of individuals and demes (and everything they subsume) is subject to a large dose of chance, owing to disasters ("bad luck") of one sort or another.

6. BIOSPHERE Last in this survey of levels is the notion that the biosphere acts as a self-regulating system—what Lovelock (1972, 1979) christened *Gaia* (summarized by Westbroek 1983). As an example of this view, the reason there is 20% oxygen in the atmosphere is said to be that the biosphere acts as a unit to keep it at that level.

Needless to report, this view of the biosphere as the quintessential "individual" has received considerable scientific skepticism (e.g. Van Valen 1983). For one thing, it considers only a minor role for traditional physical-chemical processes in setting constraints on the biosphere (Walker 1977, Schopf 1980). The concept of Gaia excludes the idea that the biosphere is simply a collection of diverse communities that happen to exist on the Earth at any given time. Rather, the view is that the biosphere is in some as yet unknown and unspecified way regulating the relative frequency and type of communities so that the biosphere itself is maintained. Evolution, for Gaia, consists of "homeostatic" mechanisms that maintain the current balance.

The Time Scale Over Which Observations Are Made

The question raised here is whether or not the time interval over which observations are made predetermines any rates that are then "observed." Traditional paleontological reliance is on intervals of resolution of no less than a few millions of years—the duration of the geologic stage. For example, Sepkoski (1975) indicated that the "linear" survivorship curves of taxa over time (Van Valen 1973a) could be generated simply from variations in the durations of the geologic stages that were sampled. And, since mean durations of stages are approximately 8 m.y., it should be no surprise that the mean duration of species is held to be approximately the same, with shorter durations where stratigraphic resolution permits finer division of geologic stages.

The quantitative basis for evaluating stratigraphic resolution has been developed for paleontological aspects by Schindel (1980, 1982; see also Dingus & Sadler 1982) and for sedimentologic aspects by Sadler (1981). Future studies on rates of evolution from specific sections cannot be carried out without regard for these analyses. Rate-related concepts are a function both of the biologic process one seeks to measure and of a stratigraphic process that determines what one actually can measure.

The work of Schindel and Sadler has been carried one step further by Gingerich (1983), who has demonstrated that rates of morphologic evolution "are inversely related to the interval of time over which they are measured." He shows that slow rates are typically due to very long periods of observation over which the data are averaged, and that in general all rates must be scaled against equivalent interval lengths before they can be compared. In unscaled measurements, vertebrates (with an average observation interval of 1.6 m.y.) have the traditionally higher rate of morphologic change over the invertebrates (average interval 7.9 m.y.). But when appropriately scaled (by using regressions to produce rates at a standard interval), the "rates of morphologic evolution of invertebrates as a whole exceed those of vertebrates over all intervals of geologic time normally sampled." This bias of unequal sample intervals needs to be added to those already discussed in evaluating species durations, in conjunction with Table 1.

LIVING FOSSILS: CONFUSION OF A SPECIATION PROBLEM WITH A DEVELOPMENTAL PROBLEM

Definitions

While in the planning stage of writing this section, I was asked by a prominent geneticist whether it was necessary: "Did any professional still believe in living fossils?" I said yes, and cited as evidence a multiauthor book [N. Eldredge & S. M. Stanley (eds.), in preparation; N. Eldredge, personal communication, May 1983], as well as the use of the term living fossil as a proposed "test" of an idea in macroevolutionary theory (Stanley 1975b, 1979). Mayr (1982, p. 1129) remarked that "the existence of 'living fossils'[,] and the occurrence of 'explosive speciation' in freshwater lakes[,] has long been accepted as evidence for drastically unlike rates of evolution and speciation." Similar credence in the belief of the reality of species as "living fossils" has been given by several prominent population geneticists (e.g. Kimura 1969, 1981, Wright 1982a, Slatkin 1983). Moreover, I replied, individuals from *Drosophila* geneticists to biochemists still cite in seminars the horseshoe crab as a surviving but extremely ancient species.

The notion of "living fossils" is also well ingrained in the popular literature, and if it is not a correct notion, some attempt should be made to set the record straight. *Argosy* (May 1969) highlighted a supposed frozen humanoid on its cover and called the beast "Living Fossil: Is this the missing link between man and the apes?" Academic Press published a book on British brachiopods, and advertised it as follows: "Today Brachiopods can be regarded as 'living fossils' found intertidally and in deeper water..." (*Nature*, 282: vii, 20/27 Dec. 1979). *Earthwatch* stated in its winter and spring issue of 1980 that "The Horseshoe Crab is called a 'living fossil' as it has not changed in over 200 million years"; and *Science News* (19 Jan. 1980) in the same year said about the same animal, "Having evolved little during the past 300 million years, it is often called a 'living fossil'." The *National Geographic* in April of 1983 advertised a television program about

Australia by stating "There's a land where some fossils have never turned to stone." And so it goes. Finally, there have been reports from time to time of bacteria that have survived for 200 m.y. or more in salt or similar deposits (most recently proposed by Deelman & De Coo 1975), although when previous claims have been closely examined, they have been thoroughly discredited by bacteriologists (De Ley 1968). There are also several popular books on living fossils (Silverberg 1966, Ley 1959, Burton 1954, Delamare-Deboutteville & Botosanéanu 1970).

But what if the "living fossil" as a problem in speciation lives largely in myth only? What does this do to our notions of species durations? How well is the concept of "living fossils" grounded in fact? The thrust of the following argument is that a "living fossil" fails as a concept related to speciation, i.e. that no definite evidence exists for any species being extremely ancient. Instead, the interesting question concerning "living fossils" is why specific traits persist. This is chiefly a problem in developmental biology, broadly conceived, and not in speciation biology. Now the evidence.

Depending upon the author, a living fossil is

1. A living species that has persisted over a very long interval of geologic time.
2. A living species that is morphologically and physiologically quite similar to a fossil species, as seen over long intervals of geologic time.
3. A living species that has a preponderance of primitive morphologic traits.
4. A living species that has one of the above, *and* a relict distribution.
5. A living species that was once thought to be extinct.
6. An extant clade of low taxonomic diversity whose species have one or more of the properties of (1.), (2.), and (3.).

Differences in concepts of a "living fossil" can best be seen in terms of the taxa that are *excluded.*

1. The well-known Viennese paleontologist Erich Thenius (1977) devoted Chapter 10 of his book on the fossil record to the topic of living fossils. He wrote (p. 172), "The longest-lived Recent species is, however, the fresh-water crustacean *Triops cancriformis,* a branchiopod which is found as early as the late Triassic period and is thought to have remained unchanged as a species ever since—that is, for 180 million years." But Thenius then adds, "However, according to their organization and their wide distribution [through Europe], the Recent branchiopods cannot be designated living fossils." According to Thenius, a living fossil must also have a relict distribution.

2. Most authors (Thenius among them) would include the brachiopod

Lingula, which is known as a genus from the Ordovician to the present. Yet Stanley (1979, p. 124) excluded lingulid brachiopods "because they contain several species today and may have undergone a rather large total number of speciation events during long intervals of persistence with little evolutionary change." On this basis, Stanley (1979) also excluded pleurotomariid gastropods and pinnid bivalves, although four years earlier he had included lingulid brachiopods and pinnid bivalves as "classic 'living fossils'" (Stanley 1975b).

3. By the criterion that a living fossil should have at least one definite fossil specimen, one should exclude the twenty-five taxa of Table 2. And by the criterion that a living fossil should have at least one fossil specimen older than the Pleistocene, one should exclude at least an additional six taxa (see next paragraph). These exclusions include several of the best publicized "living fossils," e.g. the horseshoe crab *Limulus*, the reptile *Sphenodon*, the coelacanth *Latimeria*, the mollusc *Neopilina*, the opossum *Didelphis*, the duck-billed platypus *Ornithorhynchia*, the kiwi *Apteryx*, and the okapi *Okapia*. [In modern times, Fisher (1982) showed in a phylogenetic diagram a dashed line between the Recent *Limulus polyphemus* and a common ancestor with the Upper Cretaceous *L. coffini*. *L. coffini* consists of a single mold of part of the central region of the carapace (most of the opisthosoma) but lacks the anterior cephalothorax as well as the tail spine; it was found in a concretion. It was described in 1952 by the general paleontologist J. B. Reeside and the man who found the fossil, D. V. Harris (Reeside & Harris 1952), but with all due respect, the incompleteness of the specimen renders it reasonable to query the generic assignment (D. C. Fisher, personal communication, May 1983). The specimen, in any event, seems not to have been studied since 1952. Two Triassic and one Jurassic genera are sometimes provisionally placed in *Limulus*, but the leading student of these forms writes that it would be "most appropriate for these species to go into one or more genera of their own" (D. C. Fisher, personal communication, June 1983).] In any case, there are no known fossils of the extremely abundant living species *Limulus polyphemus*.

At least one "living fossil," the gastropod *Nassarius delosi*, was called that because although it has an extensive Pleistocene record, it was known from only two living specimens (Chace 1962). Similarly, the "living fossil" snail *Clilostoma* (*Drobacia*) *maestica* is widespread in Miocene and Pleistocene beds of Transylvania but is rather rare in the modern fauna (Lupu 1966). Another gastropod, *Perplicaria clarki*, was called a "living fossil" because it has congeners in the Pliocene and Miocene (Olsson & Bergeron 1967). The freshwater fish, the bowfin *Amia calva*, is (strictly speaking) evidently only known since the Pleistocene (Boreske 1974, p. 77), but forms to be

Table 2 Twenty-five extant genera referred to as "living fossils" but for which no definite fossils are as yet known[a]

Genus[b]	Common name	Genus[b]	Common name
1. *Anaspides* (1)	syncarid crustacean	14. *Neopilina*	monoplacophora
2. *Bathynella* (1)	syncarid crustacean	15. *Nothomyrmecia* (5)	ant
3. *Echidna*	spiny anteater	16. *Peripatus*	onychophoran
4. *Hutchinsoniella* (2)	cephalocarid	17. *Platasterias* (2)	starfish
5. *Lanthanotus* (2)	anguinomorph lizard	18. *Psilotum* (2)	psilotophyta
6. *Latimeria*	coelacanth	19. *Pycnogonum* (2)	pycnogonid
7. *Leiopelma* (2)	frog	20. *Sphenodon*	New Zealand tuatara
8. *Limulus*	horseshoe crab	21. *Stephodiscus* (6)	scyphomedusa, conularid (?)
9. *Liphistius* (2)	arachnid	22. *Stylites* (2)	glossopsid microphyllopta
10. *Metacrinus* (2)	crinoid	23. *Tarsius* (7)	tarsier
11. *Nebalia* (2)	phyllocarid crustacean	24. *Thermosbaena* (1)	peracarid crustacean
12. *Neoglyphea* (3)	decapod crustacean	25. unnamed genus (8)	ostracod
13. *Neolepas* (4)	lepadomorph barnacle		

[a] Most data on stratigraphic distribution from the relevant volume of the *Treatise on Invertebrate Paleontology* or Romer (1966); for *Lanthanotus*, see Rieppel (1980).

[b] Numbers in parentheses refer to the references listed below which cite these less-well-known genera as "living fossils." Lest the reader believe I am relying too strongly on the summary of Thenius (reference 2), note that each of his three summary diagrams are reprinted in full with approval by Gordon & Jablonski (1979) in their article on living fossils in the *Encyclopedia of Paleontology* [(1) Botosanéanu & Delamare-Deboutteville 1967; (2) Thenius 1973; (3) Forest et al 1976; (4) Newman 1979; (5) Taylor 1978; (6) Werner 1967a,b; (7) Hill 1966; (8) McKenzie 1967].

compared with it ("cf" designation) occur in the Pliocene. The bivalve *Neotrigonia* (the only extant member of the family Trigonidae) goes back only to the Upper Miocene (Fleming 1964). The oldest alligator (cited by Slatkin 1983 as a living fossil) is only late Miocene in age (W. Langston, personal communication, 8 July 1983; see Webb et al 1981, p. 520). *Nautilus*, a famous "living fossil," is known only from the Oligocene to Recent (Kummel 1964). The "living fossil" chiton *Lepidopleurus* includes a large number of subgenera and has the *Treatise* range of Carboniferous to Present (Smith 1960). However, Fischer (1965) notes, "strictly speaking it did not appear until the Eocene. . . ." The bivalve genus *Nucula* may be in a similar situation. Although the genus *Nucula* is recorded from the Cretaceous to the Recent, the subgenus *Nucula* (*Nucula*) is known only from the Recent. Among vertebrate genera, only Pleistocene fossils are known for *Didelphis*, *Ornithorhynchus*, *Apteryx*, *Okapia*, and the pangolin *Manis*. The modern plant genus *Metasequoia*, allied to *Sequoia* and *Taxodium* of Cretaceous and Tertiary age, appears to be only known from Pliocene fossils (Li 1964). At least six living genera of sclerosponges have been placed in a new class of Porifera (discussed by Hutchinson 1970). The class includes the Paleozoic stromatoporiids, as well as many Paleozoic and later enigmatic genera once placed among the bryozoa or the corals. Not one of the "living fossils" mentioned in this lengthy paragraph (and the list could be longer) has an extensive fossil record.

4. By the criterion that a living fossil should have a similar physiology to ancient forms, one can exclude several taxa. The Onychophoran genus *Peripatus*, traditionally placed between the annelid and the arthropod phyla, today occurs in leaf litter and moist tropical forest regions. The fossil form typically associated with *Peripatus* is the Cambrian Burgess Shale genus *Aysheaia*, a marine taxon. [A possible onychophoran of uncertain ecologic standing has been reported from the Pennsylvanian Mazon Creek fauna (Thompson & Jones 1980).] No matter how morphologically similar to each other *Peripatus* and *Aysheaia* superficially may appear, they cannot possibly be closely related because of their vastly different physiological requirements for terrestrial versus marine life, and through these requirements to their biochemical pathways and the corresponding necessary genetic differences.

A second example of physiological incompatibility in various forms concerns the lungfish. Middle Paleozoic species are clearly marine (Campbell 1981), whereas the modern taxa are tied to freshwater (Brien 1967). A third example concerns coelacanths. Among fossil coelacanths, Devonian forms are nearly all marine, Carboniferous and Permian species include both marine and freshwater species (Forey 1981), and Triassic genera "are about 60 percent marine," as are Jurassic and Cretaceous forms

(Schaeffer 1952). The mechanism of osmoregulation in *Latimeria* appears to be by retention of urea (Griffith & Pang 1979, p. 88; but disputed by Lagios in McCosker & Lagios 1979, p. 171). If urea is retained, this is significant because urea retention in fish in general is considered "incontestible evidence of their freshwater ancestry" (Griffith & Pang 1979, p. 88).

The point of this discussion of "excludable" taxa in the past six paragraphs is not to focus on the inadequacies of any of the several concepts of living fossils as a problem in speciation, but rather to show that *the claim that "living fossils" are ancient species is completely arbitrary*, and that the concept of living fossils should not be tied to taxonomic categories such as genus or species.

There is one more definition of a living fossil, which pertains only to morphology (not speciation).

7. A relatively little morphologically modified representative of a relatively archaic lineage with little modern representation.

The archaicness of the lineage can be inferred from phylogenetic analysis, and a fossil record is not necessary (wording after D. Wake, personal communication, June 1983). If the term "living fossil" is to be used at all, then this strictly morphological sense is most appropriate. However, given the inherent uncertainty as to whether one is referring to an ancient species or to a form with conservative traits, it is perhaps best not to use the term at all.

Speciation Theory Versus Developmental Biology

Such value as exists in the idea of living fossils is in emphasizing the occurrence of specific morphologic traits that appear stable over long periods of geologic time. This, however, is a problem likely related to developmental biology in its broadest sense, and more narrowly in what has been referred to in a purely descriptive sense as "canalization" (Rendel 1967). As Waddington (1942) expressed it, "The main thesis is that developmental reactions, *as they occur in organisms submitted to natural selection*, are in general canalized.... The canalization, or perhaps it would be better to call it the buffering, of the genotype is evidenced most clearly by constancy of the wild type" (italics his). I would call the retention of morphologic traits "stasis" except for the fact that this term has been highly publicized in notions of speciation theory, where biological *species* are not meant to change over time. The phrase "stasis is data" has meaning in developmental biology, where the emphasis is on *traits*, and the phrase "stasis is bias" is appropriate in speciation theory, where the emphasis is on biological species. I share the concept of stasis of Wake et al (1983), a concept that "separates it sharply from any tie with speciation" (p. 215), and agree with their further assessment that "surely any theory of morphological evolution that has speciation as the engine of change is suspect" (p. 222).

Biochemical evidence has shown for quite some time that the rate and amount of morphologic change are not closely tied to the rate and amount of genomic change (Wilson et al 1977, Kimura 1981, Schopf 1981). This disengagement of treating change (or lack of change) in morphologic traits as a topic distinct from reproductive isolation has been given a sound basis in analyses of the genetic changes associated with morphologic change. In a series of papers, Lande (1976a,b, 1981, 1983) has concluded from several points of view that "large evolutionary changes usually occur by the accumulation of multiple genetic factors with relatively small effects" (Lande 1981). Evidence in support of this view ranges from experiments reported by King (1955, and discussion following paper by Crow & Sokal) with regard to DDT resistance, to Coyne (1983) on the genetic basis of genital morphology in *Drosophila* sibling species. Most recently, Lande (1983) has shown that major mutations carry with them such a strong deleterious pleiotropic effect that such changes are not likely to occur unless populations are strongly disturbed, as through artificial breeding in domestic lines. Much disagreement is likely to continue, however, because some apparent major morphologic changes evidently may occur with major genomic changes (Templeton 1981), but what is "major" and what is "minor" is difficult to quantify.

The broad topic of developmental biology and the retention of traits, and the relationship of this topic to related topics such as atavistic traits (Grant & Wiseman 1982) and convergence (Davis 1979), is beyond the scope of this review. However, several points should be made with regard to "living fossils."

Conceivably, anatomical traits could persist for immense periods of time because (*a*) the trait is so generalized and simple relative to the environmental gradients with which it interacts that variation and specialization is not readily achieved; (*b*) the trait is so strongly positively limited by a specific environment or physiologic aspect that deviation is unobserved; or (*c*) the trait is so restricted in its evolutionary potential by some internal developmental factor that deviation is not observed. Different restraints may apply in different cases. Consider these possibilities:

1. Generalized traits may have a limited range of possible variation. In this category might be brachiopod shells for forms that burrow into the sediment (*Lingula*). Sculpturing, ornamentation, and ridges in bottom-dwelling brachiopods are associated with water flow over and around the

shell. These features do not occur in burrowing forms, and water currents interact with the lophophore and the anterior shell margin. The lophophore is rarely, if ever, preserved in *Lingula*, and thus one customarily makes comparisons between living and fossil forms simply by reference to the common denominator of a small, unspecialized external side of the shell. [The best character for identifying modern species of *Lingula* is the arrangement of muscles (Emig 1982), and this is rarely visible in fossil material.]

As a second example, consider venation in the *Ginkgo*, long considered a "living fossil" (Major 1967), chiefly on the basis of dichotomously branching leaves known in the living *G. biloba* and in similar fossils since the Lower Cretaceous (Tralau 1968). [Andrews (1961, p. 337) wrote, however, that "many of the fossils consist of leaves that fit into an essentially continuous sequence beginning with deeply dissected ones in the early Mesozoic to the nearly entire ones of a living ginkgo. The difficulties of delimiting genera and species have perplexed many a botanist." And he adds, "In contrast to the abundance of foliage, reproductive organs are almost nonexistent in the fossil record."] Owing to the absence of dominance-subordinate type of venation (i.e. leaves with a primary vein, and secondary and tertiary veins), and to the presence in *Ginkgo* of an equivalent type of venation, leaf venation is *automatically* dichotomous. Only one major character state for this trait is possible, and species-level diagnostic leaves would not be anticipated.

2. Mechanical constraints during development may sharply limit the evolutionary potential of some traits. In *Sphenodon*, avoidance of dental occlusion, and thus tooth wear, is achieved by lining the teeth of the upper and lower jaws into nonoverlapping rows, with the result that local variation in tooth shape is inconsequential in occlusion and thus precludes the type of jaw mechanics that developed in mammals (Gorniak et al 1982). Thus, to note the presence of a "primitive" trait in *Sphenodon* may be merely to identify a particular type of morphologic trait. More generally, Gans (1983) wrote, in summarizing a great deal of work of his own and of others, that many of *Sphenodon*'s physiological and life-history specializations "represent obvious adaptations to current conditions. The tremendous diversity of adaptive patterns seen in Recent lepidosaurians makes it risky to use any small cohort of these animals to infer conditions in various extinct forms."

3. Heterochrony in development provides a third major explanation for the occurrence of a suite of "primitive traits" in a modern form. Indeed, Stanley (1975a, p. 382) wrote, "A remarkable aspect of the phylogeny of the Bivalvia is that the fossil record shows that several extant bivalve groups judged by biologists to be primitive actually evolved quite recently, long

after the class has attained great diversity and long after many advanced families were in existence." More explicitly, Bemis (1983, 1984) states that the "primitive" characters of the three modern forms of lung fish are the result of paedomorphosis and are not the direct modern descendants of an unchanging lineage. Thus the similarity between the Paleozoic and the living forms is due to convergence (Bemis 1983). The possibility that primitive traits in living forms of barnacles result from the time of expression of traits during development is also discussed by Newman (1979).

For most "living fossils," no functional analysis has been carried out. Thus, McKenzie (1967) assigned 17 marine specimens to "a new ostracod genus that both anatomically and in carapace features qualifies as a 'living fossil'." The anatomy was found to be very similar to some ancient freshwater families! This example clearly illustrates the use of the concept of "living fossils" in a purely morphological sense that is not in any way related to speciation. This interpretation also applies to the large number of fossil-resting spores of Mesozoic and Tertiary age now referred to modern dinoflagellates (Wall & Dale 1966), and to the deep-sea trench ostracod *Abyssocythereis vitjasi*, a form placed in a Jurassic-Cretaceous family (Shornikov 1975). Many of the taxa of Table 2 also lack a structural-mechanical analysis.

For other "living fossils," a partial functional analysis exists, as in the Hawaiian monk seal *Monachus schumardi*. This species retains a relatively unspecialized ear region, unspecialized posterior vena cava, primitive isolation of the obturator foramen, and primitive freedom of the fibula from the tibia (Repenning & Ray 1977). However, the ear structure and unspecialized posterior vena cava were suggested by Repenning & Ray to be adaptations to shallow, short diving (in contrast to the special ear and vena cava features found in deep-diving seals). The retention of an unfused fibula and tibia has not been studied from a functional-mechanical point of view in the monk seal. Similarly, the functional significance, if any, of a distinct foramen for the obturatus nerve is unstudied and unknown.

Every organism contains a mixture of traits that are newly evolved, plus those that are retained from an ancient lineage. This is true at the biochemical level, the physiological level, and the anatomical level. For this reason, Eldredge (1982a) noted that the term living fossil "applied loosely, could embrace nearly all extant animals and plants." In *Limulus*, physiologists do not generally consider the beast to be "primitive" (chapters in Bonaventura et al 1982). Indeed, some aspects of the anatomy are "unusually well developed." Redmond et al (1982) cite with approval Milne-Edwards' (1872) observation that (in translation) "the circulatory system of the Limulines is more perfect, and better developed, than in any other segmented animal." And Palumbi & Johnson (1982) remark on the unusual adaptation to low in vivo oxygen tension. Also quite typical of marine invertebrates are other physiological features of *Limulus* [hemocyanins discussed by Solomon et al (1982) and Bijlholt et al 1982], as well as the extent of morphologic variability (Riska 1981), the extent of hybridization among the four living species of horseshoe crabs (Sekiguchi & Sugita 1980), and the extent of genetic variability (Selander et al 1970). In fossil limulines, virtually nothing is known about the soft parts (including most of the appendages). But to judge from what is known of the living forms, *Limulus* has not lagged behind its modern marine invertebrate contemporaries in any biologically relevant fashion.

Other "living fossils" have, where examined, exhibited no biochemical traits indicative of a species "where time has stood still." There is nothing special distinguishing the DNA or RNA of *Lingula* (Shimizu & Miura 1971a,b, Shimizu 1971), the cytochrome C of *Ginkgo* (Ramshaw et al 1971), or the cuticle of *Peripatus* (Hackman & Goldberg 1975). In some biochemical traits, "living fossils" may even be precocious. For example, it was noted that "the opossum α [hemoglobin] chain has apparently evolved more rapidly than any other α chain in all cases in which a comparison is possible" (Stenzel 1974).

There is a large difference between considering "living fossils" as the problem of a biological species that is entirely primitive and has not evolved, or as the problem of the occurrence of particular "primitive" traits. In some cases these traits are a prominent aspect of a very limited fossil record. In this roundabout way, the concept of living fossils as a problem in developmental expression of traits has been tied to the problem of species stability over geologic time. These aspects must be disengaged.

Perhaps the most often-cited analysis of the rate of change of traits leading to a living fossil is that done for lungfish (Westoll 1949; see Simpson 1953). For a suite of characters, a set of ancestral conditions (and subsequent structural grades) was determined; each grade was assigned an arbitrary relative value. Through time, the morphological "score" for each genus was determined, based on the extent of retention of the primitive condition. The conclusion was that there existed a rapid early loss of primitive traits. The graph of this loss (and a similar plot for coelacanths by Schaeffer 1952) suggests that there is an initially rapid rate of acquisition of new morphologic traits, followed by a long, slow period of nearly imperceptible modification.

Although this customary explanation may be true, it appears possible that the pattern of scores follows directly from the method of analysis, for the following reasons. (*a*) By definition, the initial set of characters is considered primitive, i.e. that derived states exist and thus change must be seen; this effect is enhanced by arbitrarily adding to the score of the "presumed ancestor" to make it the highest score in 12 of the 26 characters. (*b*) There are 14 points plotted (including one for the "presumed ancestor"). Of these, 7 are in the Devonian. Because the amount of visible change is a function of the number of species available for study, it seems only reasonable that half of the change should be observed during the first 11% of the time (the length of the Devonian in Westoll's time scale), and most of the initial steep decline may be due to this effect. Three more points are added by the Middle Carboniferous, and only two more points are added during the 225 m.y. between then and the Recent. (*c*) The scale of morphologic change is not linear. Only 1 character state had as many as 7 steps, and 17 of the 26 characters had 4 or fewer steps. Changes in character states are a function of the number of specimens. The preponderance of Devonian specimens surely biases the steps in the scale of differentiation to include much more change during this early interval of time. (*d*) If the vast amount of observed changes were "late" in the history of the group, then that would in all likelihood be seen taxonomically as the splitting off of a new group, thus further resulting in character change being considered "most typical" of early periods of radiation. An example would be birds split from certain reptilian lineages. (*e*) W. Bemis (1984) has suggested that lungfish become increasingly paedomorphic toward the Recent. Thus, the available character states automatically become increasingly restricted.

For these five reasons, it seems inevitable that in Westoll's example, characters will appear to change considerably in the early part of the lineage relative to the later part of the lineage. Thus, despite a long tradition (Simpson 1953, 1983) these data on lungfish cannot be used to support the view that taxa of lungfish are inordinately slowly evolving. A thorough simulation analysis of the problem of the rate of character change has been begun by Derstler (1982), who also notes the bias (in early echinoderms) of an association between taxonomic diversity and high rate of morphologic change. (One *must* normalize for diversity, perhaps by expressing the results as change per species, i.e. change per whatever taxonomic unit is being compared.)

Counterexamples

Insects provide the most detailed evidence of absence of significant morphological change for geologically long intervals of time, although no taxa are inferred to have persisted for tens to hundreds of millions of years. Over the late Pleistocene, Coope (1978, p. 183) stated:

> . . . an empirical view of the fossil record shows that there is no evidence of any morphological evolution [in insects] during the last half million years at least. In other words insect species have remained immutable regardless of the vicissitudes of the

> Quaternary climate. Certainly for the past few glacial/interglacial cycles . . . almost all our fossil insects match their modern counterparts with extraordinary exactness even down to the intimate internal sclerites of the genitalia. Though the vast majority of insect fossils are Coleoptera, because the robustness of their exoskeleton lends itself to easy preservation, a broad spectrum of insect orders is represented in these fossil assemblages although structural frailty makes identification in many other groups a rather rare occurrence.

Coope (1970, 1979) reviewed a large amount of evidence that indicated that insect fossils 0.5–1.6 m.y. old were "identical" with living species. Older specimens of the late Miocene (~6 m.y. ago) "inevitably display subtle features which place them outside the range of variation of their modern counterparts . . ." (Matthews 1980). Unless speciation is related to karyotypic or other genomic changes, or to behavior or biochemical specialization (not unreasonable alternatives), these studies of fossil insects appear to be the best available information on actual measurement of species duration in morphologically complex forms.

The two species (of which I am aware) with the best claim to a truly long duration appear to be the branchiopod *Triops cancriformis* (late Triassic to present; see Tasch 1969) and the noncoccolithophore nannolith *Braarudosphaera bigelowi*, known from the late Jurassic to the present (Percival & Fischer 1977, p. 17). *T. cancriformis* is most notable for an anterior-posterior elongated dorsal carapace that varies, however, sufficiently in size as to be an inconsistent taxonomic character (Longhurst 1955). Portions of the carapace are chiefly what occurs in the fossil material, which is "represented by some 400 carapace fragments, without marks of body parts, 70 carapace-and-body fragments, 30 fragments of abdomen with impressions of furcal setae, eggs, [and] carapace margins" (Tasch 1969, p. R135). Although the material appears relatively extensive, it seems not to have been closely examined since its discovery and initial description in the 1930s (Trusheim 1938). The reconstruction of the ventral surface is especially sketchy (Trusheim 1938, Figure 7). No forms are known of *T. cancriformis* between the late Triassic and the present. *B. bigelowi* is a very simple nannolith composed of 5 interlocking crystals (Haq 1978, Figure 21). Forms referable to this species occur abundantly in Mesozoic and Tertiary sections. A variety of *genera* are reported to have long geologic ranges (e.g. the ostracod *Bairdia*: Ordovician-Recent; the brachiopod *Lingula*: Silurian-Recent; and the bivalve *Modiolus*: Devonian-Recent), and those who wish to defend the notion of living fossils as a problem in speciation theory (i.e. in species persistence) might possibly use *species* in these genera to develop a case, but this has not yet been done.

Habitats

The notion of living fossils has also spilled over into the idea that there are some habitats that have more than their share of these organisms. Such claims were originally made by Darwin (1859, p. 111), who attributed the persistence of "living fossils" (a term he coined) to conditions of less severe competition. On the other hand, some have emphasized the generality in form brought about by (it is claimed) "stressful environments" (Vermeij 1978, p. 180). The sulfide biome has been suggested to contain "some primary elements and functional relationships of the oldest biosystem on earth which preceded the aerobic biosphere" (Fenchel & Riedl 1970, p. 266). The marine cave fauna (at Bermuda) is said to include some species that date from the separation of the African and the American continents (Iliffe et al 1983). The interstitial sand grain fauna has been inferred to harbor a large number of species with ages "50 to 200 million years or more" (Sterrer 1973, p. 214); and cool-climate, ammonium-rich soils may include primitive nitrogen-loving taxa (Galston 1978, Hutchinson 1970). Tradition has also supported the view that the deep sea is a "marginal environment" that contains a larger-than-normal share of relect forms (Vermeij 1978), but the primary test of this idea concluded that "the shallow-water marine, the fresh-water, and the terrestrial environments have more archaic kinds of life than the abyssal marine environment" (Menzies & Imbrie 1958, p. 208). Indeed, what is clear is that if one wishes, one can find "living fossils" in virtually every habitat (most of which are included in the summary of Delamare-Deboutteville & Botosanéanu 1970), and thus the existence of "living fossils" is largely habitat-independent.

Demise of Bradytely?

"Bradytelic species"—the jargon phrase for the population of presumed extremely slowly evolving species—has been part of the conventional wisdom of paleontology for longer than the 40 years since the phrase was introduced (Simpson 1944). Why has this notion persisted for so long?

The chief scientific reason as to why "living fossils" have been part of our heritage is that prior to an understanding of the ubiquity of high levels of genetic variability (reviewed by Lewontin 1974, Coyne 1982), it was possible to *think* that some species indeed had "stood still" genomically. The demise of one third of Simpson's evolutionary rate triumvirate of bradytely, tachytely, and horotely is first and foremost a casualty of modern population genetics, now supplanted by molecular biology (Schopf 1981). It is not longer scientifically possible to believe that any group of organisms is genomically at an evolutionary standstill.

A second reason for the persistence of the belief in "living fossils" was the traditional view asserted by taxonomic paleontologists that since neontologists use morphology, and paleontologists do the same, then whatever either group calls a species is likely to be the same sort of entity. Thus, if no or slight change was seen in "hard parts," then that meant the species persisted. I dealt above (Table 1; also Schopf 1982) with the fallacy of this view, a view that I myself shared prior to working with live material. As neontologists delve more deeply into systematics related to "soft parts" (including chromosomal and genomic differences), paleontologists are left further behind. This is not to say that paleontologists are less-able scientists, or anything of that sort. It is to say that the limitations of material permit different types of judgments to be made in living versus fossil specimens. Paleontological judgments are quite useful for biostratigraphic purposes, where, for example, one correlates strata in New York with strata in Estonia. Paleontological judgments are not so useful for evolutionary purposes, where one must deal with the equivalent of biological species in order to make assessments of rates of evolutionary change.

The problem of how "canalization" of morphological features occurs in development appears to be an unsolved and at present intractable problem, though no doubt it will soon yield to a molecular accounting. Enough cases of canalization are known for there to be no question that there is something to be explained. This, however, in no way detracts from what we can say in the most positive manner about living fossils, namely, that the occurrence of classical "living fossils" as a problem in speciation theory is without scientific standing. As a first approximation it seems most reasonable to refer all species to a single distribution of rates, within which there is variation, as in any biological or chemical process.

CONCLUSIONS

The topic "Rates of Evolution," with emphasis at the species level, seems not to have been previously considered by any of the Annual Reviews series. Commonly cited reviews of rates of evolution in animals (Simpson 1944, 1949, 1953) and in plants (Stebbins 1947, 1949, 1950) were published more than a quarter of a century ago.

The present effort reflects new information obtained during this time, and thus differs from these previous works in major ways. Simpson delineated independent groups of species by their rate of evolution—those that were bradytelic, those that were horotelic, and those that were tachytelic. Within each group, there were taxa that were bradytelic, etc. Thus he wrote that "a relatively fast bradytelic rate might conceivably be faster than a relatively slow horotelic rate" (Simpson 1949, p. 221). These rates were, in all cases, largely determined by a literal reading of the fossil record, without benefit of our present understanding of the ubiquity of genomic variability and our present appreciation of the various biases that affected previous estimates of species durations.

The present contribution differs from Simpson's in that attention is drawn to the likelihood that there is but a single population of rates, and that biases of various sorts are what is responsible for the vast majority of the slower-appearing rates of evolution. In the present review, the aim of the section on "living fossils" is to show that the concept of "living fossil" fails as a concept in speciation theory. A separate population of extremely slowly evolving species does not exist. Whatever value there may be in the concept of "living fossils" resides in considering why specific traits may not change over long intervals of geologic time. This, however, is a problem in developmental biology (broadly conceived), rather than in speciation theory.

The present paper also differs appreciably in emphasis from Stebbins' work. He has given considerable attention to environmental factors in speciation, as has been common among many authors. He wrote, for example, "We can therefore postulate that evolution will be slowest in those organisms which demand relatively little of their environment, and progressively more rapid as the minimum requirements for existence become greater and greater" (Stebbins 1949, p. 239). The present paper [and, to a greater extent, a companion one (Schopf 1984)] plays down the environmental aspects per se, and instead emphasizes the continued change in the genome itself. The growth in ecologic information (which is a counterpart to an increase in understanding of genetic variability) came through an appreciation that species diversity in the deep sea (a "stable and uniform habitat") is equivalently high to the diversity of coral reefs or tropical rain forests ("variable habitats").

Thus, seen in an historical perspective, a 1984 understanding of "rates of evolution" reflects a heightened awareness of the role of genomic change in the evolutionary process. Although, in a sense, genomic rates of change only set the boundaries on speciation rates, it is now realized that genomic rates may change at a rapid enough rate to drive the process of speciation to an extent not envisaged 25 years ago. The mean rate of speciation is probably not as short as 10^3 to 10^4 yr, or else we would see a much greater turnover of taxa in both the modern and fossil worlds. However a mean rate in the range of 10^5 to 10^6 yr seems plausible if one takes into account information on both paleontologic biases and on genomic change.

Acknowledgments

During the course of preparing this paper, I have greatly benefited from discussions with Jeff Mitton, Antoni Hoffman, Guy Bush, David Wake, Alan Templeton, Jack Sepkoski, Teresa Bone, Dan Fisher, James Shapiro, and students in my course on Molecular Biology and Evolution. I am, however, most of all indebted to the flexibility of the University of Chicago, whose traditions for encouraging a person to explore his or her intellectual interests have allowed me to study for nearly 15 years, and in very time-consuming ways, areas that are quite foreign to traditional geology departments.

Literature Cited

Aarssen, L. W., Turkington, R. 1983. What is community evolution? *Evol. Theory* 6: 211–17

Andrews, H. N. Jr. 1961. *Studies in Paleobotany*. New York: Wiley. 487 pp.

Beermann, S. 1977. The diminution of heterochromatic chromosomal segments in *Cyclops* (Crustacea, Copepoda). *Chromosoma* 60: 297–344

Bemis, W. E. 1983. *Studies on the functional and evolutionary morphology of lepidosirenid lungfish (Pisces: Dipnoi)*. PhD dissertation. Univ. Calif., Berkeley

Bemis, W. E. 1984. Paedomorphosis and the evolution of the Dipnoi. *Paleobiology*. In press

Bender, W., Akam, M., Karch, F., Beachy, P. A., Peifer, M., et al. 1983. Molecular genetics of the bithorax complex in *Drosophila melanogaster*. *Science* 221: 23–29

Bijlholt, M. M. C., van Heel, M. G., van Bruggen, E. F. J. 1982. Comparison of 4 × 6-meric hemocynanins from three different arthropods using computer alignment and correspondence analysis. *J. Mol. Biol.* 161: 139–53

Bingham, P. M., Kidwell, M. G., Rubin, G. M. 1982. The molecular basis of P-M hybrid dysgenesis: the role of the P element, a P-strain-specific transposon family. *Cell* 29: 995–1004

Bold, H. C., Alexopoulos, C. J., Delevoryas, T. 1980. *Morphology of Plants and Fungi*. New York: Harper & Row. 819 pp.

Bonaventura, J., Bonaventura, C., Tesh, S., eds. 1982. *Physiology and Biology of Horseshoe Crabs*. New York: Liss. 313 pp.

Boreske, J. R. Jr. 1974. A review of the North American fossil amiid fishes. *Mus. Comp. Zool. Bull.* 146: 1–87

Botosanéanu, L., Delamare-Deboutteville, C. 1967. Fossiles vivantes des eaux souterraines. *Sciences (Paris)* 52: 17–22

Bregliano, J.-C., Kidwell, M. G. 1983. Hybrid dysgenesis determinants. See Shapiro 1983a, pp. 363–410

Brien, P. 1967. The African protoptera: living fossils. *Afr. Wildl.* 21: 213–33

Britten, R. J. 1982. Genomic alterations in evolution. In *Evolution and Development (Dahlem Konferenzen)*, ed. J. T. Bonner, pp. 215–35. Berlin: Springer-Verlag

Brown, W. M., George, M. Jr., Wilson, A. C. 1979. Rapid evolution of animal mitochondrial DNA. *Proc. Natl. Acad. Sci. USA* 76: 1967–71

Brown, W. M., Prager, E. M., Wang, A., Wilson, A. C. 1982. Mitochondrial DNA sequences of primates: tempo and mode of evolution. *J. Mol. Evol.* 18: 225–39

Burton, M. 1954. *Living Fossils*. London: Thames & Hudson. 282 pp.

Bush, G. L. 1975. Modes of animal speciation. *Ann. Rev. Ecol. Syst.* 6: 339–64

Bush, G. L. 1981. Stasipatric speciation and rapid evolution in animals. In *Evolution and Speciation, Essays in Honor of M. J. D. White*, ed. W. R. Atchley, D. S. Woodruff, pp. 201–18. London: Cambridge Univ. Press

Bush, G. L., Case, S. M., Wilson, A. C., Patton, J. L. 1977. Rapid speciation and chromosomal evolution in mammals. *Proc. Natl. Acad. Sci. USA* 74: 3942–46

Buzzati-Traverso, A. A. 1955. Populations in time and space: synthesis. *Cold Spring Harbor Symp. Quant. Biol.* 20: 300–2

Campbell, K. S. W. 1981. Lungfishes—alive and extinct. *Field Mus. Nat. Hist. Bull.* 52: 3–5

Cavalier-Smith, T. 1982. Skeletal DNA and the evolution of genome size. *Ann. Rev. Biophys. Bioeng.* 11: 273–302

Chace, E. P. 1962. A living fossil. *Veliger* 4: 162

Cherry, L. M., Case, S. M., Kunkel, J. G., Wyles, J. S., Wilson, A. C. 1982. Body shape metrics and organismal evolution. *Evolution* 36: 914–33

Cisne, J. L., Chandlee, G. O., Rabe, B. D., Cohen, J. A. 1982. Clinal variation, episodic evolution, and possible parapatric speciation: the trilobite *Flexicalymene senaria* along an Ordovician depth gradient. *Lethaia* 15: 325–41

Coope, G. R. 1970. Interpretations of Quaternary insect fossils. *Ann. Rev. Entomol.* 15: 97–120

Coope, G. R. 1978. Constancy of insect species versus inconstancy of Quaternary environments. In *Diversity of Insect Faunas*, ed. L. A. Mound, N. Waloff, pp. 176–87. Oxford: Blackwell

Coope, G. R. 1979. Late Cenozoic fossil coleoptera: evolution, biogeography, and ecology. *Ann. Rev. Ecol. Syst.* 10: 247–68

Coyne, J. A. 1982. Gel electrophoresis and

cryptic protein variation. *Isozymes: Curr. Top. Biol. Med. Res.* 6:1–32

Coyne, J. A. 1983. Genetic basis of differences in genital morphology among three sibling species of *Drosophila. Evolution* 37: In press

Cullum, J., Saedler, H. 1981. DNA rearrangements and evolution. In *Molecular and Cellular Aspects of Microbial Evolution, Symp. Soc. Gen. Microbiol., 32nd*, ed. M. J. Carlile, J. F. Collins, B. E. B. Moseley, pp. 131–50. Cambridge Univ. Press

Czelusniak, J., Goodman, M., Hewett-Emmett, D., Weiss, M. L., Venta, P. J., Tashian, R. E. 1982. Phylogenetic origins and adaptive evolution of avian and mammalian haemoglobin genes. *Nature* 298: 297–300

Darwin, C. 1859. *The Origin of Species by Means of Natural Selection.* (Doubleday & Co. reprint; New York). 517 pp.

Davis, G. M. 1979. The origin and evolution of the gastropod family Pomatiopsidae, with emphasis on the Mekong River Triculinae. *Acad. Nat. Sci. Philadelphia Monogr. 20.* 120 pp.

Dawid, I., Britten, R. J., Davidson, E. H., Dover, G. A., Gallwitz, D. F., et al. 1982. Genomic change and morphologic evolution, group report. In *Evolution and Development* (*Dahlem Konferenzen*), ed. J. T. Bonner, pp. 19–39. Berlin: Springer-Verlag

Deelman, J. C., De Coo, J. C. M. 1975. Microbial cryptobiosis and recycling: a conservation mechanism in evolution? *Mod. Geol.* 5:185–89

Delamare-Deboutteville, C., Botosanéanu, L. 1970. *Formes Primitives Vivantes.* Paris: Herman. 232 pp.

De Ley, J. 1968. Molecular biology and bacterial phylogeny. *Evol. Biol.* 2:103–56

Demerec, M. 1950. Reaction of populations of unicellular organisms to extreme changes in environment. *Am. Nat.* 84:5–16

Derstler, K. 1982. Estimating the rate of morphological change in fossil groups. *Proc. North Am. Paleontol. Conv., 3rd*, 1:131–36

Dingus, L., Sadler, P. M. 1982. The effects of stratigraphic completeness on estimates of evolutionary rates. *Syst. Zool.* 31:400–12

Dobzhansky, T. 1935. A critique of the species concept in biology. *Philos. Sci.* 2:344–55

Dobzhansky, T. 1937. *Genetics and the Origin of Species.* New York: Columbia Univ. Press. 364 pp.

Dobzhansky, T. 1955. A review of some fundamental concepts and problems of population genetics. *Cold Spring Harbor Symp. Quant. Biol.* 20:1–15

Doolittle, W. F. 1982. Selfish DNA after fourteen months. In *Genome Evolution*, ed. G. A. Dover, R. B. Flavell, pp. 3–28. New York: Academic

Dover, G. 1978. DNA conservation and speciation: adaptive or accidental. *Nature* 272:123–24

Dunbar, M. J. 1960. The evolution of stability in marine environments: natural selection at the level of the ecosystem. *Am. Nat.* 94:129–36

Echols, H., Lu, C., Burgers, P. M. J. 1983. Mutator strains of *E. coli mut*D and *dna*Q, with defective exonucleolytic editing by DNA polymerase III holoenzyme. *Proc. Natl. Acad. Sci. USA* 80:2189–92

Ehrlich, P. R. 1980. Should we worry about the extinction of other species? *Lindbergh Lecture Series in Ecology, Marine Biological Laboratory, Woods Hole, Mass.* 12 pp. (Privately printed)

Eldredge, N. 1982a. Living fossils. In *McGraw-Hill Encyclopedia of Science and Technology*, 7:783. New York: McGraw-Hill. 5th ed.

Eldredge, N. 1982b. Phenomenological levels and evolutionary rates. *Syst. Zool.* 31: 338–47

Emig, C. C. 1982. Taxonomie du genre *Lingula* (Brachiopodes, Inarticulés). *Mus. Natl. Hist. Nat. Bull., Paris, 4th Ser.* 4(A):337–67

Endler, J. A. 1977. *Geographic Variation, Speciation, and Clines.* Princeton Univ. Press. 246 pp.

Fenchel, T. M., Riedl, R. J. 1970. The sulfide system: a new biotic community underneath the oxidized layer of marine sand bottoms. *Mar. Biol.* 7:255–68

Fischer, H. J. L. 1965. Remarques sur la repartition du genre *Lepidopleurus. J. Conchyliol.* 105:143–48

Fisher, D. C. 1982. Phylogenetic and macroevolutionary patterns within the Xiphosurida. *Proc. North Am. Paleontol. Conv., 3rd*, 1:175–80

Fleming, C. A. 1964. History of the bivalve family trigoniidae in the south-west Pacific. *Austral. J. Sci.* 26:196–204

Forest, J., Laurent, M. de S., Chace, F. A. Jr. 1976. *Neoglyphea inopinata*: a crustacean "living fossil" from the Philippines. *Science* 192:884

Forey, P. J. 1981. The coelacanth *Rhabdoderma* in the Carboniferous of the British Isles. *Palaeontology* 24:203–29

Futuyma, D. J., Slatkin, M. 1983. Epilogue: The study of coevolution. In *Coevolution*, ed. D. J. Futuyma, M. Slatkin. pp. 459–64. Sunderland, MA: Sinauer Assoc.

Galston, A. W. 1978. A living fossil. *Nat. Hist.* 87:42, 44

Gans, C. 1983. Is *Sphenodon punctatus* a

maladapted relict? In *Advances in Herpetology and Evolutionary Biology: Essays in Honor of Ernest E. Williams*, ed. G. J. Rodin, K. Miyata, pp. 613–20. Cambridge, Mass: Mus. Comp. Zool.

Gingerich, P. D. 1983. Rates of evolution: effects of time and temporal scaling. *Science* 222: 159–61

Glickman, B. W. 1982. Methylation-instructed mismatch correction as a post-replication error avoidance mechanism in *Escherichia coli*. In *Molecular and Cellular Mechanisms of Mutagenesis*, ed. J. F. Lemontt, W. M. Generoso, pp. 65–87. New York: Plenum

Gooch, J. L. 1975. Mechanisms of evolution and population genetics. In *Marine Ecology*, ed. O. Kinne, 2(1): 349–409. New York: Wiley

Goodman, M. 1981. Decoding the pattern of protein evolution. *Prog. Biophys. Mol. Biol.* 37: 105–64

Gordon, T., Jablonski, D. 1979. Living fossil. In *The Encyclopedia of Paleontology*, ed. R. W. Fairbridge, D. Jablonski, pp. 419–22. Stroudsburg, Pa: Dowden, Hutchinson & Ross

Gorniak, G. C., Rosenberg, H. I., Gans, C. 1982. Mastication in the tuatara, *Sphenodon punctatus* (Reptilia: Rhynchocephalia): structure and activity of the motor system. *J. Morphol.* 171: 321–53

Gould, S. J. 1982. Darwinism and the expansion of evolutionary theory. *Science* 216: 380–87

Grant, B., Wiseman, L. L. 1982. Fossil genes: scarce as hen's teeth? *Science* 215: 698–99

Grant, V. 1971. *Plant Speciation*. New York: Columbia Univ. Press. 435 pp.

Griffith, R. W., Pang, P. K. T. 1979. Mechanisms of osmoregulation in the coelacanth: evolutionary implications. See McCosker & Lagios 1979, pp. 79–93

Gupta, A. J., Lewontin, R. C. 1982. A study of reaction norms in natural populations of *Drosophila pseudoobscura*. *Evolution* 36: 934–48

Hackman, R. H., Goldberg, M. 1975. *Peripatus*: its affinities and its cuticle. *Science* 190: 582–83

Hansen, T. A. 1982. Modes of larval development in early Tertiary neogastropods. *Paleobiology* 8: 367–77

Haq, B. U. 1978. Calcareous nannoplankton. In *Introduction to Marine Micropaleontology*, ed. B. U. Haq, A. Boersma, pp. 79–107. New York: Elsevier

Hill, O. 1966. The spectral tarsier. *Med. Biol. Illus.* 16: 182–86

Hoffman, A. 1979. Community paleoecology as an epiphenomenal science. *Paleobiology* 5: 357–79

Hoffman, A., Szubzda-Studencka, B. 1982. Bivalve species duration and ecologic characteristics in the Badenian (Miocene) marine sandy facies of Poland. *Neus. Jahrb. Geol. Paläontol. Abh.* 163: 122–35

Hübscher, U. 1983. DNA polymerases in prokaryotes and eukaryotes: mode of action and biological implications. *Experientia* 39: 1–25

Hutchinson, G. E. 1970. Living fossils. *Am. Sci.* 58: 531–35

Iliffe, T. M., Hart, C. W. Jr., Manning, R. B. 1983. Biogeography and the caves of Bermuda. *Nature* 302: 141–42

Iltis, H. H. 1957. Studies in the Capparidaceae. III. Evolution and phylogeny of the Western North American Cleomoideae. *Ann. Mo. Bot. Gard.* 44: 77–119

Jablonski, D., Lutz, R. A. 1983. Larval ecology of marine benthic invertebrates: paleobiological implications. *Biol. Rev. Cambridge Philos. Soc.* 58: 21–89

Jacob, F. 1977. Evolution and tinkering. *Science* 196: 1161–66

Jacob, F. 1983. Molecular tinkering in evolution. In *Evolution from Molecules to Men*, ed. D. S. Bendall, pp. 131–44. Cambridge Univ. Press

Jepsen, G. L., Simpson, G. G., Mayr, E. 1949. *Genetics, Paleontology and Evolution*. Princeton Univ. Press. 475 pp. Republished in paperback by Atheneum (1963)

Kauffman, E. G. 1977. Evolutionary rates and biostratigraphy. In *Concepts and Methods of Biostratigraphy*, ed. E. G. Kauffman, J. E. Hazel, pp. 109–41. Stroudsberg, Pa: Dowden, Hutchinson & Ross

Kidwell, M. G. 1983. Evolution of hybrid dysgenesis determinants in *Drosophila melanogaster*. *Proc. Natl. Acad. Sci. USA* 80: 1655–59

Kimura, M. 1967. On the evolutionary adjustment of spontaneous mutation rates. *Genet. Res.* 9: 23–34

Kimura, M. 1969. The rate of molecular evolution considered from the standpoint of population genetics. *Proc. Natl. Acad. Sci. USA* 63: 1181–88

Kimura, M. 1981. Possibility of extensive neutral evolution under stabilizing selection with special reference to nonrandom usage of synonymous codons. *Proc. Natl. Acad. Sci. USA* 78: 5773–77

King, J. C. 1955. Evidence for the integration of the gene pool from studies of DDT resistance in *Drosophila*. *Cold Spring Harbor Symp. Quant. Biol.* 20: 311–17

Kummel, B. 1964. Nautiloides—Nautilida. In *Treatise on Invertebrate Paleontology, Part K*, ed. R. C. Moore, pp. 383–466. Lawrence, Kans: Geol. Soc. Am.

Lande, R. 1976a. Natural selection and random genetic drift in phenotypic evolution. *Evolution* 30: 314–34

Lande, R. 1976b. The maintenance of genetic

variability by mutation in a polygenic character with linked loci. *Genet. Res.* 26:221–35

Lande, R. 1981. The minimum number of genes contributing to quantitative variation between and within populations. *Genetics* 99:541–53

Lande, R. 1983. The response to selection on major and minor mutations affecting a metrical trait. *Heredity* 50:47–65

Lewontin, R. C. 1957. The adaptations of populations to varying environments. *Cold Spring Harbor Symp. Quant. Biol.* 22:395–408

Lewontin, R. C. 1974. *The Genetic Basis of Evolutionary Change.* New York: Columbia Univ. Press. 346 pp.

Lewontin, R. C. 1981. Introduction: the scientific work of Th. Dobzhansky. In *Dobzhansky's Genetics of Natural Populations I-XLIII*, ed. R. C. Lewontin, J. A. Moore, W. B. Provine, B. Wallace, pp. 93–115. New York: Columbia Univ. Press

Ley, W. 1959. *Exotic Zoology.* New York: Viking. 468 pp.

Li, H.-L. 1964. *Metasequoia*, a living fossil. *Am. Sci.* 52:93–109

Liao, L. W., Rosenzweig, B., Hirsh, D. 1983. Analysis of a transposable element in *Caenorhabditis elegans. Proc. Natl. Acad. Sci. USA* 80:3585–89

Loeb, L. A., Kunkel, T. A. 1982. Fidelity of DNA synthesis. *Ann. Rev. Biochem.* 52:429–57

Longhurst, A. R. 1955. A review of the Notostraca. *Br. Mus.* (*Nat. Hist.*) *Zool. Bull.* 3(1):1–57

Lovelock, J. E. 1972. Gaia as seen through the atmosphere. *Atmos. Environ.* 6:579–80

Lovelock, J. E. 1979. *Gaia, a New Look at Life on Earth.* Oxford Univ. Press. 157 pp.

Lupu, D. 1966. Un fossile vivant de la faune de la Roumanie: *Chilostoma* (*Drobacia*) *maeotica* Wenz (Gastropoda-Pulmonata). *Trav. Mus. Hist. Nat. "Grigore Antipa"* 6:31–37. Bucharest

MacBryde, B., Altevogt, R. 1977. Endangered plant species. In *McGraw-Hill Encyclopedia of Science and Technology Yearbook*, pp. 81–91. New York: McGraw-Hill

Maguire, M. P. 1983. Chromosome behavior at premeiotic mitosis in maize. *J. Hered.* 74:93–96

Major, R. T. 1967. The Ginkgo, the most ancient living tree. *Science* 157:1270–73

Margalef, R. 1968. *Perspectives in Ecological Theory.* Univ. Chicago Press. 111 pp.

Margulis, L., Schwartz, K. V. 1982. *Five Kingdoms.* San Francisco: Freeman. 338 pp.

Matthews, J. V. Jr. 1980. Tertiary land bridges and their climate: backdrop for development of the present Canadian insect fauna. *Can. Entomol.* 112:1089–1103

Maynard Smith, J. 1982. Overview—unsolved evolutionary problems. In *Genome Evolution*, ed. G. A. Dover, R. B. Flavell, pp. 375–82. New York: Academic

Mayr, E. 1963. *Animal Species and Evolution.* Cambridge, Mass: Belknap. 797 pp.

Mayr, E. 1982. Speciation and macroevolution. *Evolution* 36:1119–32

McCosker, J. E., Lagios, M. D., eds. 1979. *The Biology and Physiology of the Living Coelacanth, Occas. Pap. Calif. Acad. Sci. No. 134.* 176 pp.

McKenzie, K. G. 1967. Ostrocod "living fossils": new finds in the Pacific. *Science* 155:1005

Menzies, R. J., Imbrie, J. 1958. On the antiquity of the deep sea bottom fauna. *Oikos* 9:192–210

Milne-Edwards, A. 1872. Recherches sur l'anatomie des Limules. *Ann. Sci. Nat.* 17(4):1–67

Miyata, T., Hayashida, H., Kikuno, R., Hasegawa, M., Kobayashi, M., Koike, K. 1982. Molecular clock of silent substitution: at least six-fold preponderance of silent changes in mitochondrial genes over those in nuclear genes. *J. Mol. Evol.* 19:28–35

Modolell, J., Bender, W., Meselson, M. 1983. *Drosophila melanogaster* mutations suppressible by the suppressor of Hairy-wing are insertions of a 7.3-kilobase mobile element. *Proc. Natl. Acad. Sci. USA* 80:1678–82

Montgomery, E. A., Langley, C. H. 1983. Transposable elements in Mendelian populations. II. Distribution of three copia-like elements in a natural population of *Drosophila melanogaster. Genetics* 104:473–83

Moore, W. S. 1977. An evaluation of narrow hybrid zones in vertebrates. *Q. Rev. Biol.* 52:263–77

Nagylaki, T. 1983. Evolution of a finite population under gene conversion. *Proc. Natl. Acad. Sci. USA* 80:6278–81

Neel, J. V. 1983. Frequency of spontaneous and induced "point" mutations in higher eukaryotes. *J. Hered.* 74:2–15

Newman, W. A. 1979. A new scalpellid (Cirripedia): a Mesozoic relic living near an abyssal hydrothermal spring. *Trans. San Diego Soc. Nat. Hist.* 19:153–67

Odum, E. P. 1959. *Fundamentals of Ecology.* Philadelphia: Saunders. 546 pp. 2nd ed.

Odum, E. P. 1969. The strategy of ecosystem development. *Science* 164:262–70

Olsson, A. A., Bergeron, E. 1967. *Perplicaria clarki* Maxwell Smith, a living fossil. *Veliger* 9:411–12

Palumbi, S. R., Johnson, B. A. 1982. A note on the influence of life-history stage on metabolic adaptation: the responses of

Limulus eggs and larvae to hypoxia. See Bonaventura et al 1982, pp. 115–31

Percival, S. F. Jr., Fischer, A. G. 1977. Changes in calcareous nannoplankton in the Cretaceous-Tertiary biotic crisis at Zumaya, Spain. *Evol. Theory* 2:1–35

Ramshaw, J. A. M., Richardson, R., Boulter, D. 1971. The amino-acid sequence of the cytochrome *c* of *Ginkgo biloba* L. *Eur. J. Biochem.* 23:475–83

Ramshaw, J. A. M., Coyne, J. A., Lewontin, R. C. 1979. The sensitivity of gel electrophoresis as a detector of genetic variation. *Genetics* 93:1019–37

Raup, D. M. 1975. Taxonomic survivorship curves and Van Valen's Law. *Paleobiology* 1:82–96

Raup, D. M. 1978. Cohort analysis of generic survivorship. *Paleobiology* 4:1–15

Raup, D. M. 1981. Extinction: bad genes or bad luck? *Acta Geol. Hisp.* 16:25–33

Raup, D. M., Gould, S. J., Schopf, T. J. M., Simberloff, D. S. 1973. Stochastic models of phylogeny and the evolution of diversity. *J. Geol.* 81:525–42

Redmond, J. R., Jorgensen, D. D., Bourne, G. B. 1982. Circulatory physiology of *Limulus*. See Bonaventura et al 1982, pp. 133–46

Reeside, J. B. Jr., Harris, D. V. 1952. A Cretaceous horseshoe crab from Colorado. *J. Wash. Acad. Sci.* 42:174–78

Rendel, J. M. 1967. *Canalisation and Gene Control.* London: Logos. 166 pp.

Repenning, C. A., Ray, C. E. 1977. The origin of the Hawaiian monk seal. *Proc. Biol. Soc. Wash.* 89:667–88

Riede, I., Renz, M. 1983. Study on the somatic pairing of polytene chromosomes. *Chromosoma* 88:116–23

Rieppel, O. 1980. The phylogeny of Anguinomorph lizards. *Denkschriften der Schweizerischen Naturforschenden Gesellschaft* 94. (*Mém. Soc. Helv. Sci. Nat.* 94:1–86)

Ripley, L. S. 1983. The specificity of infidelity of DNA polymerase. In *Induced Mutagenesis*, ed. C. W. Lawrence, pp. 83–113. New York: Plenum

Ripley, L. S., Glickman, B. W. 1983. Unique self-complementarity of palindromic sequences provides DNA structural intermediates for mutation. *Cold Spring Harbor Symp. Quant. Biol.* 47:851–61

Ripley, L. S., Shoemaker, N. B. 1983. A major role for bacteriophage T4 DNA polymerase in frameshift mutagenesis. *Genetics* 103:353–66

Riska, B. 1981. Morphological variation in the horseshoe crab *Limulus polyphemus*. *Evolution* 35:647–58

Roeder, G. S., Fink, G. R. 1983. Transposable elements in yeast. See Shapiro 1983a, pp. 299–328

Roeder, G. S., Farabaugh, P. J., Chaleff, D. T., Fink, G. R. 1980. The origins of gene instability in yeast. *Science* 209:1375–91

Romer, A. S. 1966. *Vertebrate Paleontology.* Univ. Chicago Press. 468 pp.

Rose, M. R., Doolittle, W. F. 1983. Molecular biological mechanisms of speciation. *Science* 220:157–62

Rubin, G. M. 1983. Dispersed repetitive DNAs in *Drosophila*. See Shapiro 1983a, pp. 329–61

Runnegar, B. 1982. A molecular-clock date for the origin of the animal phyla. *Lethaia* 15:199–205

Sadler, P. M. 1981. Sediment accumulation rates and the completeness of stratigraphic sections. *J. Geol.* 89:569–84

Schaeffer, B. 1952. Rates of evolution in the coelacanth and dipnoan fishes. *Evolution* 6:101–11

Schindel, D. E. 1980. Microstratigraphic sampling and the limits of paleontologic resolution. *Paleobiology* 6:408–26

Schindel, D. E. 1982. Resolution analysis: a new approach to the gaps in the fossil record. *Paleobiology* 8:340–53

Schopf, T. J. M. 1972. Varieties of paleobiologic experience. In *Models in Paleobiology*, ed. T. J. M. Schopf, pp. 8–25. San Francisco: Freeman

Schopf, T. J. M. 1977a. Patterns and themes of evolution among the bryozoa. In *Patterns of Evolution*, ed. A. Hallam, pp. 159–207. Amsterdam: Elsevier

Schopf, T. J. M. 1977b. Patterns of evolution: a summary and discussion. In *Patterns of Evolution*, ed. A. Hallam, pp. 547–61. Amsterdam: Elsevier

Schopf, T. J. M. 1979. Evolving paleontological views on deterministic and stochastic approaches. *Paleobiology* 5:337–52

Schopf, T. J. M. 1980. *Paleoceanography.* Cambridge, Mass: Harvard Univ. Press. 341 pp.

Schopf, T. J. M. 1981. Evidence from findings of molecular biology with regard to the rapidity of genomic change: implications for species durations. In *Paleobotany, Paleoecology and Evolution*, ed. K. J. Niklas, 1:135–92. New York: Praeger

Schopf, T. J. M. 1982. A critical assessment of punctuated equilibria. I. Duration of taxa. *Evolution* 36:1144–57

Schopf, T. J. M. 1983. Extinction of the dinosaurs: a 1982 understanding. *Geol. Soc. Am. Spec. Pap. No. 190*, pp. 415–22

Schopf, T. J. M. 1984. Climate is only half the story in the evolution of organisms through time. In *Fossils and Climate*, ed. P. J. Brenchley. New York: Wiley. In press

Schopf, T. J. M., Raup, D. M., Gould, S. J., Simberloff, D. S. 1975. Genomic versus morphologic rates of evolution: influence of morphologic complexity. *Paleobiology* 1:63–70

Schopf, T. J. M., Fisher, J. B., Smith, C. A. F. III. 1978. Is the marine latitudinal diversity gradient merely another example of the species area curve? In *Marine Organisms: Genetics, Ecology and Evolution*, ed. B. Battaglia, J. A. Beardmore, pp. 365–85. New York: Plenum

Sekiguchi, K., Sugita, H. 1980. Systematics and hybridization in the four living species of horseshoe crabs. *Evolution* 34:712–18

Selander, R. K., Yang, S. H., Lewontin, R. C., Johnson, W. E. 1970. Genetic variation in the horseshoe crab (*Limulus polyphemus*), a phylogenetic "relic." *Evolution* 24:402–14

Sepkoski, J. J. Jr. 1975. Stratigraphic biases in the analysis of taxonomic survivorship. *Paleobiology* 1:343–55

Shapiro, J. A., ed. 1983a. *Mobile Genetic Elements*. New York: Academic. 680 pp.

Shapiro, J. A. 1983b. Variation as a genetic engineering process. In *Evolution from Molecules to Men*, ed. D. S. Bendall, pp. 253–70. Cambridge Univ. Press

Shimizu, N. 1971. Studies on nucleic acids of living fossils. III. A classification of transfer ribonucleic acids by elution profiles on gel filtration and sedimentation profiles on sucrose density gradient. *J. Biochem.* 69: 761–70

Shimizu, N., Miura, K.-I. 1971a. Studies on nucleic acids of living fossils. I. Isolation and characterization of DNA and some RNA components from the brachiopod *Lingula*. *Biochim. Biophys. Acta* 232:271–77

Shimizu, N., Miura, K.-I. 1971b. Studies of nucleic acids of living fossils. II. Transfer RNA from the brachiopod *Lingula*. *Biochim. Biophys. Acta* 232:278–88

Shornikov, E. I. A. 1975. A "living fossil," representative of Protocytherini (Ostracoda), from the Kurilo-Kamchatka deep. *Zool. Zh.* 54:517–25 (In Russian)

Silverberg, R. 1966. *Forgotten by Time: A Book of Living Fossils*. New York: Crowell. 215 pp.

Simpson, G. G. 1944. *Tempo and Mode in Evolution*. New York: Columbia Univ. Press. 237 pp.

Simpson, G. G. 1949. Rates of evolution in animals. See Jepsen et al 1949, pp. 205–28

Simpson, G. G. 1952. How many species? *Evolution* 6:342

Simpson, G. G. 1953. *The Major Features of Evolution*. New York: Columbia Univ. Press. 434 pp.

Simpson, G. G. 1983. *Fossils and the History of Life*. New York: Freeman. 239 pp.

Slatkin, M. 1983. Genetic background. In *Coevolution*, ed. D. J. Futuyma, M. Slatkin, pp. 14–32. Sunderland, Mass: Sinauer Assoc.

Smith, A. G. 1960. Amphineura. In *Treatise on Invertebrate Paleontology, Part I*, ed. R. C. Moore, pp. 41–76. Lawrence, Kans: Geol. Soc. Am.

Solomon, E. I., Eickman, N. C., Himmelwright, R. S., Hwang, Y. T., Plon, S. E., Wilcox, D. E. 1982. The nature of the binuclear copper site in *Limulus* and other hemocyanins. See Bonaventura et al 1982, pp. 189–230

Stanley, S. M. 1975a. Adaptive themes in the evolution of the Bivalvia (Mollusca). *Ann. Rev. Earth Planet. Sci.* 3:361–85

Stanley, S. M. 1975b. A theory of evolution above the species level. *Proc. Natl. Acad. Sci. USA* 72:646–50

Stanley, S. M. 1976. Stability of species in geologic time. *Science* 192:267–69

Stanley, S. M. 1979. *Macroevolution*. San Francisco: Freeman. 332 pp.

Stebbins, G. L. Jr. 1947. Evidence on rates of evolution from the distribution of existing and fossil plant species. *Ecol. Monogr.* 17:149–58

Stebbins, G. L. Jr. 1949. Rates of evolution in plants. See Jepsen et al 1949, pp. 229–42

Stebbins, G. L. Jr. 1950. *Variation and Evolution in Plants*. New York: Columbia Univ. Press. 643 pp.

Stenzel, P. 1974. Opossum Hb chain sequence and neutral mutation theory. *Nature* 252:62–63

Sterrer, W. 1973. Plate tectonics as a mechanism for dispersal and speciation in interstitial sand fauna. *Neth. J. Sea Res.* 7: 200–22

Strong, D. R., Simberloff, D., Abele, L. G., Thistle, A. B., eds. 1984. *Ecological Communities: Conceptual Issues and the Evidence*. Princeton Univ. Press. In press

Sturtevant, A. H. 1937. I. On the effects of selection on mutation rate. *Q. Rev. Biol.* 12:464–67

Tasch, P. 1969. Branchiopoda. In *Treatise on Invertebrate Paleontology, Part R*, ed. R. C. Moore, pp. 128–91. Lawrence, Kans: Geol. Soc. Am.

Taylor, R. W. 1978. *Nothomyrmecia macrops*: a living-fossil ant rediscovered. *Science* 201:979–85

Templeton, A. R. 1981. Mechanisms of speciation—a population genetic approach. *Ann. Rev. Ecol. Syst.* 12:23–48

Templeton, A. R. 1982. Genetic architectures of speciation. In *Mechanisms of Speciation*, ed. C. Barigozzi, pp. 105–21. New York: Liss

Templeton, A. R. 1983. A population genetic overview of speciation in mammals. *Acta Zool. Fenn.* In press

Templeton, A. R. 1984. *Mechanisms of*

Speciation. Reading, Mass: Addison-Wesley. In press

Templeton, A. R., Johnston, J. S. 1982. Life history evolution under pleiotropy and K-selection in a natural population of *Drosophila mercatorum.* In *Ecological Genetics and Evolution*, pp. 225–39. Acad. Press Austral.

Thenius, E. 1973. *Fossils and the Life of the Past.* New York: Springer-Verlag. 194 pp.

Thompson, I., Jones, D. S. 1980. A possible onychophoran from the Middle Pennsylvanian Mazon Creek beds of northern Illinois. *J. Paleontol.* 54:588–96

Thompson, J. N. Jr., Woodruff, R. C. 1978. Mutator genes—pacemakers of evolution. *Nature* 274:317–21

Tralau, H. 1968. Evolutionary trends in the genus *Ginkgo. Lethaia* 1:63–101

Trusheim, F. 1938. Triopsiden (Crust. Phyll.) aus dem Keuper Frankens. *Paläontol. Z.* 19:198–216

Van Valen, L. 1969. Variation genetics of extinct animals. *Am. Nat.* 103:193–224

Van Valen, L. 1973a. A new evolutionary law. *Evol. Theory* 1:1–30

Van Valen, L. 1973b. Are categories in different phyla comparable? *Taxon* 22:233–73

Van Valen, L. 1980. Patch selection, benefactors, and a revitalization of ecology. *Evol. Theory.* 4:231–33

Van Valen, L. 1982. Why misunderstand one evolutionary half of biology? In *Conceptual Issues in Ecology*, ed. E. Saarinen, pp. 323–43

Van Valen, L. 1983. How pervasive is coevolution? In *Coevolution*, ed. M. H. Nitecki, pp. 1–19

Vermeij, G. J. 1978. *Biogeography and Adaptation.* Cambridge, Mass: Harvard Univ. Press. 332 pp.

Vrba, E. S. 1980. Evolution, species and fossils: how does life evolve? *S. Afr. J. Sci.* 76:61–84

Vuilleumier, F. 1984. Fossil evidence on the development of South American avifaunas. *Proc. Int. Ornithol. Congr., 18th, Moscow. 1982.* In press

Waddington, C. H. 1942. Canalization of development and the inheritance of acquired characters. *Nature* 150:563–65

Wake, D. B., Roth, G., Wake, M. H. 1983. On the problem of stasis in organismal evolution. *J. Theor. Biol.* 101:211–24

Walker, J. C. G. 1977. *Evolution of the Atmosphere.* New York: Macmillan. 318 pp.

Wall, D., Dale, B. 1966. "Living fossils" in western Atlantic plankton. *Nature* 211: 1025–26

Wallace, B. 1977. Recent thoughts on gene control and the fitness of populations. *Genetika* 9:323–34

Wallace, B. 1978. Population size, environment, and the maintenance of laboratory cultures of *Drosophila melanogaster. Genetika* 10:9–16

Webb, S. D., MacFadden, B. J., Baskin, J. A. 1981. Geology and paleontology of the Love Bone Bed from the late Miocene of Florida. *Am. J. Sci.* 281:513–44

Wei, K.-Y., Kennett, J. P. 1983. Nonconstant extinction rates of Neogene planktonic foraminifera. *Nature* 305:218–20

Weiner, A. M., Denison, R. A. 1983. Either gene amplification or gene conversion may maintain the homogeneity of the multigene family encoding human U1 small nuclear RNA. *Cold Spring Harbor Symp. Quant. Biol.* 47:1141–49

Werner, B. 1967a. Der Polyp *Stephanoscyphus*—ein lebendes Fossil? *Umschau* 15:495–97

Werner, B. 1967b. *Stephanoscyphus* Allman (Scyphozoa Coronatae), ein rezenter Vertreter der Conulata? *Paläontol. Z.* 41:137–53

Westbroek, P. 1983. Life as a geological force: new opportunities for paleontology? *Paleobiology* 9:91–96

Westoll, T. S. 1949. On the evolution of the Dipnoi. See Jepsen et al 1949, pp. 121–84

White, M. J. D. 1969. Chromosomal rearrangements and speciation in animals. *Ann. Rev. Genet.* 3:75–98

White, M. J. D. 1978. *Modes of Speciation.* San Francisco: Freeman. 455 pp.

Whitney, J. B. III, Lamoreux, M. L. 1982. Transposable elements controlling genetic instabilities in mammals. *J. Hered.* 73:12–18

Williamson, V. M. 1983. Transposable elements in yeast. *Int. Rev. Cytol.* 83:1–25

Wilson, A. C., Carlson, S. S., White, T. J. 1977. Biochemical evolution. *Ann. Rev. Biochem.* 46:573–639

Wilson, D. S. 1976. Evolution on the level of communities. *Science* 192:1358–60

Wilson, D. S. 1980. *The Natural Selection of Populations and Communities.* Menlo Park, Calif: Benjamin/Cummings. 186 pp.

Wilson, E. O. 1973. Group selection and its significance for ecology. *BioScience* 23: 631–38

Wilson, M. V. H. 1983. Is there a characteristic rate of radiation for the insects? *Paleobiology* 9:79–85

Woodruff, R. C., Thompson, J. N. Jr. 1980. Hybrid release of mutator activity and the genetic structure of natural populations. *Evol. Biol.* 12:129–62

Wright, S. 1982a. Dobzhansky's genetics of natural populations. *Evolution* 36:1102–6

Wright, S. 1982b. Character change, speciation, and the higher taxa. *Evolution* 36:427–43

Wynne-Edwards, V. C. 1962. *Animal Dispersion in Relation to Social Behaviour.* Edinburgh: Oliver & Boyd. 653 pp.

Ann. Rev. Earth Planet. Sci. 1984. 12: 293–305

PHYSICS AND CHEMISTRY OF BIOMINERALIZATION

Keith E. Chave

Department of Oceanography and Hawaii Institute of Geophysics, University of Hawaii, Honolulu, Hawaii 96822

INTRODUCTION

Biomineralization is many different things to different organisms. It is the formation of spicules in the tissues of coccolithophorids, flowering plants, sponges, alcyonarians, holothurians, and tunicates; the formation of a box to live in for diatoms, foraminifera, radiolarians, mollusks, annelids, and barnacles; and the formation of an articulated skeleton for echinoderms, arthropods, and vertebrates. It is also sometimes pathologic.

There are many theories concerning the physical and chemical processes of biomineralization, but there are few facts. And there are many strange experiments and measurements carried out in the name of biomineralization.

Four common biominerals are precipitated—calcite, aragonite, apatite, and opal. In addition, there are a wide variety of exotic biominerals, including calcium, strontium and iron oxides, hydroxides, fluorides, sulfates, and oxalates. The common biominerals are chemically and physically complex phases. The apatite of bones, teeth, and inarticulate brachiopods is not, as commonly indicated, hydroxyapatite, $Ca_5(PO_4)_3OH$. Rather, it contains substantial amounts of carbonate (probably in several different lattice positions), alkalis, alkaline earths, halogens, and water or some other configuration of hydrogen and oxygen. Generally, it has a Ca/P ratio well below the 1.67 value for hydroxyapatite.

Calcite biominerals contain up to 30 mol% Mg substituting for Ca in solid solution (Chave 1952). Biogenic aragonite has Sr/Ca ratios as high as 12.5×10^{-3} (Dodd 1967). Biogenic opal contains variable amounts of bound water ranging from 3 to 13% (Lewin 1962).

All of these variations in apatite, calcite, aragonite, and opal affect the

0084–6597/84/0515–0293$02.00

chemical properties of the mineral and how the mineral behaves in the biomineralization process.

Crystal size also affects the properties of minerals even when they are pure. Very small crystals less than one micrometer in size, which are often found as biominerals, have significantly greater total free energies than larger crystals because surface free energies play a greater role as the surface-to-volume ratio of the crystal increases. This effect is discussed theoretically by Garside (1982) in terms of nucleation, and for real calcite crystals by Chave & Schmalz (1966) in terms of solubility.

Thus, the Fleisch (1982) statement, for instance, that blood ultrafiltrate in mammals is supersaturated with respect to hydroxyapatite may not be too significant in biomineralization.

It is generally agreed that enzymes play a role in biomineralization. But what kind of enzymes are required—stimulatory or inhibitory? The principal enzymes of biomineralization mentioned in the literature for the past 30 years are carbonic anhydrase and alkaline phosphatase. Carbonic anhydrase speeds equilibrium reactions in the CO_2-H_2O system, whereas alkaline phosphatase decouples inorganic phosphorus from organophosphorus compounds. Both enzymes commonly occur at sites of carbonate and phosphate mineralization, and also at many noncalcifying sites and in noncalcifying organisms. (Spinach is particularly rich in carbonic anhydrase.) Both enzymes are potentially stimulatory to the biomineralization process in that they can increase the concentration of mineral anions at mineralizing sites.

An alternative model for enzyme activity in biomineralization is the inhibitory enzyme model. In this model, the medium in which the mineralizing organism or organ lives is supersaturated with respect to the mineral phase in question, and the function of the enzyme is to prevent mineralization or to allow mineralization only at specific sites.

Near-surface seawater, where immense amounts of calcification take place, is supersaturated with respect to calcite, aragonite, and, according to Schoonmaker (1981), with respect to up to 10 to 14 mol% Mg-calcites depending upon latitude. Furthermore, if the supersaturation of blood ultrafiltrate with respect to hydroxyapatite (Fleisch 1982) has bearing on bone apatite, then this system too, and perhaps all vertebrate systems, needs inhibition, rather than stimulation, of mineralization.

One way the inhibitory enzyme model could work is that the cell, organ, or body secretes an enzyme that couples Ca to organic molecules, keeping the solution undersaturated. In the absence of the enzyme, as regulated by vital processes in normal organisms, the Ca is decoupled, supersaturating the solution and allowing calcification to proceed where hard parts should occur.

Some evidence supports the inhibitory enzyme model. On the basis that damage or disease of a tissue would destroy its enzyme system function, in mammals such damage to tissues or organs bathed by large quantities of body fluids, such as arteries or kidneys, produces immediate calcification.

In mollusks, G. Bevelander (personal communications) has shown that shell regeneration in the pen shell (*Pinna*) is much more rapid than normal shell growth. Injury to the calcitic and aragonitic shell near the beak results in only rapid calcite regeneration. This could be interpreted as a breakdown of an inhibitory enzyme system due to damage to the mantle.

A third possible piece of evidence for enzyme inhibition comes from K. Chave (unpublished). Experiments on calcification of the green alga *Halimeda* used the antibiotic tetracycline to indicate sites of calcification. Tetracycline is somehow incorporated with calcium at these sites. It also fluoresces strongly under UV light, so that its location is easily seen under a UV microscope.

In experiments with several species of *Halimeda*, tetracycline uptake was observed at several places on the thallus. In agreement with Goreau's (1963) observation, based on ^{45}Ca uptake experiments, it was noted that *Halimeda* calcified both in the light and the dark, and light : dark ratios ranged from 0.71 to 1.36 (i.e. more calcification in the dark to more in the light). It was also found that in the light, tetracycline occurred at the growing tips, and in the dark it occurred diffusely in the older segments. This perhaps explains why calcification and photosynthesis are loosely coupled in the algae. The two unexpected sites of tetracycline uptake (calcification) were at locations where the thallus was damaged: where it was cut to separate it from the substrate, and where it was picked up to place it in the tetracycline-spiked seawater.

In the remainder of this review, individual groups of organisms are discussed in terms of what is known about mechanisms, rates, and periodicities of biomineralization.

PLANT BIOMINERALIZATION

Calcification takes place in marine and freshwater macroalgae, coccolithophorids, and perhaps bacteria and fungi, while silification occurs in diatoms and some flowering plants. Exotic calcium minerals are reported scattered throughout the plant kingdom.

Marine Macroalgae

Dawson (1966) listed 11 genera of calcareous green algae (Chlorophyta), 38 genera of calcareous red algae (Rhodophyta), and 1 genus of calcareous brown algae (Phaeophyta), with a combined total of almost 700 species. He notes that most of the greens and browns are tropical, whereas the reds are more commonly temperate.

The mineralogy of these plants is as follows:

1. Chlorophyta—aragonite
2. Rhodophyta
   Nemaliales—aragonite
   Cryptonemiales—Mg-calcite
3. Phaeophyta—aragonite

Borowitzka (1982) distinguished five types of calcification in the algae—extracellular, intercellular, sheath, cell wall, and intracellular. Extracellular calcification occurs in freshwater algae and the marine brown, *Padina.* Intercellular calcification is found in most of the aragonitic greens and reds as typified by *Halimeda* and *Liagora.* Sheath calcification (i.e. aragonite deposition outside the cell wall, but in an organic sheath of unknown composition) takes place in *Penicillus* and *Udotea.* Calcification of cell walls happens in the coralline reds—articulated *Corallina* and crustose *Lithothamnium.* Intracellular calcification occurs in the planktonic, unicellular coccolithophorids.

MECHANISMS OF CALCIFICATION There are several models of mechanisms of marine algal calcification. Each model involves increasing the concentration of either Ca or CO_3 at the mineralization site. The two principal ones are the photosynthesis model and the matrix model. The photosynthesis model involves the uptake of CO_2 during photosynthesis and the resultant increase in pH and CO_3^{-2} that increases the supersaturation of the fluid with respect to $CaCO_3$. The matrix model involves a sugar or lipid in the cell wall attracting or fixing Ca to the mineralization site. There is a strong suggestion that organic material in the cell wall influences $CaCO_3$ mineralogy in coralline algae (Borowitzka 1977).

These models, no matter how elegantly presented, suffer from several inconsistencies. A major inconsistency is the presence of a noncalcified alga growing next to the calcified one, with both photosynthesizing at the same rate and apparently having the same cell-wall composition. Why are they not both calcified? Another inconsistency is that the seawater in which the algae live is supersaturated with respect to most skeletal carbonate phases. Increasing the degree of supersaturation should not affect anything except, perhaps, the kinetics of precipitation.

The final sentence in Borowitzka (1981) may be true for all benthic algae: "However, no meaningful model for calcification in the coralline algae can as yet be formulated."

RATES OF CALCIFICATION Rates of calcification have been measured in such a variety of ways that meaningful comparisons between them are impossible. Rates have been measured by (*a*) ^{45}Ca uptake (Goreau 1963); (*b*) ^{14}C uptake (Borowitzka & Larkum 1976); (*c*) linear growth (Chave & Wheeler 1965, Agegian 1981); (*d*) increase in thallus weight (Smith 1970); (*e*) changes in total alkalinity (Smith 1973, Borowitzka & Larkum 1976); (*f*) increase in thallus size (Adey 1970, Adey & Vassar 1975); and (*g*) combinations of these.

Comparisons between measurements of 7–20 mm/yr (Agegian 1981, *Porolithon*), 1948 ± 490 μg Ca/mg N/hr (Goreau 1963, *Galaxaura*), 200 cpm/unit surface area/hr (Smith & Roth 1979, *Bossiella*), and 45×10^3 cpm ^{45}Ca/mg algae (Pearse 1972, *Bossiella*) are, to say the least, unhandy.

PERIODICITY OF CALCIFICATION The articulated green and red algae do not appear to have any periodicity in calcification. Segments are added to the thallus and are calcified more or less randomly in time.

Periodicity of calcification has been demonstrated for encrusting and articulated coralline algae (Johansen 1981). Chave & Wheeler (1965) showed seasonal changes in the magnesium content of the calcite of *Clathromorphum compactum* from Maine, with higher magnesium contents in the summer. Agegian (1981) described monthly bands of X-ray-dense carbonate in *Porolithon gardineri* from the Hawaiian Islands. She also described daily or half-daily primary cellular bands.

Coccolithophorids

Coccolithophorids (Pyrmnesiophceae) are golden brown algae that inhabit the euphotic zone of marine waters. They are unicellular and may possess two flagella. In culture they may or may not bear calcitic coccoliths at the surface of the cell. It is not known if naked cells occur in nature. The cells range in size from 3 to 25 μm, while the coccoliths range from 1 to 15 μm across. The number of coccoliths per cell is a function of size and age. Parke & Adams (1960) found that the mother cells of *Coccolithus pelagicus* bore 25 to 30 coccoliths, while daughters bore only 8 to 17 per cell.

Coccoliths of *Cricosphaera carterae*, for example, are formed within the cells in the Golgi apparatus and later positioned outside the cell membrane, adhering closely to the body. Crystallization of the calcite takes place on an organic matrix that controls the size and shape of the coccolith (Outka & Williams 1971).

Crenshaw (1964) concluded that coccolith calcite in *C. carterae* utilized HCO_3^- produced by the uptake of CO_2 in photosynthesis. In the absence of light, no recalcification occurs. Similar calcification schemes are reported for other species of coccolithophorids.

Cricosphaera carterae tolerates a wide range of salinities (Paasche 1968) and is therefore useful for experiments involving varied environmental chemistry. Blackwelder et al (1976) showed that cells could be decalcified by lowering the culture pH to between 5.9 and 5.3 by bubbling with CO_2 and then allowing the cells to recalcify in solution in which divalent cations were varied. They found that recalcification took place rapidly at calcium concentrations of 10^{-2} M but not at all at 10^{-3} and 10^{-4} M. When small amounts of strontium were added to the low-calcium medium, however, calcification occurred. The coccoliths formed were pure calcite. Sikes & Wilbur (1980) suggest that the effect of strontium on *C. carlera* is not on calcification, but rather on the attachment of the coccoliths to the cell.

Reduction of magnesium in the culture medium substantially reduced the amount of coccolith production. For unknown reasons, at Mg concentrations of 4.2×10^{-2} and 0 M, the coccoliths formed were 100% calcite, while at Mg concentrations of 1.4×10^{-4} and 4.2×10^{-6} M, the coccoliths formed were 60% calcite and 40% aragonite.

In an interesting paper, Sikes & Wilbur (1982) discussed the possible functions of coccoliths. There is the obvious disadvantage that coccoliths weight an otherwise buoyant cell, increasing its sinking rate from the photic zone. Calcified *Coccolithus huxleyi* sinks in seawater at a rate of about 1.3 m/day (Eppley et al 1967).

That calcite provides a reservoir of carbon for photosynthesis does not seem sufficient reason for elaborate and complex coccoliths on the cell surface. Furthermore, this function is not supported by stable carbon-isotope measurements.

Coccolithophorids have no organic cell wall. Sikes & Wilbur explored the possibility that the coccoliths function as a cell wall, regulating the potential difference across the cell "membrane" and affecting ion transport. This function is not supported by their experimental work. Naked and calcified cells behave the same way.

They also studied experimentally whether coccoliths inhibit predation by copepods. Calcified and uncalcified cells are consumed at the same rate. The regulatory effects of coccoliths on the osmotic properties of the cells were also rejected.

Finally, Sikes & Wilbur explored the effects of coccoliths on salinity tolerance of coccolithophorids; they find that with *Coccolithus huxleyi* at salinities below about 25‰, cell growth is inhibited even though, unexpectedly, the percentage of calcified cells is increased. It appears that this is not the solution to the problem of coccolith function, because *C. huxleyi* is an open ocean form and in nature probably never encounters seawater with a salinity below about 33‰.

In summary, coccolithophorids are near-ideal experimental organisms. They are hardy, fast growing, fast calcifying, and beautiful. However, they cannot be studied under natural conditions in the sea, for they are always mixed with other phytoplankton. A comparison of the behavior of naked and calcified cells, a common practice in coccolithophoridology, may not be valid, because naked cells may not occur in nature. Finally, there are relatively few coccolithophorid cultures in the world, and most of them are quite old and perhaps have undergone mutations.

Diatoms

Diatoms deposit opal, $SiO_2 \cdot nH_2O$, in their cell walls in an architecturally elegant morphology. This frustule is composed of two parts, not firmly attached to each other, fitting into each other like a box. Diatoms grow in any lighted aquatic environment.

Silicification of marine diatoms is unlike biomineralization of other groups considered, in that the environment (surface seawater) is always highly undersaturated with respect to opal (Hurd 1972).

Major silicification takes place during cell division; the most active areas are the new cytoplasmic surfaces of the daughters. Silification continues more slowly after cell division, in at least some species, causing a thickening of the frustule wall (Lewin 1962).

Mineralization is thought to occur first as an active transport of $Si(OH)_4$ across the cell membrane with the aid of monovalent ions, such as Na^+ or K^+. The silicic acid is then polymerized with or without effects of organosilicon compounds, and finally it is deposited within the cell wall as opal (Sullivan 1976).

The water content of the biogenic opal ranges up to $\sim 13\%$. Aluminum plays some role in frustule growth and preservation. Menzel et al (1963) showed that the addition of aqueous Al to Sargasso Sea water stimulated the growth of diatoms, and Lewin (1961) demonstrated that diatom frustules treated with Al-containing solutions were more resistant to solution in water. It is possible that living diatoms require Al in or on their frustules in order to prevent their rapid dissolution in seawater.

CORAL MINERALIZATION

Much of what we know about the process of calcification in corals stems from two sources: Thomas F. Goreau, of the University of the West Indies, and his family; and Leonard Muscatine, of the University of California, Los Angeles, and his students.

Mechanisms of Calcification

It is generally agreed that zooxanthellae play a significant role in calcification of corals. The question is, what role?

T. F. Goreau (1959) was the first to demonstrate satisfactorily that corals with symbiotic algae (zooxanthellae) calcified more rapidly in the light than in the dark, thus implicating the zooxanthellae in the calcification process. His experiments, which used the uptake of ^{45}Ca, also implicated carbonic anhydrase in the calcification process, in that calcification was slowed in the presence of an inhibitor of this enzyme, perhaps by rapid removal of CO_2 from calcifying sites. In this and later papers, T. F. Goreau and coworkers tended to look at calcification of corals as a chemical system, ignoring the organic matrix.

N. I. Goreau & Hayes (1977) distinguished two skeleton-forming processes—calcification and skeletogenesis. Calcification is an intracellular process that precedes skeletogenesis and involves deposition of aragonite crystals within the Golgi apparatus through the binding of Ca to this organelle. These "calcification crystals" are then expelled to extracellular sites, where they act as nucleation sites for epitaxial growth during skeletogenesis.

T. J. Goreau (1977), using stable carbon-isotope data, modeled nine carbon pools involved in calcification and photosynthesis in corals. He concludes that one half to two thirds of the carbon used in these processes comes from respiration of the zooxanthellae, with the remainder coming from seawater.

Muscatine (1973) proposed that calcification in corals may be limited by synthesis of skeletal organic matrix and that the zooxanthellae supply more components of this organic substrate. Vandermeulen & Muscatine (1973) attempted to test this hypothesis by adding glycerol, alanine, and glucose to the seawater in which corals were growing, but the results were not conclusive.

Clausen & Roth (1975) studied the effect of temperature on calcification of *Pocillopora damicornis.* They found two temperatures at which the maximum rate of calcification occurred—27 and 31°C—and suggested that there are either two isoenzymes or two metabolic pathways involved in the process.

Chalker (1977) found daily rhythms in calcification of *Acropora cervicornis*, with maxima occurring at sunrise and sunset. The rhythm persisted for at least one day in total darkness and correlated with a daily rhythm in the photosynthetic capacity of the symbiotic alga *Gimnodinium microadriaticum* taken from the coral.

On the subject of mechanisms of coral calcification, Buddemeier & Kinzie (1976) concluded, "Although a number of competent people have devoted considerable attention to the subject for nearly two decades, we have neither a theoretical, mechanistic understanding nor any very significant predictive ability."

Rates of Calcification

Rates of calcification of corals have been measured by at least as many methods as were previously discussed for the calcareous algae and have been reported in just as many different units. Buddemeier & Kinzie (1976) considered calcification separately on different time scales. Short-term measurements (radioisotope uptake studies) have the highest variability, typically in the range of 20–30% but often higher. Intermediate-term measurements (weeks to months) can also be quite variable; some of this variability is a seasonal signal. Long-term measurements (years or more) are more reliable because of the natural damping of shorter-term oscillations in the growth rate.

Chave et al (1972) summarized the long-term measurements of earlier workers and concluded that corals, other than *Acropora*, calcify at an average rate of 10^4 g $CaCO_3/m^2/yr$. The average *Acropora* rate is 10^5 g $CaCO_3/m^2/yr$.

Periodicity of Calcification

Periodic growth patterns based on visual inspection have long been recognized in corals. Notable in this respect are the many works of T. Y. H. Ma between 1933 and 1960. However, rapid developments in this field of coral research came about following the publication of Knutson et al (1972). In it, they described high- and low-density bands, perpendicular to the growth direction, seen by X-radiography. Yearly periodicity was determined from radioactive bands in Eniwetok corals that were formed at the time of well-dated bomb tests.

MOLLUSK BIOMINERALIZATION

Mollusk shells are generally composed of calcite, aragonite, or both minerals. The egg case of the cephalopod *Argonauta* is Mg-calcite, and vaterite is reported in the shell of the pearl oyster *Pinctada*. Amorphous $CaCO_3$ is also mentioned in the mollusk literature.

Mechanisms of Calcification

Mollusk shells are with few exceptions external. Shells are formed by a layer of epithelial cells of the mantle. The mineralizing medium is the extrapallial fluid between the mantle and the shell. This fluid is isolated from the surrounding water, completely or partially, by the mantle and its proximity to the shell.

The dissolved inorganic components of the extrapallial fluid in marine mollusks are similar to seawater (Crenshaw 1972a, Wada & Fujinuki 1976). The fluid, however, contains various little-known organic materials that may complex Ca, so that the saturation state with respect to calcite or aragonite cannot be calculated.

The pH of extrapallial fluid has been monitored by implantation of pH electrodes through the shell. In marine pelecypods, the pH ranges from about 7.0 to 8.0, with the low value occurring when the shell is closed. The extrapallial fluid appears to change from aerobic to anaerobic with the opening and closing of the shell, while at the same time changing from shell deposition to shell solution (at least in some species).

An organic matrix is associated with biomineralization in all organisms except, perhaps, sponges and echinoderms (Watabe 1981). In mollusks this matrix, into which the carbonate minerals are embedded, is composed largely of protein and polysaccharides. Its components are both water soluble and insoluble. The insoluble matrix is called conchiolin. There are numerous studies of the amino acid composition and X-ray diffraction-determined structure of conchiolin, yet there appears to be no correlation between these properties of the matrix and the microstructure or mineralogy of the shell. Nevertheless, the idea persists that the organic matrix is an ordered structure, a template that controls crystal growth epitaxially.

The water-soluble matrix comprises 20–50% of the organic matter in molluscan shells. The soluble matrices of bivalves and cephalopods are similar and are distinctly different from gastropods (Crenshaw 1982). Soluble matrices from clams and oysters will selectively bind Ca (Crenshaw 1972b, Wheeler et al 1981). Several functions of the soluble matrix have been suggested. Wheeler et al (1981) believe that, in solution, it binds Ca and thus prevents precipitation, whereas others believe that it is active in nucleation of $CaCO_3$ crystals (Crenshaw & Ristedt 1976).

Crenshaw (1980) concluded that "our knowledge of the processes involved in shell formation is very incomplete."

Rates of Calcification

Chave (1967) reported that nonlarval pelecypods, from all latitudes, grow in length between about 2 and 10 mm/month, although with high variability, probably because of environmental factors. Vermeij (1980) emphasized high variability of shell growth rate in gastropods, which he attributed to food supply, crowding, exposure, and genetics. He notes growth rates in the same general range as pelecypods.

Periodicity of Calcification

The most thorough documentation of periodicity in mollusks is the work of George R. Clark II on the Pectinidae (Clark 1968, 1974). He described two distinct types of external growth lines. The "annual ring" is a narrow band of light-colored shell, formed during very slow, or stopped, winter growth. A wider, darker band between the annual rings represents more rapid growth during the rest of the year. Both of these bands have large numbers of fine growth lines that are probably daily, although Clark notes that they occasionally miss days.

OSTRACOD BIOMINERALIZATION

Turpen & Angell (1971) have made the only comprehensive study of the calcification of ostracods. From studies of *Heterocypris* sp., they concluded that the process is simple, requiring no specialized cells, or passage through, or storage of Ca in the body. Unlike other crustaceans, there is no resorption of Ca before molting.

Rates of calcification are rapid, with only a few hours needed to produce new valves after molting. Molting (i.e. shedding fully calcified valves into the environment) occurs every 24–48 hr in larval (instar) stages 1–5, every 24–72 hr in stages 5–7, and every 48–96 hr in stages 7–9.

BARNACLE BIOMINERALIZATION

Barnacles have a complex shell composed of a variety of wall plates, opercular valves, and a basal disk that may or may not be calcified. Growth in height occurs by calcification at the basal edge of the wall plates, and in width by calcification at the margins of the wall plates. Complex calcification patterns take place in the opercular valves.

Bourget & Crisp (1975) described the deposition of the shell by the epithelium of the mantle underlying the plates and base, presumably in much the same manner as in the mollusks. Barnacles are able to build a calcitic shell while still periodically molting their chitinous exoskeleton.

Barnacles exhibit two types of periodicity in shell-growth features. These are growth bands, separated by growth lines, and growth striae, separated by growth ridges (Bourget 1982). The bands and lines are internal to the shell plates and are best seen in thin section. It appears that these bands were first observed by Darwin (1854); two bands per day, up to 20 μm thick, have been observed to form in common barnacle species (Bourget & Crisp 1975). The banding is probably due to tidal immersion and emersion.

Growth striae and ridges are external features, most prominent on opercular plates. These features are associated with periods of molting.

Killingley & Newman (1982) have described an ^{18}O barnacle paleotemperature equation.

VERTEBRATE BIOMINERALIZATION

Fleisch (1982) summarized the knowledge of mineralization of bones and cartilage in about seven pages, without a single reference. This may be too much space, because nearly every topic mentioned is a controversy with contradictory evidence. His abstract summarized quite succinctly the state of knowledge in the field: "Despite intensive investigations, normal and abnormal calcification mechanisms are still poorly understood. The great number of theories which have been and are still being proposed reflect the complexity of the subject and the uncertainty of our knowledge."

Need I say more?

Literature Cited

Adey, W. 1970. The effect of light and temperature on growth rates in boreal-subarctic crustose corallines. *J. Phycol.* 6:269–76

Adey, W., Vassar, J. M. 1975. Colonization, succession, and growth rates of tropical crustose coralline algae (Rhodophyta, Cryptonemiales). *Phycologia* 14:55–69

Agegian, C. R. 1981. Growth of the branched coralline alga, *Porolithon gardineri* (Foslie) in the Hawaiian Archipelago. *Proc. Int. Coral Reef Symp., 4th,* 2:419–24

Blackwelder, P. L., Weiss, R. E., Wilbur, K. M. 1976. Effect of calcium, strontium, and magnesium on the coccolithophorid *Cricosphaera (Hymanomonas) carterae.* I. Calcification. *Mar. Biol.* 34:11–16

Borowitzka, M. A. 1977. Algal calcification. *Oceanogr. Mar. Biol. Ann. Rev.* 15:189–223

Borowitzka, M. A. 1981. Photosynthesis and calcification in the articulated coralline algae *Amphiroa anceps* and *A. foliacea. Mar. Biol.* 62:17–23

Borowitzka, M. A. 1982. Morphological and cytological aspects of algal calcification. *Int. Rev. Cytol.* 74:127–62

Borowitzka, M. A., Larkum, A. W. D. 1976. Calcification in the green alga *Halimeda.* II. The exchange of Ca^{2+} and the occurrence of age gradients in calcification and photosynthesis. *J. Exp. Bot.* 27:864–78

Bourget, E. 1982. Barnacle shell growth and its relationship to environmental factors. In *Skeletal Growth of Aquatic Organisms,* ed. D. C. Rhoads, R. A. Lutz, pp. 469–92. New York: Plenum. 750 pp.

Bourget, E., Crisp, D. J. 1975. An analysis of the growth bands and ridges of barnacle shell plates. *J. Mar. Biol. Assoc. UK* 55:439–61

Buddemeier, R. W., Kinzie, R. A. III. 1976. Coral growth. *Oceanogr. Mar. Biol. Ann. Rev.* 14:183–225

Chalker, B. E. 1977. Daily variation in the calcification capacity of *Acropora cervicornis. Proc. Int. Coral Reef Symp., 3rd,* 2:417–23

Chave, K. E. 1952. A solid solution between calcite and dolomite. *J. Geol.* 60:190–92

Chave, K. E. 1967. Recent carbonate sediments—an unconventional view. *J. Geol. Educ.* 15:200–4

Chave, K. E., Wheeler, B. O. 1965. Mineralogic changes during growth in the red alga *Clathromorphum compactum. Science* 147:621

Chave, K. E., Schmalz, R. F. 1966. Carbonate-seawater interactions. *Geochim. Cosmochim. Acta* 30:1037–48

Chave, K. E., Smith, S. V., Roy, K. J. 1972. Carbonate production by coral reefs. *Mar. Geol.* 12:123–40

Clark, G. R. II. 1968. Mollusk shell: daily growth lines. *Science* 161:800–2

Clark, G. R. II. 1974. Growth lines in invertebrate skeletons. *Ann. Rev. Earth Planet. Sci.* 2:77–99

Clausen, C. D., Roth, A. A. 1975. Effect of temperature and temperature adaptation on calcification rate in the hermatypic coral *Pocillopora damicornis. Mar. Biol.* 33:93–100

Crenshaw, M. A. 1964. *Coccolith formation by two marine coccolithophorids,* Coccolithus huxleyi *and* Hymenomonas sp. PhD thesis. Duke Univ., Durham, N.C. 74 pp.

Crenshaw, M. A. 1972a. The inorganic composition of molluscan extrapallial fluid. *Biol. Bull.* 143:506–12

Crenshaw, M. A. 1972b. The soluble matrix from *Mercenaria mercenaria* shell. *Biomineral. Res. Rep.* 6:6–11

Crenshaw, M. A. 1980. Mechanisms of shell formation and dissolution. In *Skeletal Growth of Aquatic Organisms,* eds. D. C. Rhoads, R. A. Lutz, pp. 115–32. New York: Plenum. 750 pp.

Crenshaw, M. A. 1982. Mechanisms of normal biological mineralization of calcium carbonates. In *Biological Mineralization and Demineralization,* ed. G. H. Nancollas, pp. 243–58. New York: Springer-Verlag. 415 pp.

Crenshaw, M. A., Ristedt, M. 1976. The histochemical localization of reactive groups in the septal nacre from *Nautilus pompilius* L. In *Mechanisms of Mineralization in the Invertebrates and Plants,* eds. N. Watabe, K. M. Wilbur, pp. 335–67. Columbia: Univ. S.C. Press. 435 pp.

Darwin, C. 1854. *A Monograph of the Subclass Cirripedia,* Vol. 2. R. Soc. London. 684 pp.

Dawson, E. Y. 1966. *Marine Botany.* New York: Holt, Rinehart & Winston. 371 pp.

Dodd, J. R. 1967. Magnesium and strontium in calcareous skeletons. *J. Paleontol.* 41:1313–29

Eppley, R. W., Holmes, R. W., Strickland, J. D. 1967. Sinking rates of marine phytoplankton measured with a fluorometer. *J. Exp. Mar. Biol. Ecol.* 1:191–208

Fleisch, H. 1982. Mechanisms of normal

mineralization in bone and cartilage. In *Biological Mineralization and Demineralization*, ed. G. H. Nancollas, pp. 233–41. New York: Springer-Verlag. 415 pp.

Garside, J. 1982. Nucleation. In *Biological Mineralization and Demineralization*, ed. G. H. Nancollas, pp. 23–26. New York: Springer-Verlag. 415 pp.

Goreau, N. I., Hayes, R. L. 1977. Nucleation catalysis in coral skeletogenesis. *Proc. Int. Coral Reef Symp., 3rd*, 2:439–45

Goreau, T. F. 1959. The physiology of skeleton formation in corals. I. A method for measuring the rate of calcium deposition by corals under different conditions. *Biol. Bull.* 116:59–75

Goreau, T. F. 1963. Calcium carbonate deposition by coralline algae and corals in relation to their roles as reef-builders. *Ann. NY Acad. Sci.* 109:127–67

Goreau, T. J. 1977. Carbon metabolism in calcifying and photosynthetic organisms: theoretical models based on stable isotope data. *Proc. Int. Coral Reef Symp., 3rd*, 2:396–401

Hurd, D. C. 1972. Factors affecting the solution rate of biogenic opal in seawater. *Earth Planet. Sci. Lett.* 15:411–17

Johansen, H. W. 1981. *Coralline Algae, A First Synthesis*. Boca Raton, Fla: CRC Press. 239 pp.

Killingley, J. S., Newman, W. A. 1982. ^{18}O fractionation in barnacle calcite: a barnacle paleotemperature equation. *J. Mar. Res.* 40:893–902

Knutson, D. W., Buddemeier, R. W., Smith, S. V. 1972. Coral chronometers: seasonal growth bands in reef corals. *Science* 177:270–72

Lewin, J. C. 1961. The dissolution of silica from diatom walls. *Geochim. Cosmochim. Acta* 21:182–98

Lewin, J. C. 1962. Silicification. In *Physiology and Biochemistry of Algae*, ed. R. A. Lewin, pp. 445–55. New York: Academic. 929 pp.

Menzel, D. W., Hulbert, E. M., Ryther, J. H. 1963. The effects of enriching Sargasso Sea water on the production and species composition of the phytoplankton. *Deep-Sea Res.* 6:351–66

Muscatine, L. 1973. Nutrition in corals. In *Biology and Geology of Coral Reefs*, ed. O. A. Jones, R. Endean, 2:271–324. New York: Academic

Outka, D. E., Williams, D. C. 1971. Sequential coccolith morphogenesis in *Hymenomonas carterae*. *J. Protozool.* 18: 285–97

Paasche, E. 1968. Biology and physiology of coccolithophorids. *Ann. Rev. Microbiol.* 22:77–86

Parke, M., Adams, I. 1960. The motile (*Crystalolithms hyalinus* Gaarder and Markali) and non-motile phases in the life history of *Coccolithus pelagicus* (Wallich) Shiller. *J. Mar. Biol. Assoc. UK* 39:263–74

Pearse, V. B. 1972. Radioisotopic study of calcification in the articulated coralline alga *Bossiella orbigniana*. *J. Phycol.* 8:88–97

Schoonmaker, J. E. 1981. *Magnesian calcite-seawater reactions: solubility and recrystallization behavior*. PhD thesis. Northwestern Univ., Evanston, Ill. 263 pp.

Sikes, C. S., Wilbur, K. M. 1980. Calcification by coccolithophorids: effects of pH and Sr. *J. Phycol.* 16:433–36

Sikes, C. S., Wilbur, K. M. 1982. Functions of coccolith formation. *Limnol. Oceanogr.* 27: 18–26

Smith, S. V. 1970. Calcium carbonate budget of the southern California borderland. *Hawaii Inst. Geophys. Rep. 70-11*. 174 pp.

Smith, S. V. 1973. Carbon dioxide dynamics: a record of organic carbon production, respiration and calcification in the Eniwetok reef flat community. *Limnol. Oceanogr.* 18:106–20

Smith, A. D., Roth, A. A. 1979. Effect of carbon dioxide concentration on calcification in the red coralline alga *Bossiella orbigniana*. *Mar. Biol.* 52:217–25

Sullivan, D. W. 1976. Diatom mineralization of silicic acid. I. $Si(OH)_4$ transport characteristics in *Navicula pelliculosa*. *J. Phycol.* 12:390–96

Turpen, J. B., Angell, R. W. 1971. Aspects of molting and calcification in the ostracod *Heterocypris*. *Biol. Bull.* 140:331–38

Vandermeulen, J. H., Muscatine, L. 1973. Influence of symbiotic algae on calcification in reef corals: critique and progress report. In *Symbiosis in the Sea.*, ed. W. B. Vernberg, pp. 1–19. Columbia: Univ. S.C. Press

Vermeij, G. J. 1980. Gastropod shell growth rate, allometry, and adult size: environmental implications. In *Skeletal Growth of Aquatic Organisms*, eds. D. C. Rhoads, R. A. Lutz, pp. 379–94. New York: Plenum. 750 pp.

Wada, K., Fujinuki, T. 1976. Biomineralization in bivalve molluscs with emphasis on the chemical composition of the extrapallial fluid. In *The Mechanisms of Mineralization in the Invertebrates and Plants*, eds. N. Watabe, K. M. Wilbur, pp. 175–90. Columbia: Univ. S.C. Press. 435 pp.

Watabe, N. 1981. Crystal growth of calcium carbonate in invertebrates. In *Progress in Crystal Growth and Characterization*, ed. B. Ramplin, 4:48–93. Oxford: Pergamon

Wheeler, A. P., George, J. W., Evans, C. A. 1981. Control of calcium carbonate nucleation and crystal growth by soluble matrix of oyster shell. *Science* 212:1397–98

Ann. Rev. Earth Planet. Sci. 1984. 12: 307–35

THE ORDOVICIAN SYSTEM, PROGRESS AND PROBLEMS

Reuben James Ross Jr.

Department of Geology, Colorado School of Mines, Golden, Colorado 80401

INTRODUCTION

A Superlative System and Period

The Ordovician Period was approximately 80 m.y. in duration, the longest of any Paleozoic period and equivalent in span to the Cretaceous. The period probably began 510–515 m.y. ago and ended close to 435 m.y. ago.

The period witnessed great volcanicity (notably in active belts bordering tectonic plates), widespread inundations of continents, and explosive organic evolution. Despite the extinction of many taxa at the end of the Cambrian, trilobites remained abundant and increasingly diverse. There was a great increase in numbers of diverse nautiloid, conodont, and bivalve taxa. The appearance and expansion of graptolites among the protochordates was paralleled by that of clitambonitid, triplesiid, strophomenid, pentamerid, rhynchonellid, and spiriferid brachiopods. The Ordovician witnessed not only the beginnings of rugose and tabulate corals, but also the flourishing of bryozoans and three major types of stromatoporoids. As noted by James (1982), with these frame builders came structures that could be called reefs.

Global Correlation Charts

Any review of the Ordovician must take advantage of the extensive references and summaries now available with correlation charts published by the International Union of Geological Sciences (IUGS) and sponsored by the Subcommission on Ordovician Stratigraphy. One of the prime purposes of the charts is to lay a foundation for recognizing possible global events during the period. Already published by the IUGS are charts for *China* (Sheng 1980); the *Middle East* (Dean 1980); *Australia, New Zealand,*

0084–6597/84/0515–0307$02.00

and Antarctica (Webby et al 1981); *Canada* (Barnes et al 1981), *southwestern Europe* (France, Spain, and Portugal) (Hammann et al 1982); and the *United States of America* (Ross et al 1982b). Extensive correlation charts for component tectonic elements within the Soviet Union are cited in what follows. Here I attempt no synthesizing or comparative chart; that will be fitting only after the remaining Ordovician charts are published.

Born in Conflict

It is well known that the Ordovician System was born in conflict and nurtured in compromise. C. H. Holland (1974) has given us a sympathetic account of the unfortunate and bitter disagreement that marred the later lives of two good friends: Professor Adam Sedgwick of Cambridge University, founder of the Cambrian System, and Sir Roderick I. Murchison of the Geological Survey of Great Britain, founder of the Silurian System. The upper part of Sedgwick's Cambrian and the lower part of Murchison's Silurian overlapped stratigraphically. The overlapping parts, the subject of heated feuding for more than 35 years, became the basis for the compromise proposed by Professor Charles Lapworth in 1879—the Ordovician System.

The Ordovician was accepted in Scandinavia almost immediately. The US Geological Survey (USGS), under the directorship of Charles D. Walcott, became the first national survey to recognize the system in its 24th Annual Report (p. 25), published in 1903.

The Highest is Ordovician

Yin & Kuo (1978) reported on the successful geologic and mountaineering exploration of the high Himalayas in the vicinity of Mount Jolmo Lungma (Mount Everest) in May 1975. Geologic mapping was undertaken, and a cross section through the peak was constructed. Following earlier workers, they recognized an upper Jolmo Lungma Formation and a lower Yellow Band, both of Ordovician age, resting on gneisses and schists of Cambrian and Precambrian age and capping both Jolmo Lungma and Mount Changtse. These same units have been mapped at localities to the north of the peak and at lower elevations. At Chaya, and along the Chiuhala and Chienchin Rivers, all less than 35 km north of Everest, brachiopods and other fossils have been collected and are clearly of Whiterock age. Farther to the west around Nyalam, this same fauna is present in the Lower Chiatsun Group (Sheng 1980, chart column II). We infer that the highest place on Earth is not only Ordovician, but also Whiterockian, essentially equivalent to the upper Arenig and Llanvirn of the British standard series and to the upper Antelope Valley Limestone of the Toquima Range, Nevada.

Oldest Articulated Fish Are Ordovician

Disarticulated elements attributed to fish have been reported from Cambrian strata, but the oldest fossils of nearly complete fish are found in the Darriwilian Stairway Sandstone of the Amadeus Basin of Australia (Ritchie & Gilbert-Tomlinson 1977; see Webby et al 1981, column 8). These remarkable fossils occur as imprints in shallow-water marine sandstones and show exceptionally fine details of armor plating and ornament. They antedate the ostracoderms of the Harding Sandstone of Colorado, which are usually found as separated bony plates. But both occurrences are in very similar sediments (Fischer 1978), reinforcing the belief that the earliest fish were marine.

First Antarctic Ordovician

T. O. Wright early in 1983 collected trilobites from an outcrop in north Victoria Land, Antarctica, while participating in West Germany's project GANOVEX. Genera include *Harpides* (?), *Pseudokainella*, a saukiid resembling *Andersonnella* and *Sinosaukia* but differing from both, and *Tsinania*. Similar assemblages have been reported from the Middle East, Kazakhstan, North China, and western Queensland; some of these have been listed as early Ordovician in age, others as very late Cambrian. From the same collection, John Repetski has identified the first true conodonts of early Ordovician age from the South Polar continent. More complete discussions will appear in *Geology* (Wright et al 1984) and in the *Geologische Jahrbuch*, *GANOVEX III* special issue.[1]

SERIES

The Original Series

The Ordovician System is British in origin. Its pieces were assembled by British stratigraphers and paleontologists, who established the succession of stratigraphic units that they had found and studied in Wales, Shropshire, the Lake District, and southern Scotland (Williams et al 1972). The pieces were fossiliferous rocks and were called series. Originally, the series were constructional units, with which the system was built. Later, they became time-rock units.

There are six series in the British succession. They are utilized around the world as a standard. Almost always their international use is based on graptolites, which is a paradox because the dating of five of the six series was

[1] C. F. Burrett and R. H. Findley have recently informed me that they have found Tremadocian conodonts from another sample collected at the same time as Wright's. Their report is to be published in *Nature* (Burrett & Findley 1984).

originally based on shelly fossils (Table 1). Only the Llanvirn Series was based on graptolites. Murchison's type Llandeilo (Bassett 1972, pp. 29–30, 32, Figure 5, column W9) was based entirely on shelly fossils, but the section near Builth also contained graptolites and became the basis for equating the zone of *Glyptograptus teretiusculus* with the bottom of the Llandeilo.

Nonetheless, the British series are almost always cited with corresponding graptolite zones. This practice seems to be derived from Table B of Elles & Wood (1914, p. 526).

In discussing classification of the Ordovician System based on the British standard series, Skevington (1969) deplored the fact that most of the fossils of the series are shelly forms and that it is impossible (with the exception of the Llanvirn) to define the series on the basis of graptolites, "which, in the majority of cases, they do not contain. Rather, the units of Series category are better retained for the shelly facies only. . . ." Too few scholars outside the British Isles are aware of this deficiency.

Table 1 British Ordovician series, their principal fossil faunas, and selected references

British Series	Principle fauna	Selected references
Ashgill	Trilobites and brachiopods.	Ingham 1966, 1970, 1977, Ingham & Wright 1970, 1972, Temple 1968 (part)
Caradoc	Trilobites and brachiopods. Graptolites very rare. Conodonts disappointing.	Bancroft 1945, Dean 1958, 1960a,b, 1961, 1962a,b, 1963, 1972, Savage & Bassett, in preparation
Llandeilo	Brachiopods, trilobites, and graptolites. *Glyptograptus teretiusculus*	R. Addison, in Bassett 1972, pp. 35–36, Bassett 1972, Elles 1940, Hughes 1969, 1971, Toghill 1970, Williams 1948, 1949, 1953
Llanvirn	Graptolites. *Didymograptus murchisoni* ?"*D. bifidus*"?, not Hall	Bassett 1972, pp. 30, 38, Elles 1904, 1940, p. 397, Hicks 1875, Hopkinson & Lapworth 1875, C. J. Jenkins, in preparation
Arenig	Trilobites, brachiopods?, graptolites. *D. hirundo* *D. deflexus*	Whittington 1966b, Bates 1969, Skevington 1969, p. 169, Bassett 1972, p. 23
Tremadoc	Trilobites, but graptolites limited to lowest part. At base, *Dictyonema flabelliforme*.	Whitworth 1969, Skevington 1969, p. 163, Cowie et al 1972, p. 11

Australian Practice

In Australia (Webby et al 1981), Ordovician time is divided into 10 stages based on graptolites. Because the graptolites are more closely related to those of North America (Berry, in Ross & Berry 1963, pp. 69–71), these stages can be correlated with the British series with difficulty. They differ from the British series because they are based on Australian stratigraphic rock successions, not because the graptolites differ. VandenBerg (1981), in reviewing the basis for each of the stages, indicated the type section or area for each of the graptolite zones. Each of the Australian stages is properly based on a time-rock unit, so that the original series designations of T. S. Hall could still be applied to them. Comparison of the British series and the Australian stages (series?) is made by Webby et al (1981).

These graptolitic stages are readily applicable in Victoria, New South Wales, and New Zealand, but conodonts, trilobites, brachiopods, corals, and stromatoporoids must be used for correlations in the vast areas of Western Australia, Northern Territory, and Queensland. Furthermore, the shelly faunas are more readily correlated with those of North America, Kazakhstan, or the Russian platform than with those of the British series.

Webby (1978) has presented a meaty, well-illustrated account of the Ordovician history of the Australian platform and its margins. In our search for worldwide events we may note that there is no unconformity in the lower or middle Ordovician not attributed to local orogeny, and only a suggestion that the late Ordovician unconformity may have been related to a lowering of sea level.

Series in the USSR

Although stratigraphic practice in the Soviet Union has been undergoing revision, it is important for Westerners to understand some differences in the use and meaning of words. The Russian *svita,* or suite, is close in meaning to a formation, but not to a purely lithogenetic formation; the meaning is closer to one of the biostratigraphic formations of the Cincinnati area of 50 years ago.

The word *otdel* is translated by the Anglo-American as *series*, and to him a series is a building block of which one constructs a system. But the literal meaning betrays the Russian concept of a series as quite the reverse. The Russian and many of his European colleagues start with a complete system and then divide it into parts, preferably lower, middle, and upper series.

Gorizont (= horizon) is used as a regional stage. It is always a time unit, but inasmuch as it is usually named for a specific *seriya* or *svita*, it may be equivalent to a time-rock unit.

A comparison of correlation charts for widely separated structural mountain belts, such as those of northeastern USSR (Chugaeva 1976, Oradovskaya 1970), the Urals (Breivel et al 1980), and eastern Kazakhstan (Nikitin 1972, Shligin & Bandaletov 1976), shows under the column *YARUS* (= stage) the British series, adjoined by the inevitable graptolite zones. Under *gorizont* are regional stages that I would consider provincial series. Each "horizon" may, but in some cases does not, correlate precisely with some other horizon in one of the major regions. Each horizon seems to have been established on a combined lithogenetic and biostratigraphic unit, much as were the British series, and each therefore reflects some aspect of provincial geologic history.

Soviet stratigraphers have separated subparallel facies zones within each structural belt, so that the graptolite-bearing strata tend to be segregated from those enclosing shelly fossils. The latter are more likely to compare with North American, Australian, or Balto-Scandinavian faunas than with those of the British series.

Ordovician Series in China

The Ordovician of China is divided into three series: lower Ichangian Series, middle Neichiashanian Series, and upper Chientangkiangian Series. According to Sheng (1980), the boundary between the middle and upper series falls between the zones of *Climacograptus peltifer* and of *Climacograptus wilsoni* (a level within the Balto-Scandian zone of *D. multidens* in Figure 2), equivalent to the base of the Pagoda Limestone. This boundary was placed above the Pagoda Limestone by Lu (1959), presumably mid-Caradocian and above the zone of *Dicranograptus clingani.* Mu (1974) has presented a third scheme, based entirely on graptolite zones and seemingly without regard to lithologic units, that divides the system into Lower, Middle, and Upper parts. These do not agree with the divisions of either of the other authors.

Another stratigraphic chart, constructed by Lu et al (1976), reconciles the stratigraphic occurrences of a variety of fossil taxa. It appears to have marked differences with the chart of Sheng (1980). The differences not only involve the relatively unimportant question of the exact boundaries of a middle Ordovician series, but more importantly the stratigraphic level of the major unconformity accounting for the absence of middle and upper Ordovician rocks across the whole North China platform. Sheng (1980, Table 1) has demonstrated that three schemes of ages (= stages) portray stratigraphy in South China, in western Hubei and Ichang Gorges, and in North China, reflecting the differing geologic histories of these regions.

Ordovician Series in Southwestern Europe

In southwestern Europe the mainly siliciclastic and volcanic Ordovician is classified according to both British and Bohemian series (Hammann et al 1982). There are no natural divisions of stratigraphic units that coincide with series boundaries. The abundance of limestone in the late Ordovician throughout the Iberian Peninsula is unexpected for a time correlative with and a place believed to have been adjacent to a supposed center of glaciation in Morocco and Algeria. Chamositic iron ore in sandstone and quartzite of Arenig age has its counterpart in the Wabana deposits beneath Conception Bay, Newfoundland (Barnes et al 1981, column 83). These ores have been discussed by Van Houten & Bhattacharyya (1982).

Near and Middle Eastern Ordovician Deposits

The Ordovician of the Near and Middle East (Dean 1980) is devoid of carbonate rocks and is dated by graptolites, trilobites, and trace fossils, relative to the British series. The best record is found in the Arenig and Caradoc, with other series only sporadically represented. Late Caradoc diamictites in Saudi Arabia have been interpreted as tillites, but if they are of glacial origin they may have been dropstone deposits.

North American Series and Stages

The Ordovician stratigraphy of Canada and of the United States has been summarized by Barnes et al (1981) and by Ross et al (1982b), respectively. Of the two, the Canadian summary, with accompanying correlation chart, is the more conservative, considering the system to be composed of a lower Canadian Series, a middle Champlainian Series, and an upper Cincinnatian Series. The US chart breaks with some recent usage and with much tradition to update the series.

IBEX SERIES, FORMERLY CANADIAN Ross and Hintze (Ross et al 1982b, pp. 4–12, sheet 1) have noted that the modern concept of the Canadian is very different from J. D. Dana's original, that it has been stripped of its original type section, and that there have been a succession of substitute patchwork type and reference sections. The stratigraphic section in the Ibex Hills, Millard County, west-central Utah, is without peer for continuous exposure, abundance and variety of fossils, and accessibility. The Ibex succession not only equals the range of the Canadian, but it is also so complete at the base as to be a serious candidate for stratotype of the Cambrian-Ordovician boundary. The section is continuous into the overlying beds of Whiterock age. Traditionalists may wish to continue the use of the term Canadian, but the Ibex section is already well established as the standard for the Lower Ordovician in the United States. To emphasize its importance, it is urged that the lower series henceforth be called the *Ibex Series*.

WHITEROCK SERIES Based on sections in and faunas from Nevada, Cooper (1956, pp. 7–8, chart 1) erected the Whiterock Stage as the oldest in the middle Ordovician Mohawk Series. Recognizing that the old Chazy Stage was unsatisfactory, he also proposed the Marmor to take its place. Eventually it became evident that much of the Whiterock was of the same age as both the Chazy Group and the Marmor (Ross et al 1982b, pp. 10–12, column 50). Because the Chazy Group is bounded above and below by unconformities and is unsatisfactory as a type for a stage, the Whiterock has been extended to include all the Chazy. The *extended* Whiterock is now considered a series and represents approximately 15 m.y.

The Whiterock Series has particular tectonic significance because it was deposited peripheral to the North American continent when the interior was being eroded and deeply weathered (Ross 1976). The Chazy Group is best considered a shallow-water, partly reef facies of the Whiterock. Equivalents of the Whiterock are recognized widely on the basis of brachiopods and trilobites (Ross & Ingham 1970), including the Himalayan exposures.

MOHAWK SERIES The Black River and Trenton stages have traditionally composed the Mohawk Series in New York. Unfortunately, the Black River beds rest on Precambrian crystalline rocks (Fisher, in Ross et al 1982b, pp. 44, 46). Therefore, a new type locality for this series has been designated at the base of the conodont zone of *Prioniodus gerdae* near the base of the Elway Formation at the Lay School, Hogskin Valley, Tennessee (Bergstrom, in Ross et al 1982b, p. 12). The Mohawk Series is the result of renewed marine invasion of the North American continent.

The stages of the Mohawk Series have suffered added confusion because of the discovery that the upper Trenton and the Eden Stage of the Cincinnati Series are correlative. Despite obvious regional preferences, i.e. New York vs Ohio, for purposes of the latest correlation chart the Eden has been maintained intact (Bergstrom, in Ross et al 1982b, pp. 12–13, sheet 1).

CINCINNATI SERIES Rocks of Cincinnatian age provide evidence of the greatest marine submergence of the North American continent in Paleozoic times. Furthermore, in Nevada, Oklahoma, Missouri, Arkansas, Illinois, and the Williston Basin in North Dakota and Montana (Ross et al 1982b), there is evidence that the submergence continued into the Silurian.

Summary of Series

In the United Kingdom and in the United States, Ordovician series are based on lithogenetic units that have assumed temporal significance. In the Soviet Union, comparable units are called "horizons." The "horizons" differ from one major tectonic belt to another, are not coincidently correlative from area to area, and are based as much on shelly faunas as on graptolites. In all three countries, such units not only provide a means of measuring relative time, but also an insight into provincial depositional and tectonic history. Similarly, Sheng (1980, Table 1) demonstrated that three differing schemes portray stratigraphy in South China, in western Hubei and Yichang Gorges, and in North China, reflecting differing tectonic behaviors.

When the British series are used abroad, they seldom have any lithogenetic significance and are, in fact, time units. With few exceptions, they are correlated on the basis of graptolites into a black shale and chert facies, from which correlations into the carbonate facies may be impossible or very difficult. This misfortune will be rectified by improved biostratigraphic discipline, probably starting with the use of conodonts. There is no reason why dating should not be done with brachiopods or trilobites.

Jaanusson (1960) reviewed the confusion occasioned by the attempt to use the British standard series in continental Europe. Reasonably, Jaanusson attributed our inability to find suitable worldwide series to "extensive paleozoogeographical differentiation," to which I would add the basic differences in tectonic behavior and resulting patterns of sedimentation. Series are time-rock units and by their fundamental nature reflect geologic history, particularly in their type areas. By the same fundamental nature, they cannot reflect geologic history in areas far distant from the type areas. We must take stock of Jaanusson's (1960, p. 79) recommendation that series be provincial. But to do away with the international use of the British series (as stages) would be unthinkable.

Lower, Middle, and Upper

Jaanusson (1960) also called attention to the misunderstanding inherent in the use of the terms "Lower," "Middle," and "Upper," which have conflicting meanings in different continents, regions, and countries of the world. These three terms can be used informally (not capitalized) to convey the uncertainty or ignorance intrinsic in their use. There is no reason to formalize ignorance. We are in danger of having the use of these three terms dictated internationally for the sake of consistency, concerning which Emerson commented, "Consistency is the hobgoblin of little minds."

SYSTEM BOUNDARIES

Cambrian-Ordovician Boundary

The founding of the Ordovician System was not without its problems. At the outset there was ambiguity about the lower boundary (Williams 1972a, pp. 2–3), caused by some unfortunate stratigraphy not of Lapworth's doing. This led Lapworth to alter his original suggestion that the base of the system should coincide with the base of the Arenig Series; instead, he included part of the Tremadoc Series (Table 1). Henningsmoen (1973) has presented a superb review of the history, the stratigraphic alternatives, and the biostratigraphic principles concerned with the lower boundary. The Cambrian-Ordovician Boundary Working Group (IUGS) has voted formally to include the entire Tremadoc Series in the Ordovician System (J. Miller, written communication, May 1983), agreeing with virtually universal practice. But the working group has yet to choose formally the precise zone on which or the stratotype section in which the boundary is to be pegged. Bassett & Dean (1982) have edited a fine collection of 15 papers dealing with fossil faunas and potential stratotype sections from widely separated parts of the world. The only major area not represented is China, but representatives of the working group examined candidate sections in China in October and November 1983.

Ordovician-Silurian Boundary

Over the past 10 years the Ordovician-Silurian Boundary Working Group, chaired until autumn of 1981 by R. B. Rickards, produced over 50 reports on potential sites for a boundary stratotype—a remarkable accomplishment. In these reports, objective biostratigraphic data from around the world were presented. By 1979 the number of candidate sections had been narrowed considerably.

In August of 1979 an international field meeting was arranged by Dr. B. S. Sokolov (Koren et al 1979) to examine the exposures of the boundary along Mirny Creek, a tributary of the Kolyma River, in northeastern USSR. In the summer of that year, a meeting of the working group inspected the field relationships of the base of the type section of the Llandovery Series in Wales and the base of the Birkhill Shale at Dob's Linn, east of Moffat, Southern Uplands, Scotland. Both of these stratigraphic units immediately overlie the Ordovician-Silurian boundary. By tradition these were the original sections of Murchison (for shelly fossils) and of Lapworth (for graptolites). During the meeting in the United Kingdom, it was decided that the Dob's Linn section should take precedence over all those that had been examined already, but that a section on Anticosti

Island, Quebec, should be examined at a field conference in 1981. Regrettably, although Chinese representatives were present, no effort was made to nominate a Chinese section.

The Anticosti Island section, in the carbonate facies, produced almost unanimous enthusiasm among those attending the 1981 field conference. Exposure and accessibility were found to be excellent. The fauna is abundant, and the proposed level of the boundary can be pegged precisely. Although the formal proposal of McCracken & Barnes (1981) defines the boundary on conodonts, it was obvious that brachiopods and trilobites could support the choice of horizon. Unfortunately, only two specialists on graptolites attended the field meeting, and only three members represented areas outside North America.

Despite the excellence of the Anticosti section, the voting members of the Boundary Working Group have chosen the Dob's Linn section as the stratotype section. This section has two advantages. (*a*) It is the original boundary section on which Lapworth based his upper Ordovician and lower Silurian graptolite zones. (*b*) Fission tracks in zircon crystals from bentonite beds in the zone of *Monograptus cyphus*, a very short distance above the base of the Birkhill Shale, provide an isotopic date of approximately 437 m.y. (Ross et al 1982a, Ross & Naeser 1984). Few intersystem boundaries can be dated this closely. Whether the base of the zone of *Glyptograptus persculptus* or of the zone of *Akidograptus acuminatus* will mark the boundary remains to be decided.

The graptolitophilic choice must not be permitted to deny the use of the section in the upper Ellis Bay Formation, on the west side of Ellis Bay (McCracken & Barnes 1981) on Anticosti Island, as a parastratotype for students of almost every other kind of fossil. Nor should the graptolitic section in the Wufeng Shale at Huanghuachang, north of Yichang, be overlooked (Wang et al 1982). That and a more westerly section in the Yangtze Gorges were visited by representatives of the Ordovician Subcommission in 1978 and found to be rich in graptolites and structurally little disturbed.

ISOTOPIC DATING

The Ordovician and Silurian systems are isotopically dated more thoroughly than any other Paleozoic systems. Fission-track dating (Ross et al 1982a, Ross & Naeser 1984) gives a good indication of the relative lengths of the British standard series of both systems. The early Silurian zone of *Monograptus cyphus* has yielded a date of 437 ± 10 m.y. using fission tracks (Ross et al 1982a, p. 144) and 433 ± 3 m.y. based on the K/Ar method

(Lanphere et al 1977). Another sample, from the late Ordovician zone of *Dicellograptus anceps*, gave a fission-track date of 434 ± 12 m.y. Compston et al (1983), using an ion microprobe, have analyzed zircons from the same sample of bentonite from the Birkhill Shale that produced the fission-track date on the zone of *M. cyphus*. They conclude that the age may be closer to 431 m.y. Despite large analytical errors, it would seem that the Ordovician-Silurian boundary is close to 435 m.y. ago.

The oldest Ordovician fission-track date is from a tuff in the Llyfnant Flags of early Arenig age and is 493 ± 11 m.y. From the late Tremadocian Rhobell volcanics a date of 508 ± 11 m.y. has been obtained, using the K/Ar method, by Kokelar et al (1982). Two fission-track dates on Middle Cambrian bentonites from the Grand Canyon (Ross & Naeser 1984)—535 ± 12 m.y. for the *Bathyuriscus-Bolaspidella* zone and 563 ± 12 m.y. for the *Glossopleura* zone—give support to placing the Cambrian-Ordovician boundary at 510–515 m.y. (Figure 1).

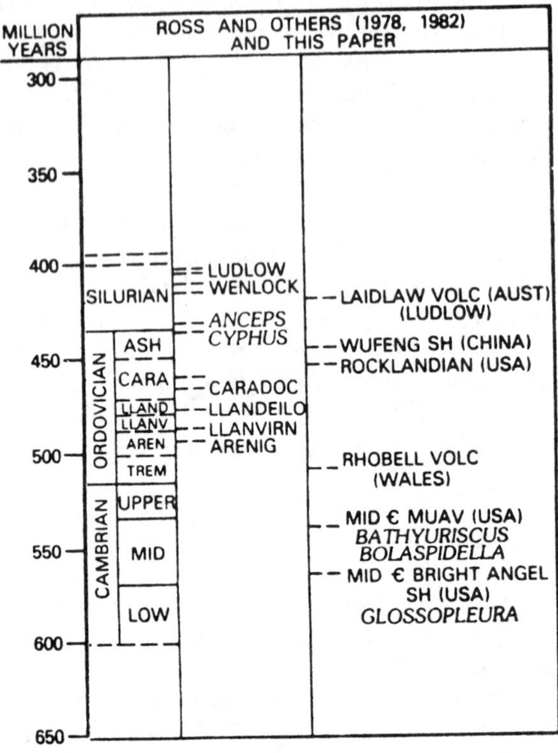

Figure 1 Isotopic dates significant in gauging the age and duration of the Ordovician Period (modified from Ross & Naeser 1984).

WIDESPREAD VOLCANISM

An incomplete review of the Ordovician provincial series of the world indicates that volcanism was widespread. It is evident in tectonically complex linear belts, especially those bordering the stable platforms in the Soviet Union. One may speculate that much of the volcanism took place along subduction zones and that the Iapetus Ocean was only one of several that closed or were closing in Ordovician time.

Europe

Tuffs, ashes, bentonitic ashes, and related intrusive and extrusive rocks abound in Wales from the Lleyn Peninsula to the Berwyn Hills and from St. David's to Builth (Bassett 1972). The Ordovician volcanic rocks along the Welsh Borders from South Shropshire to the Breidden Hills range in age from the lower Llanvirn Stapeley volcanic group to the Caradoc Acton Scott beds (Dean 1972, Ross et al 1982a). In Yorkshire and the Lake District of northern England (Ingham & Wright 1972), the thick rhyolites, andesites, and tuffs of the Borrowdale volcanic complex are considered to be of Caradocian age, while the Cautley volcanic beds are late Ashgillian. In Scotland (Whittington 1972) and in Ireland (Williams 1972b), volcanic rocks are Arenigian to Silurian in age, if one includes the bentonites interbedded in the highest Ordovician Hartfell Shale and the lowest Silurian Birkhill Shale at Dob's Linn, east of Moffat.

In southern France (Hammann et al 1982, p. 18, columns 11–14), rhyolitic and andesitic tuffs and flows are exposed from the Montagne Noire to the eastern Pyrenees. Far to the north in Sweden, bentonite in the Caradocian Dalby Limestone is evidence of a distant volcanic source. Similarly, bentonite in the middle Ordovician of Latvia (Ulst et al 1982, Figure 48) probably originated in volcanic areas bordering the Russian platform.

Examples in the Soviet Union

In the folded and thrust belt of the Urals (Breivel et al 1980), the elongate Sakmaro-Lemvinskaya zone is characterized by great thicknesses of middle Ordovician tuff, basalt, porphyry, spillite, and diabase. In the upper Ordovician, volcanic rocks are mixed with carbonates. In the upper Kuyagach Suite of Arenig age, tuffs and other effusive rocks are intertongued with limestone.

To the east in the drainage basin of the Rassokha River, northwest of the Omulev Mountains, of northeastern USSR, Oradovskaya (1970) and Chugaeva (1976) report volcanic rocks from the upper Arenig to the top of the system. In this complex north-south belt, along the Serechen River, middle and upper Ordovician basalt, andesite, dacite, trachytic lavas, and volcanogenic sandstone aggregates 1560 m in thickness.

Nikitin (1972) described five north-south zones in southeastern Kazakhstan, of which the Stepnyak-Betpakdala zone and the Chirghiz-Tarbagati zone are characterized by volcanic rocks throughout the Ordovician section. Only upper Ordovician rocks are known in the Dzungaro-Balkhash zone, and most of them are volcanic.

Mongolia

There are no complete Ordovician sections in Mongolia. In the Agach-ula section in the Kobdinskaya district of southwestern Mongolia, Arenigian, Llanvirnian, and lower Llandeilian rocks are composed mainly of siltstone, sandstone, shales, and tuffs (Rozman et al 1981).

Thick Volcanics in China

A cursory examination of the Ordovician correlation chart for China (Sheng 1980, column IV) reveals that the thickest Ordovician section is in the north slope of the Chilien Shan (Qilian Shan). Over 2000 m of interbedded limestone, shale, chert, and a variety of volcanic rocks range in age from bottom to top of the system. Graptolites and certain trilobites suggest that this was an outer-shelf environment, possibly a continental margin.

The discovery of a datable bentonite layer in the Wufeng Shale of Ashgill age (Ross & Naeser 1984) in the Yangtze Gorges suggests that more volcanics will eventually be found.

Australian Volcanics

The Ordovician sections in New South Wales and east-central Queensland include great thicknesses of andesite and tuffs along the north-south fold and thrust belt of eastern Australia (Webby et al 1981, columns 34–53, 58). The oldest are the Mount Windsor volcanics of Queensland (column 58), but ages range throughout the Ordovician.

Arenig Volcanics in Argentina

In the Famatina and Puna regions of northwestern Argentina, Ordovician volcanics range in composition from rhyolitic to andesitic, occurring as tuffs, flows, breccias, ignimbrites, and a few spillites (Acenolaza & Toselli 1981, 1982). Graptolites indicate that these are of Arenig age. Ordovician volcanics are also known in southern Bolivia and southern Peru.

North America

Bentonitic ashes are widespread in the middle Ordovician of the Mississippi Valley, Kentucky, Tennessee, New York, Maine, Ontario, Quebec, and Newfoundland. Their sources may now be in the British Isles or in easternmost Canada, but they probably were related to subduction along the former eastern margin of the continent. Williams & Hatcher (1982) preview a more explicit portrayal of the volcanic and tectonic history of the Appalachian orogen, an intriguing story of suspect terrains. Volcanic rocks are widespread in eastern Canada (Barnes et al 1981, columns 64, 65, 69, 76, 79, 80, 82). The composition of the tholeiitic amphibolites and greenstones of the middle Ordovician Ammonoosuc volcanics of New Hampshire and Vermont led Aleinikoff (1977) to conclude that they were partly of abyssal oceanic and partly of island-arc affinity.

In the western United States, greenstones and pillow basalts of the middle Ordovician Valmy Formation in Nevada were erupted along the supposed trailing border of the continent. Similarly, in British Columbia and Yukon (Douglas et al 1969, Figure VIII-9), volcanics are interbedded with shales west of the carbonate belt. Ophiolitic suites are present in the Klamath Mountains of northern California, indicating that volcanism was active wherever that terrane was in the Ordovician (Ross et al 1982b, column 8, pp. 30–33).

SEA-LEVEL CHANGES

In much of the cratonic part of North America, an unconformity exists above units of Ibex (= Canadian) age (Ross et al 1982b, sheets 1–3). This would be approximately equivalent to the conodont zone of *Oepikodus evae* and to the Arenig graptolite zone of *Didymograptus hirundo*. An upper limit on the unconformity is placed in the southern Appalachians in the Lenoir Limestone of late Whiterock age at the base of the zone of *Pygodus serra*. That conodont zone is approximately equal to the Llanvirn graptolite zone of *Didymograptus murchisoni*. In the northern midcontinent region, the St. Peter Sandstone of approximate Llandeilo age rests above the unconformity. This unconformity of late Arenig through Llanvirn age is the basis for the sea-level drop postulated by Vail et al (1977, p. 84, Figure 1), supposedly the most prominent of the Ordovician sea-level changes.

Examination of the Chinese sections (Sheng 1980) indicates that in the North China platform there is a post-Llandeilo unconformity, lasting throughout the period. In northeastern USSR (Chugaeva 1976), Tremadoc and most of the lower Arenig sediments are missing in the terrigenous, volcanic, and transitional zones, but a complete Ordovician section is present in the carbonate facies. This suggests that the unconformity resulted from local, very early Ordovician tectonic activity, not from a sea-level change.

No unconformities are evident in Kazakhstan except in a volcanic-rich section in the Kazik Range of central-east pre-Balkhash (Shligin & Bandaletov 1976, Table 5, column 32). This unconformity is Llanvirn in age and is probably tectonic in origin.

On the other hand, on the Siberian platform (Chugaeva 1976, p. 287) there is a post-Arenig/pre-Llandeilo break similar to south-central North America. The Latvian sections (Ulst et al 1982, Table 23, p. 170) contain three Tremadoc and Arenig unconformities and one at the Arenig-Llanvirn boundary. There is no reason to associate these minor breaks with local tectonism, nor do they correlate with breaks elsewhere to suggest universal sea-level changes.

In Australia (Webby et al 1981), there is an unconformity above the Darriwillian (post-Llanvirn) and below the Eastonian (lower Caradoc) in New South Wales. But in Western Australia the upper limit is Silurian, and in the Northern Territory it is Devonian to Cretaceous. Dating of the discontinuity in New South Wales does overlap that in the United States, but their coincidence is not impressive.

Locally in Spain and Portugal, the Caradoc section is missing (Hammann et al 1982). The positions of unconformities in the British Isles are erratic. In Wales they are locally present in the Llanvirn, Llandeilo, upper Caradoc, and lower Ashgill (Williams et al 1972, Figure 6). Along the Welsh Borderland and in northern England (Williams et al 1972, Figures 7, 8), it is more a matter of determining how much Ordovician section is present than what is missing locally. In the United Kingdom, tectonism must have been far more important than sea level in determining the position of unconformities during Ordovician time.

Evidence for a late Arenig to late Llanvirn drop in sea level, as proposed by Vail et al (1977, p. 84), is shaky at best. There is ample indication of local tectonic unrest during the Ordovician if one couples the evidence of local unconformities with that of volcanism. Only an infinitesimal percentage of that evidence has been assembled here, but it is enough to suggest that preoccupation with the behavior of the Atlantic Ocean and its predecessor(s?) may blind us to the existence of other oceans and subduction zones active during the Ordovician.

ORDOVICIAN CLIMATES

North African Evidence

Ordovician climates, particularly the supposedly late Ordovician glaciation in Gondwana, have received much attention of late. Spjeldnaes (1981) has written a very thorough review of early Paleozoic climates, with special reference to glaciations. The beautifully illustrated work of Beuf et al (1971), following their work (1966) on the extent of Silurian glaciation in the Sahara, leaves no doubt that there are glacial features and deposits of early Paleozoic age in North Africa. However, there is no mention of an Ordovician fossil in the 1971 report. In fact, there is very little biostratigraphy to support claims of Ordovician, rather than Silurian, glaciation. If one checks the references cited by Spjeldnaes (1981, pp. 230–35), one will understand why Spjeldnaes himself (p. 234) has written, "The dating of the Ordovician glaciation is still somewhat doubtful." Legrand (1974, 1981a,b) commented on the tectonic instability of the great area of Algeria and Morocco in the Ordovician and Silurian, and noted that the youngest Ordovician fossils beneath glacial deposits are Caradocian in age and that the oldest Silurian fossils above the deposits are late Llandoverian, a span of 30 m.y. Destombes (1976, p. 413) has recorded Hirnantian (latest Ordovician) brachiopods associated with glacial deposits, but only in the Anti-Atlas Mountains of southern Morocco. Dean (1980, p. 8) cites glacially derived, probably dropstone, deposits in Saudi Arabia of Caradocian age.

A variety of stratigraphic and biostratigraphic phenomena have been attributed to this late Ordovician or early Silurian glaciation. The supposed drop in sea level that resulted in deposition of middle Silurian sediments on late Ordovician or older strata in cratonic North America is one of these phenomena, but whether there is a worldwide unconformity at this stratigraphic level remains to be documented. The question of how much of any such change in sea level might be related to spreading rates of continental plates, rather than to accumulations of ice, must be considered. But Thompson & Diecchio (1982) have already shown that in the eastern United States alone there was no such unconformity because of the rapid accumulation of the Queenston Delta, which spread westward onto the craton from the crumpled eastern continental margin.

Drastic changes in the composition of brachiopod faunas are supposed to have been caused by the glaciation (Sheehan 1973), particularly in North America. However, Amsden (1974, 1980, 1982, 1983) has found no marked differences between latest Ordovician and earliest Silurian faunas in Oklahoma, despite an interruption of the stratigraphic sequence by beds inhospitable to brachiopods and one lower Silurian unconformity. It is possible that faunas found closer to the cratonal center might seem to demonstrate a greater change because the entire lower Silurian is missing (see also Ross et al 1982b, column 40). Future studies in other lands are needed to assess faunal changes, their magnitude, and the reasons for the changes.

Earliest Land Plants from Libya

According to Gray et al (1982), the discovery of spores from Caradocian subsurface rocks of Libya provides the oldest evidence of land plants. They speculate that land plants may have radiated from a North African center. When in the future more is known about these primitive plants, they may provide improved information regarding the rigors of a polar climate approximately 460 m.y. ago. In the same paper (Figures 2–6), they show similar spores from the late Ordovician Elkhorn Formation of Kentucky and the early Silurian of New York and Libya, suggesting a biostratigraphic succession in which increased size may be important.

Strother & Traverse (1983) are critical of this report and its lack of proper taxonomy. They claim that similar evidence of Ordovician land plants is far more plentiful and widespread than realized by Gray et al, who (1983) reply to the charge. These published discussions and accompanying references are informative sources for students of land-plant origins.

The Empty Arctic

As noted by Ziegler et al (1979), Ross (1975), Ormiston & Ross (1979), and Spjeldnaes (1981), there is no record of any land area around the North Pole during the Ordovician. Presumably the northern ocean was open. Without obstruction to oceanic circulation or to mixing with equatorial waters, the climate of the northern hemisphere may have been mild all the way to the pole. Surely there was a very different weather pattern from what we experience today. Warm-water life should have been able to flourish far north of present latitudinal limits.

BIOSTRATIGRAPHIC CORRELATION

Lapworth's Legacy, Concurrent Ranges

In addition to the Ordovician System, Lapworth's most valuable legacy is exemplified by his use of overlapping stratigraphic ranges of many graptolite taxa to define zones. It is the overlapping ranges based on measured sections—the *concurrent ranges* of several taxa—that define the standard graptolite zones and give to them a stable utility not yet equaled by other kinds of assemblage and range zones.

Facies Faunas

Biostratigraphic correlation within the Ordovician seems to be good so long as measured sections have been the basis for determining ranges of various taxa and all the components of the fauna are taken into account. The greatest problems are met when the contemporaneity of coeval faunal facies is not recognized, and the facies are concluded to be sequential. Jaanusson (1976) studied the changes in faunal distribution around the Baltic within the Viruan Series (middle Ordovician), noting that some were related to lithologic changes, while the reason for others was obscure. He also found that no single factor, such as water depth, could explain the distributions of all elements of a fauna. In a far more complex tectonic setting, Jaanusson & Bergstrom (1980) were able to delineate three faunal facies belts in the Appalachian Mountains of essentially the same age as those in the Baltic study. Brachiopods fell into three belts parallel to and bounded by the major thrust faults. Conodont distributions paralleled those of the brachiopods, as did some of the trilobites. But inasmuch as correlative lithologies do not precisely parallel the thrusts, the control on faunal distributions is not evident. Yet, so different are faunas of the more widely separated belts that one might easily conclude that they are of different ages.

Faunal developments and worldwide distributions of fossil taxa throughout the Ordovician have been reviewed by Jaanusson (1979). It has become customary to think of early and middle Ordovician faunas as having been provincial, while late Ordovician faunas were cosmopolitan. Graptolite faunas supposedly exemplify such distributions, although trilobites (Whittington & Hughes 1972, Ross 1975) have been similarly cited. Jaanusson (1979), however, noted that there were several geographically distinct shelly faunal assemblages in the late Ordovician. Among these is an assemblage characterized by *Monorakos* (Ormiston & Ross 1979) and found only in the Siberian platform, northeastern USSR (Kolyma Platform), and the Seward Peninsula of Alaska.

Another assemblage, characterized by *Dicoelosia* and other brachiopods, has been reported (Ross & Dutro 1966) from Jones Ridge, close to the border between Alaska and Yukon; from Percé, Quebec (Schuchert & Cooper 1930); from the Drummuck Group near Girvan, Scotland; and from the Klamath Mountains, northern California (A. W. Potter, personal communication, and Ross et al 1982b, column 8). This brachiopod assemblage may be depth controlled and may border the former North American continent, as well as other continents, since several of its elements have been reported by Amsden (1974) from Oklahoma and by Roomusoks (1968) from Estonia. It has little in common with correlative(?) Cincinnati faunas of Ohio and Kentucky.

Ludvigsen (1975, 1978a) and Chatterton & Ludvigsen (1976), working in the southern MacKenzie Mountains (District of MacKenzie), have documented superbly preserved, silicified trilobite faunas of late Whiterock to late Mohawk age in alternating transgressive-regressive sequences. The trilobites of a late Cincinnati age trilobite and conodont assemblage from the Hanson Creek Formation of central Nevada (Ross et al 1980) are so similar to those of Ludvigsen's shelf-edge Biofacies III that they might have been considered correlative had not the associated conodonts confirmed the younger age of the Nevada occurrence. In their next more westerly exposure, the Hanson Creek beds are composed of very thinly laminated limestone-bearing graptolites and radiolarians, but no shelly fossils.

Obviously, Ludvigsen's (1978a) paleoecologic analysis is applicable to this younger fauna about 3000 km distant from the MacKenzie Mountains.

Dating Facies Faunas, and Graptolite Depth Zones

How are fossil assemblages as strongly influenced by environment as these to be correlated? Traditionally, biostratigraphers might have appealed to graptolites, but Jaanusson (1979) has noted the provincialism of these fossils.

Despite Skevington's cautioning (1969, p. 161), it is often assumed that graptolites, unlike other animals, were not affected by environmental conditions. Unfortunate disputes concerning the stratigraphic value of graptolite zonation are related in large part to the failure to recognize the possibility of facies control. In an important ongoing study of middle Ordovician stratigraphy in western New York state, Cisné & Chandlee (1982) have used numerous bentonite layers to establish time planes along which they traced graptolite populations from shallow- to deeper-water paleoenvironments. They documented the lateral change of faunas dominated by *Orthograptus* (including *O. ruedemanni* and *O. amplexicaulus*) to contemporaneous assemblages dominated by *Corynoides* (including *C. americanus* and *C. calicularis*). This discovery bears on the graptolite correlation problem discussed in the correlation chart for the United States (Ross et al 1982b, pp. 2–3, chart sheet 1 of 3). Clearly the same kind of documentation is needed for the whole stratigraphic range of graptolite zones, although bentonites are not ordinarily available to provide the necessary temporal control.

Need for Unbiased Stratigraphic Data

Concurrent range zones are applicable to any fossil group studied in the context of measured sections. The increased interest in using conodonts for purposes of correlation is a matter of discipline, addressed recently by Sweet (1982). The biostratigraphic use of microfossils, which cannot be seen by the collector, requires disciplined measurement of sections and recording of lithologies. The unbiased end result produces detailed information on stratigraphic ranges of genera and species. All too often in the past, macrofossils have been collected and lumped as the "fauna of formation X" with little regard to stratigraphic detail. Macropaleontologists who put more emphasis on biostratigraphy (for example, see Stitt 1971, 1977) will remedy this deficiency.

Pelagic Fossils—Trilobites and Conodonts(?)

The biostratigraphically useful trilobites were surely pelagic; some, like the olenids, were widespread in their distribution, while others, such as the bathyurids, preferred shallower, platform environments (Whittington

1966a, Ross 1975, Fortey 1974, 1975a,b, 1980). In these habits, conodonts may have resembled trilobites, on whom they may have preyed.

Briggs et al (1983) have discovered the first fossil conodont animal and have discussed the possible relationship of conodonts with and their similarity to Chaetognathids, or "arrow worms." If Briggs et al are correct, the conodonts may have been the carnivorous scourge of the planktonic realm in Paleozoic time. The distribution of conodonts indicates that they were as cosmopolitan as graptolites. It is known that they occur in shallow-water populations over carbonate platforms, as well as in distinct deeper(?) water assemblages, and that they can be used for accurate worldwide correlation. Conodont taxa that characterize the presumably warm, shallow-water facies [North American Midcontinent fauna (Sweet et al 1971, Barnes & Fahraeus 1975, Barnes et al 1973)] are recognized as the same fauna in Australia (Palmieri 1978; see Webby et al 1981, column 56). Although the shallow-water forms might be thought to be endemic to a continent, their distribution on continential platforms that seem to have been widely separated all through the Phanerozoic indicates that these forms were widespread.

Conodont-Graptolite Correlations

In a comprehensive review of the evidence mainly from Balto-Scandia, from the British Isles, and from North America, Bergstrom (1984) has attempted to show the equivalence of conodont and graptolite zones. This herculean effort has produced close to 80 ties between the two schemes, despite the rarity of conodonts in graptolitic shales. More ties are available in Arenig through Llandeilo time than later in the period. Furthermore, the graptolite zones of Balto-Scandia are not precisely the same as those of Britain, just as the zones of eastern North America (Riva 1969) differ from those of western North America [Berry 1960, Jackson 1964–1979 (see Barnes et al 1981, p. 19)]. The North Atlantic conodont zones, preferred by Bergstrom, are difficult to mesh with the North American cratonic zones. Despite these and other difficulties, Bergstrom's review is a tribute to his persistence (Figure 2).

Sweet (1979a, 1982) has been revising and updating the North American conodont zonal scheme using Shaw's (1964) graphic solution of concurrent ranges, as modified by Miller (1977). This is a very different method from the evolutionary lineage zones employed by Bergstrom.

Trilobites, Brachiopods, and Other Taxa

One of the most complete recent efforts to coordinate the concurrent ranges of trilobites, brachiopods, ostracodes, graptolites, and conodonts is found in a book on the Ordovician of Latvia (Ulst et al 1982). There is nothing like it for North America. To be sure, trilobite zones have been erected for the lower and lower middle Ordovician by Ross (1951) and Hintze (1952), but these are assemblage zones rather than concurrent range zones. Ross (1964, 1967, 1970) and Hintze (1979) showed the ranges of Ibexian (= Canadian of authors) and Whiterockian trilobites, brachiopods, and other macrofossils relative to stratigraphic sections in Nevada and Utah, as Harris et al (1979) and Ethington & Clark (1981) have done for conodonts. It is important to

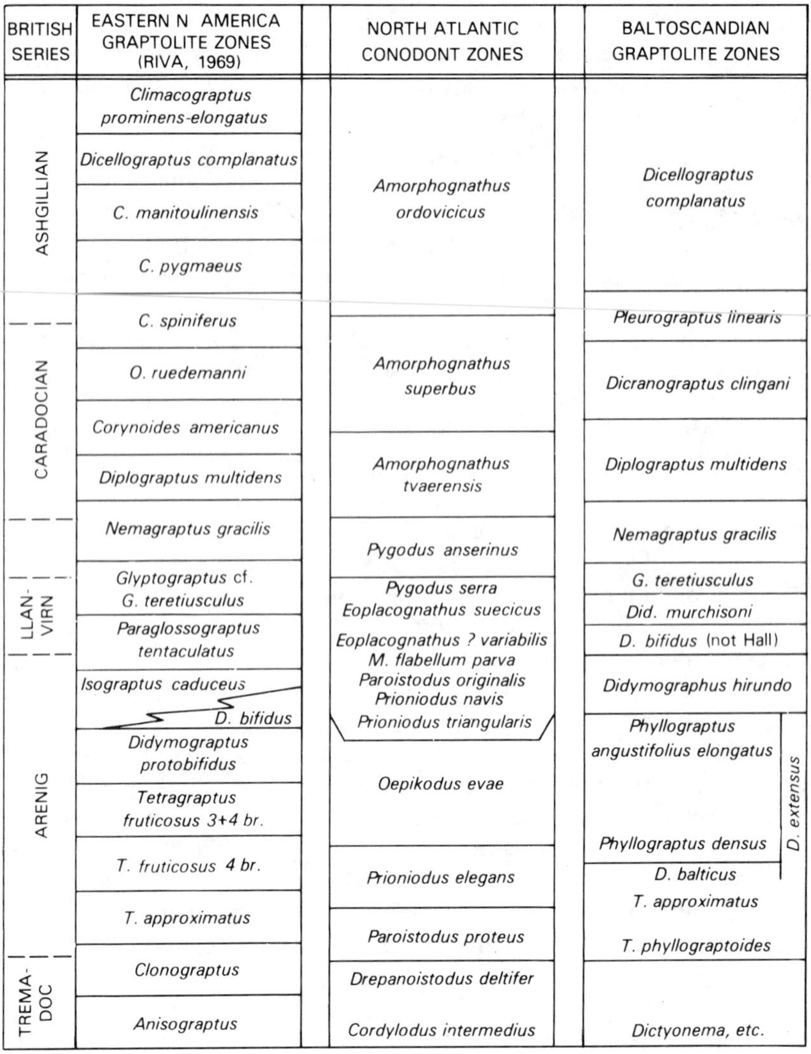

BRITISH SERIES	EASTERN N AMERICA GRAPTOLITE ZONES (RIVA, 1969)	NORTH ATLANTIC CONODONT ZONES	BALTOSCANDIAN GRAPTOLITE ZONES
ASHGILLIAN	*Climacograptus prominens-elongatus*	*Amorphognathus ordovicicus*	*Dicellograptus complanatus*
	Dicellograptus complanatus		
	C. manitoulinensis		
	C. pygmaeus		
CARADOCIAN	*C. spiniferus*		*Pleurograptus linearis*
	O. ruedemanni	*Amorphognathus superbus*	*Dicranograptus clingani*
	Corynoides americanus		
	Diplograptus multidens	*Amorphognathus tvaerensis*	*Diplograptus multidens*
	Nemagraptus gracilis	*Pygodus anserinus*	*Nemagraptus gracilis*
LLAN-VIRN	*Glyptograptus* cf. *G. teretiusculus*	*Pygodus serra* *Eoplacognathus suecicus*	*G. teretiusculus* *Did. murchisoni*
	Paraglossograptus tentaculatus	*Eoplacognathus ? variabilis*	*D. bifidus* (not Hall)
ARENIG	*Isograptus caduceus* *D. bifidus*	*M. flabellum parva* *Paroistodus originalis* *Prioniodus navis* *Prioniodus triangularis*	*Didymographus hirundo*
	Didymograptus protobifidus	*Oepikodus evae*	*Phyllograptus angustifolius elongatus* (*D. extensus*)
	Tetragraptus fruticosus 3+4 br.		
	T. fruticosus 4 br.	*Prioniodus elegans*	*Phyllograptus densus* (*D. extensus*) *D. balticus*
	T. approximatus	*Paroistodus proteus*	*T. approximatus* *T. phyllograptoides*
TREMA-DOC	*Clonograptus*	*Drepanoistodus deltifer*	*Dictyonema, etc.*
	Anisograptus	*Cordylodus intermedius*	

Figure 2 Comparison of conodont and graptolite zones (modified from Bergstrom 1984). Note that the graptolite zones of Balto-Scandia and those of northeastern North America are not precisely correlative.

note that Cooper's (1956) brachiopod assemblage zones are fundamental to the Whiterock with only minor changes.

Ludvigsen (1978b) has produced a brief, well-illustrated trilobite biostratigraphy, in which he has tabulated concurrent range zones for southern Ontario. This plus Ludvigsen's (1978a) work in the MacKenzie Mountains should form a firm base for continent-wide trilobite zones in the Mohawk and Cincinnati series.

Despite excellent reconstruction of the complex facies relationships of upper Mohawkian strata of the Cincinnati Arch (Cressman 1973), there is no study of the faunas integrated with the stratigraphy. Similarly, in spite of a complete updating of the mapping of most of the classical Eden, Maysville, and Cincinnati areas, only a few stratigraphic and faunal reports have been written (Pojeta 1979, Alberstadt 1979, Howe 1979, Ross 1979, Bell 1979, Branstrator 1979, Sweet 1979b). These cover only a few taxa and are not integrated into a stratigraphic summation. The USGS effort to map lithologies provided an important break from the earlier practice of mapping faunas, but the biostratigraphic fruits have yet to be reaped to produce a concurrent range chart for the macrofossils of the type Cincinnati Series.

Faunas Away from Home

A single trilobite, *Colpocoryphe* (a Tethyian genus), was reported from the subsurface Ordovician of Florida (Whittington 1953). It and two graptolite species are considered evidence that Florida is a remnant of western Africa left attached to North America after a late Paleozoic collision. And the asaphid fauna at Oaxaca, Mexico (Robison & Pantoja-Alor 1968), may have a like significance with reference to South America.

Similarly, the brachiopod assemblage of the Klamath Mountains, noted above, might be cited in support of the currently popular theory that calls for accretion of Ordovician ophiolitic terrain of northern California to North America long after Ordovician time. The distribution of the trilobite *Monorakos*, cited above, provides evidence of post-Ordovician joining of the Seward Peninsula to Alaska. On the other hand, probably Ordovician sediments in roof pendants of the Sierra Nevada Batholith near Convict Lake (Rinehart & Ross 1964, Ross et al 1982b, column 19) are so similar to sequences eastward in Nevada as to suggest that they were part of Ordovician North America and thus not accreted later.

SUMMARY

Before indulging in further speculation about faunal migrations, sea-level changes, or sudden extinctions due to climatic changes, it would seem wise to seek the participation of specialists outside the field of biostratigraphy. The various massifs of Soviet Asia, China, and South America have individual histories about which paleomagnetists can surely add to our enlightenment. The positions of Ordovician island arcs and the timing of volcanism allied with subductions should provide evidence of the closing or altering of several seas, accompanied by changes in oceanic currents, sea levels, and faunal distributions.

In North America, the Ordovician histories of the Reelfoot Basin and Michigan Basin (Ross et al 1982b, pp. 36–40), as well as the en echelon eastern Great Lakes and St. Lawrence estuary, need to be studied as possible sites of rift valleys and early extensional phases.

Completion of the correlation charts for South America, northern Europe, Mexico, and Africa will surely produce surprises. Extensive charts for the major tectonic terrains of the Soviet Union have been available, and several were cited in the above discussions of series and volcanism. But we must either translate them into English or learn our Russian better to fully appreciate the information therein. A similar comment is in order concerning China.

The Ordovician is only beginning to release its secrets to us and remains a fertile field for all kinds of geologic endeavor.

Literature Cited

Acenolaza, F. G., Toselli, A. J. 1981. Geologia del noroeste Argentino. *Univ. Nac. Tucuman, Fac. Cienc. Nat.* 212 pp.

Acenolaza, F. G., Toselli, A. J. 1982. Lower Ordovician vulcanism in northwest Argentina. *Int. Symp. Ordovician Syst., 4th, Palaeontol. Contrib. Univ. Oslo* 280:2 (Abstr.)

Alberstadt, L. P. 1979. The brachiopod genus *Platystrophia. US Geol. Surv. Prof. Pap. 1066-B.* 20 pp. + 7 pls.

Aleinikoff, J. N. 1977. Petrochemistry and tectonic origin of the Ammonoosuc volcanics, New Hampshire-Vermont. *Geol. Soc. Am. Bull.* 88:1546–52

Amsden, T. W. 1974. Late Ordovician and early Silurian articulate brachiopods from Oklahoma, southwestern Illinois, and eastern Missouri. *Okla. Geol. Surv. Bull. 119.* 154 pp. + 28 pls.

Amsden, T. W. 1980. Hunton Group (late Ordovician, Silurian, and early Devonian) in the Arkoma Basin of Oklahoma. *Okla. Geol. Surv. Bull. 129.* 136 pp. + 12 pls.

Amsden, T. W. 1982. Late Ordovician–Silurian brachiopod biostratigraphy, southern Midcontinent region, U.S.A. *Int. Symp. Ordovician Syst., 4th, Palaeontol. Contrib. Univ. Oslo* 280:4 (Abstr.)

Amsden, T. W. 1983. Upper Bromide Formation and Viola Group (middle and upper Ordovician) in eastern Oklahoma. *Okla. Geol. Surv. Bull. 132.* 76 pp. + 14 pls.

Bancroft, B. B. 1945. The brachiopod zonal indices of the stages Costonian to Onnian in Britain. *J. Paleontol.* 19:181–252

Barnes, C. R., Fahraeus, L. E. 1975. Provinces, communities, and the proposed nektobenthic habit of Ordovician conodontophorids. *Lethaia* 8:133–49

Barnes, C. R., Rexroad, C. B., Miller, J. F. 1973 (1972). Lower Paleozoic conodont provincialism. *Geol. Soc. Am. Spec. Paper No. 141*, pp. 157–90

Barnes, C. R., Norford, B. S., Skevington, D. 1981. The Ordovician System in Canada, correlation chart and explanatory notes. *Int. Union Geol. Sci. Publ. No. 8.* 27 pp.

Bassett, D. A. 1972. Wales. See Williams et al 1972, pp. 14–39

Bassett, M. G., ed. 1976. *The Ordovician System, Proc. Palaeontol. Assoc. Symp., Birmingham, Sept., 1974* Cardiff: Univ. Wales Press and Natl. Mus. Wales. 696 pp.

Bassett, M. G., Dean, W. T., eds. 1982. The Cambrian-Ordovician boundary: sections, fossil distributions, and correlations. *Int. Union Geol. Sci. Publ. No. 10 and Natl. Mus. Wales Geol. Ser. No. 3.* 227 pp.

Bates, D. E. B. 1969. Some aspects of the

Arenig faunas of Wales. In *The Pre-Cambrian and Lower Paleozoic Rocks of Wales*, ed. A. Wood, pp. 155–59. Cardiff: Univ. Wales Press. 461 pp.

Bell, B. M. 1979. Edrioasteroids (Echinodermata). *US Geol. Surv. Prof. Pap. 1066-E.* 7 pp. + 2 pls.

Bergstrom, S. M. 1984(?). Biostratigraphic integration of Ordovician graptolite and conodont zones—a regional review. In *Biostratigraphy, Spec. Publ. Geol. Soc. London*, ed. R. S. Rickards, C. P. Hughes, A. J. Chapman (Graptolite Working Group, Int. Paleontol. Assoc., *Proc. Int. Conf., 2nd, Cambridge, 1981*)

Berry, W. B. N. 1960. Graptolite faunas of the Marathon region, West Texas. *Univ. Tex. Bur. Econ. Geol. Publ. No. 6005.* 179 pp. + 20 pls.

Beuf, S., Biju-Duval, B., Stevaux, J., Kulbicki, G. 1966. Ampleur des glaciations "siluriennes" du Sahara: leurs influence et leurs consequences sur la sedimentation. *Rev. Inst. Fr. Pet.* 21. 363 pp.

Beuf, S., Biju-Duval, B., de Chaparal, O., Rognon, R., Gariel, O., Bennacef, A. 1971. *Les Gres du Paleozoique Inferieur au Sahara.* Paris: Technip

Branstrator, J. W. 1979. Asteroidea (Echinodermata). *US Geol. Surv. Prof. Pap. 1066-F.* 7 pp. + 2 pls.

Breivel, M. G., Papulov, G. N., Hodalevich, A. N., eds. 1980. *Report on a Unified Stratigraphic Correlation Scheme for the Urals.* Sverdlovsk: Acad. Sci. USSR and Minist. Geol. USSR

Briggs, D. E. G., Clarkson, E. N. K., Aldridge, R. J. 1983. The conodont animal. *Lethaia* 16:1–14

Burrett, C. F., Findley, R. H. 1984. Cambrian and Ordovician conodonts from the Robertson Bay Group, Antarctica, and their tectonic significance. *Nature.* In press

Chatterton, B. D. E., Ludvigsen, R. 1976. Silicified middle Ordovician trilobites from the South Nahani River area, District of MacKenzie, Canada. *Palaeontogr. Abt. A* 154. 106 pp.

Chugaeva, M. N. 1976. Ordovician in the northeastern U.S.S.R. See Bassett 1976, pp. 283–92

Cisné, J. L., Chandlee, G. O. 1982. Taconic foreland basin graptolites: age zonation, depth zonation, and use in ecostratigraphic correlation. *Lethaia* 15:343–63

Compston, W., Williams, I. S., Froude, D., Foster, J. J. 1983. U-Pb dating of zircons from lower Paleozoic tuffs using the ion microprobe. *Ann. Rep. 1982, Res. Sch. Earth Sci., Aust. Natl. Univ.*, pp. 244–46

Cooper, G. A. 1956. Chazyan and related brachiopods. *Smithson. Misc. Collect.*, Vol. 127. 1245 pp. + 269 pls.

Cowie, J. W., Rushton, A. W. A., Stubblefield, C. J. 1972. A correlation of Cambrian rocks in the British Isles. *Geol. Soc. London Spec. Rep. No. 2.* 42 pp.

Cressman, E. R. 1973. Lithostratigraphy and depositional environments of the Lexington Limestone (Ordovician) of central Kentucky. *US Geol. Surv. Prof. Pap. 768.* 61 pp. + 11 pls.

Dean, W. T. 1958. The faunal succession in the Caradoc Series of South Shropshire. *Bull. Br. Mus. (Nat. Hist.) Geol.* 3:191–231

Dean, W. T. 1960a. The Ordovician trilobite faunas of South Shropshire. I. *Bull. Br. Mus. (Nat. Hist.) Geol.* 4:71–143

Dean, W. T. 1960b. The use of shelly faunas in a comparison of the Caradoc Series in England, Wales, and parts of Scandinavia. In *Ordovician and Silurian Stratigraphy and Correlations, Int. Geol. Congr., 21st, Part 7*, pp. 82–87

Dean, W. T. 1961. The Ordovician trilobite faunas of South Shropshire. II. *Bull. Br. Mus. (Nat. Hist.) Geol.* 5:311–58

Dean, W. T. 1962a. The Ordovician trilobite faunas of South Shropshire. III. *Bull. Br. Mus. (Nat. Hist.) Geol.* 7:213–54, pls. 37–46

Dean, W. T. 1962b. The trilobites of the Caradoc Series in the Cross Fell inlier of northern England. *Bull. Br. Mus. (Nat. Hist.) Geol.* 7:65–134

Dean W. T. 1963. The Ordovician trilobite faunas of South Shropshire. IV. *Bull. Br. Mus. (Nat. Hist.) Geol.* 9:257–96

Dean, W. T. 1972. The Welsh borderland. See Williams et al 1972, pp. 39–42, fig. 7

Dean, W. T. 1980. The Ordovician System in the Near and Middle East, correlation chart and explanatory notes. *Int. Union Geol. Sci. Publ. No. 2.* 22 pp.

Destombes, J. 1976. The Ordovician of the Moroccan Anti-Atlas. See Bassett 1976, pp. 411–13

Douglas, R. J. W., Gabrielse, H., Wheeler, J. O., Stott, D. F., Belyea, H. R. 1969. Geology of western Canada. In *Geology and Economic Minerals of Canada, Geol. Surv. Can., Econ. Geol. Rep. No. 1*, pp. 367–488. 838 pp.

Elles, G. L. 1904. Some graptolite zones in the Arenig rocks of Wales. *Geol. Mag.* 41:199–211

Elles, G. L. 1940. The stratigraphy and faunal succession in the Ordovician rocks of the Builth-Llandindrod inlier, Radnorshire. *Q. J. Geol. Soc. London* 95:383–445 (for 1939)

Elles, G. L., Wood, E. M. R. 1901–1914. *A Monograph of British Graptolites.* Publ. in 10 parts by Palaeontogr. Soc. London (1914, Part 10, pp. 487–526, pls. 50–52, Vol. 67)

Ethington, R. L., Clark, D. L. 1981. Lower and middle Ordovician conodonts from the Ibex area, western Millard County, Utah. *Brigham Young Univ. Geol. Stud.* 28(2). 160 pp. + 14 pls.

Fischer, W. A. 1978. The habitat of the early vertebrates: trace and body fossil evidence from the Harding Formation (middle Ordovician), Colorado. *Mt. Geol.* 15:1–26, 9 pls.

Fortey, R. A. 1974. The Ordovician trilobites of Spitsbergen. I. Olenidae. *Norsk Polarinst. Skr. No. 160.* 129 pp. + 24 pls.

Fortey, R. A. 1975a. The Ordovician trilobites of Spitsbergen. II. Asaphidae, Nileidae, Raphiophoridae, and Telephinidae of the Valhallfona Formation. *Norsk Polarinst. Skr. No. 162.* 207 pp. + 41 pls.

Fortey, R. A. 1975b. Early Ordovician trilobite communities. *Fossils and Strata* 4: 339–60

Fortey, R. A. 1980. The Ordovician trilobites of Spitsbergen. III. Remaining trilobites of the Valhallfona Formation. *Norsk Polarinst. Skr. No. 171.* 163 pp. + 25 pls.

Gray, J., Massa, D., Boucot, A. J. 1982. Caradocian land plants microfossils from Libya. *Geology* 10:197–201

Gray, J., Massa, D., Boucot, A. J. 1983. Reply to comment by Strother & Traverse. *Geology* 11(5):317–18

Hammann, W., Robardet, M., Romano, M. 1982. The Ordovician System in southwestern Europe (France, Spain, and Portugal), correlation chart and explanatory notes. *Int. Union Geol. Soc. Publ. No. 11.* 47 pp.

Harris, A. G., Bergstrom, S. M., Ethington, R. L., Ross, R. J. Jr. 1979. Aspects of middle and upper Ordovician conodont biostratigraphy of carbonate facies in Nevada and southeast California and comparison with some Appalachian successions. *Brigham Young Univ. Geol. Stud.* 26(3):7–33, 5 pls.

Henningsmoen, G. 1973. The Cambro-Ordovician boundary. *Lethaia* 6(4):423–39

Hicks, H. 1875. On the succession of the ancient rocks in the vicinity of St. David's, Pembrokeshire, with special reference to those of the Arenig and Llandeilo Groups, and their fossil contents. *Q. J. Geol. Soc. London* 31:167–95

Hintze, L. F. 1952. Lower Ordovician trilobites from western Utah and eastern Nevada. *Utah Geol. Miner. Surv. Bull. 48.* 249 pp. + 28 pls.

Hintze, L. F. 1979. Preliminary zonations of lower Ordovician of western Utah by various taxa. *Brigham Young Univ. Res. Stud., Geol. Ser.* 26(2):13–19

Holland, C. H. 1974. The lower Paleozoic systems: an introduction. In *Cambrian of the British Isles, Norden, and Spitsbergen, Lower Paleozoic Rocks of the World*, 2:1–13. New York: Wiley. 300 pp.

Hopkinson, J., Lapworth, C. 1875. Description of the graptolites of the Arenig and Llandeilo rocks of St. David's. *Q. J. Geol. Soc. London* 31:631–72

Howe, H. J. 1979. Middle and late Ordovician Plectambonitacean, Rhynchonellacean, Syntrophiacean, Trimerellacean, and Atrypacean brachiopods. *US Geol. Surv. Prof. Pap. 1066-C.* 18 pp. + 7 pls.

Hughes, C. P. 1969. The Ordovician trilobite faunas of the Builth-Llandrindod inlier, central Wales. Part I. *Bull. Br. Mus. (Nat. Hist.) Geol.* 18(3):39–103, 14 pls.

Hughes, C. P. 1971. The Ordovician trilobite faunas of the Builth-Llandrindod inlier, central Wales. Part II. *Bull. Br. Mus. (Nat. Hist.) Geol.* 20(4):118–82, 16 pls.

Ingham, J. K. 1966. The Ordovician rocks in the Cautley and Dent districts of Westmorland and Yorkshire. *Proc. Yorks. Geol. Soc.* 35:455–505

Ingham, J. K. 1970–77. The upper Ordovician trilobites from the Cautley and Dent districts of Westmorland and Yorkshire. *Monogr. Palaeontogr. Soc. London* (1970, Part 1, pp. 1–58, pls. 1–9; 1974, Part 2, pp. 59–87, pls. 10–18; 1977, Part 3, pp. 89–121, pls. 19–27)

Ingham, J. K., Wright, A. D. 1970. A revised classification of the Ashgill Series. *Lethaia* 3:233–42

Ingham, J. K., Wright, A. D. 1972. The north of England. See Williams et al 1972, pp. 43–49

Jaanusson, V. 1960. On the series of the Ordovician System. *Int. Geol. Congr., Norden, 21st*, 7:70–81

Jaanusson, V. 1976. Faunal dynamics in the middle Ordovician (Viruan) of Balto-Scandia. See Bassett 1976, pp. 301–26

Jaanusson, V. 1979. Ordovician. In *Biogeography and Biostratigraphy, Treatise on Invertebrate Paleontology, Part A*, pp. 136–66. Lawrence: Geol. Soc. Am./Univ. Kansas Press

Jaanusson, V., Bergstrom, S. M. 1980. Middle Ordovician faunal spatial differentiation in Baltoscandia and the Appalachians. *Alcheringa* 4:89–110

James, N. P. 1982. Geologic history of reefs. *Amer. Assoc. Pet. Geol. Bull.* 66(10):1681 (Abstr.)

Kokelar, B. P., Fitch, F. J., Hooker, P. J. 1982. A new K-Ar age from uppermost Tremadoc rocks of north Wales. *Geol. Mag.* 119(2):207–11

Koren, T. N., Oradovskaya, M. M., Sobolevskaya, R. F., Chugaeva, M. N. 1979. *Field Excursion Guidebook on Tour VIII and Atlas of Paleontological Plates, 14th USSR Pac. Sci. Congr., Khabarovsk.* Magadan: North-East. Interdisciplinary Sci. Res. Inst.

Lanphere, M. A., Churkin, M., Eberlein, G. D. 1977. Radiometric age of the *Monograptus cyphus* zone in southeastern Alaska—an estimate of the age of the Ordovician-Silurian boundary. *Geol. Mag.* 114: 15–24

Legrand, P. 1974. Essai sur la paleogeographie de l'Ordovicien au Sahara Algerian. *Cie. Fr. Pet. Notes & Mem. No. 11*, pp. 121–38, 3 tables, 8 pls.

Legrand, P. 1981a. Essai sur la paleogeographie du Silurien au Sahara Algerien. *Cie. Fr. Pet. Notes & Mem. No. 16*, pp. 9–24, 3 tables, 9 pls.

Legrand, P. 1981b. Les graptolites de l'Oued in Djeranes et le probleme de la limite Ordovicienne-Silurien au Sahara-Algeria. *Int. Conf. Graptolite Work. Group, Int. Paleontol. Assoc., 2nd, Cambridge*

Lu, Y. 1959. Subdivision and correlation of Ordovician formations of South China. In *Review Papers on Basic Data and Special Topics in Chinese Geology.* Ed. Comm. Chinese Geol. 105 pp. (with English summary)

Lu, Y., Chu, C., Chien, Y., Zhou, Z., Chen, J., et al. 1976. Ordovician biostratigraphy and paleozoogeography of China. *Mem. Nanjing Inst. Geol. Paleontol. No. 7.* 83 pp. + 14 pls.

Ludvigsen, R. 1975. Ordovician formations and faunas, southern MacKenzie Mountains. *Can. J. Earth Sci.* 12(4): 663–97

Ludvigsen, R. 1978a. Middle Ordovician trilobite biofacies, southern MacKenzie Mountains. *Geol. Assoc. Can. Spec. Pap. No. 18.* 37 pp. + 3 pls.

Ludvigsen, R. 1978b. Towards an Ordovician trilobite biostratigraphy of southern Ontario. In *Geology of the Manitoulin Area, Mich. Basin Geol. Soc. Spec. Pap.*, ed. J. T. Sanford, R. E. Mosher, 3: 73–84

McCracken, A. D., Barnes, C. R. 1981. Conodont biostratigraphy and paleoecology of the Ellis Bay Formation, Anticosti Island, Quebec, with special reference to late Ordovician–early Silurian chronostratigraphy and the systematic boundary. *Geol. Surv. Can. Bull.* 329(2): 52–134, 7 pls.

Miller, F. X. 1977. The graphic correlation method in biostratigraphy. In *Concepts and Methods in Biostratigraphy*, ed. E. G. Kauffman, J. E. Hazel, pp. 165–86. Stroudsburg, Pa: Dowden, Hutchinson & Ross

Mu, A. T. 1974. Evolution, classification, and distribution of graptoloidea and graptodendroids. *Sci. Sin.* 17(2): 227–38

Nikitin, I. F. 1972. *Ordovician of Kazakhstan. Chap. 1, Stratigraphy.* Alma-Ata: Acad. Sci. Kazak. SSR. 233 pp. (English summary, pp. 234–40)

Oradovskaya, M. M. 1970. Ordovician System (pp. 80–104) and correlation chart of the Ordovician deposits of northeastern U.S.S.R. (Suppl. 2). *Geology of the USSR, Tom. 30, Northeast USSR Geological Structure. Book 1.* Moscow: NEDRA

Ormiston, A. R., Ross, R. J. Jr. 1979. *Monorakos* in the Ordovician of Alaska and its zoogeographic significance. In *Historical Biogeography, Plate Tectonics, and the Changing Environment*, ed. J. Gray, A. J. Boucot, pp. 53–59. Corvallis: Oreg. State Univ. Press

Palmieri, V. 1978. Late Ordovician conodonts from the Fork Lagoons beds, Emerald area, central Queensland. *Publ. Geol. Surv. Queensland* 369: 1–31

Pojeta, J. Jr. 1979. The Ordovician paleontology of Kentucky and nearby states. Introduction. *US Geol. Surv. Prof. Pap. 1066-A.* 48 pp.

Rinehart, C. D., Ross, D. C. 1964. Geology and mineral deposits of the Mount Morrison quadrangle, Sierra Nevada, California. *US Geol. Surv. Prof. Pap. 385.* 106 pp., geol. map

Ritchie, A., Gilbert-Tomlinson, J. 1977. First Ordovician vertebrates from the Southern Hemisphere. *Alcheringa* 1: 351–68

Riva, J. 1969. Middle and Upper Ordovician graptolite faunas of St. Lawrence lowlands of Quebec, and of Anticosti Island. *North Atlantic Geology and Continental Drift, Amer. Asso. Pet. Geol. Mem.* 12: 513–56

Robison, R. A., Pantoja-Alor, J. 1968. Tremadocian trilobites from the Nochixtlan region, Oaxaca, Mexico. *J. Paleontol.* 42(3): 767–800, pls. 97–104

Roomusoks, A. 1968. On the relation between the Ordovician brachiopod faunas of northern Estonia, Scandinavia, Bohemia, Britain, and North America. *Stratigraphy of Central European Lower Paleozoic, Int. Geol. Congr., 23rd, Sect. 9*, pp. 21–30

Ross, R. J. Jr. 1951. Stratigraphy of the Garden City Formation in northeastern Utah, and its trilobite faunas. *Peabody Mus. Nat. Hist. Yale Univ. Bull. 6.* 161 pp. + 36 pls.

Ross, R. J. Jr. 1964. Middle and lower Ordovician formations in southernmost Nevada and adjacent California. *US Geol. Surv. Bull. 1180-C.* 101 pp., charts

Ross, R. J. Jr. 1967. Some middle Ordovician brachiopods and trilobites from the Basin Ranges, western United States. *US Geol. Surv. Prof. Pap. 523-D.* 43 pp. + 10 pls.

Ross, R. J. Jr. 1970. Ordovician brachiopods, trilobites, and stratigraphy in eastern and central Nevada. *US Geol. Surv. Prof. Pap. 639.* 103 pp. + 22 pls.

Ross, R. J. Jr. 1975. Early Paleozoic trilobites, sedimentary facies, lithospheric plates, and ocean currents. *Fossils and Strata* 4:307–29

Ross, R. J. Jr. 1976. Ordovician sedimentation in the western U.S.A. See Bassett 1976, pp. 73–105

Ross, R. J. Jr. 1979. Additional trilobites from the Ordovician of Kentucky. *US Geol. Surv. Prof. Pap. 1066-D.* 13 pp. + 6 pls.

Ross, R. J. Jr., Berry, W. B. N. 1963. Ordovician graptolites of the Basin Ranges in California, Nevada, Utah, and Idaho. *US Geol. Surv. Bull. 1134.* 177 pp. + 13 pls.

Ross, R. J. Jr., Dutro, J. T. 1966. Silicified Ordovician brachiopods from east-central Alaska. *Smithson. Misc. Collect.* 149(7):1–22, 3 pls.

Ross, R. J. Jr., Ingham, J. K. 1970. Distribution of the Toquima–Table Head (middle Ordovician Whiterock) faunal realm in the Northern Hemisphere. *Geol. Soc. Am. Bull.* 81(2):393–408, figs. 1–5

Ross, R. J. Jr., Naeser, C. W. 1984. The Ordovician time scale—new refinements. In *Aspects of the Ordovician System, Paleontol. Contrib. Univ. Oslo*, ed. D. L. Bruton, 295:5–10

Ross, R. J. Jr., Naeser, C. W., Izett, G. A., Whittington, H. B., Hughes, C. P., et al. 1978. Fission-track dating of lower Paleozoic volcanic ashes in British stratotypes. *US Geol. Surv. Open-File Rep. 78-701*, pp. 363–65

Ross, R. J. Jr., Nolan, T. B., Harris, A. G. 1980. The upper Ordovician and Silurian Hanson Creek Formation of central Nevada. *US Geol. Surv. Prof. Pap. 1126-C.* 23 pp. + illus.

Ross, R. J. Jr., Naeser, C. W., Izett, G. A., Obradovich, J. D., Bassett, M. G., et al. 1982a. Fission track dating of British Ordovician and Silurian stratotypes. *Geol. Mag.* 119(2):135–53

Ross, R. J. Jr., Adler, F. J., Amsden, T. W., Bergstrom, D., Bergstrom, S. M., et al. 1982b. The Ordovician System in the United States of America, correlation chart and explanatory notes. *Int. Union Geol. Sci. Publ. No. 12.* 73 pp. + 3 sheets

Rozman, Kh. S., Bondarenko, O. B., Minzhin, Ch. 1981. Part 1, Chap. 3. Regional stratigraphic subdivision of Ordovician of Mongolia. *Atlas of Ordovician Faunas of Mongolia. Acad. Sci. U.S.S.R. Trans.* 354. 228 pp. + 48 pls.

Schuchert, C., Cooper, G. A. 1930. Upper Ordovician and lower Devonian stratigraphy and paleontology of Percé, Quebec. *Am. J. Sci., 5th Ser.* 20:161–76, 265–88, 365–93

Shaw, A. B. 1964. *Time in Stratigraphy.* New York: McGraw-Hill. 365 pp.

Sheehan, P. M. 1973. The relation of late Ordovician glaciation to the Ordovician-Silurian changeover in North American brachiopod faunas. *Lethaia* 6(2):147–54

Sheng, S. 1980. The Ordovician System in China, correlation chart and explanatory notes. *Int. Union Geol. Sci. Publ. No. 1.* 7 pp., 6 tables, chart

Shligin, E. D., Bandaletov, S. M., eds. 1976. *Resolution of the Full Boundary Commission for the Establishment of a Unified Stratigraphic Scheme for the Pre-Cambrian and Paleozoic of Eastern Kazakhstan, 1971.* Moscow: Kazak. Minist. Geol. 96 pp. + 10 charts

Skevington, D. 1969. The classification of the Ordovician System in Wales. In *The Pre-Cambrian and Lower Paleozoic Rocks of Wales*, ed. A. Wood, pp. 161–79. Cardiff: Univ. Wales Press. 461 pp.

Spjeldnaes, N. 1981. Lower Paleozoic climatology. In *Lower Paleozoic of the Middle East, Eastern and Southern Africa, and Antarctica*, ed. C. H. Holland, pp. 199–256. New York: Wiley

Stitt, J. H. 1971. Late Cambrian and earliest Ordovician trilobites, Timbered Hills and lower Arbuckle Groups, western Arbuckle Mountains, Murray County, Oklahoma. *Okla. Geol. Surv. Bull. 110.* 83 pp. + 8 pls.

Stitt, J. H. 1977. Late Cambrian and earliest Ordovician trilobites, Wichita Mountains, Oklahoma. *Okla. Geol. Surv. Bull.* 124. 79 pp. + 7 pls., incl. range chart

Strother, P. K., Traverse, A. 1983. Comment on "Caradocian land plant microfossils from Libya." *Geology* 11(5):316–17

Sweet, W. C. 1979a. Late Ordovician conodonts and biostratigraphy of the western Midcontinent province. *Brigham Young Univ. Geol. Stud.* 26(3):46–86, 10 figs.

Sweet, W. C. 1979b. Conodonts and conodont biostratigraphy of post-Tyrone Ordovician rocks of the Cincinnati region. *US Geol. Surv. Prof. Pap. 1066-G*, pp. G1–G26

Sweet, W. C. 1982. Fossils and time: an example from the North American Ordovician. In *Paleontology, Essential of Historical Geology, Proc. Int. Meet., Venice, 1981*, ed. E. Montanaro Gallitelli, pp. 309–21. Modena: STEM Mucchi. 524 pp.

Sweet, W. C., Ethington, R. L., Barnes, C. R. 1971. North American middle and upper Ordovician conodont faunas. *Geol. Soc. Am. Mem.* 127:163–94

Temple, J. T. 1968. The lower Llandovery (Silurian) brachiopods from Keisley, Westmorland. *Paleontogr. Soc. Monogr.* 58 pp. + 10 pls.

Thompson, A. M., Diecchio, R. J. 1982. Taconian clastic facies and clastic wedges in the Appalachian region, U.S.A. *Int.*

Symp. Ordovician Syst., 4th, Paleontol. Contrib. Univ. Oslo 280: 54 (Abstr.)

Toghill, P. 1970. A fauna from the Hendre Shales (Llandeilo) of the Mydrim area, Carmathenshire. *Proc. Geol. Soc. London No. 1663*, pp. 121–29

Ulst, R. J., Gailite, L. K., Yakovleva, V. I. 1982. *The Ordovician of Latvia*. Riga: Minist. Gas Prod. USSR, Mar. Geol. Geophys. 294 pp.

Vail, P. R., Mitchum, R. M. Jr., Thompson, S. III. 1977. Seismic stratigraphy and global changes of sea level. Part 4. Global cycles of relative changes of sea level. *Am. Assoc. Pet. Geol. Mem. 26*, pp. 83–97

VandenBerg, A. H. M. 1981. Victorian stages and graptolite zones. See Webby et al 1981, pp. 2–7, fig. 2

Van Houten, F. B., Bhattacharyya, D. P. 1982. Phanerozoic oolitic ironstones—geologic record and facies model. *Ann. Rev. Earth Planet. Sci.* 10: 441–57

Wang, X., Zeng, Q., Zhou, T., Ni, S., Xu, G., Li, Z. 1982. The Ordovician-Silurian boundary in the eastern Yangtze Gorges, China. *Int. Symp. Ordovician Syst., 4th, Palaeontol. Contrib. Univ. Oslo* 280: 57 (Abstr.)

Webby, B. D. 1978. History of the Ordovician continental platform and shelf margin of Australia. *J. Geol. Soc. Aust.* 25(1): 41–63

Webby, B. D., VandenBerg, A. H. M., Cooper, R. A., Banks, M. R., Burrett, C. F., et al. 1981. The Ordovician System in Australia, New Zealand and Antarctica, correlation chart and explanatory notes. *Int. Union Geol. Sci. Publ. No. 6.* 64 pp. + 4 figs., chart

Whittington, H. B. 1953. A new Ordovician trilobite from Florida. *Breviora, Harvard Mus. Comp. Zool.* 17: 1–6

Whittington, H. B. 1966a. Phylogeny and distribution of Ordovician trilobites. *J. Paleontol.* 40: 696–737

Whittington, H. B. 1966b. Trilobites of the Henllan Ash, Arenig Series, Merioneth. *Bull. Br. Mus. (Nat. Hist.) Geol.* 11: 489–505

Whittington, H. B. 1972. Scotland. See Williams et al 1972, pp. 49–53, fig. 9

Whittington, H. B., Hughes, C. P. 1972. Ordovician geography and faunal provinces deduced from trilobite distribution. *Philos. Trans. R. Soc. London Ser. B* 263: 235–78

Whitworth, P. H. 1969. The Tremadoc trilobite *Pseudokainella impar* (Salter). *Palaeontology* 12(3): 406–13, pl. 75

Williams, A. 1948. The lower Ordovician cryptolithids of the Llandeilo district. *Geol. Mag.* 85: 65–88

Williams, A. 1949. New lower Ordovician brachiopods from the Llandeilo-Llangadock district. Parts I and II. *Geol. Mag.* 86: 161–74, 226–38

Williams, A. 1953. The geology of the Llandeilo district, Carmarthenshire. *Q. J. Geol. Soc. London* 108: 177–208

Williams, A. 1972a. Introduction and general aspects of correlation. See Williams et al 1972, pp. 1–10

Williams, A. 1972b. Ireland. See Williams et al 1972, pp. 53–59

Williams, A., Strachan, I., Bassett, D. A., Dean, W. T., Ingham, J. K., et al. 1972. A correlation of Ordovician rocks in the British Isles. *Geol. Soc. London Spec. Rep. No. 3.* 74 pp.

Williams, H., Hatcher, R. D. Jr. 1982. Suspect terranes and accretionary history of the Appalachian orogen. *Geology* 10: 530–36

Wood, A. 1969. *The Pre-Cambrian and Lower Paleozoic Rocks of Wales*. Cardiff: Univ. Wales Press. 461 pp.

Wright, T. O., Ross, R. J. Jr., Repetski, J. 1984. New youngest Cambrian or earliest Ordovician marine fossils from northern Victoria Land, Antarctica. *Geology*. In preparation

Yin, C.-H., Kuo, S.-T. 1978. Stratigraphy of the Mount Jolmo Lungma and its north slope. *Sci. Sin.* 21(5): 629–44

Ziegler, A. M., Scotese, C. R., McKerrow, W. S., Johnson, M. E., Bambach, R. K. 1979. Paleozoic paleogeography. *Ann. Rev. Earth Planet. Sci.* 7: 473–502

Ann. Rev. Earth Planet. Sci. 1984. 12: 337–57

RHEOLOGICAL PROPERTIES OF MAGMAS

Alexander R. McBirney

Department of Geology, University of Oregon, Eugene, Oregon 97403

Tsutomu Murase

Department of Physics, Institute of Vocational Training, 1960 Aihara, Sagamihara 229, Japan

INTRODUCTION

Recent studies of igneous rocks have taken a refreshing new direction, mainly as the result of a greater awareness of the important role played by physical properties of magmas in determining the eruptive behavior and compositional variations of volcanic rocks. This review summarizes the present state of knowledge, along with some of the recent rheological studies having a direct bearing on interpretations of volcanic phenomena and processes of crystallization and differentiation of shallow magmatic intrusions.

The rheological properties of magmas reflect the inherent structures of molten silicates and will not be thoroughly understood until more is known about the basic nature of silicate liquids. Recent progress in this field has been ably reviewed by Hess (1980) and needs no elaboration here aside from mention of the obvious need for better laboratory measurements in order to place greater constraints on interpretations of silicate melts. For petrogenetic and volcanological processes, the most conspicuous need is for improved data on rheological properties, chiefly viscosity and yield strength, and for more precise equations for predicting these properties under a wide range of natural conditions.

Viscosity, η, is usually defined as the ratio of shear stress to strain rate and is expressed in units of poise (g cm^{-1} s^{-1}) or Pascal seconds (10 poise = 1.0 Pa s). In more general terms, it is the coefficient for transfer of

0084–6597/84/0515–0337$02.00

momentum in the equation

$$\tau = \tau_0 + \eta \left(\frac{du}{dy}\right)^n, \tag{1}$$

in which τ is the shear stress applied in the direction x parallel to the plane of flow, τ_0 is the minimum stress required to initiate permanent deformation, du/dy is the velocity gradient normal to the plane of shear, and n is a constant with a value of one or less. A fluid having no yield stress ($\tau_0 = 0$) and a direct linear relation between shear stress and rate of strain (i.e. $n = 1$) is said to be Newtonian. Many complex fluids, suspensions, and emulsions are non-Newtonian in the sense that the relation between strain rate and stress is nonlinear, and in some, viscous flow takes place only when the shear stress exceeds a finite value. Although, for convenience, magmas are commonly treated as Newtonian fluids, most are not. Below their liquidus, particularly when they are charged with phenocrysts, their rate of shear is not directly proportional to stress, i.e. $n < 1.0$ in Equation (1).

In 1967, Robson noted that some lavas have a finite yield strength ($\tau_0 > 0$) that must be exceeded for the lava to move under the force of gravity (Robson 1967). Non-Newtonian behavior of basaltic magma was demonstrated by Shaw et al (1968) in studies of the basalts of Hawaiian lava lakes and by Gauthier (1973) and Pinkerton & Sparks (1978) at Mount Etna. Subsequent workers (e.g. Hulme 1974, Sparks et al 1976, Moore et al 1978, Peterson & Tilling 1980, Cigolini et al 1983, Murase et al 1984) have shown how many morphological features of lavas can be explained in terms of these properties.

Yield strength is not strictly a threshold value below which a magma does not deform under any condition. Elastic deformation may result from any stress, with a value either greater or less than the yield strength. It differs from viscous flow in that it is recovered when stress is removed. Small stresses may cause permanent deformation in the form of creep. Thus, the yield strength of magma is more properly thought of as a transition value below which the dominant mechanism of deformation is not viscous flow but another process that, though slower, is still capable of producing permanent strain.

The different modes of permanent deformation within the ranges of temperature and stresses that prevail in crystallizing magmas can be illustrated by means of a "deformation map" of the kind devised by Ashby (1972). Figure 1 is a preliminary attempt to construct such a diagram on the basis of the sparse data now available (H. J. Melosh and A. R. McBirney, unpublished data). The fields for the dominant mechanisms of deformation are shown for different regimes of temperature and deviatoric stress; estimated strain rates are shown as contours crossing the fields. Such a

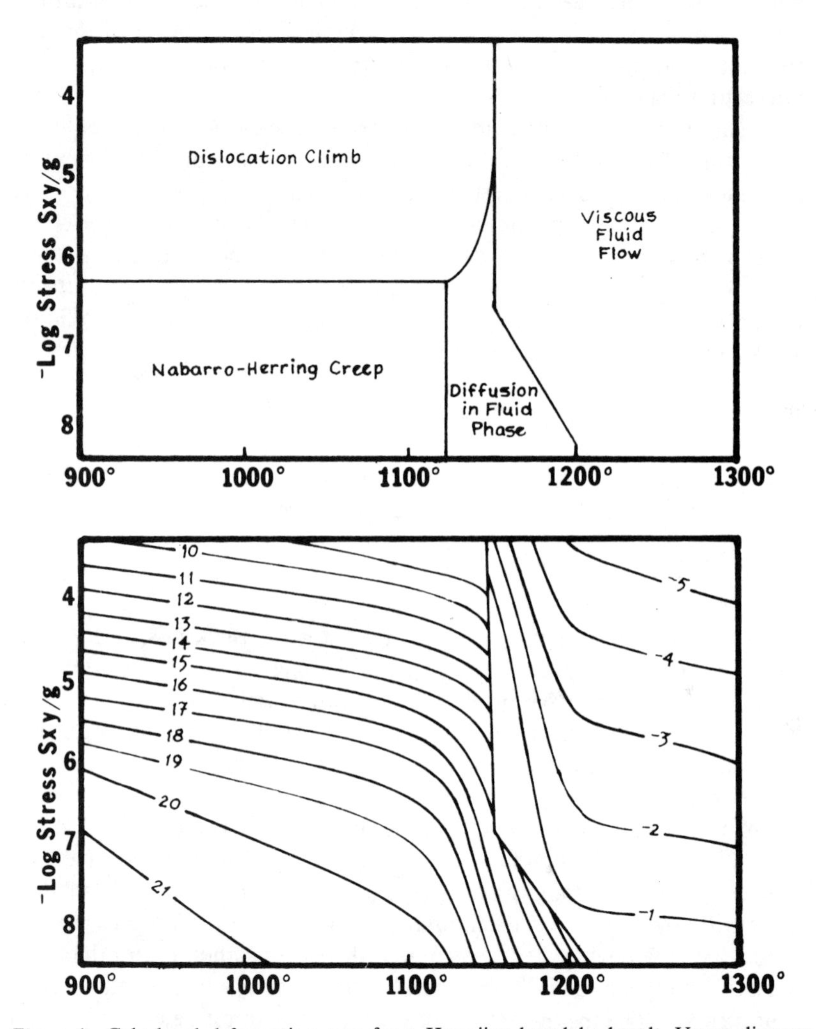

Figure 1 Calculated deformation map for a Hawaiian lava-lake basalt. Upper diagram displays the dominant mechanism of deformation in various ranges of temperature and stress. The lower diagram shows strain-rate contours in negative exponential values (cm s^{-1}). The solidus temperature is assumed to be 1100° and the liquidus 1200°C. Calculations were carried out by Jay Melosh on the basis of properties measured by T. Murase and A. R. McBirney (unpublished data). The method of calculation is similar to that of Stocker & Ashby (1973) and is designed to illustrate the forms and rates of deformation under differing deviatoric stresses over a wide range of temperatures.

diagram, if it were based on reliable data, could provide valuable insights into a number of geological processes occurring within the transition from the liquid to the solid state. Unfortunately, the details of the most interesting regions are poorly known.

In summarizing recent field and laboratory studies of the rheological properties of natural silicate melts, we treat viscosity and yield strength separately, but the two properties are closely related and are only two parts of the much broader phenomenon of deformation of igneous rocks and melts at high temperature. Although we are not concerned here with the behavior of rapidly cooled magmas and glasses, we draw attention to the important work being done in this field, most notably by Ryan and co-workers (e.g. Ryan & Sammis 1981).

FIELD MEASUREMENTS

Viscosities have been estimated from the velocities of moving lavas by using Jeffery's equation for laminar flow of an extensive sheet on a uniform slope:

$$\eta = \frac{g\rho \sin \alpha d^2}{3V} \tag{2}$$

where ρ is the density, α is the slope angle, d is the thickness, and V is the mean velocity. For flow in restricted channels, the coefficient in the denominator is replaced by 4. This equation yields only crude values because it takes no account of the fact that most of the velocity gradient in lavas may be in a relatively thin basal layer and that viscosity can vary greatly between the base, interior, and crust of the flow.

Most, if not all, lavas move by laminar flow (Einarsson 1949, Booth & Self 1973); no well-documented examples of turbulent flow have been reported, although some have been postulated for lunar basalts. Many lavas with low viscosities are thought to have a motion similar to that of a caterpillar tread, with shear distributed throughout their entire thickness, but others appear to slide on a basal layer in which most of the shear is concentrated. The mean velocity of such a flow cannot be estimated from observations of the surface. An additional problem arises in the case of non-Newtonian lavas in which the apparent viscosity may vary with shear rate. At low velocities and shear rates, the effective viscosity may be greater than it is at high velocities because the ratio of shear rate to shear stress is not linear.

The most reliable field data for lavas are those obtained from measurements made over a range of temperatures and shear stresses in the interiors of flows. Excellent opportunities for measurements of this kind have been provided by recent eruptions of the Hawaiian volcano, Kilauea, where basalt has ponded in deep pit craters and remained molten for periods of a decade or more. Viscosity is measured by drilling through the crust, inserting a vane-shear apparatus directly into the interior, and measuring the torque required for a constant rate of rotation (Shaw et al 1968).

Yield strength can be estimated from the angle of repose of flow fronts and the minimum slope on which lavas continue to move (Johnson 1970, Hulme 1974). Values obtained by these methods are useful for interpreting the properties of lavas long after they have come to rest, but they are subject to uncertainty owing to the difficulty of evaluating temperatures and time-dependent factors. The measurements may be relevant only to the crust and not to the body as a whole. Again, the best field data are those obtained from ponded lavas, such as those of the Hawaiian lava lakes. Yield strength can be estimated with the same apparatus used to measure viscosity. The value corresponds to the torque required to initiate rotation of a spindle inserted below the crust. More recently, Pinkerton & Sparks (1978) developed a portable instrument that can be used to obtain measurements from flowing lava. Measurements are made by determining the force required to push a cylindrical steel rod into the molten interior of the flow. Yield strengths measured in this way were found to be less than half those estimated from the morphology of flow fronts.

At best, however, the results of field measurements are subject to uncertainties arising mainly from the problem of taking into account all the various factors that affect rheological properties under natural conditions. Most serious, perhaps, are time-dependent factors, which may lead to widely differing values for both viscosity and yield strength, depending on how long the lava has been undisturbed before measurements are made.

LABORATORY MEASUREMENTS

Three methods have been used for laboratory studies of the rheological properties of natural silicate melts. Rotational viscometers of the kind used in the glass industry are satisfactory for measurements at large stresses and strain rates requiring only low precision. For lower stresses and strain rates, the counterbalanced sphere method of Shartsis & Spinner (1959) [Figure 2 (*left*)] is more satisfactory, especially if strain rates are measured electronically (Murase et al 1984). Densities can be measured simultaneously in the same experiment by determining the weight of the platinum sphere while suspended in the molten sample (MacKenzie 1956).

Use of the counterbalanced sphere technique is limited to samples prepared from powdered rock or glass and with low yield strengths and viscosities (below about 10^5 dyn cm^{-2} and 10^8 poise, respectively). Samples with more than 20 or 30% crystals and with high viscosities and yield

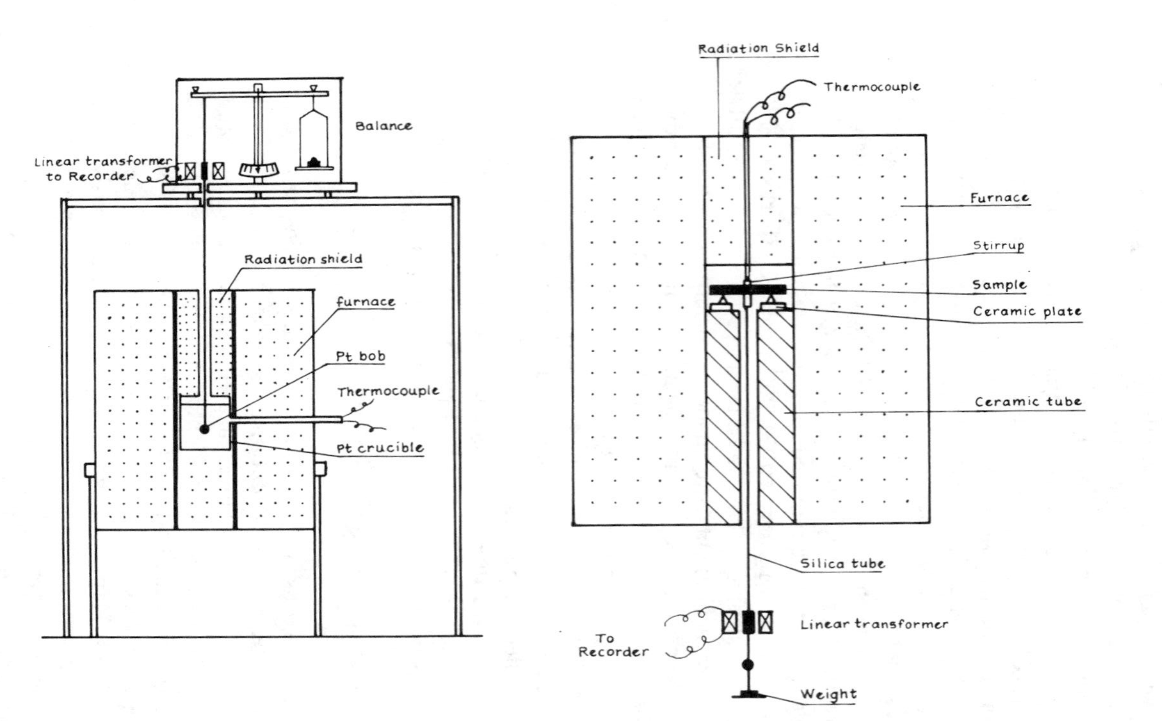

Figure 2 Two types of apparatus used for rheological measurements in the laboratory. (*Left*) In the counterbalanced sphere method, a platinum bob is suspended in the molten sample and balanced by weights on the opposing arm of an analytical balance. A platinum rod between the balance and the wire suspending the sphere in the sample reduces springlike effects in the suspension system. Displacement of the sphere is measured by a linear transformer connected through an amplifying system to a chart recorder. Corrections are made for the effect of surface tension on platinum wire by measuring the stress on the wire when suspended in the sample without a sphere attached. Loss of iron to the crucible and sphere is minimized by prior saturation of the platinum. The atmosphere of the furnace is controlled by mixtures of CO and CO_2 monitored by a zirconia cell. (*Right*) The deflected-beam method utilizes the same basic furnace to measure the bending of a slab of precisely cut rock supported between two ceramic prisms. Stress is transmitted to the center of the beam by means of a system of weights and a silica glass stirrup. As before, displacement is measured by a linear transformer, and the atmosphere of the furnace is regulated by a mixture of CO and CO_2.

strengths can be studied by observing the rate of bending of a precisely cut slab under a stress applied to its center [Figure 2 (*right*)].

In order to study the behavior of crystals with densities close to the liquid in which they are suspended, Roeder & Dixon (1977) developed an ingenious centrifuge furnace capable of generating forces equivalent to 700 times the acceleration of gravity.

These various techniques can be used in controlled atmospheres, but only at low pressures; more elaborate methods must be used at elevated pressures and volatile contents. Measurements of the latter type have been carried out by Shaw (1963), Burnham (1963), Carron (1969), Scarfe (1973), Kushiro et al (1976), and Khitarov et al (1978). Most of these studies have used a falling-sphere technique, in which the sample, enclosed in a small capsule, contains a platinum sphere or round mineral grain of known size and density, which is allowed to settle through the liquid while the sample is held at an elevated temperature and pressure.

VISCOSITY

Viscosity is sensitive to a variety of factors, including composition, temperature, pressure, volatile content, abundance and size of crystals, and thermal and mechanical history of the melt. Many of these factors are interrelated, and their individual effects are not easy to unravel.

Effects of Temperature and Composition

Typical viscosities for some common igneous compositions are shown in Figure 3. As in all determinations for silicate melts above their liquidus, these viscosities have an inverse logarithmic variation with temperature according to a relation having the form

$$\eta = A_\eta \exp E_\eta/RT, \tag{3}$$

where A_η is a constant, E_η the activation energy of viscous flow, and R the gas constant. Activation energy seems to be related to the proportion of bridging silicon and aluminum atoms as expressed by the ratio (Si + Al)/O for the melt (Euler & Winkler 1957), but only at ratios greater than about 0.4 (Murase & McBirney 1973). Structural differences reflecting the proportions of bridging and nonbridging silicon and aluminum atoms seem to provide the best empirical correlation between composition and viscosity (Murase 1962), but no simple rule has been found to account for the differing effects of individual network-modifying elements, such as Mg, Fe, Ca, K, and Na, each of which tends to lower viscosity but by slightly different amounts.

Viscosities of crystal-free liquids can be calculated from the empirical

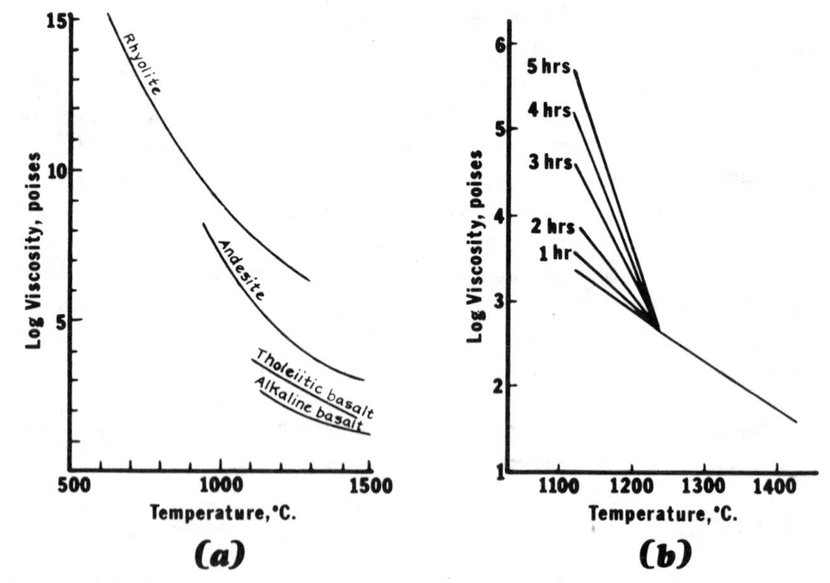

Figure 3 (*a*) Viscosities of some common igneous rocks at temperatures in and above their melting ranges (after Murase & McBirney 1973). (*b*) The viscosities shown in (*a*) are metastable below the liquidus temperatures of the respective samples. The increase that comes with time as crystals nucleate and grow is illustrated by the behavior of the same tholeiitic basalt shown in (*a*).

equations of Shaw (1972) and Bottinga & Weill (1972). Values predicted by these equations are in close accord with laboratory measurements, at least for low pressures and water contents and for temperatures above the liquidus.

Effects of Water and Load Pressure

Shaw (1963), Burnham (1963), and Carron (1969) obtained data on the effect of water on the viscosities of rhyolitic liquids, and Scarfe (1973) and Khitarov & Lebedev (1978) collected similar data for basaltic compositions (Figure 4). At constant temperature, one weight percent of water lowers the viscosity of silica-rich liquids by nearly an order of magnitude, but the effect is much less pronounced in liquids of lower silica content, probably because the polymerization of mafic melts is already low and cannot be substantially reduced by the addition of water. The effects of other volatile components, such as CO_2 and S, have yet to be explored.

Kushiro (1980) has reviewed the work that he and others have done on the effects of load pressure on the viscosity and other properties of natural silicate melts. Load pressure has been found to reduce the viscosity of basaltic and andesitic melts at constant temperatures above the liquidus. A tholeiitic basalt was found to fall from about 170 poise at atmospheric

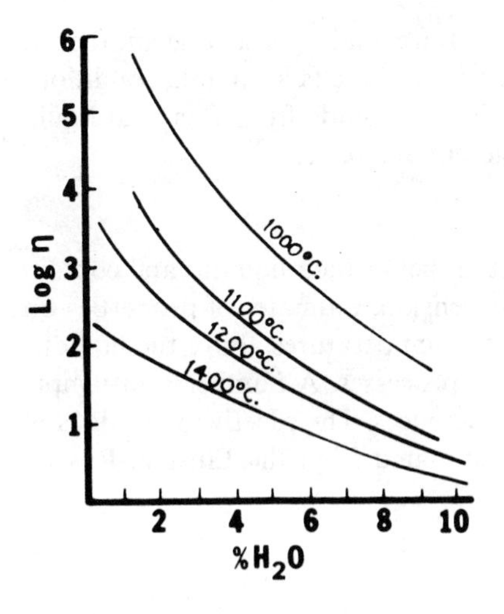

Figure 4 The effect of water on the viscosity of basaltic liquids at various temperatures (after Khitarov & Lebedev 1978).

pressure to 40, 25, and 8 poise at 15, 20, and 30 kbar, respectively. Corresponding values for an andesite from Crater Lake were about 3200 and 900 poise at 7.5 and 20 kbar (Figure 5). Since pressure also has an effect on the liquidus temperature, and since viscosity is a function of temperature

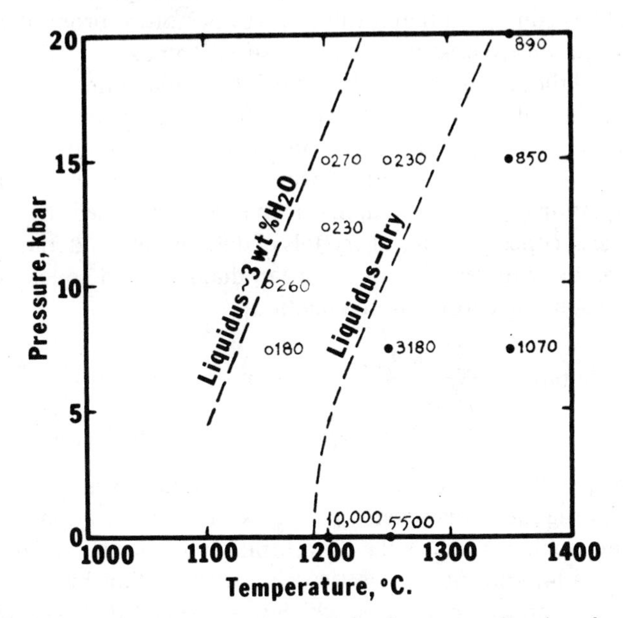

Figure 5 Viscosity of an andesite from Crater Lake, Oregon, as a function of pressure and temperature above the liquidus. Data points shown by filled circles are for water-free conditions; open circles denote samples with about 3 wt% H_2O. Corresponding liquidus curves are shown by broken lines [from data of Kushiro et al (1976) and Kushiro (1978)].

above the liquidus, the two effects of pressure tend to offset one another. It appears likely, however, that the viscosities of melts at mantle conditions are low enough to facilitate segregation of liquids from a crystal-liquid assemblage under relatively small differential stresses.

Effects of Crystal Content

Because most magmas have temperatures below their liquidus and contain differing proportions of crystals in suspension, estimates of properties on the basis of measurements carried out at temperatures above the liquidus have only limited relevance to natural processes. A number of attempts have been made to correct for these differences. The effective viscosities of crystal-rich magmas are commonly estimated from the Einstein-Roscoe equation

$$\eta = \eta_0(1 - R\phi)^{-2.5}, \tag{4}$$

in which η is the effective viscosity of a liquid containing a volume fraction Φ of suspended solids, and η_0 is the viscosity of the liquid alone. The constant R is based on the volumetric ratio of solids at maximum packing and is usually taken as 1.35 for spheres of uniform size or as 1.0 for crystals with serial sizes. Marsh (1981) has considered the packing ratios of crystals in lavas and has concluded that a value of 1.67 is more appropriate.

The large differences between measured viscosities of crystal-rich magmas and those predicted by Equation (4) are illustrated by the data for Hawaiian tholeiitic basalts and the dacitic dome of Mount St. Helens (Figure 6). The discrepancies are attributable in part to the sizes and high concentrations of crystals, both of which exceed the range of values for which Equation (4) was derived. Sherman (1968) examined the effects of a wide range of concentrations of crystals of differing size and found that the relative viscosity increases with the mean diameter and concentration of suspended solids according to the relation

$$\ln \eta_R = \ln(\eta/\eta_0) = \frac{\alpha D_m}{\left(\dfrac{\Phi_{max}}{\Phi}\right)^{1/3} - 1} - 0.15, \tag{5}$$

in which η_0 is the viscosity of the crystal-free liquid, α is a constant that varies with the mean diameter D_m, Φ_{max} is the concentration of solids at maximum packing, and Φ is the concentration of suspended particles.

Equations (4) and (5) assume that the solid fraction has no physical-chemical relation to the liquid in which it is suspended. This is obviously not the case for magmas if crystals are nucleating and growing during cooling. Under such conditions, the only limit of Φ_{max} is 1.0, when the liquid is totally crystallized at the solidus. As Φ increases, the mean diameter D_m also changes. It does not necessarily increase with cooling; in lavas the opposite is more likely to be the case because large numbers of small crystals growing from the groundmass more than offset the slight increase in size of phenocrysts.

An even greater effect arises from the fact that the composition of the liquid is not independent of Φ but changes at an accelerating rate with progressive crystallization. Thus, as the composition of the liquid fraction changes, so too does η_0, and the magnitude of this effect may be much greater than that of the suspended crystals alone.

We know of no simple way of integrating these effects, other than by empirically matching the physical constants in Equation (5) to experimentally determined values of η_0, Φ, and D_m. If the groundmass composition of the sample is determined, either by a mass-balance calculation or direct analysis, η_0 can be calculated from the equations of Shaw (1972). The value of D_m can be estimated petrographically, and Φ can be determined in the same way or by mass balance if the compositions of all phases are known.

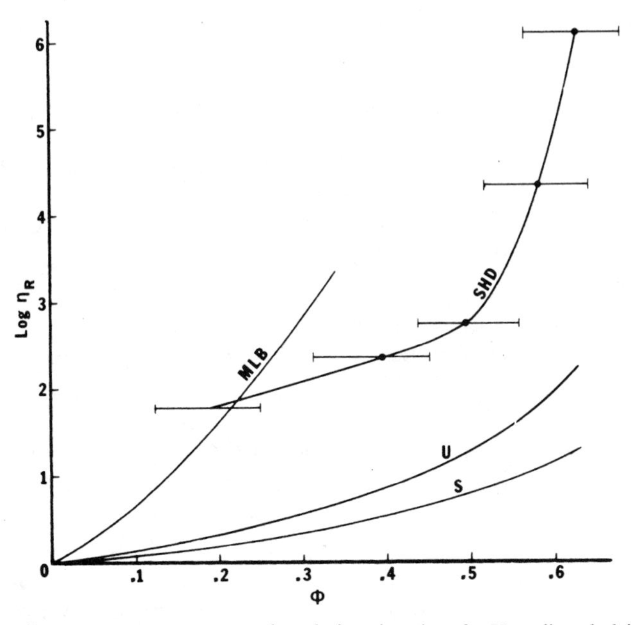

Figure 6 Effects of crystal content on the relative viscosity of a Hawaiian tholeiitic basalt (MLB) and a Mount St. Helens dacite (SHD). Relative viscosity is defined as the ratio of the viscosity of the liquid-crystal suspension to that of the crystal-free liquid at the same temperature. Curves labeled U and S are calculated for spheres of uniform and serial sizes, respectively, using Equation (4) [after Shaw (1969, Figure 4) and Murase et al (1984)].

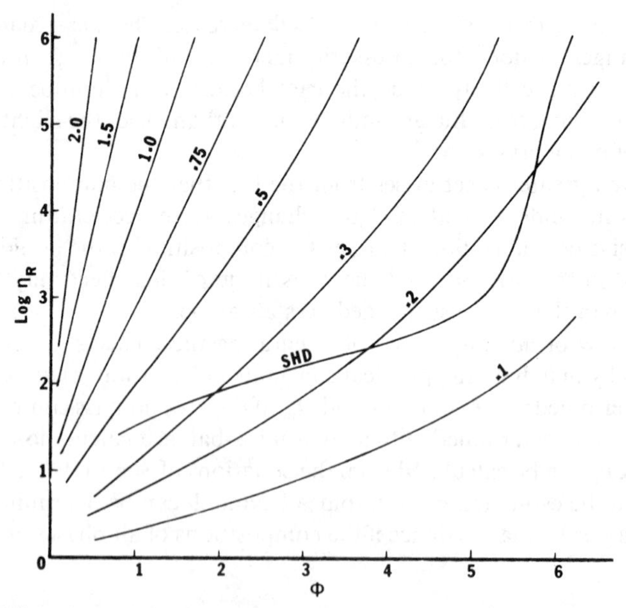

Figure 7 Effect of crystal size on the relative viscosity of a Mount St. Helens dacite. Sloping lines indicate mean diameter in millimeters. Curve labeled SHD is plotted from measured viscosities, crystal contents, and compositions of glass. It forms the basis for values used in Equation (5) and the computer program in the Appendix (data from Murase et al 1984).

Using data of these kinds for the dacite of Mount St. Helens and a Hawaiian tholeiitic basalt, we find that the best agreement with measured viscosities is obtained when $\alpha = 0.011$ and $\Phi_{max} = 1.0$. If these values are substituted into Equation (5), a graphical plot (Figure 7) can be constructed for the relative viscosity (Log η_R) as a function of the proportions (Φ) and mean diameter of crystals (in mm). A computer program for calculating effective viscosities of magmas at subliquidus temperatures is given in the Appendix.

Time-Dependent Factors

Most measurements of viscosity, either in the field or in the laboratory, show an increase of viscosity as the melt remains at subliquidus temperatures for increasing periods of time (Figure 8*a*). Most of this increase is attributable to the growth of crystals, but part may be due to the ordering and polymerization of the liquid. Samples that have remained undisturbed for many hours may revert to a lower effective viscosity if they are stirred or otherwise disturbed. In a similar way, apparent viscosities of many crystal-rich silicate liquids tend to be lower at high rates of shear. Most of this difference is due to the non-Newtonian nature of the sample, which can only be determined if rates of strain are measured at more than one rate of shear.

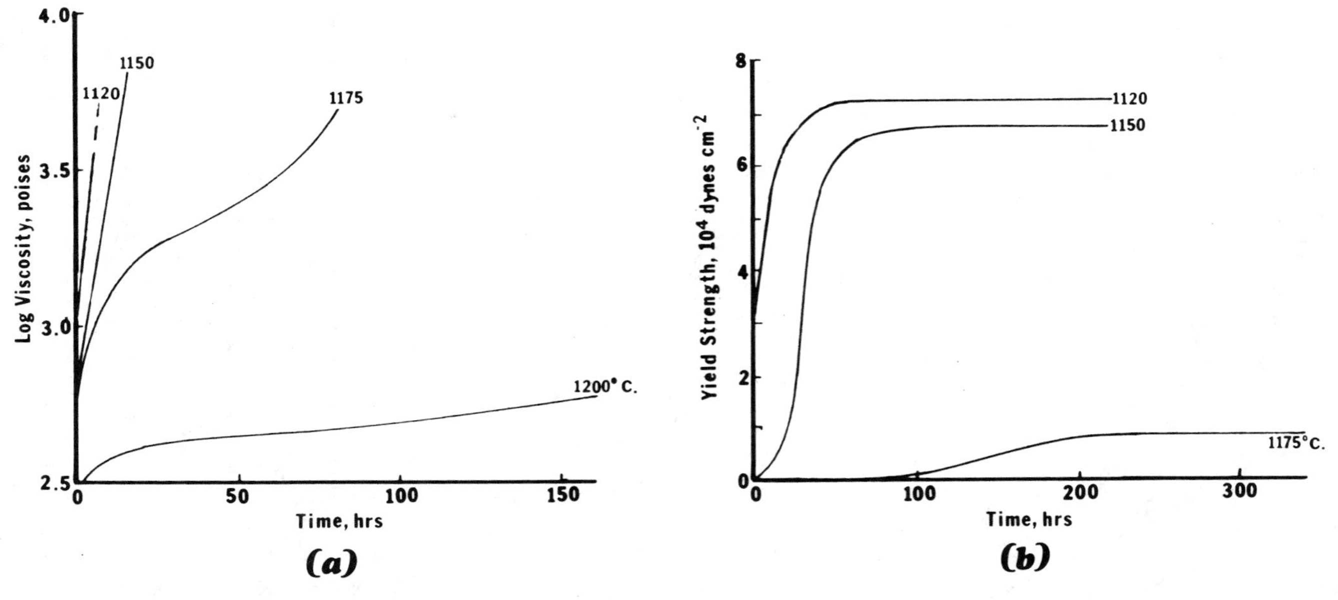

Figure 8 (*a*) A sample that was initially at a higher temperature increases in viscosity with time, as shown by the behavior a Hawaiian tholeiitic basalt from Makaopuhi lava lake. Most of the increase is due to crystallization, but a similar effect is observed after the melt is strongly stirred at constant temperature. (*b*) The same sample shows a similar increase in yield strength with time. The measurements were made in air; reliable values for yield strength have not yet been obtained under reducing conditions, but preliminary results indicate that they are much lower.

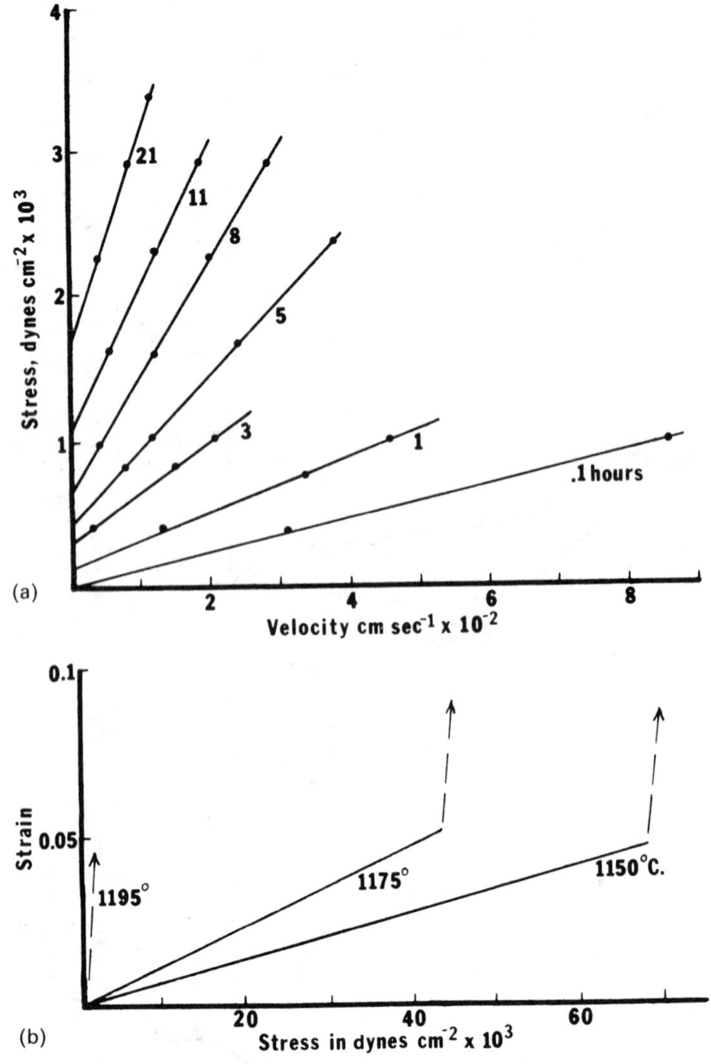

Figure 9 Yield strength can be determined by (*a*) measuring the slope of a curve for velocity of the ascending sphere as a function of the applied stress and then extrapolating to a velocity of zero to find the minimum stress at which motion begins. Alternatively, the elastic strain at low stresses (*b*) can be measured to find the stress at which the elastic limit is exceeded and permanent strain begins. In the latter method, each measurement of strain for a given stress produces a characteristic strain-time (*c*). The effect of surface tension (σ_s) is determined without a sphere attached to the wire and is subtracted to obtain the stress required to produce strain in the interior of the liquid. At low stresses (below the yield strength), elastic strain results when stress is applied, and is recovered when the stress is removed. Stresses greater than the yield strength result in viscous flow and strain that is not recovered when the stress is removed.

YIELD STRENGTH

Two methods have been used to estimate yield strength from laboratory measurements. By measuring strain rates as a function of stress and plotting one aginst the other (Figure 9*a*), one can obtain a curve that extrapolates to the stress axis to give the minimum stress required to produce measurable viscous flow. This intercept corresponds to τ_0 in Equation (1). In the second method, strain is measured as a function of stress (Figure 9*b*). Elastic strain does not increase with time after the initial loading, and it is recovered when stress is removed. At stresses greater than the yield strength, however, viscous flow results in permanent strain that continues to increase with time. A series of experiments can be run at progressively greater stresses to find where this transition occurs (Figure 9*c*).

Most laboratory measurements of yield strength show a marked increase as temperature falls below the liquidus and crystallinity increases. No unambiguous evidence has yet been reported that crystal-free liquids above their liquidus temperatures have measurable yield strengths. In the critical range of temperatures between the liquidus and solidus in which the liquid is saturated with one or more crystalline phases, the bulk crystal-liquid assemblages of most natural compositions have a yield strength that increases with both time and increasing crystallinity (Figure 8*b*).

Hulme (1974) has shown that the apparent yield strength estimated from the morphologies of arrested lava flows increases with silica content (Figure 10), but the relationship is ambiguous because the estimates do not provide a way of separating the effects of silica content from those of temperature and crystal content (Murase et al 1984). The relationship may be similar to

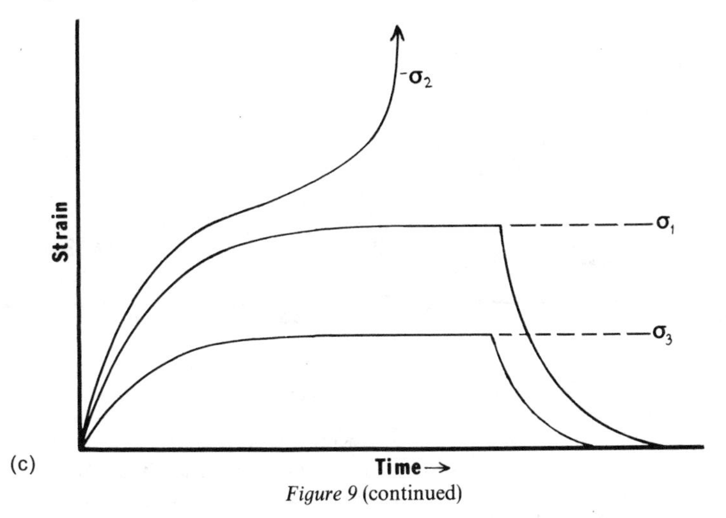

Figure 9 (continued)

that found by Marsh (1981), who noted that beyond a certain limiting viscosity the rates of cooling and crystallization are rapid compared with the rate of flow, and the magma is simply demobilized by its slow response to shear stress.

Laboratory measurements show less correlation of yield strength with silica content than with the amount of cooling below the liquidus. If silica content is a factor, it cannot be the only one, because silica-rich liquids with cation ratios of alkalies to alumina greater than 1.0 have little or no yield strength, even at temperatures well below their liquidi (T. Murase and A. R. McBirney, unpublished data). Iron may also be an important factor, particularly at a high degree of oxidation. Exploratory experiments we have carried out on basalts show that samples having high yield strengths in air have much lower values under reducing conditions. We have no explanation for this difference.

No simple relation has yet been found between yield strength and the

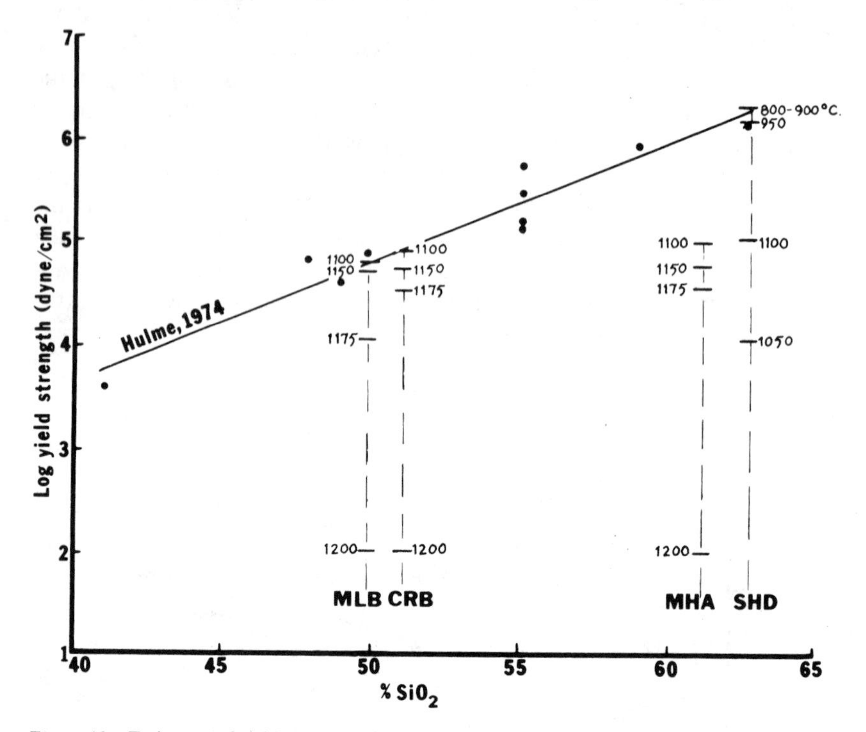

Figure 10 Estimates of yield strength of lavas of different silica contents. The sloping line has been defined by Hulme (1974) from the morphologies of arrested lava flows (solid dots). The last point at the high-SiO_2 end of the line is based on the shape of the dome of Mount St. Helens. Values for four rocks studied in the laboratory (MLB, CRB, MHA, SHD) are plotted as functions of temperature to show the increase of yield strength and the approach to Hulme's line as temperature falls and crystallization advances.

volumetric proportions of crystals. A basic question still to be resolved is whether any part of the yield strength of a crystallizing silicate is inherent in the saturated liquid and not a bulk property of the crystal-liquid suspension. The magnitude of values measured in complex alumina-silicates containing only a few percent of small crystals suggests an extensive polymerization of the liquid phase under saturated conditions. None of the techniques developed thus far are able to make this distinction.

NEEDS FOR FURTHER RESEARCH

Much remains to be done in order to quantify the relations between rheological properties of natural silicates and such factors as composition, pressure, crystal content, shear rate, and a variety of time-dependent factors. Next to nothing is known about rates and mechanisms of deformation at very low long-term stresses, and little has been done on the effects of nonuniform temperature distributions and the simultaneous transfer of heat and momentum. The early studies by Shaw (1969) of viscous heating and mechanisms of thermal feedback are of unusual geological interest but have never been expanded, mainly because of the serious experimental difficulties they entail.

Rheological studies are certain to advance rapidly as experimental techniques are improved and more detailed measurements are made, particularly in the range of conditions between the liquidus and solidus temperatures of natural silicates. Few fields of geological research hold greater potential for enhancing our understanding of basic magmatic and tectonic processes.

Acknowledgments

Grants to the senior author by NASA and the National Science Foundation (No. EAR-7909458) are gratefully acknowledged. R. S. J. Sparks and H. E. Huppert reviewed the manuscript.

Appendix

Calculation of effective viscosity of a subliquidus magma Knowing the proportions and size of phenocrysts, it is possible to calculate the effective viscosity of a magma by means of relations outlined in the text, provided the composition of the groundmass liquid can be determined. The following program is designed to do this by first estimating the compositions of crystals that would form in a liquid having the bulk composition of the magma and then by subtracting these crystals to obtain the composition of the remaining liquid.

Volumetric proportions of crystals are converted to weight proportions by use of simple density relations. The density values used to do this need not be precise because the calculation of viscosity is much less sensitive to this factor than to others that are much more loosely constrained.

Densities and compositions of crystals in equilibrium with the liquid are estimated using the equilibrium coefficients compiled by Nathan & Van Kirk (1978) with subsequent modifications (H. D. Nathan, personal communication). Plagioclase is assumed to have perfect normal zoning, with a core in equilibrium with the initial liquid and subsequent growth reflecting the changing composition of the liquid as increments of 0.1 vol% are crystallized. The loop used to achieve this effect prolongs the run-time, and if less precision is needed, the growth step can be made larger by increasing the value of the step in line 1050. All other minerals are assumed to be unzoned, with compositions in equilibrium with the original (bulk) composition.

The viscosity of the remaining (groundmass) liquid is calculated using Shaw's (1972) equations and then corrected for the effect of suspended solids by using the equation of Sherman (1968), given as Equation (5) in the text. Constants for the Sherman equation were selected to give a result conforming to experimental measurements of the dacite of Mount St. Helens and a Hawaiian tholeiitic basalt.

The program is written in BASIC for an Apple computer and, in the interest of clarity, is given here in a more expanded form than necessary. A more condensed version can be obtained from either of the authors.

Table 1 Program to calculate effective viscosity of subliquidous magmas

```
100   REM : CALC VISC OF SUBLIQUIDUS MAGMAS
110   DEF  FN A(X)  = ( INT (X * 100 + .5)) / 100
130   HOME : PRINT "ENTER OXIDES IN WT.%"
140   PRINT : INPUT "    SIO2    ";SI
150   INPUT "    TIO2    ";TI
160   INPUT "    AL2O3   ";AL
170   INPUT "    FE2O3   ";F3
180   INPUT "    FEO     ";F2
190   INPUT "    MNO     ";MN
200   INPUT "    MGO     ";MG
210   INPUT "    CAO     ";CA
220   INPUT "    NA2O    ";NA
230   INPUT "    K2O     ";K
240   INPUT "    P2O5    ";P
250   INPUT "ENTER H2O FOR VISCOSITY: ";H
255 DL = 2.2 + .027 * (F3 + F2 + TI): REM : APPROX DENSITY OF LIQUID
260 T = (SI + TI + AL + F3 + F2 + MN + MG + CA + NA + K + P) / (100 - H)
300 SI = SI / T / 60.0843
310 TI = TI / T / 79.8988
320 AL = AL / T / 50.9806
330 F3 = F3 / T / 79.8461
340 F2 = F2 / T / 71.8464
350 MN = MN / T / 70.9374
```

Table 1 (*continued*)

```
360 MG = MG / T / 40.3044
370 CA = CA / T / 56.0794
380 NA = NA / T / 30.9895
390 K = K / T / 47.0980
400 H = H / T / 18.0152
410 P = P / T / 70.9723
420 FM = F3 + F2 + MN + MG
430 TN = SI + TI + AL + FM + CA + NA + K + P + H
440  INPUT "ENTER MEAN DIAM OF XLS IN MM.: ";DM:DM = DM * 1000
500  PRINT "ENTER VOL. FRACTION OF:": PRINT
510  INPUT "    PLAGIOCLASE:    ";PF
520  INPUT "    K-FELDSPAR:     ";KF
530  INPUT "    OLIVINE:        ";OL
540  INPUT "    AUGITE:         ";AU
550  INPUT "    HYPERSTHENE:    ";HY
560  INPUT "    MT-ILM:         ";MT
570  INPUT "    QUARTZ:         ";QZ
580 PH = OL + PF + KF + AU + HY + MT + QZ
590  INPUT "ENTER TEMP, DEG C: ";TE:TE = TE + 273.15
670  IF OL = 0 THEN 760
680 MF = 3.33 * MG / F2: REM   MG/F2 OF CRYSTALS
690 DO = 4.39 - 1.17 * MF / (1 + MF):OL = OL * TN * DO / DL
700 SS = SS + OL * .333
710 SF = OL * .667
760  IF HY = 0 THEN 850
770 MF = 4.15 * MG / F2: REM   MG/F2 OF CRYSTALS
780 DH = 3.96 - .75 * MF / (1 + MF):HY = HY * TN * DH / DL
790 SS = SS + HY * .5
800 SF = SF + HY * .5
850  IF AU = 0 THEN 990
860 MF = 5.53 * MG / F2: REM   MG/F2 OF CRYSTALS
870 DA = 3.56 - .26 * MF / (1 + MF)
880 AU = AU * TN * DA / DL
900 SS = SS + AU * .5
920 SF = SF + AU * .25
940 SC = SC + AU * .25
990  IF QZ = 0 THEN 1040
1000 QZ = QZ * 2.65 * TN / DL
1020 SS = SS + QZ
1040  IF PF = 0 THEN 1140
1050  FOR Z = 0 TO PF STEP .001
1060 CN = 3.68 * (CA * AL) / (NA * SI): REM    CA/NA OF PLAG
1070 DP = 2.62 + .14 * CN / (1 + CN):PL = .001 * TN * DP / DL
1080 SS = SS + PL * .6 - PL * .2 * CN / (1 + CN)
1090 SA = SA + PL * .2 + PL * .2 * CN / (1 + CN)
1100 SC = SC + PL * .2 * CN / (1 + CN)
1110 SN = SN + PL * .2 / (1 + CN)
1120  NEXT Z
1140  IF KF = 0 THEN 1200
1150 KF = KF * TN * 2.62 / DL
1160 SS = SS + KF * .6 * TN
1170 SA = SA + KF * .2 * TN
1180 SK = KF * .2
1200  IF MT = 0 THEN 1300
1230 FT = .92 * F3 * F3 / (F2 * TI): REM   MAGNETITE/ULVOSPINEL
1240 MT = MT * TN * 5.18 / DL
1250 ST = MT * .1200 / (1 + FT)
```

Table 1 (*continued*)

```
1260 SF = SF + MT * .88
1300 SI = SI - SS
1310 TI = TI - ST
1320 AL = AL - SA
1330 FM = FM - SF
1360 CA = CA - SC
1370 NK = NA - SN + K - SK
1380 H = H / (1 - PH)
1400 TV = SI + TI + AL + FM + CA + NK + P + H:TR = TN / TV
1410 SI = SI * TR
1420 TI = TI * TR
1430 AL = AL * TR
1440 FM = FM * TR
1480 CA = CA * TR
1490 NK = NK * TR
1510 P = P * TR
1520 H = H * TR
1530 TC = SI + TI + AL + FM + CA + (NK + P) / 2 + H
2000  REM    CALCULATE VISC OF LIQ FRACTION
2010 Q1 = AL * 6.7
2020 Q2 = FM * 3.4
2030 Q3 = (CA + TI) * 4.5
2040 Q4 = NK * 1.4
2050 Q5 = H * 2
2060 QT = (Q1 + Q2 + Q3 + Q4 + Q5) * SI / TC ^ 2
2070 MU = (QT / (1 - SI / TC)) * (1E4 / TE - 1.5) - 6.4
2100  REM : CALC VISC OF LIQ + XLS
2110  IF PH = 0 THEN NE = MU: GOTO 2170
2120  IF DM = 0 THEN 2070
2130 PH = (1 / PH) ^ .333
2140 SH = .011 * DM / (PH - 1) - .15
2150 NE = MU + SH
2170  PRINT "LOG VISC OF GLASS = " FN A(MU / 2.3026)
2180  PRINT "LOG VISCOSITY = " FN A(NE / 2.3026)
3000  END
```

Literature Cited

Ashby, M. F. 1972. A first report on deformation-mechanism maps. *Acta Metall.* 20:887–97

Booth, B., Self, S. 1973. Rheological features of the 1971 Mount Etna lavas. *Philos. Trans. R. Soc. London Ser. A* 274:99–106

Bottinga, Y., Weill, D. F. 1972. The viscosity of magmatic silicate liquids: a model for calculation. *Am. J. Sci.* 272:438–75

Burnham, C. W. 1963. Viscosity of a water-rich pegmatite melt at high pressures. *Geol. Soc. Am. Spec. Pap. No. 76*, p. 26

Carron, J. 1969. Recherches sur la viscosite et les phénomènes de transport des ions alcalins dans les obsidiennes grantiques. *Trav. Lab. Geol., Éc. Norm. Super., Paris.* 112 pp.

Cigolini, C., Borgia, A., Cassertano, L. 1983. Intracrateric activity, aa–block lava, viscosity and flow dynamics: Arenal Volcano, Costa Rica. *J. Volcanol. Geotherm. Res.* In press

Einarsson, T. 1949. The eruption of Hekla, 1949–1948, Pt. IV, 3. *Soc. Sci. Isl.* 74 pp.

Euler, R., Winkler, H. G. F. 1957. Über die Viskositaten von Gesteins und Silikatschmelzen. *Glastech. Ber.* 30:325–32

Gauthier, F. 1973. Field and laboratory studies of the rheology of Mount Etna lava. *Philos. Trans. R. Soc. London Ser. A* 274:83–98

Hess, P. C. 1980. Polymerization model for silicate melts. In *Physics of Magmatic Processes*, ed. R. B. Hargraves, pp. 3–48. Princeton, NJ: Princeton Univ. Press

Hulme, G. 1974. The interpretation of lava flow morphology. *Geophys. J. R. Astron. Soc.* 39:361–83

Johnson, A. M. 1970. *Physical Processes in Geology*. San Francisco: Freeman & Cooper. 577 pp.

Khitarov, N. I., Lebedev, E. B. 1978. The peculiarities of magma rise in presence of water. *Bull. Volcanol.* 41:354–59

Khitarov, N. I., Lebedev, E. B., Dorfman, A. M., Slutsky, A. B. 1978. Viscosity of dry and water-bearing basalt melts under the pressure. *Geokhimiya* 6:900–5

Kushiro, I. 1978. Density and viscosity of hydrous calc-alkalic andesite magma at high pressures. *Carnegie Inst. Washington Yearb.* 77:675–77

Kushiro, I. 1980. Viscosity, density, and structure of silicate melts at high pressures, and their petrological applications. In *Physics of Magmatic Processes*, ed. R. B. Hargraves, pp. 93–120. Princeton, NJ: Princeton Univ. Press

Kushiro, I., Yoder, H. S. Jr., Mysen, B. O. 1976. Viscosities of basalt and andesite melts at high pressures. *J. Geophys. Res.* 81:6351–56

MacKenzie, J. D. 1956. Simultaneous measurements of density, viscosity, and electrical conductivity of melts. *Rev. Sci. Instrum.* 37:297–99

Marsh, B. D. 1981. On the crystallinity, probability of occurrence, and rheology of lava and magma. *Contrib. Mineral. Petrol.* 78:85–98

Moore, H. J., Arthur, D. W. G., Schaber, G. G. 1978. Yield strengths of flows on the Earth, Mars, and Moon. *Proc. Lunar Planet. Sci. Conf., 9th*, pp. 3351–78

Murase, T. 1962. Viscosity and related properties of volcanic rocks at 800° to 1400°C. *J. Fac. Sci. Hokkaido Univ. Ser.* 7 1:487–584

Murase, T., McBirney, A. R. 1973. Properties of some common igneous rocks and their melts at high temperatures. *Geol. Soc. Am. Bull.* 84:3563–92

Murase, T., McBirney, A. R., Melson, W. G. 1984. The viscosity of the dome of Mt. St. Helens. *J. Volcanol. Geotherm. Res.* In press

Nathan, H. D., Van Kirk, C. K. 1978. A model of magmatic crystallization. *J. Petrol.* 19:66–94

Peterson, D. W., Tilling, R. I. 1980. Transition of basaltic lavas from pahoehoe to aa, Kilauea Volcano, Hawaii. *J. Volcanol. Geotherm. Res.* 7:271–94

Pinkerton, H., Sparks, R. S. J. 1978. Field measurements of the rheology of lava. *Nature* 276:383–85

Robson, G. R. 1967. Thickness of Etnean lavas. *Nature* 216:251–52

Roeder, P. L., Dixon, J. M. 1977. A centrifuge furnace for separating phases at high temperature in experimental petrology. *Can. J. Earth Sci.* 14:1077–84

Ryan, M. P., Sammis, C. G. 1981. The glass transition in basalt. *J. Geophys. Res.* 86:9519–35

Scarfe, C. M. 1973. Viscosity of basic magmas at varying pressure. *Nature* 241:101–2

Shartsis, L., Spinner, S. 1959. Viscosity and density of molten glasses. *J. Res. US Nat. Bur. Stand.* 46:176–94

Shaw, H. R. 1963. Obsidian-H_2O viscosities at 1000 and 2000 bars in the temperature range 700° to 900°C. *J. Geophys. Res.* 68:6337–42

Shaw, H. R. 1969. Rheology of basalt in the melting range. *J. Petrol.* 10:510–35

Shaw, H. R. 1972. Viscosities of magmatic silicate liquids: an empirical method of prediction. *Am. J. Sci.* 272:870–93

Shaw, H. R., Wright, T. L., Peck, D. L., Okamura, R. 1968. The viscosity of basaltic magma: an analysis of field measurements in Makaopuhi lava lake, Hawaii. *Am. J. Sci.* 266:255–64

Sherman, P. 1968. *Emulsion Science*. New York: Academic. 351 pp.

Sparks, R. S. J., Pinkerton, H., Hulme G. 1976. Classification and formation of lava levees on Mount Etna, Sicily. *Geology* 4:269–71

Stocker, R. L., Ashby, M. F. 1973.

Ann. Rev. Earth Planet. Sci. 1984. 12: 359–81

TECTONIC PROCESSES ALONG THE FRONT OF MODERN CONVERGENT MARGINS—Research of the Past Decade[1]

Roland von Huene[2]

US Geological Survey, Menlo Park, California 94025

INTRODUCTION

At convergent margins, oceanic crust descends beneath continental crust to form a complex crustal juncture. This juncture begins at an ocean trench and can be followed deep into the Earth as a zone of high seismicity created as the two crustal plates move past one another. The places where oceanic crust descends into the Earth, known as *subduction zones*, are highly dynamic features; modern examples contain analogues for processes that created the now-static remains of ancient margins exposed on land. At the fronts of subduction zones, where the descending oceanic crust is still relatively shallow, exploration geophysical techniques can define the structure of sedimentary strata. Modern seismic reflection techniques can resolve 100-m intervals in the upper 2–6 km of a sedimentary section at trench depths. Drill cores of such sediment from trenches have been successfully recovered between about 0.5 and 1.2 km below the seafloor. This review focuses on the advances in knowledge of the fronts of modern convergent margins, where the structure and tectonic history can be so observed. The review is in part historical because new findings of the past decade became increasingly difficult to explain without extensive modifi-

[2] Prepared while a visiting researcher at the Laboratoire de Géologie Structurale, Département de Géotectonique, Université Pierre et Marie Curie, Académie de Paris, France.

cation of a widely accepted simplifying model. In this model, the continent is likened to a bulldozer blade that scrapes all sediment from the igneous oceanic crust and thus builds a continuously rising and outward-building accretionary prism of compressionally deformed oceanic sediment. To some workers, it appeared that because of the numerous exceptions, the accretionary model was seriously flawed. However, this was not the case; what was flawed was the view that a single model could be the only simplifying concept for all convergent margins.

The past decade includes a transition from a focus on testing of the accretionary model to research based more on observation. The accretionary model developed in the 1960s and refined in the 1970s provided the direction of investigation as deeper observation into subduction zones was made possible by the availability of advanced geophysical and sampling techniques.

IMPROVEMENT OF TECHNIQUES

During the past 20 years, the marine seismic reflection technique has become the most commonly applied method in marine tectonic study. Many of the seismic reflection records used in the papers published before about 5 years ago were obtained with relatively simple analog single-channel instruments. The single-channel records gradually improved from those that showed little more than bathymetry along the lower slopes of a trench, to records that showed the top of the subducted igneous oceanic crust beneath the front of the subduction zone. Although single-channel records showed simple structure at depths of 4–6 km on the shelf, the resolution deteriorated rapidly farther down the continental slope. Hyperbolic reflections or diffractions dominated the records here, and the slopes were commonly interpreted as zones of chaotic structure or mélange—hence the term "mélange wedge" used by some authors.

The far more powerful multichannel seismic reflection methods developed by the petroleum industry acquire data digitally from 24 to 96 hydrophone groups that sample each subsurface point along many ray paths. The different ray paths to a single point are velocity corrected to bring them in line, and summed to yield records with significantly greater depth of penetration and resolution of structure. The cost, however, is also about 10 to 20 times greater than the equivalent single-channel coverage. Multichannel records across trenches presented by industry scientists (Beck & Lehner 1974, Seely et al 1974) established the potential of this technique, but the high cost delayed its use in academic research for several years. Continuing improvement of multichannel records by advanced processing techniques has yielded reflection images of thrust-fault planes (Aoki et al

1983) and clear delineations of décollements 40 km long (von Huene et al 1983, Westbrook & Smith 1983) and folded sediment layers.

Another geophysical technique, multibeam sonar or swath mapping, has recently begun to advance studies of convergent margins. With swath mapping, bathymetric soundings from a 30° to 45° arc beneath the ship are received at frequent intervals, rather than the conventional reception of single soundings along the ship's path. The bathymetry from swath mapping delineates complex topography that was previously unresolvable. The increase in resolution is comparable to the increased resolution in topographic mapping on land with the introduction of photogrammetric techniques.

Combined swath mapping and seismic data show that many of the diffractions in seismic reflection records are caused by rugged topography at the fronts of convergent margins. As the seismic instrument passes from the shallow waters of a shelf into the increasingly deeper waters of the slope and trench, the signal it detects contains reflections from a correspondingly greater area of the sea floor and from an increasingly greater volume of rock in the subsurface. At the 4- to 7-km depth of ocean trenches, the seismic signal received is not only from directly below the instrument but also from topography at a 15–30° angle from the vertical. Thus, more strongly reflective features from a few kilometers on either side of the section are superimposed on features from below. Single-channel seismic records, interpreted as a two-dimensional or single plane of imaging rather than as reflections from a three dimensionally complex trench topography, fostered the view of chaotic structure. A comparison of single-channel records with the final multichannel-processed record along the same track reveals that the chaotic appearance of many earlier records was a function of the technique. By interpreting most frontal areas of convergent margins as chaotic, the structural differences were minimized, and all seismic data seemed to indicate a general similarity.

Geophysical data collection is generally a more efficient technique in exploring large areas of the oceans than is the sampling of rock, especially in deep water and far below the ocean floor. However, a geophysical study without some sampling to establish the age and type of rock generally gives a poorly constrained tectonic interpretation. Samples of deep layers are obtained either by dredging outcrops identified in seismic records or by deep-ocean drilling. Before 1977, the Deep Sea Drilling Project (DSDP) had drilled single holes above only three subduction zones. This initial experience led to a more comprehensive study under the International Program of Ocean Drilling (IPOD), and a series of holes were combined with grids of geophysical information along several transects. The results of these IPOD transects represented a turning point in research on modern convergent margins. It was discovered that many important time stratigraphic boundaries were not apparent in seismic records, even with good geophysical data, and that the tectonic history deduced from geophysical studies commonly differed greatly from that demonstrated by combined geophysical and geologic studies.

The past two years have witnessed an advance in active-margin research from an essentially two-dimensional bathymetric analysis to a three-dimensional analysis because improved geophysical methods and instruments have now become available to academic scientists. Selected images from deep below the seafloor show faults and folds, as well as tracts of mildly deformed strata. Drilling along the IPOD transects has provided important new limits on interpretation and has expanded the number of models to be considered.

Some terms applied to the features in subduction zones are used here in the following way. "Accretion" refers to the tectonic addition of material from the oceanic crust to the front of a continent. To distinguish accretion at the front of convergent margins from sedimentary accretion or the accretion of colliding lithospheric blocks, Scholl et al (1980) use the term "subduction-accretion." Accretion of materials to the base of a continent from below is referred to as "underplating." All of the material accreted is included in the term "accretionary complex." "Collision" involves the processes incurred as a positive topographic feature on the oceanic crust is brought against a margin. Thus, the addition of oceanic plateaus or other lithospheric blocks to a continent is differentiated from subduction-accretion. "Sediment subduction" refers to the passage of sedimentary layers under the front of a margin beyond the zone of frontal deformation. "Erosion," a tectonic process referred to as "subduction-erosion" by Scholl et al (1980), involves the removal of material from the front of a continental margin. Material removed by strike-slip faulting is referred to under a more general term, "tectonic erosion."

CONVERGENT MARGIN MODELS A DECADE AGO

A decade ago, Karig (1974) presented in this review series a comprehensive analysis of the general structure across a convergent margin, including the volcanic arc, forearc, trench, and oceanic crust (Figure 1). Knowledge has expanded in the ensuing 10 years to the extent that I am able to cover only the frontal part of the arc-trench system in the same amount of space. Much research along active margins then focused on developing a simplifying concept to complement ocean-floor spreading and to complete the proof of plate tectonics (e.g. Seely et al 1974, Karig & Sharman 1975, Dickinson & Seely 1979). Also, the concept of subduction-accretion seemed to have provided insight into the fundamental understanding of complex coastal fold belts along western North America (Hamilton 1969, Dickinson 1971). The papers of Hamilton, Dickinson, and Karig provided the hypothesis to be tested for the next decade.

The search for evidence of subduction and accretion along modern margins was still a topic of emphasis, but it was reaching culmination. The plate-tectonic theory, and an Earth of constant radius, required the disposal of oceanic crust at a rate equivalent to its generation; constant crustal generation at a spreading ridge required constant crustal disposal at a convergent margin, and thus the need for steady-state subduction. Evidence for constant subduction, at the rate of crustal generation, was not obtainable from the early marine seismic reflection records across convergent margins. Early single-channel records showed undeformed sediment, 1–3 km thick, ponded in the axis of trenches (Ross & Shor 1965, Scholl et al 1968, von Huene & Shor 1969). This undeformed trench sediment was 2–4 m.y. old, as extrapolated from the accumulation rates seen in conventional piston cores, which sampled only recent sediment. In corresponding forearc basins, most of the sediment also was relatively undeformed or gently tilted, with few major faults. The structures seen on the landward slopes of trenches were so poorly resolved that they could be explained by either thrusting or slumping.

About 1970, the first seismic reflection data showing a smooth continuation of oceanic crust from a trench to below the front of a margin became available (Chase & Bunce 1969, Holmes et al 1972, von Huene 1972), and the first industry multichannel data were published (Beck & Lehner 1974, Seely et al 1974). In 1970, the first deformed sediment was recovered in DSDP cores from the landward slopes of two trenches (von Huene, Kulm et al 1971, Scholl, Creager et al 1971, Karig, Ingle et al 1975). The study of DSDP cores established a 0.6-m.y. age for the oldest fill in the Aleutian

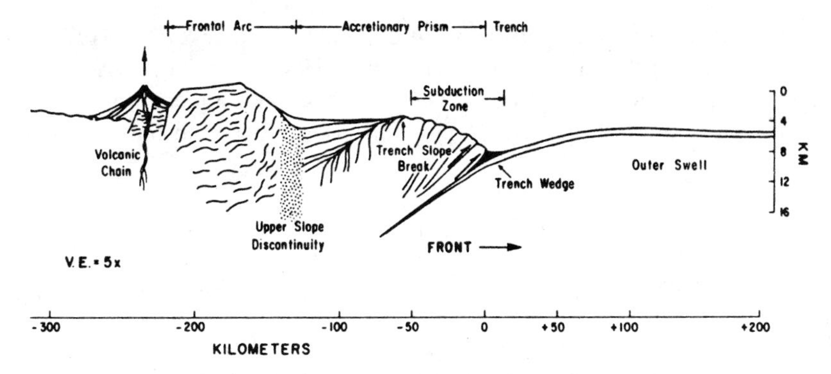

Figure 1 Basic accretionary model of a decade ago (from Karig 1974).

Trench rather than 2–4 m.y. (von Huene, Kulm et al 1971); thus, the trench sediment was far younger than the Neogene age of the arc-trench systems. This required disposal of the older trench fill. A decade ago, the observational evidence for subduction-accretion along modern margins had become convincing.

Various authors elaborated on a single basic simplifying tectonic model (Figure 1) that might explain most subduction zones (e.g. Seely et al 1974, Karig & Sharman 1975, Dickinson & Seely 1979). Steady-state subduction was accompanied by accretion, and as an oceanic plate passed the front of a continent, the sediment above the igneous basement was scraped off and accreted, commonly in multiple slices bounded by thrust faults. Variations could be explained by such subsidiary and related processes as collision, oblique subduction, and sediment distribution (Karig & Sharman 1975). Slumping, sediment subduction, and elevated pore-fluid pressure were regarded as possible, but they were thought to be secondary processes; tectonic erosion was debatable. Because tectonic erosion or any means of disposal of sediment was minimized, subduction required accretion, which in turn required constant seaward growth of a margin. Thus, an implied idea developed that all convergence results in accretion and that the two processes are inseparable.

The great depth of trenches makes it difficult to obtain cores, and only the Aleutian Trench had been drilled. Thus, few samples were available to indicate the composition of sediment in trenches. A simple model of sediment distribution and origin was inferred from the geometry of strata in seismic records. The topography-conforming sediment of a deep ocean basin was assumed to be essentially pelagic, the ponded layers in a trench axis were assumed to represent sandy turbidites, and the ponded and topography-conforming geometries of slope deposits were assumed to be mainly hemipelagic.

The accretionary model seemed to provide a simplifying concept for some highly complex structured areas on land (e.g. Blake & Jones 1974). It came to be identified as *the* plate-tectonic model. In his review, Karig (1974) stressed the accretionary "plate-tectonic" model but tempered his conclusions by noting that the structural variations in the data at that time also allowed other interpretations.

DISCOVERIES FROM MODERN STUDIES

Between 1974 and 1978, seismic reflection data and conventional sampling seemed both to confirm accretion and to indicate a greater variation in tectonism than that allowed by some secondary modifications to the basic accretionary model. The interpretation of multichannel seismic data released by Exxon (Seely et al 1974) had a significant impact. Although Seely's main premise regarding accretion was not new, this publication used multichannel data in which clear reflections from the top of igneous oceanic crust continued for tens of kilometers beneath the Guatemalan margin. The presentation of these data at meetings (later published in Seely et al 1974) was sufficiently convincing to stimulate academic funding for three multichannel seismic records off Peru in 1973 as part of the Nazca plate project. In this region, the top of the oceanic crust could be followed in multichannel records beyond the trench and beneath the front of the margin. However, Paleozoic metamorphic rocks of the Peruvian continent extended to within 15 to 60 km of the trench (Hussong et al 1976, Kulm et al 1981, Thornburg & Kulm 1981, Moberly et al 1982). The estimated volume of accreted material was far less than the probable volume of sediment input during subduction. In 1976, D. M. Hussong and coworkers formally postulated tectonic erosion by subduction, followed by Kulm et al (1977) and Coulbourn & Moberly (1977). Erosion along much of Chile had been inferred from truncated cratonic and igneous complexes (Rutland 1971).

The *Glomar Challenger* drilling of active margins under IPOD began off Japan in 1977. The IPOD drilling was preceded by the most extensive geophysical surveys of active margin areas undertaken up to that time (Nasu et al 1980). At least 4, and as many as 30, multichannel seismic records were acquired in the areas targeted for drilling (e.g. Ladd et al 1978, Shipley et al 1980, Biju-Duval et al 1982). Typical areas selected for drilling included the Japan, Nankai, Marianas, Middle America, and Barbados margins. The Peru Trench area was not drilled for logistic reasons; instead, an area off southern Mexico, suspected to contain a truncated boundary, was drilled. Thus, two transects were studied across the Middle America Trench—one that appeared to be purely accretionary and another that appeared to involve truncation. The combined multichannel seismic, lithostratigraphic, and biostratigraphic data in the transects of several drill sites greatly increased the constraints on interpretation. Accretion was dominant along the Barbados, Mexican, and Nankai margins, but sediment subduction and subduction-erosion were found to be dominant along the Japan, Mariana, and Guatemalan margins (Blanchet & Montadert 1981).

Along the IPOD Japan Trench transect (Scientific Staffs, Legs 56 & 57 1980), one highly significant finding was of the subsidence of a large landmass beneath the forearc area off northern Honshu Island, simultaneous with the present period of subduction (Figure 2). This landmass had subsided about 6 km at a point close to the present trench, and the subsidence decreased progressively landward to less than 0.5 km near the present coast (von Huene et al 1982). Then, about 5–3 m.y. ago, uplift of the forearc area began. Reconstruction of the history of the subsidence, the geology of the slope, and the forearc volcanism suggests simultaneous retreat of the landward slope of the trench, subsidence, and subduction. In addition, at least 80% of the sediment entering the trench since the late Miocene has been subducted.

Similar discoveries were made soon afterward on Leg 60 (Hussong, Uyeda et al 1983) across the Marianas forearc. Studies there showed no evidence to support development of the forearc by accretion. Instead, erosion of the frontal part of the arc has left only arc-generated rocks. Extensional structure dominates seismic records across the edge of the shelf (Mrozowski & Hayes 1980), and the recovery of some carbonate rocks on the trench slope well below the carbonate-compensation depths supports the case for subsidence accompanied by extensional faulting. The absence of much accreted sediment, despite convergence since Eocene time, indicates a net subduction of sediment.

The first of two transects across the Middle America Trench was located off southern Mexico (Watkins, Moore et al 1981). This area has well-developed landward-dipping reflections, interpreted in other areas as thrust faults (Figure 3); however, in this location they represent mainly bedding. The seaward area is an accretionary complex in which beds are tilted and faulted only at the front of the slope. Although landward tilting is complete

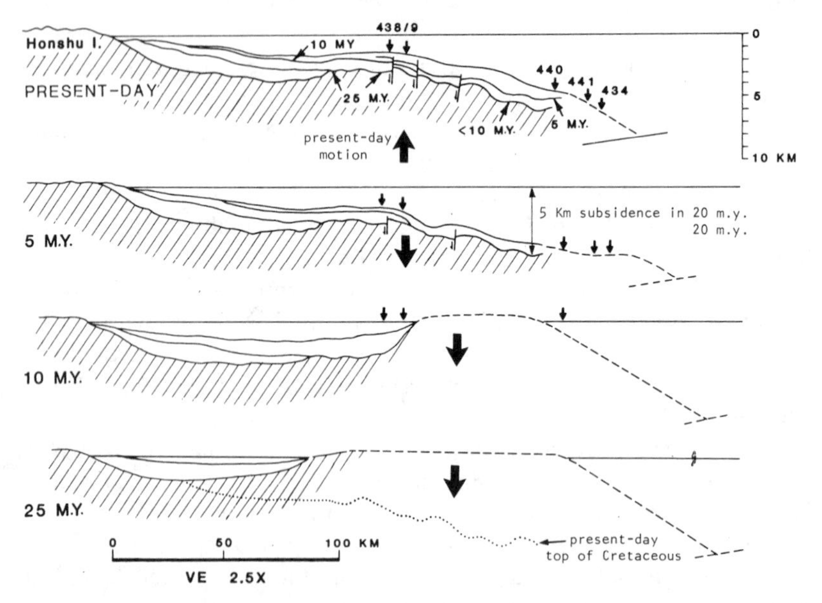

Figure 2 History of the subsidence of the Oyashio landmass off northern Honshu Island, Japan (from von Huene et al 1982).

in the first 2 or 3 km, the slope continues to be uplifted as material is added from below by underplating; sediment has been accreting for about 10 m.y. Before accretion, the Mexican part of the continent was truncated along the trench in this region, although the cause of this truncation—whether by lateral faulting or by subduction—is still debated (e.g. Karig et al 1978, Moore et al 1981, Scholl & Vallier 1983). Drilling and geophysical results suggest that one third of the incoming sediment was accreted, one third was underplated, and one third was subducted (Watkins, Moore et al 1981).

In contrast to the Mexican transect, drilling off Guatemala yielded an assemblage of ophiolitic rocks covered by slope sediment of at least early Eocene age (Aubouin, von Huene et al 1982a,b). The ophiolites were highly deformed and disrupted during a period of tectonism before the present arc-trench system developed. The total post-Eocene sedimentary input to the trench axis has been subducted. This subduction is probably aided by elevated pore-fluid pressures, as measured both near the subduction zone and high above it in the slope sediment.

The Barbados Ridge complex was drilled during an abbreviated leg (Moore, Biju-Duval et al 1982) in which the drilling time was insufficient to overcome problems and to allow penetration through the subduction zone as was planned. However, a repeated sequence of Miocene over Pliocene strata marks the first thrust faults to be clearly documented by repetition of strata at the front of a modern subduction zone. The near-lithostatic pore pressure that was measured just above the décollement that forms the master fault of the Barbados subduction zone indicates that near the drill site, the accretionary complex is essentially floating on the underlying undeformed sedimentary section (Figure 4).

Deep sampling of trench sediment along the IPOD transects did not support a simple sedimentary-facies distribution model, as inferred by some investigators a decade ago. In all but one trench, the sampled sediment fill consisted of less than 20% sand; sediment on the adjacent oceanic plate consisted of a lower pelagic and an upper hemipelagic unit; and the

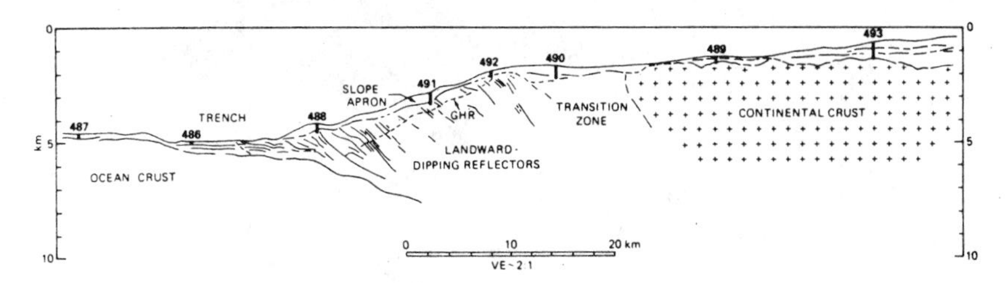

Figure 3 Generalized seismic section across the Middle America Trench off southern Mexico, showing well-developed landward-dipping reflections in the accretionary complex (from Shipley et al 1980).

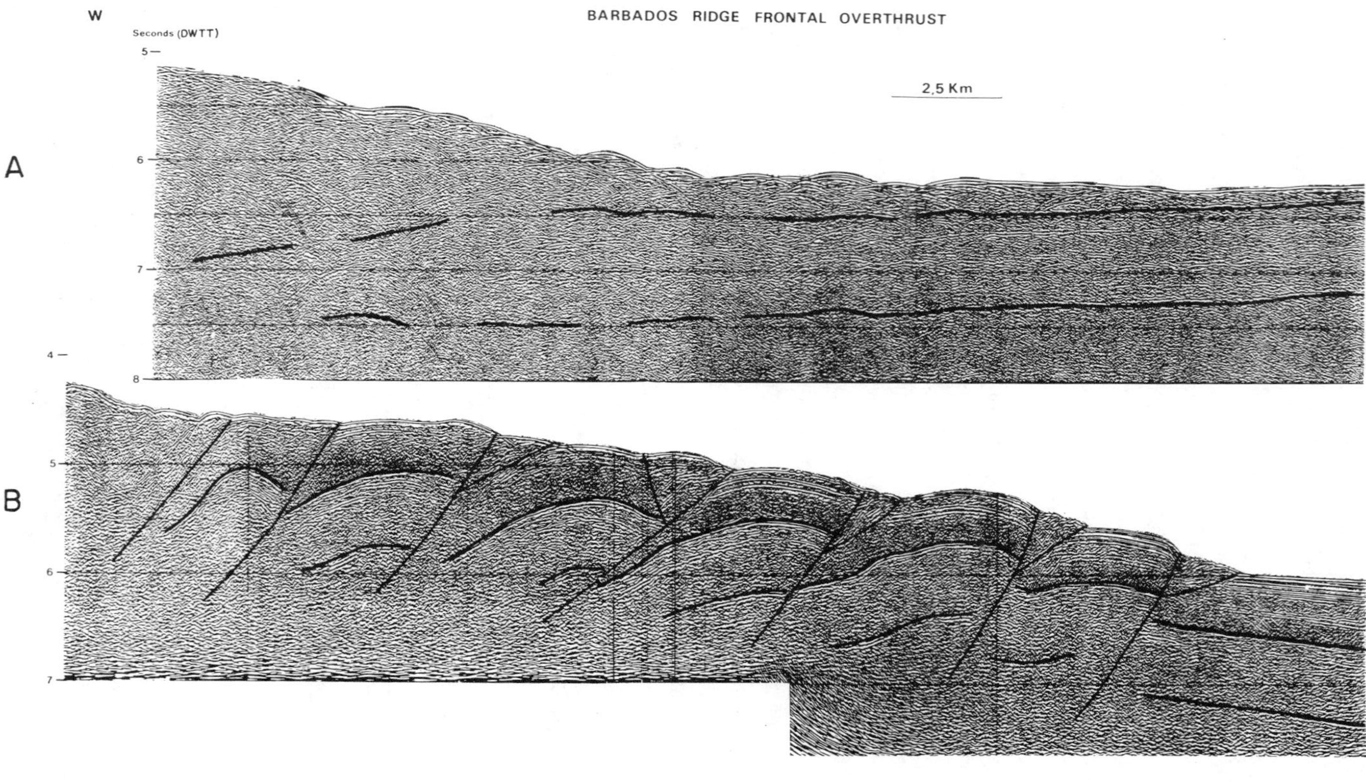

Figure 4 Two constrasting structural styles in seismic records from the northern (*A*) and southern (*B*) Lesser Antilles arc (from Biju-Duval et al 1982).

landward slopes yielded cores with a diversity of sedimentary types, from pelagic diatomaceous ooze to sandy turbidites. Thus, the distribution of sediment in each tectonic environment results from multiple processes, and no distinct lithology separates slope from trench sediment in the DSDP cores.

Elevated pore-fluid pressures were measured directly at drill sites across the Guatemalan margin and at the front of the Barbados margin. These data provided direct confirmation of elevated pore pressures that were also indicated by physical properties and drilling characteristics off Japan (Carson et al 1982), from industry drilling in forearc basins, and from suspected diapiric structures in seismic records (von Huene & Lee 1983, Westbrook & Smith 1983).

Wherever pre-Neogene rocks were recovered, more than one episode of tectonism was detected, and an earlier developed structural geometry was overprinted by a subsequent one. Thus, the structure seen in some IPOD seismic records had to be reinterpreted after drilling.

The IPOD investigations, in particular, confirmed accretion, as proposed in earlier studies, but they shifted the previous emphasis from accretion as a single fundamental process to one with a greater balance among multiple tectonic mechanisms. This shift was required to explain the following observed variations in modern convergent margin structure and history: many convergent margins lack large accretionary complexes; much of the sediment that has entered the trenches has been subducted; the presence of truncated older rocks indicates erosion of the front of many margins; convergent margins have a history not only of uplift but also of profound subsidence; and even in the frontal areas of some margins, the occurrence of more than one stress system results in a composite deformational geometry. The tectonic processes observed over the past decade indicate that the fronts of modern margins are tectonically no less diverse than are ancient thrust and fold belts on land.

STRUCTURAL DIVERSITY

Table 1 summarizes tectonic features reported from the IPOD transects and from other convergent margins, supplemented with DSDP (Aleutian) or industry (Peru) drilling information. These tectonic features include a continuum from those associated with accretion to those associated with erosion, and considered as a group, no single type of feature dominates the table. The relative frequency of tectonic features is suggested by the emphasis in reports, but in the Pacific, the lengths of margins that display Neogene erosional features (Peru-Chile, Middle America, Aleutian, Japan, Marianas) do not differ much from the lengths of the remaining convergent margins. Structural diversity is common because many of the margins with

Table 1 Summary of observations from studies of transects[a]

Tectonic features of the present period of convergence	Accretion			Sediment Subduction		Erosion			Other	
	Frontal accretion, thrusts, folds	Inferred underplating	Uplift	Décollement	Deficient mass balance	Subsidence	Truncation at start of period	Extensional structure	Mass wasting	Overpressure
Japan Trench	×		×	?	■	■	×	×	×	×
Mariana Trench					■	■	■	■	×	
Mid. Am. Trench, Guatemala				×	■	×		×	×	■
Mid. Am. Trench, Mexico	■	■	■	×	×	×	■	×		
Barbados	■	?	×	■					×	■
Aleutian Trench	■	■	×	■	×	×	×	×	■	I
Peru Trench	■			×	■	■	×	×	×	
Nankai	■	?	×	■						I

[a] Margins with a transect of holes are in the upper half of the table, whereas those sampled only at the foot of the slope are in the lower half. Rectangular vertical bars indicate features emphasized in reports, ×'s represent features that have been noted, and ?'s indicate inferences from published data. Direct measurement of overpressure is shown by a rectangular vertical bar; overpressure inferred from logs shown by ×'s; and overpressure calculated from the length of décollement by the method of Hubbert & Rubey (1959) shown by the error-bar symbols.

erosional features also have modest frontal accretionary complexes along strike. Thus, all of the features listed in Table 1 might be found along any single margin.

The structural variation between margins in various parts of the Pacific might reflect different plate motion histories or a difference in the types of marginal crust. Relative plate motion is one of the principal controls of the rate and direction in which stress is applied to the deforming sediment. The two transects across the Middle America Trench demonstrate a sharp divergence in structure and tectonic history along parts of the same trench where relative plate motion is nearly the same. These two transects could hardly contrast more in showing the dominance of accretion in one area and of nonaccretion in another area along a single convergent margin. The structural diversity along single margins indicates the importance of causes other than plate motion in defining structural style.

Studies in the Lesser Antilles (Biju-Duval et al 1982, Westbrook et al 1982) illustrate that in the south, near a large sediment source, the margin is sediment flooded and has a wide accretionary complex, whereas in the north, far from the sediment source, the margin is sediment starved and has a narrower accretionary complex. The northern and southern accretionary complexes also contrast in structural style (Figure 4). The upper part of the southern complex exhibits a repeated imbricate structure, whereas in the northern part, the thin accretionary complex is separated by a décollement from the underlying undisturbed beds. This difference suggests a tectonic influence from sediment thickness. The structural style changes suddenly from the imbricate sequence, marked by regularly spaced ridges, to a décollement, overlain by a high-frequency, weakly lineated topography in the SEABEAM bathymetry (Biju-Duval et al 1982). This situation is complicated, however, by a varying topography on the converging ocean floor. Nonetheless, the sudden change emphasizes the narrow range of conditions under which one or the other structural style will dominate.

A relatively complete set of geophysical data has been collected along the eastern Aleutian Trench (Figure 5). Off southern Kodiak Island, one record in a network of lines shows a well-developed décollement and a series of thrust packets that become increasingly deformed toward land (Figure 5*A*). About 220 km northeast, another record illustrates a mildly faulted but thickened sedimentary sequence at the front of the margin (Figure 5*B*). This thickened sediment at the front of the margin appears to be subducting. Approximately 290 km farther northeast (Figure 5*C*), in the midslope area, a broad landward-dipping sequence of reflection alternates with a seaward-dipping sequence, and so both subduction and obduction are indicated. At the northeast end of the eastern Aleutian Trench, the time stratigraphic information projected into a seismic record from an industry drill hole

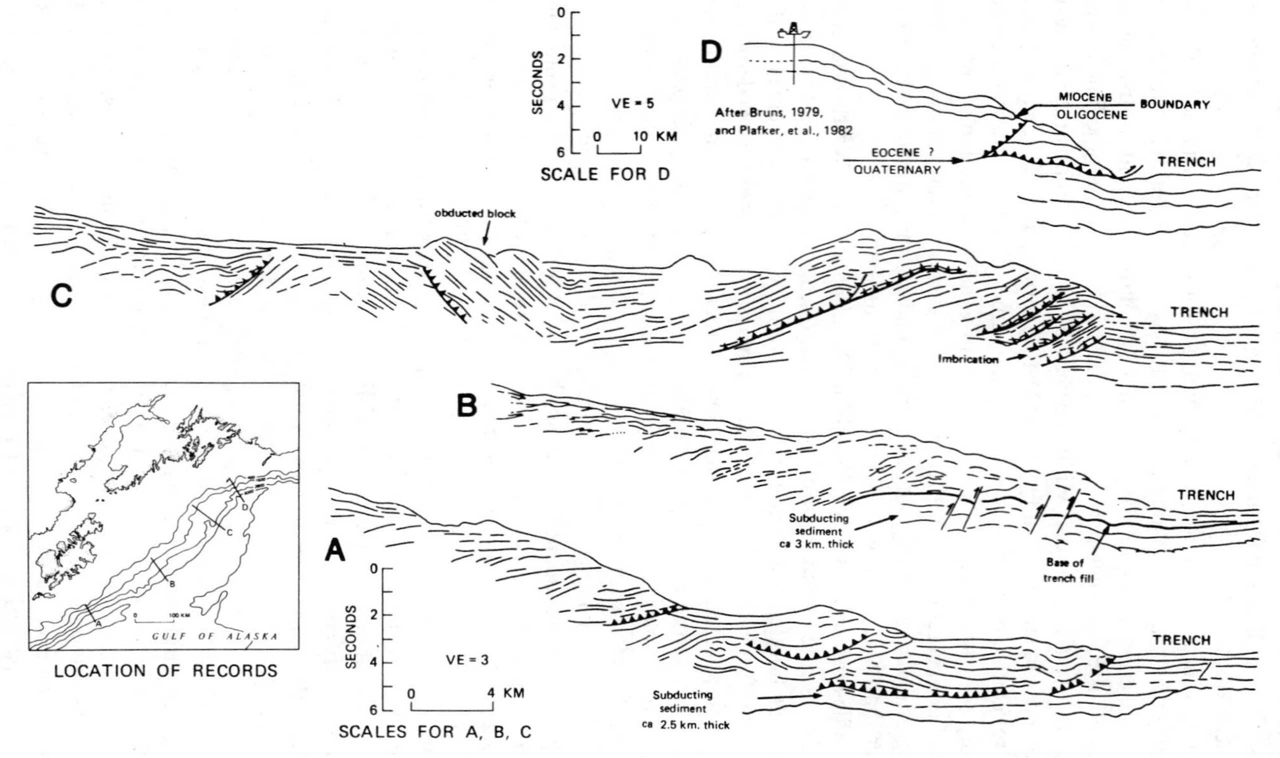

Figure 5 Seismic records across the landward slope of the eastern Aleutian Trench, showing structural diversity at intervals of hundreds of kilometers.

indicates a major décollement not obvious in the seismic data alone (Figure 5*D*). The Eocene contact can be followed downslope along reflectors that crop out near the trench axis (Plafker et al 1982). Below, reflectors can be traced to DSDP Hole 178, where they are Pleistocene. Thus, a Pleistocene and older oceanic section is being subducted beneath an Eocene and younger continental-margin section along a master décollement concordant with bedding. Possibly more significant is the question of what has happened to the Eocene section now truncated at the front of the margin. Either it has been eroded, or else it was never accreted. The latter possibility is less likely because much accreted sediment is seen just a short distance away along strike.

The four seismic sections along a 600-km-long segment of the eastern Aleutian Trench (Figure 5) show features indicating imbrication, décollement, folding, sediment subduction, tectonic erosion, and obduction. Along the Barbados and Middle America trenches, where the seismic records are fewer, similar diversity is indicated. This structural diversity, independent of changes in the rate and direction of plate motion, indicates that other causes related to the strength of material may equally affect the tectonic style. Few margins, however, are known in sufficient detail to relate such differences in structure to the causative phenomena that control structural mechanisms.

TECTONIC PROCESSES

Scholl et al (1980) called attention to four common tectonic processes that they termed "subduction-accretion," "sediment subduction," "subduction-erosion," and "tectonic kneading." Processes other than accretion had been proposed previously, but they were seldom considered in interpretations of active margins. There was little appeal at the time to consider eroded rock bodies that could not be seen, because much attention was then focused on understanding the complex rock masses in mountain belts.

Major tectonic processes at the fronts of convergent margins (Table 2) can be grouped under the three descriptive headings in Table 1, which resemble those of Scholl et al (1980). The headings along the horizontal axis span a continuous spectrum of material transfer between accretion and erosion, whereas the vertical ordering is intended to represent a spectrum from strong coupling along the top to near-decoupling along the bottom of the table.

Coupling, as well as the rate and direction of convergence between the upper and lower plates, controls the stresses transmitted across a subduction zone. Coupling is greatly affected by the topographic roughness of a subducting plate, the thickness and strength of sediment entering a

Table 2 Tectonic processes, ordered from accretion to erosion along the horizontal axis, and from strong to weak coupling on the vertical axis

Accretion	Sediment subduction	Erosion	
Accretion by *tectonic addition* of sediment and crustal rock to the front of a margin; associated with compressional structure	Sediment *trapped in pockets* of igneous ocean floor and thus guarded from offscraping	Erosion from *abrasion* by drag of positive features against the underside of the upper plate	Strong coupling
Accretion by *underplating*: subducted sediment detached from the descending plate and added from below to the upper plate	Sediment subduction by the passage of *deep-ocean sediment* beyond the front of the margin, commonly beneath a décollement	Erosion from *fragmentation* of the underside of the upper plate by hydrofracture and transport of slurry with subducting matter down a subduction zone	
Accretion by *deposition* of sediment and products of mass wasting that become tectonically incorporated into an accretionary complex	Sediment subduction including the upper *terrigenous debris* ponded in a trench axis	Erosion by *failure* of margin front through mass wasting, with gravity transport to trench floor and incorporation into subducting sediment	Weak coupling

subduction zone, and the configuration of the Wadati-Benioff zone. The coupling from these causes, however, depends on the distribution of strength in rocks and sediment of the subduction zone, which depends, in turn, on the level of pore-fluid pressure. Elevated fluid pressure decreases shear strength and thus resistance to shear (Hubbert & Rubey 1959). If pore fluid in a subduction zone were elevated from hydrostatic to lithostatic pressures (a pressure equal to the mass of the overlying materials), the upper plate would nearly float upon the lower plate, and the stress transmitted across the subduction zone or the coupling would be low. Direct measurements of near-lithostatic pressure were made at the front of the Barbados and Guatemalan subduction zones (Moore, Biju-Duval et al 1982, Aubouin, von Huene et al 1982b). Many more such measurements would have been made were it not for their difficulty and high cost.

Elevated pore-fluid pressure across the forearc areas of subduction zones is further indicated by pressure measurements in industry drill holes on the shelf, by observations of diapiric structure, and from simple model studies (von Huene 1984). Modeling by Davis et al (1983) suggests that elevated pore-fluid pressure controls the angle of failure of convergent margin slopes, an idea supported by measurements of Poisson's ratio in slope sediment along the Japan Trench (Nagumo et al 1980), which indicate very low shear strength. The indications of elevated pore-fluid pressure suggest that the frontal areas of most subduction zones have high pressures. Local differences in pore-fluid pressure provide a likely explanation for structural variations because of the resulting differences in strength of material and transmission of stress.

Elevated pore-fluid pressure is a function of the rate at which horizontal and vertical stresses are applied to the materials in a subduction zone and the ability of fluids to drain from it. When two plates converge, horizontal pressure results from the convergence, and vertical pressure occurs due to the increasing thickness of rocks and sediment. If the convergence of two plates is very slow, the contained fluids have more time to drain from the material and to release pressure continuously, thus keeping pressures low. An increase in the rate of convergence causes an elevation of pressure. Rising pressure with increasing convergence rate will weaken the material until the increasing stress exceeds the strength of the material. Once the rock begins to fail, the style of deformation will reflect more the differences in strength and less the rate of convergence. When pore pressure is elevated to near-lithostatic levels, further increase in the rate of convergence no longer influences the deformational style because the increased shear can no longer be transmitted across the subduction zone in highly weakened material. Thus, there can be one regime of low convergence rates and low pore pressure where structural style is dominantly controlled by the convergence rate, and another regime of higher convergence rates and excess pore pressure where structural style is dominantly controlled by the pore-pressure distribution. This difference in coupling is useful in recognizing related tectonic processes at the fronts of convergent margins.

Some of the tectonic processes listed in Table 2 have been observed, but others are only conceptual, partly because they concern rock bodies that have been destroyed. Accretion is the best known process in subduction zones because it conserves material to provide a record of past convergence. Where clearly displayed in seismic records across modern subduction zones, accretion is active mainly at the fronts of these zones. Examples of "classical" accretion are seen in records from the frontal parts of the Nankai Trough (Aoki et al 1983), the Barbados forearc (Biju-Duval et al 1982), the Middle America Trench off Mexico (Shipley et al 1980), the Java Trench (Lehner et al 1983a), and the Aleutian Trench (von Huene et al 1983) (Figure 5*A*). These examples show active deformation along the lower slope.

Underplating has been inferred to explain the uplift, without much further deformation of the upper 2 to 6 km, seen in many midslope and upper-slope areas. It requires the detachment of sediment from the subducting plate beneath the accretionary complex. Such detachment may be indicated by the down-stepping of a décollement, as interpreted off Barbados (Westbrook & Smith 1983). A hypothetical detachment may occur when the water bound in clay minerals is released from subducted sediment and oceanic crust as the temperatures reach 82°C and higher (von Huene & Lee 1983). Temperatures above 82°C in the subducted sediment can begin below the midslope, as extrapolated from measured thermal gradients. The water released at that depth will form a layer of overpressured pore fluid that provides a low-strength, low-friction glide plane near the base of the sedimentary sequence and thereby decouples the overlying sediment. Underplating is difficult to observe because it occurs at depths that cannot be resolved well in most seismic records and that are too deep to drill. In ancient fold belts, the underplated rocks probably have a structure similar to that of the adjacent thrust and imbricate rocks and could be recognized only by a detailed time stratigraphy. Thus, underplating has probably gone unrecognized in ancient exposed accretionary complexes where diagenesis has destroyed microfossils and, thus, the ability to discriminate small differences in age between rock masses.

Underplating is closely related to sediment subduction. The case for sediment subduction became convincing when sufficient constraints were afforded by DSDP sampling to show large deficiencies of accreted sediment at the fronts of four subduction zones. The subsequent clear delineation of thick subducting sedimentary sections in several seismic records now leaves little doubt regarding the importance and common occurrence of this process (Biju-Duval et al 1982, Aoki et al 1983, Lehner et al 1983a, von Huene et al 1983; see also Figure 5*A*). A related form of sediment subduction involves the sediment trapped in topographic pockets (Hilde & Sharman 1978, Schweller & Kulm 1978). Oceanic sediment appears to be subducted to levels of magma generation, as indicated by Be^{10} in some island-arc lavas (Brown et al 1982). This isotope is generated by cosmic rays in the upper atmosphere, and it settles into the deep-ocean sediment that is carried into a subduction zone. The concentration of Be^{10} in island-arc lava is direct evidence of sediment subduction far beyond the front of a convergent margin.

Sediment subduction commonly involves an extensive décollement. Seismic records show clear décollement surfaces 30 and 40 km long below which the strata are continuous (Aoki et al 1983, Lehner et al 1983a,b, von Huene et al 1983, Westbrook et al 1982). Such structure requires considerable reduction of friction, which is easiest to explain by overpressured pore fluids (Hubbert & Rubey 1959, Westbrook & Smith 1983). Critical to the position of the décollement, the amount of sediment subducted, and thus the structural style of a margin, are the composition, thickness, and history of the sediment entering a subduction zone. Such factors establish the distribution of pore-fluid pressure in the subducted sediment and the position of the plane of least shear strength. Simple modeling of pore-fluid conditions in a subduction zone indicates an inevitable elevation of pore pressure unless subduction rates are very low and sediment permeability is high (von Huene 1984, von Huene & Lee 1983). The subduction of all sediment, as seen along the IPOD transect in the Middle America Trench off Guatemala, is well documented but remarkably unexplained (Aubouin, von Huene et al 1982a).

The truncation of structure by tectonic erosion includes strike-slip faulting and subduction. Strike-slip faulting is an important tectonic component along margins where the compressional stress is not normal to the regional trend. The mechanisms of tectonic erosion by subduction (subduction-erosion) are hypothetical because erosion is evidenced by the absence of material or by massive subsidence requiring subcrustal thinning. An abrasion of the base of the upper plate by the seafloor topography is one proposed mechanism (Hilde & Sharman 1978, Schweller & Kulm 1978). Another abrasive mechanism that has been proposed requires sufficient coupling to drag slices of the upper plate into the subducting material (Coulbourn & Moberly 1977). A third proposed mechanism, operating in an environment of decoupling, relies on upward migration of the water released during subduction to soften the base of the upper plate (Murauchi & Ludwig 1980) or to fragment the base by hydrofracting (von Huene & Lee 1983), in which the resulting slurry becomes part of the subducting material. A fourth mechanism of erosion is by mass wasting of the front of the margin, with subsequent accumulation of debris in the trench axis and subduction of this debris. Such a mechanism may be exemplified at the front of the northern Japan Trench, where failure of the slope has caused a large debris flow in one area but the trench is empty at the base of failed slopes nearby (von Huene et al 1982, Lehner et al 1983b).

CONCLUSIONS

The possibility that a single model could explain the fronts of most modern convergent margins was very appealing because it offered a simplifying hypothesis paralleling the plate-tectonic theory. The findings of the past decade, however, have shown the necessity of considering additional mechanisms in interpretation, and the topic has now grown more comprehensive; the basic concept of accretion, with relatively little change, is the best understood of multiple-end-member models.

Once the materially conservative end-member process of accretion was coupled with the materially nonconservative end-member concept of erosion, the episodic history of accretionary complexes could be made consistent with the required plate-tectonic history of constant crustal disposal equal to crustal generation. The time stratigraphic record along convergent margins can be broken by tectonic erosion and sediment subduction, and no longer do gaps in accretion require a change in plate motion.

The differences in tectonic processes along the fronts of convergent margins may be largely influenced by changes in the strength of rock masses from complex pore-water pressure distributions. Subduction environments contain a great variety of boundary conditions, determined by such factors as sediment grain size and composition, sediment thickness, folds, faults, and fractures, to name only a few. In an environment of relatively rapid loading and overpressure, the strength of a deforming rock and sediment mass will vary within very broad limits in comparison with that of a normally pressured undeformed sedimentary sequence. Thus, great structural diversity is possible within a fixed setting of constant plate convergence but varying sediment supply and ocean-basin topography.

The tectonic processes at the fronts of convergent margins may be unique to those areas. Landward, beyond the front of a margin, the continental crust thickens, the lithostatic loads increase, and as sediment dewaters, the grain-to-grain contact also increases. Such trends must lead to environments of greater friction farther down the subduction zone, where earthquakes become more numerous.

Modern convergent margins were traditionally studied with widely spaced seismic records across the structural grain. This procedure was essentially a two-dimensional approach. Lately, more attention has been given to the third dimension, with swath mapping and gridded surveys of seismic records. Continued three-dimensional studies with state-of-the-art techniques will be an important part of future convergent margin research. Moreover, the temporal history provided by DSDP sampling has greatly advanced knowledge in this field and is also of great importance to future research. Aubouin has likened convergent margin geophysical data without temporal control to a tectonic map or cross section without a legend of rock units.

ACKNOWLEDGMENTS

Reviews and constructive comments by T. R. Bruns, D. S. Cowan, J. C. Moore, and A. J. Stevenson significantly improved the manuscript.

Literature Cited

Aoki, Y., Tamano, T., Kato, S. 1983. Detailed structure of the Nankai Trough from migrated seismic sections. *Am. Assoc. Pet. Geol. Mem.* 34:309–24

Aubouin, J., von Huene, R., Azema, J., Blackington, G., Carter, J. A., et al. 1982a. *Initial Reports of the Deep Sea Drilling Project*, Vol. 67. Washington DC: GPO

Aubouin, J., von Huene, R., Baltuck, M., Arnott, R., Bourgois, J., et al. 1982b. Leg 84 of the Deep Sea Drilling Project, subduction without accretion, Middle America Trench off Guatemala. *Nature* 297:458–60

Beck, R. H., Lehner, P. 1974. Oceans new frontier in exploration. *Am. Assoc. Pet. Geol. Bull.* 58:376–95

Biju-Duval, B., Lequellec, P., Mascle, A., Renard, V., Valery, P. 1982. Multibeam bathymetric survey and high-resolution seismic investigations on the Barbados Ridge complex (eastern Caribbean): a key to the knowledge and interpretation of an accretionary wedge. *Tectonophysics* 86: 275–304

Blake, M. C. Jr., Jones, D. L. 1974. Origin of the Franciscan mélanges in northern California. *Soc. Econ. Paleontol. Mineral. Spec. Publ.* 19:255–63

Blanchet, R., Montadert, L., conveners. 1981 *Colloq. C3, Int. Geol. Congr., 26th. Evolution of Active Margins in the Light of Deep Boreholes in the Pacific. Oceanol. Acta*, Vol. 4 (Suppl.). 294 pp.

Brown, L., Klein, J., Middleton, R., Pavich, M., Sacks, I., Tera, F. 1982. Beryllium-10 as a geochemical and geophysical probe. *Carnegie Inst. Washington Yearb.* 81:464–70

Carson, B., von Huene, R., Arthur, M. 1982. Small-scale deformational structures and physical properties related to convergence in Japan Trench slope sediment. *Tectonics* 1(3):277–302

Chase, R. L., Bunce, E. T. 1969. Underthrusting of the eastern margin of the Antilles by the floor of the western North Atlantic Ocean, and the origin of the Barbados Ridge. *J. Geophys. Res.* 74:1413–20

Coulbourn, W. T., Moberly, R. 1977. Structural evolution of forearc basins off southern Peru and northern Chile. *Can. J. Earth Sci.* 14:102–16

Davis, D., Suppe, J., Dahlen, F. A. 1983. Mechanics of fold-and-thrust belts and accretionary wedges. *J. Geophys. Res.* 88:1153–72

Dickinson, W. R. 1971. Plate tectonics in geologic history. *Science* 174:107–13

Dickinson, W. R., Seely, D. R. 1979. Structure and stratigraphy of forearc regions. *Am. Assoc. Pet. Geol. Bull.* 63:2–31

Hamilton, W. 1969. Mesozoic California and underflow of Pacific mantle. *Geol. Soc. Am. Bull.* 80:2409–30

Hilde, T. W. C., Sharman, G. F. 1978. Fault structure of the descending plate and its influence on the subduction process. *Trans. Am. Geophys. Union* 59:1182

Holmes, M. L., von Huene, R., McManus, D. A. 1972. Seismic reflection evidence supporting underthrusting beneath the Aleutian Arc near Amchitka Island. *J. Geophys. Res.* 77:959–64

Hubbert, M. K., Rubey, W. W. 1959. Role of fluid pressure in mechanics of overthrust faulting. *Geol. Soc. Am. Bull.* 70:115–66

Hussong, D. M., Edwards, P. B., Johnson, S. H., Campbell, J. F., Sutton, G. H. 1976.

Crustal structure of the Peru-Chile Trench. In *The Geophysics of the Pacific Ocean Basin and Its Margin*, ed. G. Sutton, et al. *Geophys. Monogr. Am. Geophys. Union* 19:17–86

Hussong, D., Uyeda, S., et al. 1983. In *Initial Reports of the Deep Sea Drilling Project*, 60:909–29. Washington DC: GPO

Karig, D. E. 1974. Evolution of arc systems in the western Pacific. *Ann. Rev. Earth Planet Sci.* 2:51–75

Karig, D. E., Sharman, G. F. 1975. Subduction and accretion in trenches. *Geol. Soc. Am. Bull.* 86:377–89

Karig, D. E., Ingle, J. C. Jr., et al. 1975. *Initial Reports of the Deep Sea Drilling Project*, Vol. 31. Washington DC: GPO

Karig, D. E., Cardwell, R. K., Moore, G. F., Moore, D. G. 1978. Late Cenozoic subduction and continental truncation along the northern Middle America Trench. *Geol. Soc. Am. Bull.* 89:265–76

Kulm, L. D., Schweller, W. J., Masias, A. 1977. A preliminary analysis of the geotectonic processes of the Andean continental margin, 6° to 45°S. In *Problems in the Evolution of Island Arcs, Deep Sea Trenches and Back-Arc Basins*, ed. M. Talwani, W. C. Pitman III, *Maurice Ewing Symp.* 1:285–301. Washington DC: Am. Geophys. Union

Kulm, L. D., Prince, R. A., French, W., Johnson, S., Masias, A. 1981. Crustal structure and tectonics of the central Peru continental margin and trench. In *Nazca Plate: Crustal Formation and Andean Convergence*, ed. L. D. Kulm, et al. *Geol. Soc. Am. Mem.* 154:445–68

Ladd, J. W., Ibrahim, A. K., McMillen, K. J., et al. 1978. *Tectonics of the Middle America Trench offshore Guatemala.* Presented at Int. Symp. on the February 4, 1976, Guatemala Earthquake and the Reconstruction Process, Agency Int. Dev., Guatemala City

Lehner, P., Doujt, H., Bakker, C., Allenbalm, P., Gueneau, J. 1983a. Java Trench. In *Seismic Expression of Structural Styles—A Picture and Work Atlas*, ed. A. W. Bally. *Am. Assoc. Pet. Geol. Stud. Geol.* 15:123-3–36

Lehner, P., Doujt, H., Bakker, C., Allenbalm, P., Gueneau, J. 1983b. Japan Trench. In *Seismic Expression of Structural Styles—A Picture and Work Atlas*, ed. A. W. Bally. *Am. Assoc. Pet. Geol. Stud. Geol.* 15:121-1–16

Moberly, R., Sheperd, G. L., Coulbourn, W. T. 1982. Forearc and other basins, continental margin of northern and southern Peru and adjacent Ecuador and Chile. In *Trench-Forearc Geology*, ed. J. K. Leggett. *Geol. Soc. London Spec. Publ.* 10:171–90

Moore, J. C., Watkins, J. S., Shipley, T. S. 1981. Summary of accretionary processes, Deep Sea Drilling Project Leg 66: offscraping, underplating, and deformation of the slope apron. In *Initial Reports of the Deep Sea Drilling Project*, Vol. 66. Washington DC: GPO

Moore, J. C., Biju-Duval, B., Bergen, J. A., Blackington, G., Claypool, G. E., et al. 1982. Offscraping and underthrusting of sediment at the deformation front of the Barbados Ridge: results from Leg 78A DSDP. *Geol. Soc. Am. Bull.* 93:1065–76

Mrozowski, C. L., Hayes, D. E. 1980. A seismic reflection study of faulting in the Mariana fore-arc. In *The Tectonic and Geologic Evolution of Southeast Asian Seas and Islands*, ed. D. E. Hayes. *Am. Geophys. Union Monogr.* 23:223–34

Murauchi, S., Ludwig, W. J. 1980. Crustal structure of the Japan Trench; the effect of subduction of ocean crust. In *Initial Reports of the Deep Sea Drilling Project*, 56–57(1):463–69. Washington DC: GPO

Nagumo, S., Kasahara, J., Koresawa, S. 1980. Airgun seismic refraction survey near sites 441 and 434 (J-1A), 438 and 439 (J-12), and proposed site J-2B: Legs 56 and 67, Deep Sea Drilling Project. In *Initial Reports of the Deep Sea Drilling Project*, 56–57(1):459–62. Washington DC: GPO

Nasu, N., von Huene, R., Ishiwada, Y., Langseth, M., Bruns, T., Honza, E. 1980. Interpretation of multichannel seismic reflection data, Legs 56 & 57, Japan Trench transect, Deep Sea Drilling Project. In *Initial Reports of the Deep Sea Drilling Project*, 56–57(1):489–504. Washington DC: GPO

Plafker, G., Bruns, T. R., Winkler, G. R., Tysdale, R. G. 1982. Cross section of the eastern Aleutian Arc from Mount Spurr to the Aleutian Trench near Middleton Island, Alaska. *Geol. Soc. Am. Map Chart Ser. MC-28-P*

Ross, D. A., Shor, G. G. Jr. 1965. Reflection profiles across the Middle America Trench. *J. Geophys. Res.* 70:5551–72

Rutland, R. W. R. 1971. Andean orogeny and ocean floor spreading. *Nature* 233:252–55

Scholl, D. W., Vallier, T. L. 1983. Subduction and the rock record of the Pacific margins. *Expanding Earth Symp., Sydney, 1981*, ed. S. W. Carey, pp. 235–45. Univ. Tasmania, Austral.

Scholl, D. W., von Huene, R., Ridlon, J. B. 1968. Spreading of the ocean floor—undeformed sediments in the Peru-Chile Trench. *Science* 159:869–71

Scholl, D. W., Creager, J., et al. 1971. Deep Sea Drilling Project Leg 19. *Geotimes* 16:12–15

Scholl, D. W., von Huene, R., Vallier, T. L.,

Howell, D. G. 1980. Sedimentary masses and concepts about tectonic processes at underthrust ocean margins. *Geology* 28:271–91

Scientific Staff, Legs 56 & 57. 1980. *Initial Reports of the Deep Sea Drilling Project*, Vol. 56–57. Washington DC: GPO

Schweller, W. J., Kulm, L. D. 1978. Extensional rupture of oceanic crust in the Chile Trench. *Mar. Geol.* 28:271–91

Seely, D. R., Vail, P. R., Walton, G. G. 1974. Trench slope model. In *The Geology of Continental Margins*, ed. C. A. Burk, D. L. Drake, pp. 249–60. New York: Springer-Verlag

Shipley, T. H., McMillen, K. J., Watkins, J. S., Moore, J. C., Sandoval-Ochoa, J. H., Worzel, J. L. 1980. Continental margin and lower slope structures of the Middle America Trench near Acapulco (Mexico). *Mar. Geol.* 35:65–82

Thornburg, T., Kulm, L. D. 1981. Sedimentary basins of the Peru continental margin: structure, stratigraphy, and Cenozoic tectonics from 6° to 16°S latitude. In *Nazca Plate: Crustal Formation and Andean Convergence*, ed. L. D. Kulm, et al. *Geol. Soc. Am. Mem.* 154:423–44

von Huene, R. 1972. Structure of the continental margin and tectonism at the eastern Aleutian trench. *Geol. Soc. Am. Bull.* 83:3613–26

von Huene, R. 1984. Structural diversity along modern convergent margins and the role of overpressured pore fluids in subduction zones. *Bull. Soc. Géol. Fr.* 21: In press

von Huene, R., Lee, H. 1983. The possible significance of pore fluid pressure in subduction zones. *Am. Assoc. Pet. Geol. Mem.* 34:781–91

von Huene, R., Shor, G. G. Jr. 1969. The structure and tectonic history of the eastern Aleutian Trench. *Geol. Soc. Am. Bull.* 80:1889–1902

von Huene, R., Kulm, L. D., et al. 1971. Deep Sea Drilling Project Leg 18. *Geotimes* Oct: 12–15

von Huene, R., Langseth, M., Nasu, N., Okada, H. 1982. A summary of Cenozoic tectonic history along IPOD Japan Trench transect. *Geol. Soc. Am. Bull.* 93:829–46

von Huene, R., Miller, J., Fisher, M., Smith, G. 1983. An eastern Aleutian Trench seismic record. In *Seismic Expression of Structural Styles—A Picture and Work Atlas*, ed. A. W. Bally. *Am. Assoc. Pet. Geol. Stud. Geol.* 15: In press

Watkins, J. S., Moore, J. C., et al. 1981. Accretion, underplating, subduction, and tectonic evolution, Middle America Trench, southern Mexico: results of DSDP Leg 66. *Oceanol. Acta* 4:213–24 (Suppl.)

Westbrook, G. K., Smith, M. J. 1983. Long décollements and mud volcanoes: evidence from the Barbados Ridge complex for the role of high pore-fluid pressure in the development of an accretionary complex. *Geology* 11:279–83

Westbrook, G. K., Smith, M. J., Peacock, J. H., Poulter, M. J. 1982. Extensive underthrusting of undeformed sediment beneath the accretionary complex of the Lesser Antilles subduction zone. *Nature* 300:625–28

Ann. Rev. Earth Planet. Sci. 1984. 12: 383–409

COOLING HISTORIES FROM $^{40}Ar/^{39}Ar$ AGE SPECTRA: Implications for Precambrian Plate Tectonics

Derek York

Department of Physics, University of Toronto, Toronto,
Ontario M5S 1A7, Canada

Introduction

While a great deal of effort is currently being expended by Earth scientists in mapping out the present state of the Earth, the ultimate aim of the Earth sciences must be to explain how such a temporary state has been reached and from what sort of beginnings. A key parameter in unraveling such a complex physical and biological evolution is, undoubtedly, the thermal history of the system involved. The total thermal history of a body such as the Earth is, of course, a complex interweaving of many local thermal histories. The present article describes some recent progress that has been made in the mapping out of some local thermal histories with the aid of diffusion-based interpretations using the $^{40}Ar/^{39}Ar$ method of dating. The significance of these studies for the larger scale becomes apparent as the examples follow. We discuss the following types of thermal histories:

1. The simplest ideal, in which a sample effectively is formed in an instant, cools "geologically instantaneously," and is never subsequently disturbed. The Texas tektites are used to illustrate this rarely found case.
2. The simplest disturbance of the ideal; that is, a sample that is formed and cools instantaneously, as in (1), and experiences later a very brief thermal pulse in an otherwise peaceful history. This represents the K-Ar (i.e. $^{40}Ar/^{39}Ar$) equivalent of the Wetherill (1956) episodic lead-loss model in U-Pb dating. This is exemplified by meteorite data (Turner 1969) and terrestrial results (Harrison & McDougall 1980b, Hall & York 1982).

0084–6597/84/0515–0383$02.00

3. The case of very slow postorogenic cooling. We look here at recent attempts to quantify such histories (Berger & York 1981a, Harrison & McDougall 1980a). This yields important information on the time scale of vertical tectonics. In addition, it is shown how the combination of palaeomagnetism data with isotopically derived cooling curves provides information on Precambrian plate motions that is exceedingly difficult (if not impossible) to derive in any other way.

We begin with a brief explanation of the "step-heating" approach to $^{40}Ar/^{39}Ar$ dating. With this method, "age spectra" are derived, and it is the interpretation of such spectra that gives us thermal histories.

From the beginning, it was realized by Merrihue & Turner (1966) that the results of step-heating experiments contained "the exciting possibility of deducing information of the past thermal history of the sample." Since much $^{40}Ar/^{39}Ar$ modeling sprang from meteorite studies, it is necessary to discuss these to some degree.

$^{40}Ar/^{39}Ar$ *Chronometry*

The $^{40}Ar/^{39}Ar$ dating technique (Merrihue 1965, Merrihue & Turner 1966) is an improvement of the traditional K-Ar method (Dalrymple & Lanphere 1969, York & Farquhar 1972). In the latter, the ^{40}Ar and ^{40}K concentrations in a sample are measured in quite different experiments, and the age is calculated from the ensuing $^{40}Ar/^{40}K$ ratio. The $^{40}Ar/^{39}Ar$ approach differs in two ways. Firstly, the daughter/parent ratio is measured in one and the same experiment. Secondly, an *age spectrum* is derived for the sample instead of a single age. We reiterate, it is the age spectrum that holds the key to unraveling thermal histories. These two improvements are achieved in the following way.

The mineral (or rock) to be dated is irradiated in a nuclear reactor. Fast neutrons convert some ^{39}K to ^{39}Ar. The irradiated sample is then gradually heated to melting temperature in a vacuum in a series of steps. Gas fractions are collected at each step, and the ratio $^{40}Ar/^{39}Ar$ is measured mass spectrometrically. Since the ^{39}Ar is proportional to the ^{39}K, and since ^{39}K is proportional to ^{40}K, the ^{39}Ar is proportional to the ^{40}K in the sample. The measured $^{40}Ar/^{39}Ar$ ratio is therefore proportional to the $^{40}Ar/^{40}K$ ratio and is consequently a measure of age. If the sample, then, is fused after eight temperature steps, eight ages will have been found, one per step, and an age spectrum will result on a plot of fraction age versus temperature of the step (or, more commonly, versus fraction of ^{39}Ar released; see Figure 1).

1. IDEALLY SIMPLE THERMAL HISTORIES AND ASSOCIATED AGE SPECTRA The ideal materials for this category should be the tektites. They are formed in very short-lived, high-temperature events. After rapid cooling to the glassy

state, they have frequently undergone little or no subsequent significant thermal disturbance. The tight clustering of the various tektite K-Ar ages about the respective means for the strewn fields (Reynolds 1960, Gentner & Zähringer 1960, Gentner et al 1963, Zähringer 1963, Schaeffer 1966) supports the idea of a simple thermal history. In $^{40}Ar/^{39}Ar$ step-heating analyses of the Texas tektites (bediasites) (Bottomley & York, in preparation) the results agree with earlier conventional K-Ar dates but additionally display the absolutely flat age spectra that we would expect for all samples of category (1.) (Figure 1).

Such perfect plateaus, indicative of ideally simple histories, are rarely found outside the tektites, however, although some have been observed in stone meteorites (Turner et al 1978). The vast majority of spectra from terrestrial samples presumed to have had an undisturbed history display low ages in the first few small steps in the low-temperature end of the spectrum. This is presumably indicative of a slight loss of argon from crystal sites of low argon retentivity.

2. SINGLE-PULSE THERMAL HISTORIES AND RELATED AGE SPECTRA The first detailed interpretation of age spectra in terms of a single-pulse disturbance of the ideal was made by Turner (Turner et al 1966, Turner 1968) during his early studies of stone meteorites. Using the assumption that radiogenic

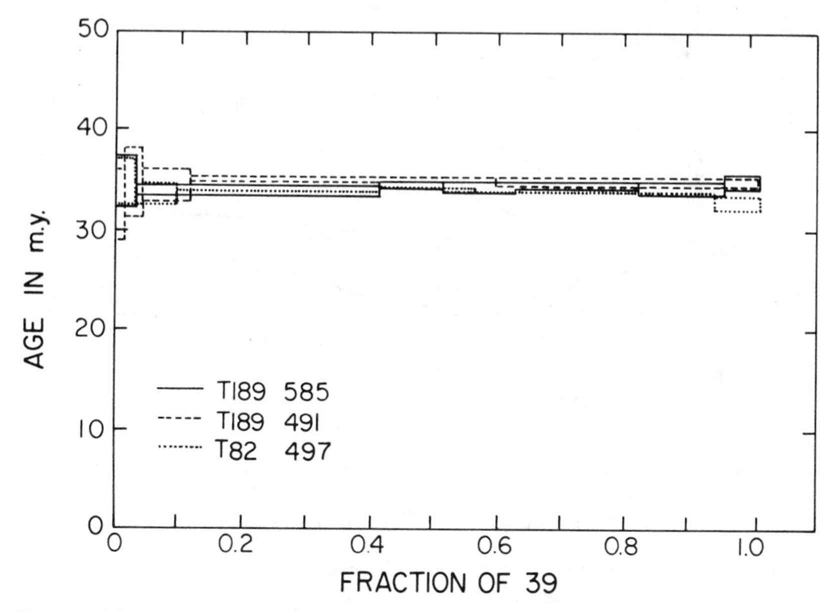

Figure 1 Three Texas tektites show essentially the ideal flat age spectrum to be expected from their simple thermal histories (Bottomley & York, in preparation).

argon loss during the brief pulse was governed by simple volume diffusion from spheres, Turner derived the spectra that should be found in a present-day analysis of such samples (Figure 2). As may be seen from the figure, two key features emerged from the calculations. Firstly, regardless of the fraction of radiogenic argon assumed to have been lost during the metamorphic pulse, the lowest temperature fraction in the age spectrum will give (within experimental errors) an upper limit to the time of occurrence of the pulse. The smaller the fraction of gas extracted in the first step, the more closely will the age of this step approach that of the actual pulse (Figure 3).

The second important feature of Turner's model curves (Figure 2) is the fact that a reasonable approximation to a high-temperature age plateau would be seen in the spectrum if argon losses during the pulse were less than

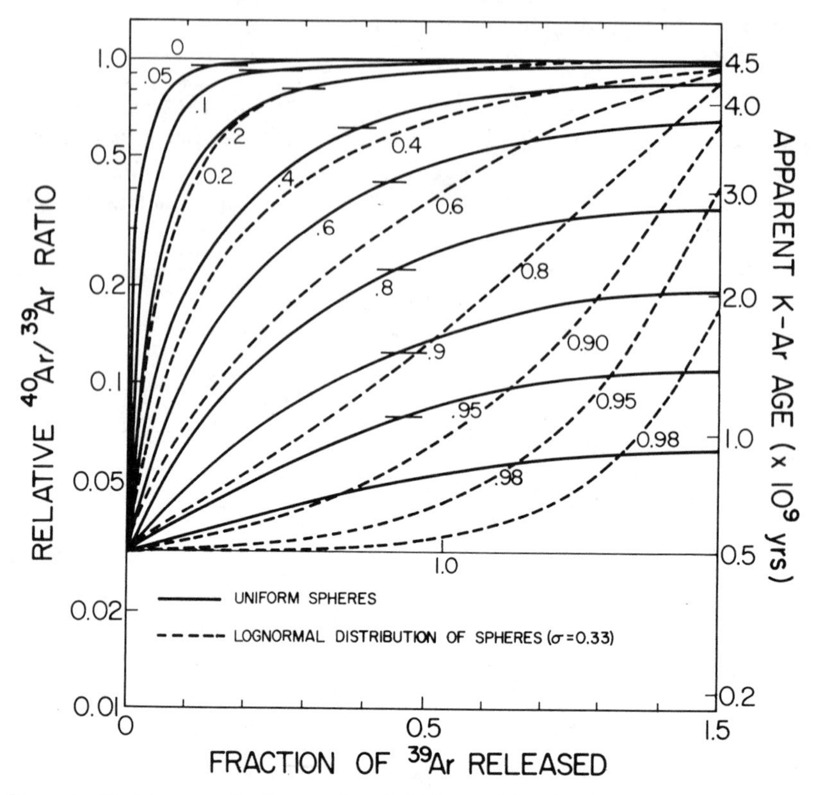

Figure 2 Model age spectra for samples of initial age 4.5 b.y. that have suffered, to differing degrees, a degassing event 500 m.y. ago (Turner 1969). The numbers on the curves indicate the different fractions of argon lost in the event. The solid curves are for a uniform collection of spheres. The dashed curves are for a log-normal distribution of grain sizes.

about 25%. If these fractional losses are small, then the curves show that the plateau region will give a lower limit that is very close to the time when the system became closed to argon diffusion. However, it is equally clear from the model curves that the apparent plateau falls further and further below the true initial age the larger the argon loss during the metamorphic episode (see also Huneke 1976). To calculate both the times of formation and disturbance of the sample, it is obviously necessary to find the model curve that fits the experimentally derived age spectrum.

Turner (1969) examined the low-temperature portions of a number of stone meteorite experimental age spectra and found that eight hypersthene chondrites recorded major outgassing events in the range 300–550 m.y. ago. The weighted mean of the six highest of these was 500 ± 30 m.y. ago, which agreed well with a value of 520 ± 60 m.y. ago deduced by Heymann (1967) from U, Th-He data for the date of a major outgassing event among the hypersthene chondrites. More recent studies of $^{40}Ar/^{39}Ar$ spectra for meteorites have been given by Bogard et al (1976), Bogard & Hirsch (1980), Wang et al (1980), and Harrison & Wang (1981). In the latter two papers, it

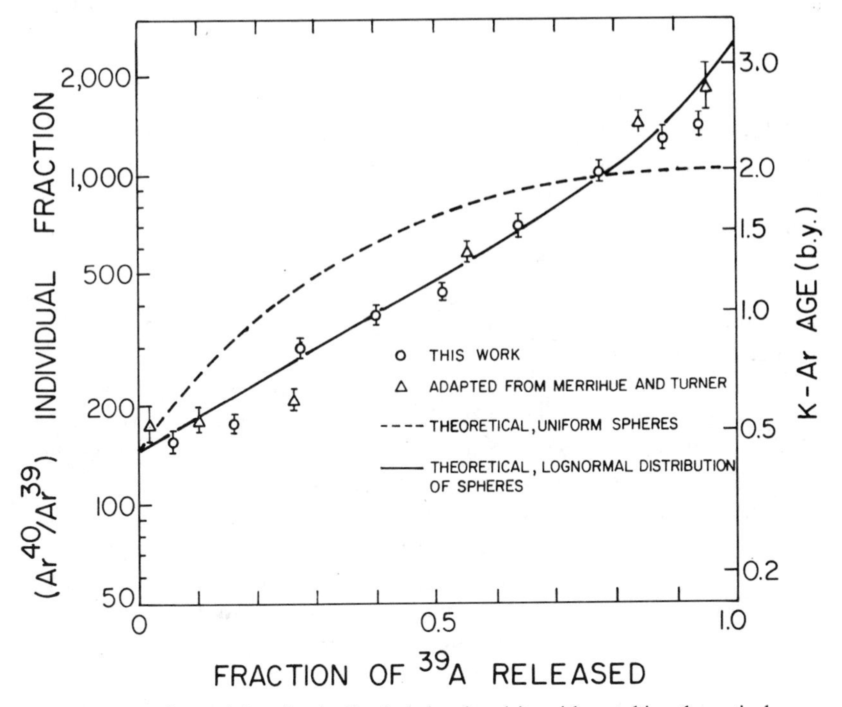

Figure 3 Experimental data for the Bruderheim chondrite with matching theoretical curve (for a log-normal distribution of grains), indicating a degassing age of 495 ± 30 m.y. (Turner et al 1966).

was argued that the age spectra recorded two outgassing events for the Kirin chondrite at 2.2 b.y. and 0.53 b.y. ago. Estimates were made of the temperatures experienced during the events and of the relative positioning of the fragments in an earlier parental body. Turner (1979) meanwhile presented some results of a Monte Carlo fragmentation model of meteorite production and discussed the statistical distribution of outgassing ages that could be expected.

Less episodic modeling of this type has been done with terrestrial samples. A notable exception is contained in the study of an intrusive contact in northwest Nelson, New Zealand, by Harrison & McDougall (1980b), where reasonable agreement between a known episodic event and the expected Turner profile (Figure 2) was found in some hornblendes (Figure 12). An interesting episodic modeling study was also reported by Gillespie et al (1983), who showed how the age of a young volcanic rock could be reliably found from the age spectrum of a partially degassed xenolith.

It is, of course, highly desirable to match the Turner profiles to the experimental data by a statistical method, rather than by eye, in order to extract the best estimates of original age, time of disturbance, and fraction of argon lost during the disturbance. Hall & York (1982) presented a least-squares approach to this problem (Figure 4).

A further interesting point was made by Turner (1969) in his paper on episodic loss from hypersthene chondrites. He noted that if a mineral were slowly cooling after formation and thereafter was undisturbed, the concentration profile of radiogenic ^{40}Ar in the mineral at a later date would not be a simple step-function with zero concentration at the boundary and uniform distribution across the grain. The concentration profile would, in fact, look "superficially similar" to that produced by an episodic loss. Turner did no detailed analytical modeling but concluded that since the Bjurböle chondrite showed an age spectrum that was essentially flat throughout (within ± 20 m.y.), the lower limit for the cooling rate of the Bjurböle parent body was 0.5–1°C/m.y. Berger & York (1981a) echoed Turner's comments in suggesting that the fine structure in a Grenville Province K-feldspar spectrum (see Figure 8) might reflect such extremely slow cooling. In a discussion of K-feldspars, Harrison & McDougall (1982) gave a diagram of the radiogenic argon concentration gradient that should be found in a plane sheet after slow cooling. Figure 5 shows the measured age spectrum of a biotite from the La Encrucijada pluton in Venezuela (Onstott et al 1983). Superimposed on this is a theoretically derived age spectrum (dashed lines) that a very slowly cooled biotite would be expected to display. A remarkably good fit can be seen, corresponding to a cooling rate of 0.43°C/m.y. To show the sensitivity of the shape of the age spectrum to the initial cooling rate, two theoretically derived smooth spectra are shown enveloping the experimental curve. These correspond to cooling rates of 0.72 and 0.30°C/m.y. It is clear from the graph (Figure 5) that if such a sample were indeed to experience such a history, its subsequent $^{40}Ar/^{39}Ar$ age spectrum would constrain the estimated cooling rate very precisely. The data cannot be fitted as well with the episodic-loss model using a single episode. To obtain a reasonable fit, at least two disturbances are required.

3. VERY SLOW POSTOROGENIC COOLING AND RELATED AGE SPECTRA The first routinized approach to estimating temperatures and corresponding times for very slowly cooling terrestrial materials was developed by York and Berger (Buchan et al 1977, Berger & York 1981a). Further applications were made and the approach refined by York and a series of coworkers (Berger & York 1979, Berger et al 1979, Berger & York 1981b, Lopez-Martinez & York 1983, Onstott et al 1983, Berger et al 1978). The Toronto work (Buchan et al 1977) introduced the concept of coupling the diffusion parameters of minerals (calculable from $^{40}Ar/^{39}Ar$ step-heating data) with

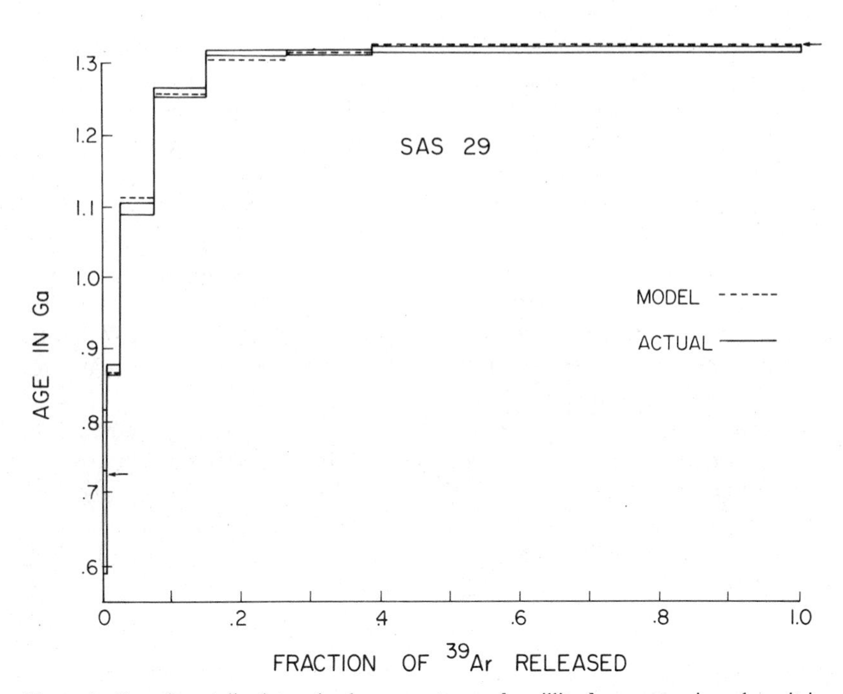

Figure 4 Experimentally determined age spectrum of an illite from a uranium deposit in northern Saskatchewan. The best-fitting least-squares matching spectrum was derived by Hall & York (1982). A continuous laser was used in the laboratory step-heating experiment (York et al 1981).

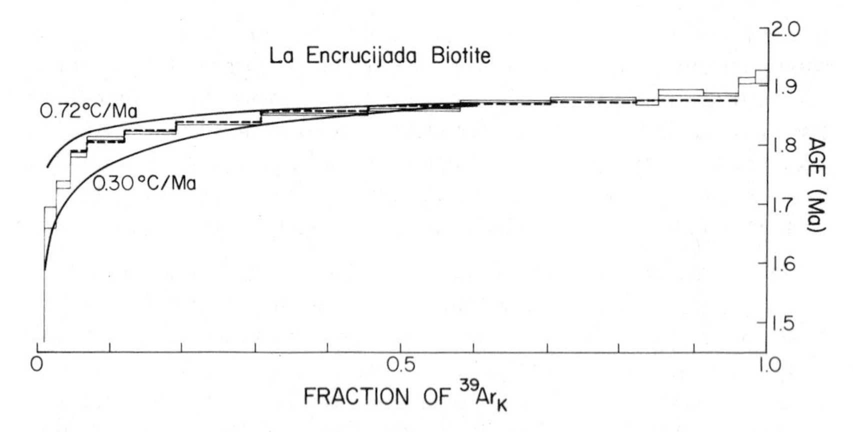

Figure 5 The $^{40}Ar/^{39}Ar$ age spectrum of biotite from the La Encrucijada intrusive in Venezuela, measured by T. Onstott. The matching theoretical stepped spectrum (derived by the author) is what would be found for a sample that cooled at a very slow rate (0.43°C/m.y.). The two continuous enveloping spectra were derived by C. M. Hall and correspond to cooling rates of 0.72 and 0.30°C/m.y.

Dodson's equation for isotopic blocking temperature (Dodson 1973), thereby calculating *simultaneously* for a mineral both the *time* when the mineral became a closed system for argon diffusion and the *temperature* at which this cessation of diffusion occurred. By performing such analyses for different mineral types from the same very slowly cooled rock body, one then obtained a series of times (ages) and corresponding temperatures. When these data were plotted on a graph of temperature versus time, it was possible to draw a cooling curve through the experimentally derived points. The first such thermal history curve derived in this way was given for a region of the Grenville Province (Canada) by Berger et al (1978), who adopted the principles, as well as some time-temperature data for hornblende and biotite, found in Buchan et al (1977).

Turner et al (1978) used a conceptually similar approach to calculate the argon blocking temperatures of meteorites using diffusion parameters from $^{40}Ar/^{39}Ar$ step-heating experiments. Mineral separates were not analyzed, the diffusion parameters being measured on whole-rock specimens.

We now consider the blocking phenomenon in a little more detail, giving a sketch of Dodson's blocking-temperature theory, followed by some applications.

Blocking-Temperature Considerations

Suppose the minerals in a deeply buried rock body are cooling very slowly following an orogenic episode. Furthermore, suppose that argon loss from these minerals is governed by the usual thermally activated diffusion process involving a diffusion coefficient $D = D_0 \exp(-E/RT)$, where D_0 is a frequency factor, E an activation energy for argon diffusion, T the absolute temperature, and R the gas constant. Then obviously D is a very strong function of temperature. For instance, if a given mineral has an activation energy $E = 50$ kcal mole^{-1}, its diffusion coefficient will change by a factor of approximately four as the temperature falls from 500°C to 470°C. Thus, a drop of a mere 30°C would change the average distance diffused by an atom in a given time by about a factor of two. Obviously, this means that such a mineral would make the transition from an open to a closed system for argon diffusion over a few tens of degrees. To a good approximation, therefore, it becomes meaningful to speak of a specific "blocking temperature" below which argon is no longer totally lost from the crystal but is henceforth totally retained (assuming no later reheating).

What Dodson (1973) did was to carry out successfully an elaborate, remarkable analysis of the interplay of daughter production by radioactive decay and daughter loss by diffusion as a crystal cools through its blocking temperature. The result was an expression for the blocking temperature given by

$$\frac{E}{RT_B} = \ln\left(A\frac{D_0}{a^2}\frac{RT_B^2}{E(-\dot{T})_B}\right), \tag{1}$$

where T_B is the blocking temperature; E the activation energy; R the gas constant; D_0 the frequency factor; a a dimension of the crystal; A a geometrical factor that is equal to 55, 27, or 8.7 for a sphere, cylinder, or sheet, respectively; and $(-\dot{T})_B$ the cooling rate at blocking. Dodson's analysis, of course, was carried out for any radioactive parent-daughter system, not merely for the K-Ar system.

The concept of isotopic blocking temperature was well known before Dodson's analysis and had been widely used (e.g. Anders 1963, Harper 1964, Jäger 1965, Armstrong 1966, Macintyre et al 1967). His discussion, however, was the first comprehensive one and contained a remarkable result. We have seen above that a mineral does not actually go from being a totally open system to a completely closed one at a single temperature at just one moment in time. There is a range of temperature (and therefore of time) over which blocking occurs. It might seem, therefore, that to talk of a specific blocking temperature (and corresponding time) is an idealization. And in some ways, this is correct. What Dodson showed, however, was that if a mineral crystallized (say, 2 b.y. ago), experienced slow cooling (say, at 3°C/m.y. for 100 m.y.), and subsequently, following uplift and rapid cooling, remained a completely closed system, then Equation (1) gives a definite single temperature [T_B in Equation (1)] that corresponds *exactly* (limited only by experimental errors) to the age calculated for that same mineral by a conventional K-Ar analysis. That is, at the time given by the K-Ar age of the sample, the temperature in the cooling body was precisely that given by T_B in Dodson's formula. This is true, even though the sample did not in some magical way snap shut at the calculated T_B value. Dodson defined the T_B value thus found as the blocking temperature. He actually used the term "closure temperature," but because in magnetic theory essentially the same concept has long been known as "blocking temperature," we use the latter expression. In fact, Dodson's equation, with the appropriate interpretation, applies equally well to magnetic blocking (York 1978a,b).

York and Berger (Buchan et al 1977) realized that Dodson's formula could be routinely combined with the $^{40}Ar/^{39}Ar$ step-heating data from samples that had experienced the appropriate slow-cooling history in order to calculate the temperature T_B to accompany the age found from the same data. They then set out to analyze different mineral species from a single body. Since different mineral types were known to have different diffusion parameters (Mussett 1969), it was hoped that the minerals with the highest values of T_B would exhibit the greatest ages, while those with lower T_B values should have correspondingly lower ages. Their results for the Bark Lake Diorite are shown in Figure 8 and are discussed later.

The T_B values are derived from the $^{40}Ar/^{39}Ar$ analytical data by calculating the values of E and D_0/a^2 to be substituted into Dodson's formula. To do this, an Arrhenius plot of $\ln D/a^2$ versus $1000/T$ must be constructed with the values of D/a^2 at each temperature (i.e. for each step in the $^{40}Ar/^{39}Ar$ experiment) calculated using the approach of Fechtig & Kalbitzer (1966). If the diffusion loss during the experiment obeys the expression $D = D_0 \exp(-E/RT)$, then the data should describe a straight line on the Arrhenius plot. The slope of the line gives the activation energy (E) and the y-intercept yields the frequency factor (D_0/a^2). Some data for hornblende, biotite, and feldspar (Berger & York 1981a) are shown in Figure 6.

When E and D_0/a^2 have been found for a mineral from the Arrhenius plot, the appropriate T_B value is then calculated from Equation (1). In doing this, two points should be noted. Firstly, the cooling rate at the time of blocking, $(-\dot{T})_B$, must be estimated. This, however, is no real problem, since the temperature T_B is not very sensitive to errors in assumed cooling rate because of the logarithmic nature of Equation (1). Furthermore, after a cooling curve has been constructed from the data for several minerals, a better estimate of cooling rate can be found from the slope of the cooling curve at the appropriate point. This new estimate of $(-\dot{T})_B$ may be then reinserted into Equation (1) and a better value for T_B calculated.

The second point to note is that Equation (1) is an implicit equation for T_B. A preliminary approximation for T_B must be chosen, inserted into the

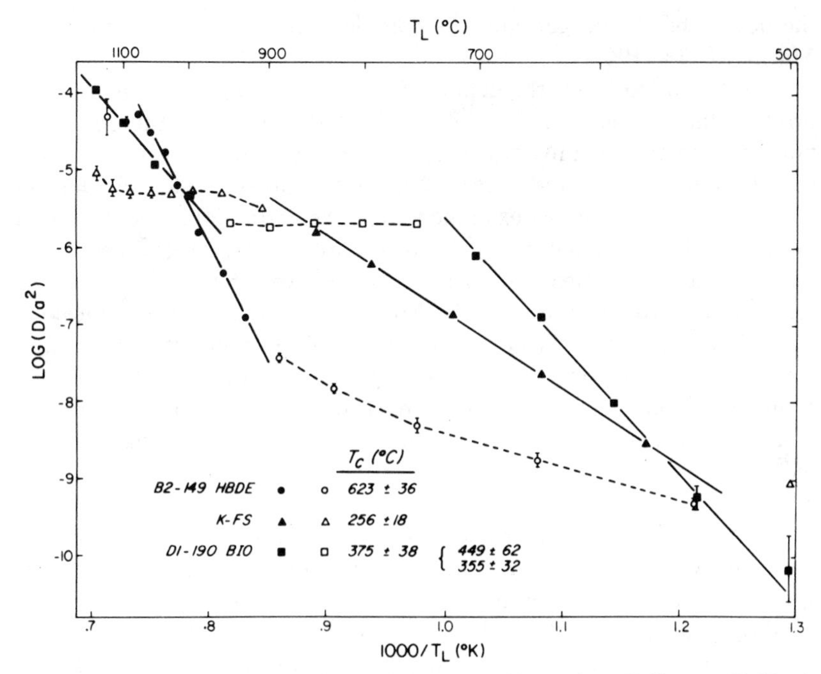

Figure 6 Arrhenius plot for hornblende, biotite, and K-feldspar from Haliburton Highlands intrusives. The blocking temperatures shown were calculated only from the solid data points. Note the kink in the biotite plot beginning close to 700°C (Berger & York 1981a).

right-hand side, and the equation solved for T_B. This new value is reinserted into the right-hand side of Equation (1) and a better T_B found, and so on. Usually one iteration is enough to calculate an accurate solution. The concepts are set out in detail in Berger & York (1981a), where the approach of simultaneously estimating a time (i.e. a date) and a corresponding temperature is called "thermochronometry." A thermochronometric study of a region of the southwest Grenville Province is described in the next section.

Postorogenic Cooling of the Grenville

The Grenville Province is a belt of predominantly high-grade metamorphic rocks in southern Canada, with exposures stretching from Ontario to the Atlantic coast (Figure 7). Ages from the metamorphic and intrusive rocks fall generally in a very wide range from about 1400 to 800 my. ago (Stockwell & Williams 1964, Silver & Lumbers 1965, Macintyre et al 1967, Krogh & Hurley 1968). In general, the higher ages are from U-Pb and Rb-Sr analyses. The majority of K-Ar analyses on micas are below 1000 m.y. in age, and it appears that these dates reflect the final slow cooling and uplift of the deeply buried orogen after its complex early history (Harper 1967, Macintyre et al 1967).

In order to work out the details of this final phase of one area of the orogen's thermal history, Berger & York (1981a) carried out thermochronometric analyses on mineral separates from three mafic intrusive bodies in the Haliburton Highlands area of the Grenville Province in Ontario (Figure 7). The intrusives examined were the Bark Lake and Dudmon dioritic and the Glamorgan gabbroic-anorthositic bodies. These were selected because of their importance for unraveling the paleomagnetic record of the Grenville (Buchan & Dunlop 1976). While it was critical for elucidating the tectonic history of the Grenville Province, the paleomagnetic evidence could only be clearly interpreted if the thermal history of the magnetized rocks involved was known. How the thermochronometry and paleomagnetism finally meshed, and the significance of this conjunction for the tectonic history of the Grenville Province, is described two sections below.

The age spectra for coexisting hornblende, biotite, potassium feldspar, and plagioclase from the Bark Lake diorite are shown in Figure 8. The spectra are striking because although the minerals all come from a single hand specimen, they record three quite different ages. The hornblende, biotite, and potassium feldspar all show clear plateaus in their age spectra, but the plateaus correspond to ages of about 970, 900, and 810 m.y. ago, respectively. Berger & York interpreted these results as being due to the

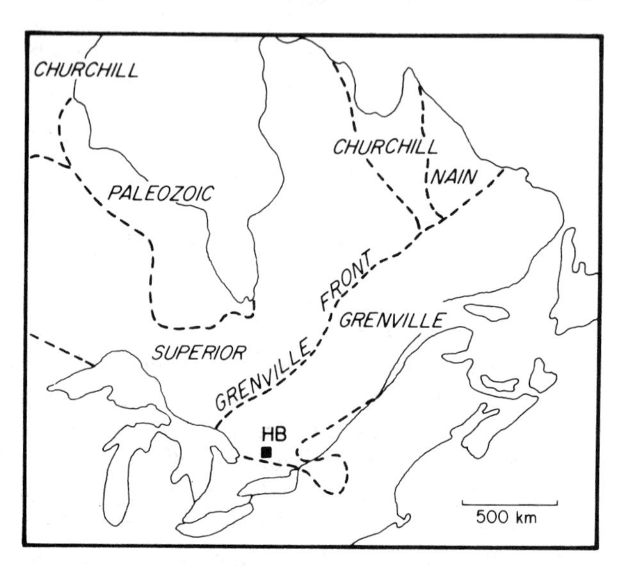

Figure 7 Location map of the Haliburton (HB) intrusives in the Grenville Province.

slow cooling of this part of the orogen after the peak of the Grenville metamorphism had passed. In this interpretation, prior to about 970 m.y. ago, all three minerals were completely open systems for argon retention. Then, as cooling proceeded, at about 970 m.y. ago, the hornblende blocking temperature was reached, and the hornblende argon clock was switched on. Meanwhile, the biotite and potassium feldspar still accumulated no argon. After another 70 m.y. of cooling, the biotite blocking temperature was reached and this mineral began to record time. Finally, after another 90 m.y., the potassium-feldspar clock was switched on at about 810 m.y. ago.

To estimate the actual blocking temperatures at which these different clocks switched on, Arrhenius plots were constructed (Figure 6), as referred to earlier. Values of E and D_0/a^2 were found, and Dodson's equation for T_B was solved for each mineral. The blocking temperatures calculated were 623 ± 36°C (1σ) for the hornblende, 352 ± 22°C for the biotite, and 230 ± 18°C for the potassium feldspar.

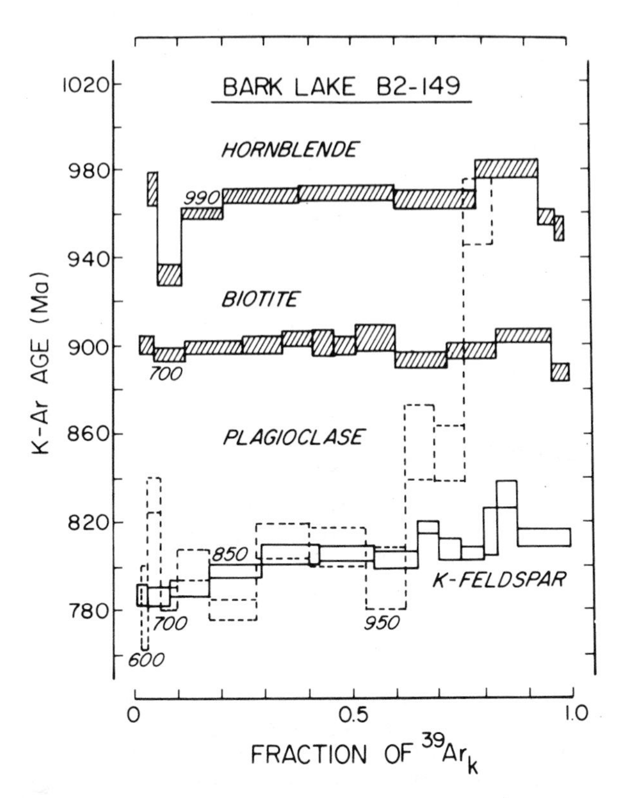

Figure 8 Age spectra from a single hand specimen of the Bark Lake diorite. Note that the plateau portion of the plagioclase overlaps that of the K-feldspar (Berger & York 1981a).

Several other hornblendes, biotites, and plagioclases from Bark Lake and the Glamorgan body were analyzed, along with a biotite from the Dudmon diorite. The data (age and corresponding temperature) for all the samples are shown in Figure 9, but before discussing the cooling curve drawn through them, it is necessary to discuss some of the decisions that had to be made during the calculations.

Firstly, since the fractions in an age spectrum will not all lie along a perfect plateau, and since not all the data points in the associated Arrhenius plot will lie (within error) on a single straight line, it is necessary to establish minimum criteria that the data must satisfy in order to qualify as "reliable" blocking temperatures. Berger & York (1981a) adopted two criteria: "Firstly, at least five points (in the Arrhenius plot) should correspond to a 'reliable' plateau in an age-spectrum plot. Secondly, the same five points should lie on a statistically well-defined straight line in an Arrhenius diagram." A reliable plateau in the age spectrum was defined as "any sequence of five or more heating steps where all dates or all but one agree

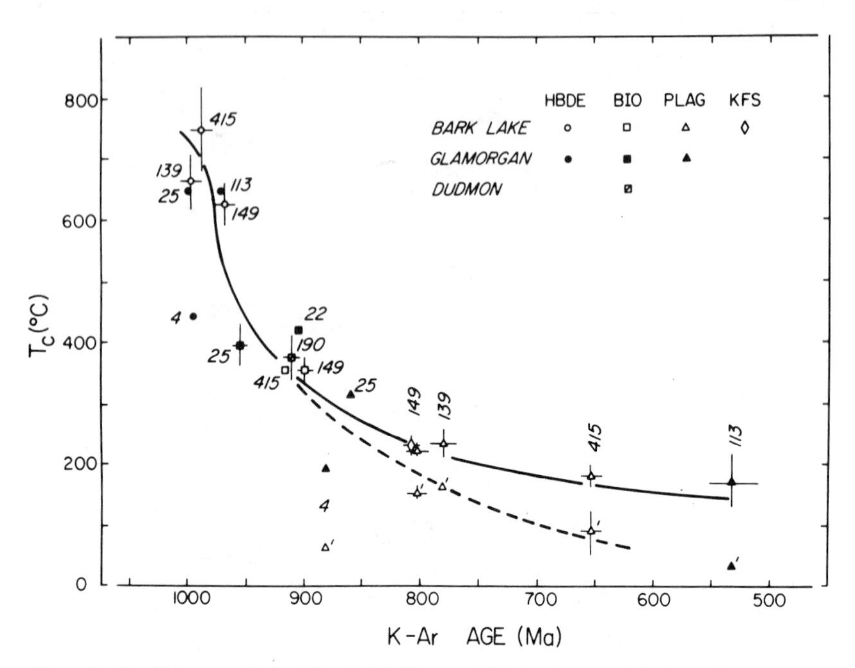

Figure 9 Cooling curve derived for the Haliburton Highlands area of the Grenville Province. The solid curve is preferred to the alternative late-stage dashed segment. The preferred curve uses the feldspar blocking temperatures calculated with the "multi-component" approach described in the text. The overall accuracy of the curve remains to be determined, but there can be little doubt that the sequence of hornblende, biotite, and feldspar ages outlines the eventual cooling history of an orogen from high to low temperature (Berger & York 1981a).

within 2σ error limits." Using these criteria, 11 out of 18 samples were considered to yield reliable blocking temperature values. The data points for these eleven are shown in Figure 9 with error bars. The unreliable estimates are displayed without them.

A second problem is presented by the structure of the biotite Arrhenius plots. Instead of defining a single straight line, all the samples displayed essentially two subparallel straight line segments separated by a kink that begins at 700°C (see Figure 6). Berger & York (1981a) considered that this discontinuity is probably the result of a breakdown of the biotites during the heating in a vacuum. Interlayer spacing may be affected as water of composition is driven off. The authors took weighted means of the two blocking temperatures that could be calculated for the two linear segments. It seems, however, that the lower-temperature line would be the better choice, since it is derived prior to the onset of an episode of instability in the vacuum heating.

It should be noted, however, that while all the Haliburton biotites showed kinks in their Arrhenius plots, not all biotites display this characteristic 700°C break. Onstott et al (1983), in a thermochronometric study of rocks from the Venezuelan Shield, found several biotites that displayed straight lines on Arrhenius plots over a wider range of temperature. These micas showed no sign in their Arrhenius plots of structural change during vacuum heating until temperatures of 800°C had been reached. Microprobe measurements at Princeton are currently being performed on these samples (from both Ontario and Venezuela) to determine what chemical parameters control this variation in behavior.

A third problem arose in the interpretation of the plagioclase age spectra, which were generally saddle- or U-shaped. Usually, there was an approximation to a plateau at the bottom of the U in the laboratory temperature range $\simeq$700–1000°C. Since in the case of the samples from the single hand specimen from Bark Lake, this plagioclase plateau portion coincides in age with the K-feldspar plateau (Figure 8), Berger & York (1981a) treated the plagioclase segment as though it came from a distinct phase. In essence, they calculated E and D_0/a^2 from the plateau region as though it were a separate mineral. This was called the "multi-component" approach. When this was done for the Bark Lake plagioclase, a blocking temperature of 225 ± 8°C [for $(-\dot{T})_B = 0.5$°C/m.y.] was found, which agreed with the value of 230 ± 18°C found for the coexisting K-feldspar. Equality of ages and equality of blocking temperatures were thus obtained for the two feldspar types. This multi-component approach was used with the remaining plagioclases.

The Haliburton Highlands cooling curve finally obtained is shown in Figure 9. It is the first such curve deduced from a self-contained single isotopic system. Its closeness to reality remains to be determined. It does indicate, however, that about 1000 m.y. ago the temperature in these rocks was about 700°C. At 975 m.y. ago, the cooling was fairly rapid at about 20°C/m.y. By 925 m.y. ago, the cooling rate had fallen to about 2°C/m.y., and 200 m.y. later an extremely slow cooling rate of about 0.3°C/m.y. existed. The concavity of the curve suggests that the initially rapid uplift was followed by a considerable slowing down with a consequently long cooling tail.

Thermal Modeling

Obviously, such a cooling curve would provide a powerful constraint on models of the thermal behavior of the Earth during and following orogeny (Vitorello & Pollack 1980, England 1980). Such modeling, however, has not yet been done for the Grenville Province. Useful studies combining thermal modeling and isotope measurements have been made by Clark & Jäger (1969) and Wagner et al (1977) on Alpine rocks. These investigations do not involve $^{40}Ar/^{39}Ar$ measurements and are on an orogenic belt in a very different stage of development from that seen in the Grenville; thus, they are not discussed here.

More recently, Harrison & McDougall (1980a) showed how it was possible to fit a theoretically derived cooling curve to isotopic age and blocking-temperature data. These workers synthesized the age data (most of which were not $^{40}Ar/^{39}Ar$) for a variety of minerals with their estimated blocking temperatures and deduced the thermal history of the Separation Point Batholith (shown in Figure 10). This Cretaceous intrusion occurs in northwest Nelson, New Zealand.

The theoretical model curve found to approximate the experimental isotopic data illustrates the enormous potential power of isotopic measurements in this context. If the blocking-temperature estimates can be verified, they will obviously provide enormously important information on tectonothermal processes when combined with the corresponding age measurements.

Other similar thermal-history studies by Harrison and coworkers may be found in Harrison & McDougall (1981), Harrison et al (1979), and Harrison & Clarke (1979), in which the mathematics of the thermal modeling is described.

Implications for Precambrian Plate Tectonics

The revolution in our understanding of the kinematics of the behavior of the Earth's surface and the unraveling of the details of recent plate motions have been based on the magnetic record in the ocean floors (Vine & Matthews 1963, Morley & Larochelle 1964). Unfortunately, because of the

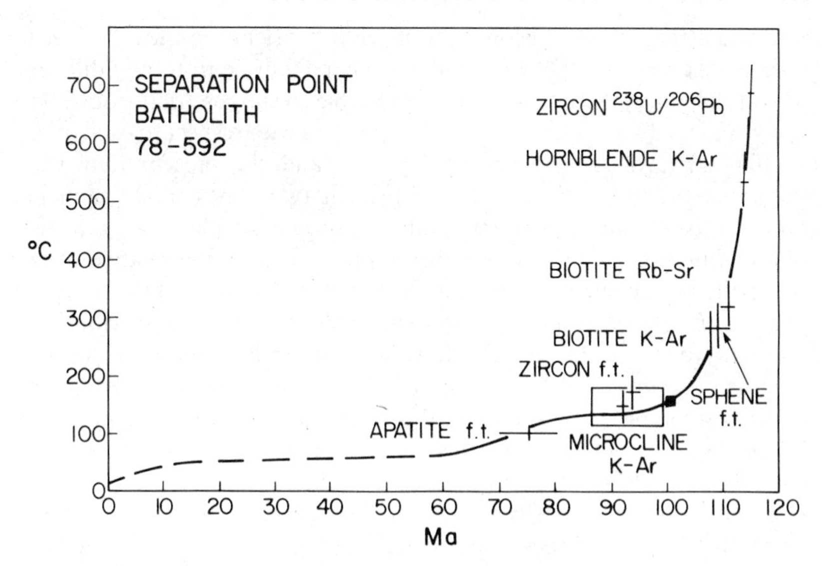

Figure 10 The cooling curve of the Separation Point Batholith. Its derivation differs from the curve in Figure 9 in that it is not based on $^{40}Ar/^{39}Ar$ thermochronometry. K-Ar, Rb-Sr, U-Pb, and fission-track age and blocking temperature estimates have been synthesized for matching with the solid curve, which was deduced from a thermal model (Harrison & McDougall 1980a).

subduction of ocean floors, this record scarcely goes back 200 m.y., a mere 5% of the Earth's history. No vestige remains in the ocean floors of the record of the first 95% of our planet's surface behavior. It will, therefore, be extremely difficult to work out the outlines, let alone the details, of plate motions during the vast history of the Precambrian. So far as I am aware, by far the most rewarding approach will be to retread in and extend the footsteps left in the 1950s by Runcorn, Irving, and others (Runcorn 1956, Irving 1964), whose work on apparent polar wander paths recorded in rocks less than 400 m.y. in age led to the revival of interest in continental drift that precipitated the plate tectonic revolution.

The principle of the approach is simple. The practice is hard. Imagine two continental segments that at this moment may be different parts of the same large continent or may be fragments of two currently widely geographically separated continents. Suppose the two segments both contain igneous rocks spanning a given time interval, say 2.0 to 1.5 b.y. ago. Then suppose we determine the apparent polar wander path (APWP) recorded in these rocks on one of the continental segments and compare this curve with that derived from the corresponding rocks on the second continent. If the two curves are similar and fall on top of each other, within experimental error, then it is extremely likely that the two continental segments have maintained their present geographical relative orientation for 2.0 b.y. If, however, the two APWPs are similar in shape, but lie in totally different parts of the globe, then it would be reasonable to suppose that from 2.0 to 1.5 b.y. ago the two continental pieces were fixed with respect to each other, but that somewhere between 1.5 b.y. ago and the present they were mutually reoriented. By moving the present two continental fragments about relative to each other on the globe, we should be able to bring the two APWPs into coincidence. When this happens, it fixes the relative orientations the two segments had in the time interval 2.0 to 1.5 b.y. ago.

Such an approach to Precambrian plate tectonics is obviously a potentially powerful technique. However, everything hinges on the construction of accurate APWPs. Such a curve is a line (or swath) drawn on the globe (or some projection) linking the successive apparent pole positions (Figure 11). The direction of the curve is from older to younger. To construct an APWP for Precambrian rocks, then, requires accurate measurements of pole positions and of the times when these were frozen into the rocks. Both of these requirements are difficult, even for Paleozoic age rocks. For Precambrian rocks, it becomes very difficult to satisfy them. The vast majority of Precambrian rocks have been metamorphosed and/or altered to some extent, and this makes it extremely difficult to recover accurate original directions of magnetization. Given, however, that reliable Precambrian poles can be determined, the problem remains as to how one can measure the time at which the rocks involved were magnetized. If one is studying basic dykes, for instance, and if these are fresh and unmetamorphosed, then this is a relatively straightforward problem (Gates & Hurley 1973). The rapid cooling of dykes will usually mean that the magnetization and the switching on of all the various isotopic clocks will have effectively occurred simultaneously. The age of crystallization will be essentially the age of magnetization, and any careful age measurement would, in principal, date the magnetization. However, pristine basic dykes only sample Precambrian time in a limited, uneven way and are not uniformly distributed in space. Paleomagnetists have therefore had to turn to the very slowly cooled and often metamorphosed rocks deep within an orogen to find Precambrian poles. In such cases, the magnetic poles in the rocks will have been acquired long after the time of original crystallization or metamorphism (i.e. sometimes well over 100 m.y. after). If one finds several different magnetic poles in a single specimen of slowly cooled igneous or metamorphic rock (say, from the Grenville Province), then it is of little use (magnetically speaking) to know the Rb-Sr whole rock age or the U-Pb zircon age of the body. In a slowly cooling orogen, such as the Grenville, the episodes of magnetization will postdate crystallization by extremely important time intervals.

To date the magnetic pole (or poles) recorded in a slowly cooled orogenic rock, we need two pieces of information. Firstly, we require the temperature at which the magnetic component was acquired, the so-called magnetic blocking temperature of the magnetization. Secondly, we need the later stages of the thermal history of the orogen in that area. Given these items, then the date of the magnetization is found simply by reading off the cooling curve the time when the temperature of the orogen was equal to the magnetic blocking temperature of the magnetic phase being dated. That this was the approach to take was emphasized by Ueno et al (1975). It was the hope of implementing this method in the Haliburton Highlands area of the Grenville Province that led Berger and York to develop the thermochronometric method of combining the diffusion treatment of their data with Dodson's equation (Buchan et al 1977, Dodson 1973).

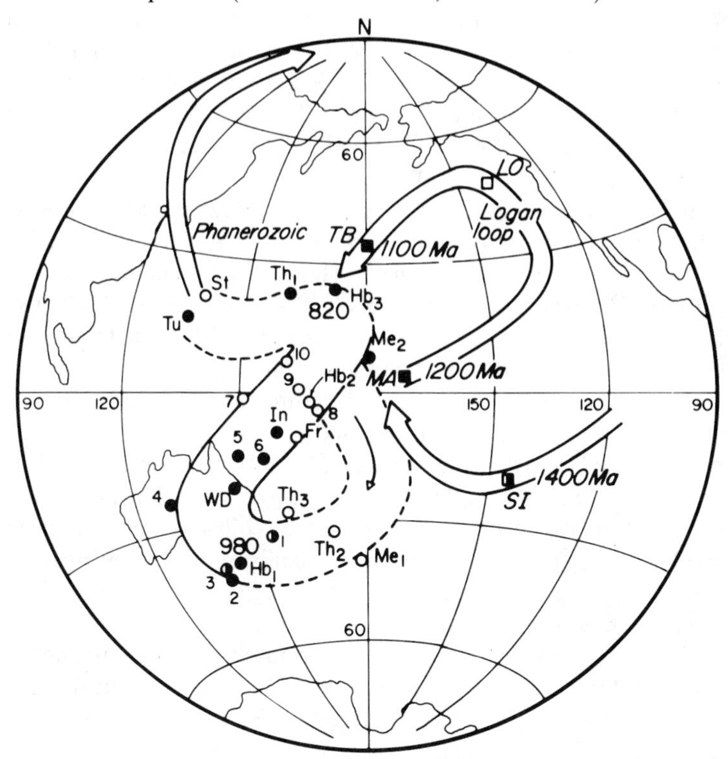

Figure 11 Grenville paleopoles fall largely to the south and west of the 1400–1100 m.y. ago segment of the Interior Laurentia apparent polar wander path (APWP). The Haliburton Highlands thermochronometry showed that the Grenville poles are almost certainly significantly younger than the Interior Laurentia poles and cannot be used to support the idea of a collision between a Grenville plate and Interior Laurentia. Berger et al (1979) linked the Grenville poles with the Interior Laurentia APWP in the so-called Grenville Loop.

How the cooling curve of Figure 9 was used to clarify the Grenville paleomagnetic record and what this means for Grenville tectonics are briefly described in the next section.

Grenville Tectonics and Paleomagnetism

As paleopoles were acquired from an increasing number of Grenville rocks during the late 1960s and the 1970s (Hargraves & Burt 1967, Palmer & Carmichael 1973, Buchan & Dunlop 1973, Irving et al 1974b), it became apparent that the Grenville poles did not fall on the APWP for the period 1300–1100 m.y. ago, which had been determined from rocks (mostly rapidly cooled) from the rest of North America, i.e. from Interior Laurentia (Figure 11). It was therefore suggested that if the Grenville poles were of this age (1100–1300 m.y. ago), then the present positioning of the Grenville Province with respect to Interior Laurentia could not be that which existed 1100–1300 m.y. ago. In fact, from the distribution of the Grenville poles with respect to the Interior Laurentia APWP, it was possible to suggest that the paleomagnetic data were a record of the convergence and collision of the Grenville plate with Interior Laurentia in this time interval (Irving et al 1974a, Buchan & Dunlop 1976).

With the derivation of the cooling curve of Figure 9, it became possible to test this very attractive hypothesis. Buchan & Dunlop (1973, 1976) had found two paleopoles recorded in the Haliburton Highlands rocks. They had also determined the temperatures (520–650°C and $\simeq$250°C) at which these two poles were acquired (their field magnetic blocking temperatures). Berger et al (1979) therefore read off Figure 9 the times when this part of the Grenville passed through these temperatures and concluded that these paleopoles were acquired at about 980 ± 10 and $\lesssim 820$ m.y. ago. This showed that the Grenville poles were significantly younger than the Interior Laurentia poles with which they were being compared. The paleomagnetic data did not therefore record a continental collision that produced the Grenville orogeny (see Figure 11). Such a collision (Dewey & Burke 1973) may well have occurred. If it did, however, the presently known paleomagnetic record does not show it.

Dunlop and his coworkers have tried to find Grenville paleopoles that have survived the intense thermal history of the orogeny, but so far to no avail. By examining rocks from the Hastings Basin, a region of low-grade metamorphism, they hope to find a magnetic window through the orogeny. However, the most promising candidate so far—the magnetism in the Cordova Gabbro—has failed the thermochronometric test. An anomalous pole found in this body could be interpreted to be a record of a small ocean that existed between the Grenville and Interior Laurentia plates over 1200 m.y. ago. The closure of such a hypothesized ocean could be speculated to have been the cause of the Grenville orogeny (Dunlop et al 1980). However, detailed $^{40}Ar/^{39}Ar$ studies of hornblendes and a plagioclase from the gabbro showed that a very mild thermal pulse had reset the very sensitive plagioclase clock approximately 450 m.y. ago (Lopez-Martinez & York 1983). Strikingly, the anomalous Cordova pole falls near the 450 m.y. ago position on the North American APWP. This is therefore obviously either a remarkable coincidence, or else the thermal pulse that reset the plagioclase clock also produced the anomalous Cordova pole, which is not therefore a record of a pre-Grenville ocean.

Limitations

From our preceding discussions, it is obvious that the step-heating approach to $^{40}Ar/^{39}Ar$ dating has enormous potential for the elucidation of the thermal histories of planetary material. However, it should be emphasized that much remains to be done to verify the various thermal histories that have been deduced.

In the simple episodic-loss models referred to, for example, it is extremely difficult to verify the postulated "major meteorite fragmentation event (or events) within the last 500 m.y." (Turner 1979) used in the interpretation of Figure 3. Turner et al (1979) reported that brief (15 min) high-temperature (1000–1100°C) heating of the Barwell chondrite in the laboratory did in fact set the low-temperature fractions in the age spectrum to zero, in keeping with the episodic-loss model. However, heating to 1300°C did not fully reset the clock, which allows the possibility of erroneously high ages in the low-temperature portion of the spectrum, perhaps because of argon redistribution during partial melting. In commenting on these data, Turner (1979) said, "The young outgassing ages (300–500 m.y.) inferred for the hypersthene chondrites (Turner 1969) appear to be a marginal case requiring more detailed investigation."

Several encouraging results, however, were found by Harrison & McDougall (1980b) for hornblendes from the Rameka Gabbro (emplacement age $\simeq$367 m.y. ago), which had been unquestionably (from the direct dating of the later intrusion) reheated about 114 m.y. ago by the Separation Point Batholith. It can be seen in Figure 12 how reasonably an episodic-loss model curve fits the data for one sample. The model curve shown corresponds to the hornblende having been switched on as a clock 367 m.y. ago and then having lost 31% of its radiogenic ^{40}Ar 114 m.y. ago. The fit is not perfect, but there is a general consistency between the idealized model curve and the experimental data.

As for the thermochronometry, much verification and development remain to be done. We have already seen that frequently the age spectra and/or the associated Arrhenius plots are disturbed and that selection criteria are required. More fundamentally important than this is the question of the validity of taking the diffusion parameters, deduced from relatively brief heating experiments carried out with the sample in vacuo, and applying them to the field situation. In the field, the minerals were often cooling extraordinarily slowly under elevated pressure and were possibly surrounded by fluid and vapor phases (including argon isotopes). Berger & York (1981a) took the pragmatic approach that since the $^{40}Ar/^{39}Ar$ step-heating runs always contain diffusion information, it would be negligent not to solve the simple equations involved and routinely calculate blocking temperatures. The resulting thermal histories could then eventually be tested internally and externally for consistency. However, the uncertainties surrounding the use of vacuum-heating data remain.

It has long been known (Evernden et al 1960) that different laboratory diffusion behavior is observed depending upon the experimental conditions. Giletti (1974) has emphasized the differences observed in diffusion parameters depending on whether the minerals are heated hydrothermally

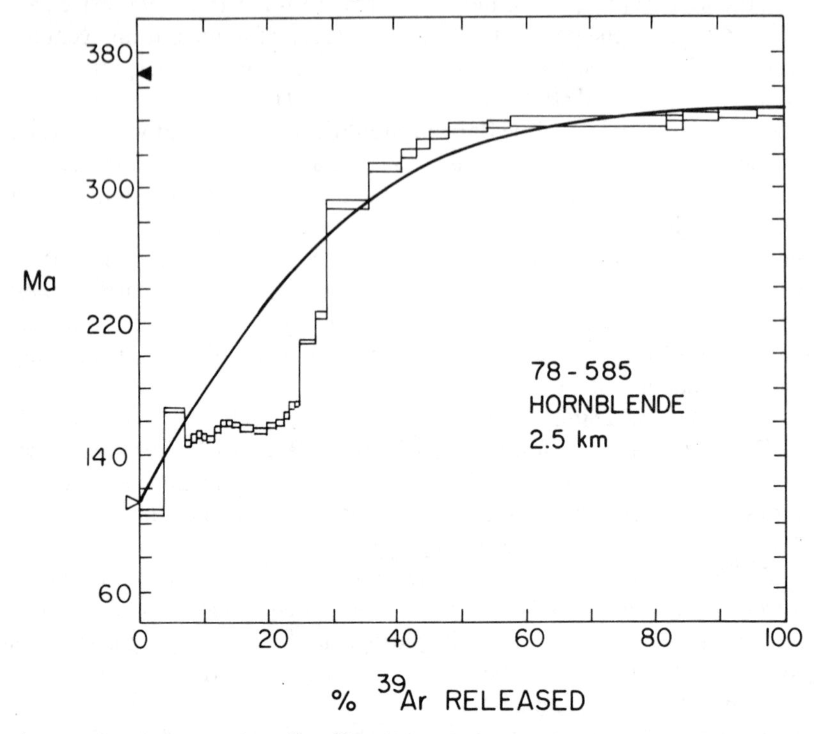

Figure 12 Age spectrum from $^{40}Ar/^{39}Ar$ dating of a Rameka hornblende. The solid curve is the theoretical spectrum corresponding to formation of the sample 367 m.y. ago, followed by degassing of 31% of the radiogenic argon present 114 m.y. ago (Harrison & McDougall 1980b).

or in vacuo. Difficulties in interpretation are obviously introduced if the mineral undergoes a phase transition or loss of volatiles during vacuum heating. This introduces questions in the handling of hornblende and biotite vacuum-generated data.

As Harrison & McDougall (1980b) have noted, Berger & York (1981a) calculated hornblende blocking-temperature values from relatively high-temperature laboratory data. If the hornblendes had lost water before reaching these temperatures, then diffusion parameters might well have been significantly altered as a consequence. This is, of course, quite possible, and only much further research can clarify how best to perform laboratory diffusion experiments to mimic accurately the field experience. Only limited data on hornblende are available from hydrothermal studies. Harrison (1981) reported some data for two hornblendes. The results are very useful, but they also illustrate the difficulties that remain to be overcome before the correct parameters for thermochronometry are available. Firstly, because of experimental requirements, the data were all taken (at $\geq 750°C$) above the range at which the field blocking would occur. What the hydrothermal data would show for this lower-temperature range is not known. To seek some answer to this, Harrison (1981) plotted three lower-temperature data points derived from a field-based study. While these points fell on the continuation of the hydrothermal line, it is not clear what the significance of this is, since the positioning of these three points is model dependent.

The extrapolation of the hydrothermal data to lower temperatures is further made very difficult with hornblende, because even at 750°C, an extremely small gas loss ($<5\%$) must be measured. Uncertainties in the diffusion coefficients at these temperatures are at least a factor of four. An additional complication arises with the interpretation of such low fractional losses of argon in hydrothermal studies. These minute losses will be controlled by the concentration gradients in the boundary regions of the crystals. In the vast majority of crystals, these are very different from the ideal horizontal gradients assumed so far in hydrothermal experiments. When small losses are occurring, grain boundary diffusion may also be a complicating factor.

As noted earlier, the data from vacuum-heated biotites often show a kink beginning at about 700°C on the Arrhenius plot. While hydrothermal data are lacking, it is possible to speculate that perhaps the correct biotite blocking temperatures are derivable from the data obtained in vacuum-heating experiments below 700°C. Or it may turn out that those biotites that march straight through (for $\geq 100°C$) the 700°C stage are the only reliable ones to use. It will be extremely useful to carry out hydrothermal studies on these two types of biotites, which show such different behavior when heated in vacuo.

Much new hydrothermal experimentation is evidently needed. In particular, some new approaches are required to extend these measurements much nearer to the lower temperatures at which the minerals presumably blocked in the field.

It was noted by Harrison & McDougall (1982) that feldspar thermochronometry developed by Berger & York (Berger et al 1978, Berger & York 1981a) was likely to be less subject to problems with vacuum-breakdown of the minerals being studied because of their anhydrous nature. Berger & York had shown the importance of plagioclase and K-feldspar for fleshing out the lower-temperature end of the cooling curves (see Figure 9) in the approximate temperature range 100–250°C. Harrison & McDougall (1982) deduced a blocking temperature of 132 ± 13°C from three samples of microcline from the Separation Point Batholith and came out in support of at least this aspect of thermochronometry, stating that "these internally consistent observations suggest that microcline geothermometry using kinetic data obtained via the step-heating experiment may yield meaningful temperature information although we reiterate that temperature control of these extractions was poor." They also found quite good linear correlations on Arrhenius plots of four anorthoclases, noting that the activation energies of these homogeneous feldspars were much higher ($\simeq 40$ kcal mole^{-1}) than for the perthitic microcline ($\simeq 29$ kcal mole^{-1}).

Conclusions

Age spectra from $^{40}Ar/^{39}Ar$ dating come in many shapes. Some are very simple to interpret, and the same interpretation will be found by almost all analysts. Many are complex, and interpretations will be more contentious, sometimes impossible. There can be little doubt, however, that complex though they may sometimes be, the various spectra are far from a jumble of random shapes. A particular mineral type tends to be characterized by a particular type of spectrum in a given context. The $^{40}Ar/^{39}Ar$ spectra in general undoubtedly reflect the thermal history of the sample coupled with its chemical environment [for instance, were there locally high ambient partial pressures of argon? (Harrison & McDougall 1981)]. The challenge is to extract these histories from the spectra. The attempts to do this, described earlier, represent only the beginning. Spectra are being matched to the output of fairly simple models that involve either simple thermal episodes (Turner 1969, 1979) or straightforward slow-cooling histories (Berger & York 1981a). When a much better understanding of the field and laboratory diffusion processes has been attained, it is very probable that quite detailed, reliable thermal histories will be extracted routinely from age spectra by inverse theory (Albarede 1978).

The $^{40}Ar/^{39}Ar$ blocking temperatures of hornblende, mica, and feldspars span the range of magnetic blocking temperatures usually found. This means that even in its present, primitive, embryonic state, $^{40}Ar/^{39}Ar$ thermochronometry is a necessary adjunct of paleomagnetic studies. This is particularly so in the Precambrian, where because of complex and often protracted thermal histories, times of magnetization must be expected frequently to postdate times of intrusion or metamorphism by geologically important time intervals.

ACKNOWLEDGMENTS

My views on blocking phenomena and argon age spectra have been strongly influenced over the years by discussions with Drs. C. T. Harper, G. W. Berger, C. M. Hall, and T. C. Onstott.

Literature Cited

Albarede, F. 1978. The recovery of spatial isotope distributions from stepwise degassing data. *Earth Planet. Sci. Lett.* 39:387–97

Anders, E. 1963. Meteorite ages. In *The Moon, Meteorites and Comets*, ed. B. M. Middlehurst, G. P. Kuiper, pp. 402–95. Univ. Chicago Press. 810 pp.

Armstrong, R. L. 1966. K-Ar dating of plutonic and volcanic rocks in orogenic belts. In *Potassium-Argon Dating*, ed. O. A. Schaeffer, J. Zähringer, pp. 117–33. Berlin: Springer-Verlag. 234 pp.

Berger, G. W., York, D. 1981a. Geothermometry from $^{40}Ar/^{39}Ar$ dating experiments. *Geochim. Cosmochim. Acta* 45:795–811

Berger, G. W., York, D. 1981b. $^{40}Ar/^{39}Ar$ dating of the Thanet gabbro, Ontario: looking through the Grenvillian metamorphic veil and implications for paleomagnetism. *Can. J. Earth Sci.* 18:266–73

Berger, G. W., York, D., Dunlop, D. J., Buchan, K. L. 1978. $^{40}Ar/^{39}Ar$ dating of Precambrian apparent polar wander paths. *Short Pap. Int. Conf., 4th, Geochron. Cosmochron. Isot. Geol. Surv. Open File Rep. 78-70*, pp. 30–32

Berger, G. W., York, D., Dunlop, D. J. 1979. Calibration of Grenvillian paleopoles by $^{40}Ar/^{39}Ar$ dating. *Nature* 277:46–48

Bogard, D. D., Hirsch, W. C. 1980. $^{40}Ar/^{39}Ar$ dating, Ar diffusion properties and cooling rate determinations of severely shocked chondrites. *Geochim. Cosmochim. Acta* 44:1667–82

Bogard, D. D., Husain, L., Wright, R. J. 1976. $^{40}Ar/^{39}Ar$ dating of collisional events in chondrite parent bodies. *J. Geophys. Res.* 81:5664–78

Buchan, K. L., Dunlop, D. J. 1973. Magnetization episodes and tectonics of the Grenville Province. *Nature* 246:28–30

Buchan, K. L., Dunlop, D. J. 1976. Paleomagnetism of the Haliburton intrusions: superimposed magnetizations, metamorphism and tectonics in the late Precambrian. *J. Geophys. Res.* 81:2951–67

Buchan, K. L., Berger, G. W., McWilliams, M. O., York, D., Dunlop, D. J. 1977. Thermal overprinting of natural remanent magnetization and K/Ar ages in metamorphic rocks. *J. Geomagn. Geoelectr.* 29:401–10

Clark, S. P. Jr., Jäger, E. 1969. Denudation rates in the Alps from geochronologic and heat flow data. *Am. J. Sci.* 267:1143–60

Dalrymple, G. B., Lanphere, M. A. 1969. *Potassium-Argon Dating*. San Francisco: Freeman. 258 pp.

Dewey, J. F., Burke, K. C. A. 1973. Tibetan, Variscan and Precambrian basement reactivation: products of continental collision. *J. Geol.* 81:683–92

Dodson, M. H. 1973. Closure temperature in cooling geochronological and petrological systems. *Contrib. Mineral. Petrol.* 40:259–74

Dunlop, D. J., York, D., Berger, G. W., Buchan, K. L., Stirling, J. M. 1980. The Grenville Province: a paleomagnetic case-study of Precambrian continental drift. In *The Continental Crust and its Mineral Deposits*, ed. D. W. Strangway, pp. 487–502. *Geol. Assoc. Can. Spec. Pap. No. 20*

England, P. 1980. Heat flow and deep structure of the continents. *Nature* 285:611–12

Evernden, J. F., Curtis, G. H., Kistler, R. W., Obradovich, J. 1960. Argon diffusion in glauconite, microcline, sanidine, leucite, and phlogopite. *Am. J. Sci.* 258:583–604

Fechtig, H., Kalbitzer, S. 1966. The diffusion of argon in potassium-bearing solids. In *Potassium-Argon Dating*, ed. O. A. Schaeffer, J. Zähringer, pp. 68–107. Berlin: Springer-Verlag. 234 pp.

Gates, T. M., Hurley, P. M. 1973. Evaluation of Rb/Sr dating methods applied to the Matachewan, Abitibi, MacKenzie and Sudbury dike swarms in Canada. *Can. J. Earth Sci.* 10:900–19

Gentner, W., Zähringer, J. 1960. Das Kalium-Argon-Alter von Tektiten. *Z. Naturforsch.* 15a:93–99

Gentner, W., Lippolt, H. J., Schaeffer, O. A. 1973. Argonbestimungen an Kalium-mineralien. XI. Die Kalium-Argon-Alter des Gläser des Nördlinger Rieses und der bömisch-mährischen Tektite. *Geochim. Cosmochim. Acta* 27:191–200

Giletti, B. J. 1974. Diffusion related to geochronology. In *Geochemical Transport and Kinetics*, ed. A. W. Hofmann, B. J. Giletti, H. S. Yoder Jr., R. A. Yund, pp. 61–76. Carnegie Inst. Washington. 353 pp.

Gillespie, A. R., Huneke, J. C., Wasserburg, G. T. 1983. Eruption age of a Pleistocene basalt from $^{40}Ar/^{39}Ar$ analysis of partially degassed xenoliths. *J. Geophys. Res.* 88: 4997–5008

Hall, C. M., York, D. 1982. Least-squares estimation of formation and metamorphic ages from $^{40}Ar/^{39}Ar$ age spectra. *Eos, Trans. Am. Geophys. Union* 63:453–54

Hargraves, R. B., Burt, D. W. 1967. Paleomagnetism of the Allard Lake anorthosite suite. *Can. J. Earth Sci.* 4:357–69

Harper, C. T. 1964. Potassium-argon ages of slates and their geological significance. *Nature* 203:468–70

Harper, C. T. 1967. On the interpretation of potassium-argon ages from Precambrian shields and Phanerozoic orogens. *Earth Planet. Sci. Lett.* 3:128–32

Harrison, T. M. 1981. Diffusion of ^{40}Ar in hornblende. *Contrib. Mineral. Petrol.* 78: 324–31

Harrison, T. M., Clarke, G. K. C. 1979. A model of igneous intrusion and uplift as applied to Quotoon Pluton, B.C. *Can. J. Earth Sci.* 16:411–20

Harrison, T. M., McDougall, I. 1980a. Investigations of an intrusive contact, northwest Nelson, New Zealand. I. Thermal, chronological and isotopic constraints. *Geochim. Cosmochim. Acta* 44:1985–2003

Harrison, T. M., McDougall, I. 1980b. Investigations of an intrusive contact, northwest Nelson, New Zealand. II. Diffusion of radiogenic and excess ^{40}Ar in hornblende revealed by $^{40}Ar/^{39}Ar$ age spectrum analysis. *Geochim. Cosmochim Acta* 44:2005–20

Harrison, T. M., McDougall, I. 1981. Excess ^{40}Ar in metamorphic rocks from Broken Hill, New South Wales: implications for $^{40}Ar/^{39}Ar$ age spectra and the thermal history of the region. *Earth Planet. Sci. Lett.* 55:123–49

Harrison, T. M., McDougall, I. 1982. The thermal significance of potassium feldspar K-Ar ages inferred from $^{40}Ar/^{39}Ar$ age spectrum results. *Geochim. Cosmochim. Acta* 46:1811–20

Harrison, T. M., Wang, S. 1981. Further $^{40}Ar/^{39}Ar$ evidence for the multicollisional heating of the Kirin chondrite. *Geochim. Cosmochim. Acta* 45:2513–17

Harrison, T. M., Armstrong, R. L., Naeser, C. W., Harakal, J. E. 1979. Geochronology and thermal history of the Coast Plutonic Complex, near Prince Rupert, B.C. *Can. J. Earth Sci.* 16:400–10

Heymann, D. 1967. On the origin of hypersthene chondrites: ages and shock effects of black chondrites. *Icarus* 6:189–221

Huneke, J. C. 1976. Diffusion artifacts in dating by step-wise thermal release of rare gases. *Earth Planet. Sci. Lett.* 28:407–17

Irving, E. 1964. *Palaeomagnetism.* New York: Wiley. 399 pp.

Irving, E., Emslie, R. F., Ueno, H. 1974a. Upper Proterozoic poles from Laurentia and the history of the Grenville structural province. *J. Geophys. Res.* 79:5491–5502

Irving, E., Park, J. K., Emslie, R. F. 1974b. Paleomagnetism of the Morin complex. *J. Geophys. Res.* 79:5482–90

Jäger, E. 1965. Rb-Sr age determinations on minerals and rocks from the Alps. In *Géochronologie Absolute, Cent. Natl. Rech. Sci. No. 151*, pp. 191–201

Krogh, T. E., Hurley, P. M. 1968. Strontium isotope variation and whole-rock isochron studies, Grenville Province of Ontario. *J. Geophys. Res.* 73:7107–25

Lopez-Martinez, M., York, D. 1983. Further chronometric unravelling of the age and palaeomagnetic record of the southwest Grenville Province. *Can. J. Earth Sci.* 20:953–60

Macintyre, R. M., York, D., Moorhouse, W. W. 1967. Potassium-argon age determinations in the Madoc-Bancroft area in the Grenville Province of the Canadian Shield. *Can. J. Earth Sci.* 4:815–28

Merrihue, C. M. 1965. Trace-element determinations and potassium-argon dating by mass spectroscopy of neutron-irradiated samples. *Trans. Am. Geophys. Union* 46:125

Merrihue, C. M., Turner, G. 1966. Potassium-argon dating by activation with fast neutrons. *J. Geophys. Res.* 71:2852–57

Morley, L. W., Larochelle, A. 1964. Palaeomagnetism as a means of dating geological events. In *Geochronology in*

Canada, ed. F. F. Osborne, pp. 39–51. *R. Soc. Can. Spec. Publ. No. 8*

Mussett, A. E. 1969. Diffusion measurements and the potassium-argon method of dating. *Geophys. J. R. Astron. Soc.* 18:257–303

Onstott, T. C., Hargraves, R. B., York, D., Hall, C. M. 1983. Constraints on the motions of South American and African Shields during the Proterozoic. I. $^{40}Ar/^{39}Ar$ and paleomagnetic correlations between Venezuela and Liberia. *Geol. Soc. Am. Bull.* In press

Palmer, H. C., Carmichael, C. M. 1973. Paleomagnetism of some Grenville Province rocks. *Can. J. Earth Sci.* 10:1175–90

Reynolds, J. H. 1960. Rare gases in tektites. *Geochim. Cosmochim. Acta* 20:101–14

Runcorn, S. K. 1956. Paleomagnetic comparisons between Europe and North America. *Proc. Geol. Assoc. Can.* 8:77–85

Schaeffer, O. A. 1966. Tektites. In *Potassium-Argon Dating*, ed. O. A. Schaeffer, J. Zähringer, pp. 162–73. Berlin: Springer-Verlag. 234 pp.

Silver, L. T., Lumbers, S. B. 1965. Geochronologic studies in the Bancroft-Madoc area of the Grenville Province, Ontario, Canada. *Geol. Soc. Am., Program 1965 Ann. Meeting*, p. 153. (Abstr.)

Stockwell, C. H., Williams, H. 1964. Fourth report on structural provinces, orogenies, and time-classification of rocks of the Canadian Precambrian Shield. *Geol. Surv. Can. Pap. 64-17, Part 2*

Turner, G. 1968. The distribution of potassium and argon in meteorites. In *Origin and Distribution of the Elements*, ed. L. H. Ahrens, pp. 387–97. London: Pergamon. 1178 pp.

Turner, G. 1969. Thermal histories of meteorites by the ^{39}Ar-^{40}Ar method. In *Meteoritic Research*, ed. P. M. Millman, pp. 407–17. Dordrecht: Reidel. 940 pp.

Turner, G. 1979. A Monte Carlo fragmentation model for the production of meteorites: implications for gas retention ages. *Proc. Lunar Planet. Sci. Conf., 10th*, pp. 1917–41

Turner, G., Miller, J. A., Grasty, R. L. 1966. Thermal history of the Bruderheim meteorite. *Earth Planet. Sci. Lett.* 1:155–57

Turner, G., Enright, M. C., Cadogan, P. H. 1978. The early history of chondrite parent bodies inferred from ^{40}Ar-^{39}Ar ages. *Proc. Lunar Planet. Sci. Conf., 9th*, pp. 989–1025

Turner, G., Enright, M. C., Hennessy, J. 1979. Diffusive loss of argon from chondritic meteorites. *Meteoritics* 13:648–49

Ueno, H., Irving, E., McNutt, R. H. 1975. Paleomagnetism of the Whitestone Anorthosite and Diorite, the Grenville polar track, and relative motions of the Laurentian and Baltic Shields. *Can. J. Earth Sci.* 12:209–26

Vine, F. J., Matthews, D. H. 1963. Magnetic anomalies over ocean ridges. *Nature* 199:947–49

Vitorello, I., Pollack, H. N. 1980. On the variation of continental heat flow with age and the thermal evolution of continents. *J. Geophys. Res.* 85:983–95

Wagner, G. A., Reimer, G. M., Jäger, E. 1977. Cooling ages derived by apatite fission-track, mica Rb-Sr and K-Ar dating: the uplift and cooling history of the Central Alps. *Mem. Padova 30.* 27 pp.

Wang, S., McDougall, I., Tetley, N., Harrison, T. M. 1980. $^{40}Ar/^{39}Ar$ age and thermal history of the Kirin chondrite. *Earth Planet. Sci. Lett.* 49:117–31

Wetherill, G. W. 1956. Discordant uranium-lead ages. *Trans. Am. Geophys. Union* 37: 320–26

York, D. 1978a. A formula describing both magnetic and isotopic blocking temperatures. *Earth Planet. Sci. Lett.* 39:89–93

York, D. 1978b. Magnetic blocking temperature. *Earth Planet. Sci. Lett.* 39:94–97

York, D., Farquhar, R. M. 1972. *The Earth's Age and Geochronology.* Oxford: Pergamon. 178 pp.

York, D., Hall, C. M., Yanase, Y., Hanes, J. A., Kenyon, W. J. 1981. $^{40}Ar/^{39}Ar$ dating of terrestrial minerals with a continuous laser. *Geophys. Res. Lett.* 8:1136–38

Zähringer, J. 1963. K-Ar measurements of tektites. In *Radioactive Dating*, pp. 289–305. Vienna: Int. At. Energy Agency

Ann. Rev. Earth Planet. Sci. 1984. 12: 411–43

TECTONICS OF VENUS

Roger J. Phillips

Department of Geological Sciences, Southern Methodist University, Dallas, Texas 75275

Michael C. Malin

Department of Geology, Arizona State University, Tempe, Arizona 85281

INTRODUCTION

Insight into the tectonic evolution and present state of the Earth has been gained over the past century through geological, geophysical, and geochemical/petrological investigations at a variety of spatial scales. At one time, specific mechanisms were considered mostly on a regional to subregional basis, and the discipline was more properly termed "structural geology." More recently, investigations of terrestrial tectonism have concentrated on a global framework provided by the concepts of seafloor spreading and continental drift. The basic tenets of the plate tectonic hypothesis have been confirmed by detailed investigations in paleomagnetism and seismology.

What chance do we have of unraveling the tectonics of other planets, where the number of detailed and useful data sets is extremely limited? Compounding this basic problem are the unique conditions of each planet: each exhibits such different characteristics as to present a different challenge in tectonic interpretation. For example, the Moon is, with the exception of the Earth, the planet for which we have acquired the largest number of useful data sets. A combination of data from photogeology, petrology, radiometric age determinations, seismology, and gravity and heat-flow studies shows the Moon to have been in a state of tectonic quiescence over the last 4 b.y., with only minor regional deformation of the lithosphere. Signs of the more vigorous tectonics that may have existed early in the Moon's history, when the planet was thermally robust, have been

0084-6597/84/0515-0411$02.00

obliterated by the intense meteoritic impact that persisted to about 3.9 b.y. ago. A similar statement can be made for Mercury, although signs of global tectonism, possibly due to despinning and cooling, are evident.

The tectonics of Mars has been addressed principally through photogeologic interpretation, aided by low-resolution elevation and gravity data. Photographic resolution is better than one kilometer over the entire planet, permitting detailed searches for landforms with tectonic implications. Mars clearly reveals volcanic-tectonic events that took place since the end of heavy meteoritic bombardment. In particular, the Tharsis region displays the dominant tectonic features of Mars: grabens, fractures, and enormous shield volcanos. These features have been interpreted in terms of the stress fields associated with the topography and the gravity field of the planet (Banerdt et al 1982, Willemann & Turcotte 1982). However, the origin of the great elevation of Tharsis, largely responsible for the lithospheric stress field, is enigmatic (e.g. Solomon & Head 1982a) and will likely remain so, since the more detailed data sets that are available for the Earth are lacking for Mars.

Two simple concepts are helpful in approaching the question of the tectonics of planets: (*a*) The global tectonics of a terrestrial planet reflects the style in which the outer rigid shell—the lithosphere—is able to expel the planet's internal heat. (*b*) The history of a planet's heat flux—its thermal history—is dependent on a planet's size, or the ratio of volume to surface area. Small planets should reach thermal robustness early in their history and then decline rapidly, while larger planets should remain thermally active for longer periods of time (Kaula 1975). This appears to be the case when we compare the Moon to Mars to Earth.

An axiom to point (*b*) above is that planets of approximately the same size and mass—such as Venus and Earth—can be expected to have similar histories of heat flux from their interior. Then, given point (*a*), a reasonable prediction is that Venus and Earth ought to have the same tectonic history, and specifically, Venus ought to be presently in a state of plate tectonics.

A plate tectonic comparison of Earth and Venus is particularly valuable because it offers an opportunity to evaluate the effects of those variables that are different for the two planets. The concepts and axioms discussed above are, of course, simplistic, and terrestrial plate tectonics is controlled by more than just the quantity of heat coming out of the Earth's interior. The importance of other geophysical variables in controlling plate tectonics is a source of considerable disagreement. For example, two additional factors are the role of oceanic lithospheric buoyancy and the role of a low-viscosity zone beneath the lithosphere.

Venus, then, offers the potential to aid in a greater understanding of the mechanisms of the Earth's tectonics. But our present understanding of venusian tectonics rests on data sets that are woefully inadequate compared with the information that was available to verify the plate tectonic hypothesis on Earth.

Surface Observations

Observations of the surface of Venus come from three main sources: Earth-based radars, Soviet Venera surface landers, and the US Pioneer Venus (PV) orbital radar altimeter.

Earth-based radar data are of three types—topographic profiles, reflectivity maps, and topographic maps. Topographic profiles around the equator (with spatial resolutions of about 50 km and vertical resolutions of about 150 m relative and 300 m absolute) provide some information on regional and global elevation variations (Pettengill 1978, Campbell et al 1972, Golovkov et al 1976). These equatorial data reveal variations as great as 6.5 km over distances of thousands of kilometers, relief contrasts of 4.5 km in as little as 100–200 km, and a number of areas with variations of up to 2 km in distances approaching 600 km. Radar reflectivity maps portrayed as photographic images (Rogers & Ingalls 1969, Goldstein & Rumsey 1970, Rumsey et al 1974, Goldstein et al 1976, 1978, Campbell et al 1976, 1979, Campbell & Burns 1980) can be used to study surface roughness and, in some instances, the physiography of surface features. It is from these data that many of the geological interpretations of the surface features are derived. Topographic maps acquired simultaneously with the high-resolution reflectivity maps through bistatic and tristatic radar observations (Rumsey et al 1974, Goldstein et al 1976) have somewhat lower spatial resolutions, with a vertical resolution of several hundred meters.

The Venera 8, 9, 10, 13, and 14 lander spacecraft returned observations of surface chemistry and, except for Venera 8, images of the surrounding terrain. From gamma-ray spectrometry on Venera 8, 9, and 10 (Surkov 1977) and X-ray fluorescence spectrometry on Venera 13 and 14 (Surkov et al 1983), the surface materials sampled appear generally basaltic in composition. Rocks and fines have low albedos, consistent with their mafic composition.

Orbital radar observations have greatly advanced our knowledge of Venus. The PV spacecraft acquired altimetry over 93% of the planet at horizontal resolutions of 25–100 km and a standard error of measurement in the vertical dimension of 200 m (Pettengill et al 1980). These data, along with reflectivity measurements, provide the basis for global studies of the physiography and its geologic interpretation (Masursky et al 1980, McGill et al 1983). Topography on Venus is generally confined to values within a few kilometers of the mean planetary radius (6051.9 km). The histogram of percent surface area as a function of altitude is unimodal and quite distinct from that of the Earth, which is bimodal. The terrestrial histogram reflects the differences between high-standing continental and low-standing oceanic crust. Venus has high-standing areas, but they occupy a relatively small fraction of surface area and merge smoothly with low-lying areas; thus they do not form a distinct mode in the histogram.

The geography of Venus is relatively simple (Figures 1 and 2). There are two continental-sized elevated areas: Aphrodite Terra, about the size of Africa and situated near the equator in the eastern hemisphere; and Ishtar Terra, about the size of Greenland and situated on the prime meridian at 70°N. Beta, Phoebe, Atla, Metis, and Alpha regione are the largest of several dozen smaller, elevated regions scattered over the surface of Venus. Beta and Phoebe, located near 270°W and about 25° north and south of the equator, respectively, are the most prominent. A small number of basins exist and are no more than about 2 km below the mean elevation. A few (Guinevere Planitia, Aino Planitia) are distinctly elongate, with the linear dimension several thousands of kilometers in length. Atalanta Planitia is nearly circular (about a thousand kilometers in diameter).

Gravity Data

Variations in the gravity field of a planet are an important means to test tectonic hypotheses. Both the magnetic and gravity fields were mapped by the PV orbiter, but the former displays no contribution by a component intrinsic to the planet (Russell et al 1980) and thus has little bearing on tectonic discussions (but it may have implications for thermal history). The gravity field has been mapped by time differentiating the variations in spacecraft velocity implied by the Doppler shift of a radio communication signal. This yields an estimate for the component of acceleration (gravity) in the line-of-sight (LOS) between the spacecraft and Earth. Systematic variations in the gravity field were first discussed by Phillips et al (1979). A strong correlation between gravity variations and topography was reported by Phillips & Lambeck (1980) and subsequently by others (Sjogren et al 1980, Reasenberg et al 1982, Esposito et al 1982, Sjogren et al 1983). This strong correlation of gravity to topography (Figure 3) is in marked contrast to the Earth, where the correlation is very weak at long wavelengths (Phillips & Lambeck 1980).

There are several ways to characterize a quantitative relationship between gravity and topography. In terrestrial applications the most common measure is the *spectral admittance function*, the ratio of gravity to topography as a function of wavelength (McKenzie & Bowin 1976). On a spatial basis, the slope of the linear regression function on a scatter diagram of gravity vs topography defines a *spatial admittance function* (e.g. Phillips & Saunders 1975). The spectral approach is perhaps the more useful

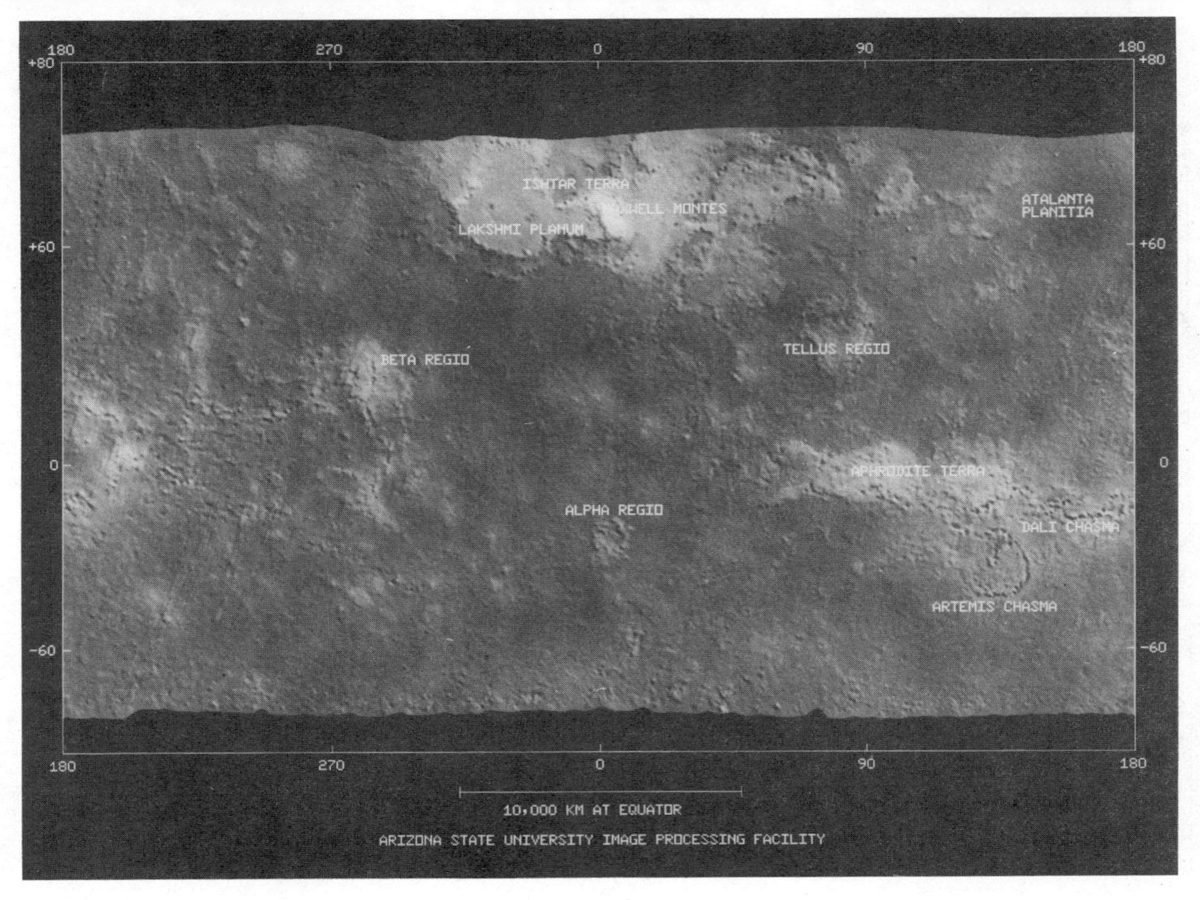

Figure 1 Shaded relief image of Venus, showing major physiographic features. Brightness of image directly corresponds with elevation.

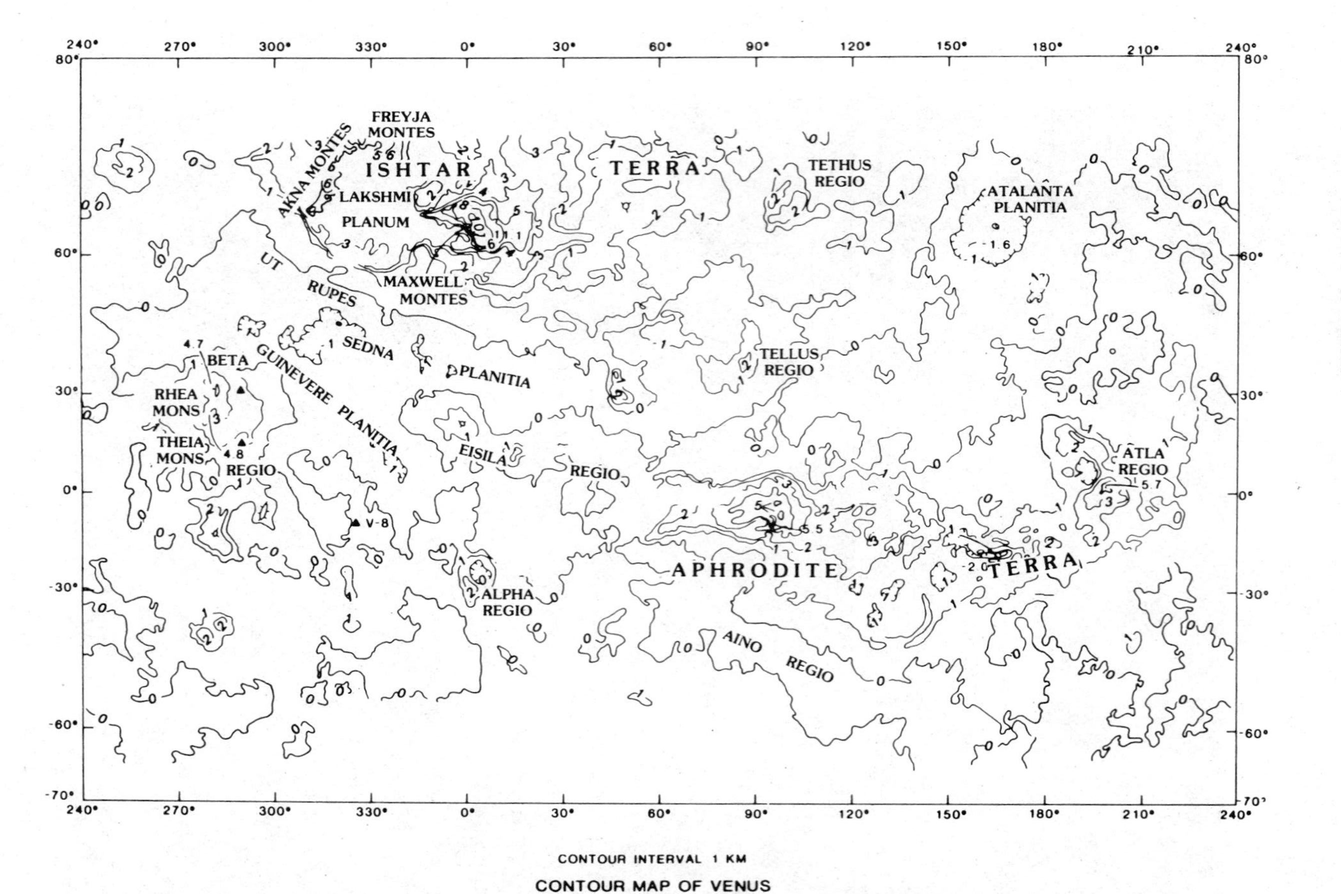

Figure 2 Topographic contour map of Venus, showing some of the major place names (modified from Masursky et al 1980).

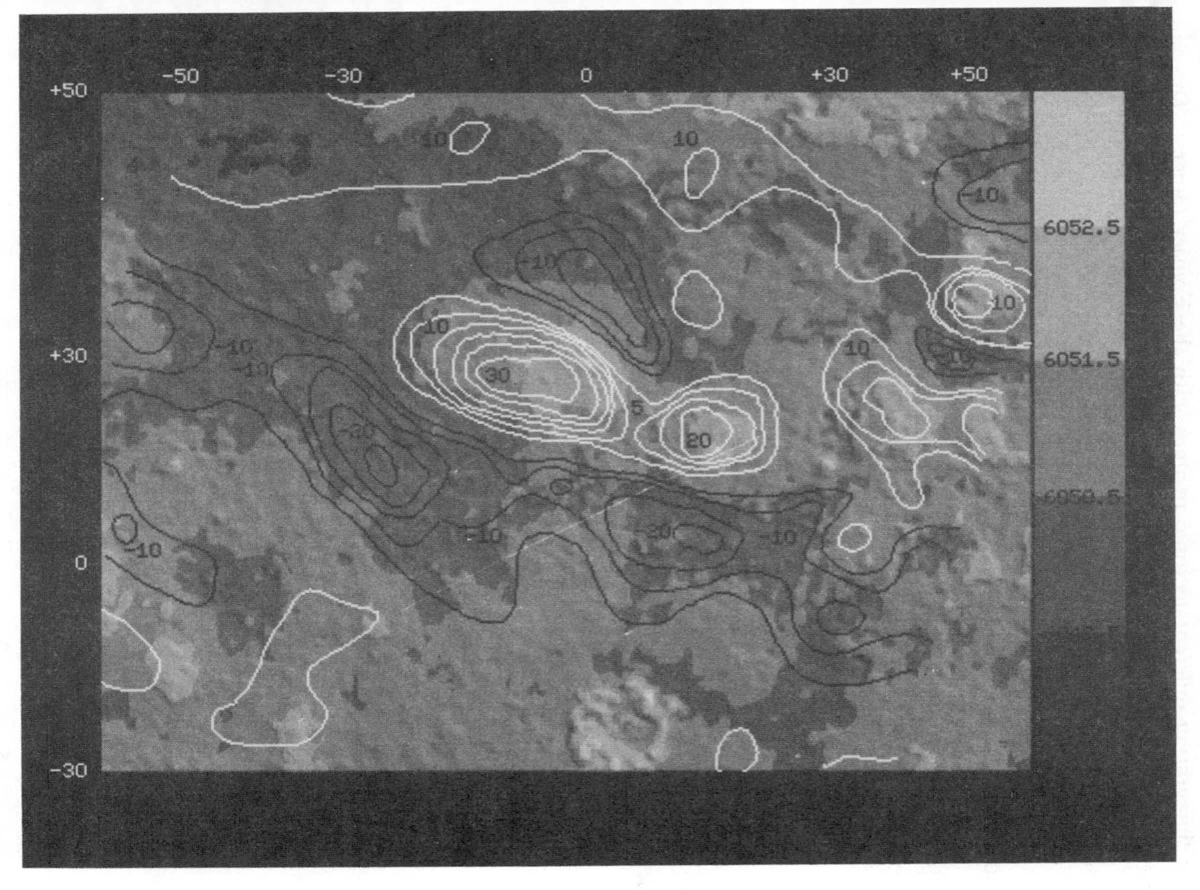

Figure 3 LOS gravity contours (in mgal) superimposed on a shaded relief map for a portion of the equatorial region of Venus. The brighter shades correspond to higher elevation; a clear correlation between gravity and topography is observed (modified from Phillips & Lambeck 1980).

technique, because interpretative geophysical models are wavelength dependent. However, with our limited knowledge of Venus, we run the risk of integrating different tectonic regimes with the Fourier integral. The spatial admittance estimate represents a more cautious alternative, although the characteristic wavelength must often be crudely estimated and assigned to topographic features in order to carry out interpretation, thus weakening its reliability.

Figure 4 (from Sjogren et al 1983) shows spatial admittance estimates for a number of geographic regions of Venus. All regions except Tellus Regio show a good correlation of gravity to topography; the low resolution of gravity data at the latitude of Tellus Regio may preclude a reasonable estimate of the level of correlation there.

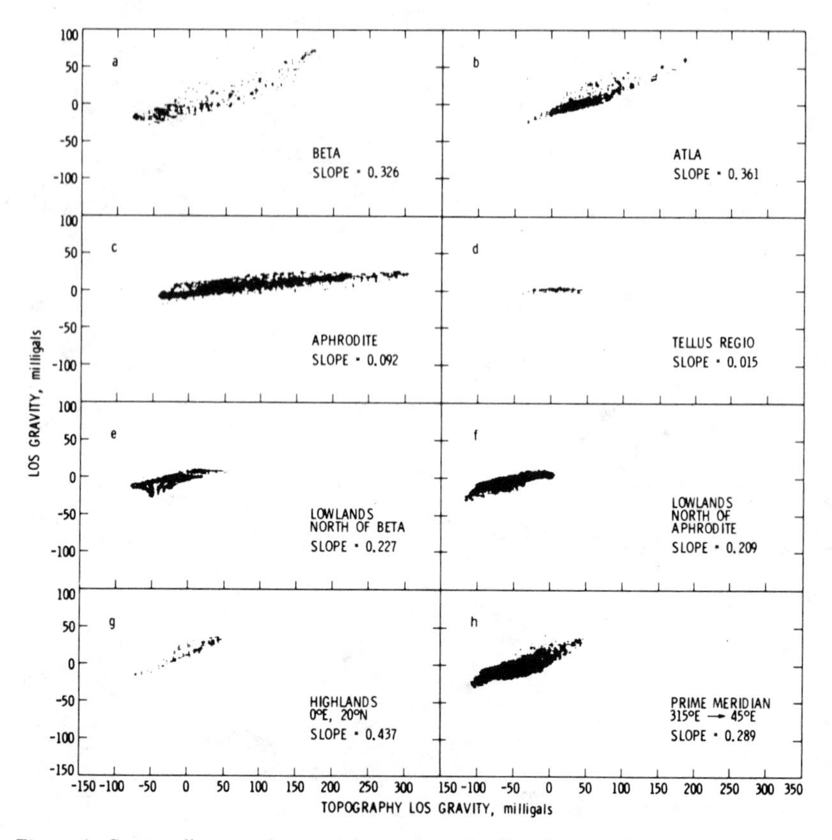

Figure 4 Scatter diagrams for spatial sampling of selected areas of Venus, showing the correlation of observed LOS gravity with the LOS gravity that would be due to the topography alone (from Sjogren et al 1983).

If a characteristic wavelength ($\lambda = 2\pi/k$) is assigned to a topographic feature, then the regression slope ψ is related to wavelength by

$$\psi \approx 1 - \exp(-kd), \tag{1}$$

where d is termed the "depth of compensation" and is the depth of a density interface that provides a mass balance with the topography. In the classic sense, this would be the depth of Airy isostatic compensation; for Earth, much of the continental topography is compensated by Airy isostasy arising at the crust-mantle boundary. We take d as a conceptually convenient alternative to ψ, although no particular type of compensation, or even complete compensation, is implied when estimates of d are discussed. It is clear that the greater the value of ψ, then either the shorter the wavelength of the topographic feature, and/or the greater the depth of compensation, and/or the less the degree of compensation.

Phillips & Malin (1983) modeled the Beta Regio gravity data and obtained a direct estimate of $d \approx 400$ km. This is consistent with the spatial admittance estimate by Esposito et al (1982) of $\psi = 0.44$ if the dimension of Beta Regio is $\lambda/2$. [The Espositio et al (1982) value of $\psi = 0.44$ is in agreement with the slope through the right portion of the scatter diagram of Figure 4*a*; this is appropriate, since the 0.44 value is associated with the main elevated region of Beta Regio.] The admittance estimates in Figure 4 imply at least a factor of three variation in d. If we discount the Tellus Regio result, the smallest value of d is found for Aphrodite. The three highest admittances are associated with Beta Regio, Atla Regio, and the highlands at 0°E, 20°N. For Atla, d is about the same as Beta (approximately 400 km), and for the highlands at 0°E, 20°N, d is $\approx$100–200 km because the larger admittance estimate is more than offset by the smaller wavelength of the feature. Thus, Beta and Atla have the greatest depths of apparent compensation of the regions sampled on Venus. [A preliminary estimate for Ishtar Terra is $d \approx 170$ km (W. L. Sjogren, personal communication, 1983).]

At least two alternative interpretations to that of a single interface depth of compensation are reasonable. First, the topography could be compensated over a vertical interval (i.e. as in the concept of Pratt compensation); in this model the compensation zones under Beta and Atla extend to at least 1000 km depth (Phillips & Malin 1983). Second, the spatial admittances can be interpreted in terms of the flexural rigidity of the lithosphere. Phillips et al (1981) find a flexural solution for Aphrodite that yields an elastic lithospheric thickness in excess of 100 km. Corresponding calculations for Atla and Beta regione lead to even greater thicknesses. As is discussed later, these values are probably implausible, given the thermal state of the interior of Venus. More likely a compensation solution is correct, and the appropriate line of inquiry is whether the compensation is passive (isostasy) or maintained dynamically by thermal processes in the venusian mantle.

In summary, it is seen that Venus, unlike the Earth, shows a high correlation at long wavelengths between gravity and topography, although the character of the correlation shows high variability over the various geographic regions of the planet. Both the correlation and its variability must be considered in any tectonic model.

OBSERVATIONS OF TECTONIC FEATURES

Owing to the low spatial resolution of Earth-based and orbital radar reflectivity and altimetry observations, a discussion of tectonic features on Venus must by necessity be based partly on informed speculation. A hierarchy of confidence can be built based on plausible interpretations of these available data. The evidence is reasonably strong for volcanism and highly suggestive of extensional tectonism. It is less suggestive but permissive of compressional tectonism. A discussion of this hierarchy is presented in the remainder of this section.

Volcanism

Volcanism, after impact cratering, is the most widespread geological phenomenon in the solar system. Since volcanism is a natural consequence of planetary thermal evolution, it would be extremely perplexing if a planet the size of Venus did not exhibit volcanic landforms. Thus, the identification of such features was an obvious goal of investigations of Earth-based radar images of modest to (relatively) high spatial resolution ($\sim$5–20 km). The acquisition of these images in the mid-1970s (Goldstein et al 1976, 1978, Jurgens et al 1980, Campbell et al 1976, 1979) permitted such studies, and several examples of potential volcanic features were identified (Malin & Saunders 1977, Saunders & Malin 1977), although with some uncertainty.

The most prominent and least equivocal evidence of volcanism is the summits of Beta Regio, now called Rhea and Theia mons. Radar reflectivity and altimetry maps acquired at the Goldstone Deep Space Network's tracking station (Goldstein et al 1978) showed the area to be broadly elevated, with a high point nearly 10 km above the lowest area in the surrounding region. (PV altimetry later showed this figure to be an overestimate.) High reflectivity indicated rough surfaces, and a small ($\sim$60 $\times$ 90 km) depression was seen near the highest peak of the region. Saunders & Malin (1977) interpreted this feature as a volcano, similar in scale to several martian volcanos. Subsequent data acquired by Venera 9 and 10 (Surkov 1977) on the K, U, and Th content of rocks in the vicinity of

Beta Regio suggested basaltic composition, as did the major-element chemistry of soils as measured by Venera 13 and 14 to the south, near Phoebe Regio. Pioneer Venus altimetry (Masursky et al 1980) and Earth-based radar reflectivity maps (Campbell et al 1979) have further supported the volcanic interpretation. Beta Regio itself may represent an area of crustal doming and rifting, with which the volcanism may be associated (Figure 5 herein, McGill et al 1981).

Other possible volcanos can be identified in the Goldstone data. Visible in one image is a 200-km dome with a summit crater defined by multiple escarpments (suggesting terracing), steep inner slopes, and shallow outer slopes. In another image, a cluster of mountains (some over 1 km in height) is seen, isolated on an otherwise smooth and featureless plain (Malin & Saunders 1977). Owing to the difficulty in creating such groups of peaks by processes other than volcanism, these features, too, are considered volcanic. The numerous, singly occurring, peaks (40–80 km diameter, 1–2 km high)

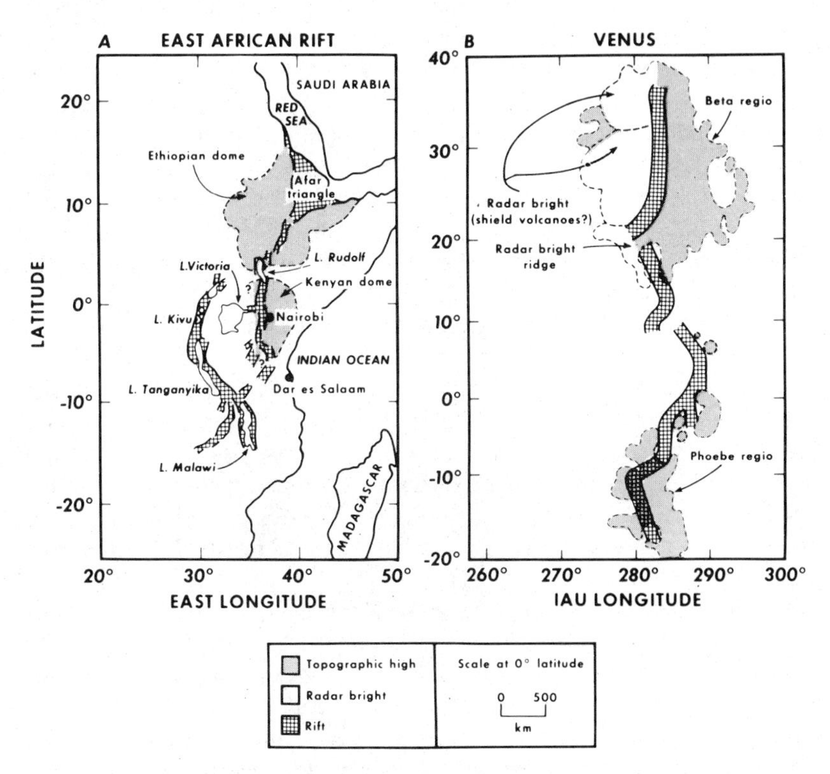

Figure 5 Comparison of East African Rift System with proposed rift system associated with the Beta Regio and Phoebe Regio regions of Venus. Areas that are radar bright and elevated have been proposed as shield volcanos (modified from McGill et al 1981).

seen in many areas on Venus are also believed to be volcanos (Masursky et al 1980).

Extensional Tectonism

Rifts, defined as elongate depressions overlying places where the lithosphere has ruptured in extension, are widespread on Earth. The discovery in 1975 of a large trough on Venus (Figure 6 herein, Goldstein et al 1976) and its interpretation as a rift (Malin & Saunders 1977) are prima facie evidence of extensional tectonism if the interpretation is correct. Subsequent discovery of many such features in the PV altimetry data (Pettengill et al 1979a,b, 1980) and analysis of their patterns and morphology (Masursky et al 1980, McGill et al 1981, 1983, Schaber 1982) have further supported the arguments for extension.

The origin of the extensional stresses is, however, of great uncertainty. Many rifts appear along or near the crests of regional domes, the best example being Devana Chasma, the axial trough of Beta Regio (Figure 5). Rifting of elevated regions is a natural consequence of the tensional stress associated with high-standing topography (Artyushkov 1974). Other rifts on Venus have no such association with elevation (e.g. Artemis Chasma south of Aphrodite Terra; see Figure 1). Within a given area, most rifts appear systematically oriented with respect to one another (e.g. subparallel). Schaber (1982) recognizes a global relationship of three irregular and discontinuous, yet clearly associated, zones of rift development (Figure 7). The largest stretches from Beta Regio west along a great circle to the western end of Aphrodite Terra, a distance of over 20,000 km. If the subdued linear-trending high topography west of Aphrodite Terra and east of Beta Regio is included, then this zone is truly circumglobal (Phillips et al 1981). The second trends northwest from Themis Regio through Alta Regio, a distance of 14,000 km; the third zone trends roughly north-south 8,000 km between Beta and Phoebe regione. Although the pattern does not extend to all areas of the planet, extensional stresses appear to be of global scale, but are most intense on a regional scale. The elevated areas encompassing these three tectonic regions have been termed the "Equatorial Highlands" by Phillips et al (1981).

Schaber (1982) attaches particular significance to the two intersections of these three tectonic zones. One intersection is at Beta Regio, for which we have already presented the evidence of volcanism; a second intersection is at Atla Regio, where neither Earth-based radar nor Venera surface-sampling data are available to support a volcanic interpretation. Lightning discharges associated with Beta and Atla provide tantalizing but equivocal evidence for active volcanism (Ksanfomality et al 1983).

The most convincing piece of evidence for anomalies at Schaber's

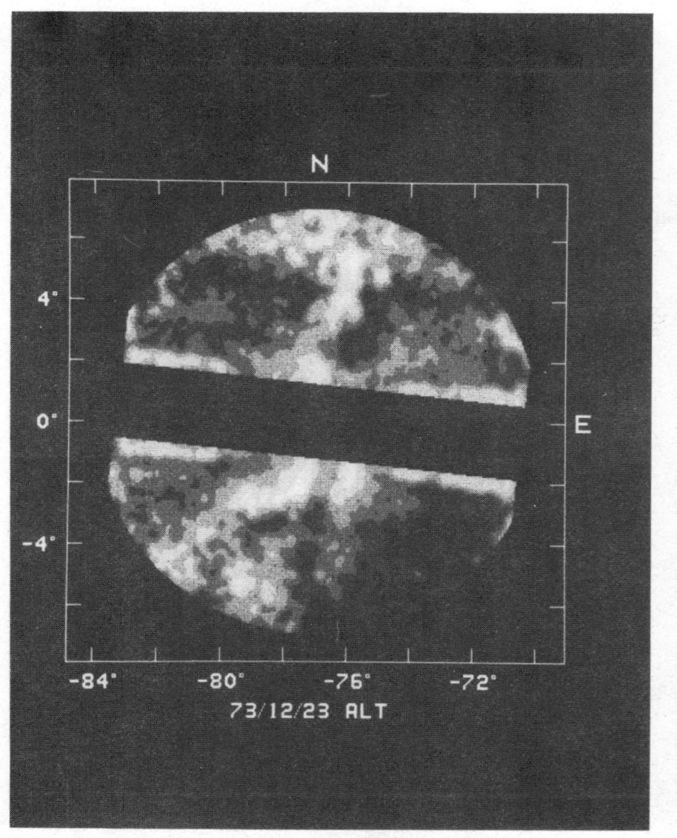

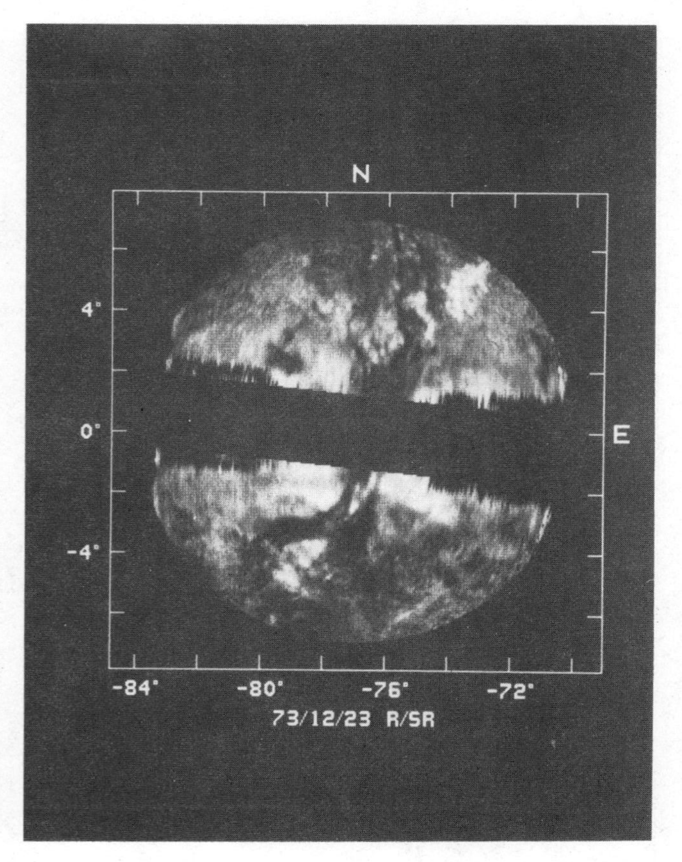

Figure 6 Radar brightness image (*right*) and elevation image (*left*; the higher elevations are brighter) from Goldstone radar data. The topographic trough was interpreted as a rift by Malin & Saunders (1977).

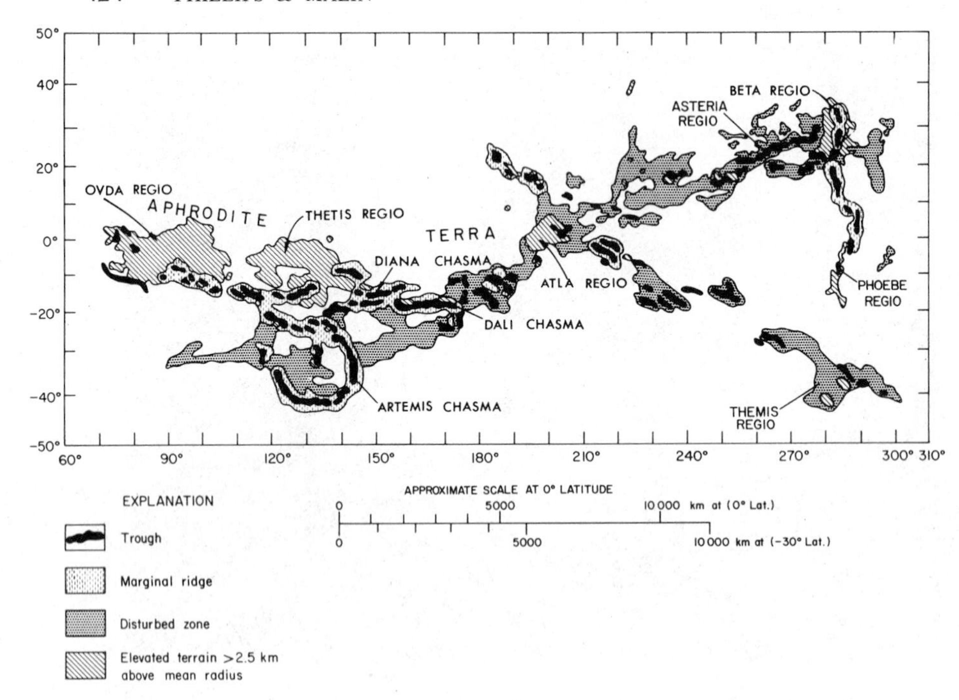

Figure 7 Tectonic sketch map showing three disturbed, rifted zones between Beta Regio, Themis Regio, and Aphrodite Terra (from Schaber 1982).

tectonic intersections are the great depths of compensation of the topography, which (as is argued below) are most likely maintained by thermal processes in the mantle. The picture that is now envisioned is that extensional tectonism is concentrated in the Equatorial Highlands of Venus; presently, the most thermally active areas appear to be concentrated at the intersections of major extensional lineaments, where the lithosphere is expected to be weakest and most subject to penetration by mantle magmas.

Compressional Tectonism

On Earth the most prominent features produced by compressive stresses at converging plate boundaries are mountain ranges. Mountain ranges on Venus have been observed in Earth-based radar profiles (e.g. Campbell et al 1972) and images (e.g. Malin & Saunders 1977), mostly as relatively small and low-standing regions with limited areal extent. PV altimetry data are unable to resolve these smaller areas, but when combined with Earth-based radar images acquired at the Arecibo Observatory, they have shown that there are only three major mountain ranges on Venus, all associated with

Ishtar Terra (Masursky et al 1980, Campbell et al 1979). Akna, Freyja, and Maxwell montes rise as much as 10 km above the mean venusian datum and over 5 km above the mean altitude of Ishtar Terra (Figure 2). Campbell et al (1983) have described these mountain ranges in detail based on 3–6 km resolution radar imagery (Figures 8 and 9). Akna Montes, for example, are a few hundred kilometers in width and extend for almost 1000 km in an arcuate band along the western margin of Lakshmi Planum (Figure 9). The peak radar returns coincide with the main axis of the mountains as determined by PV altimetry. Two broad bands of radar backscatter along this axis are 15–50 km in width and are separated by distances of 25–40 km.

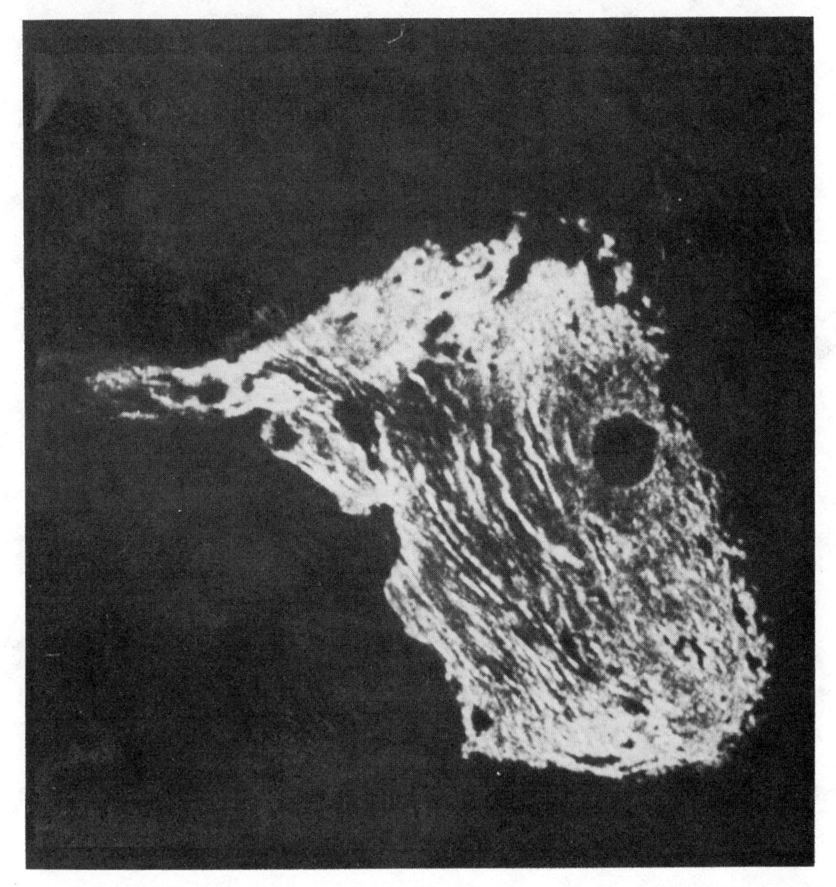

Figure 8 Radar image, taken from the Arecibo Observatory in Puerto Rico, of Maxwell Montes. Banded terrain development can be seen in the central and western regions. The width of Maxwell is about 1500 km (image courtesy of J. Head, Brown University, and D. B. Campbell, Arecibo Observatory).

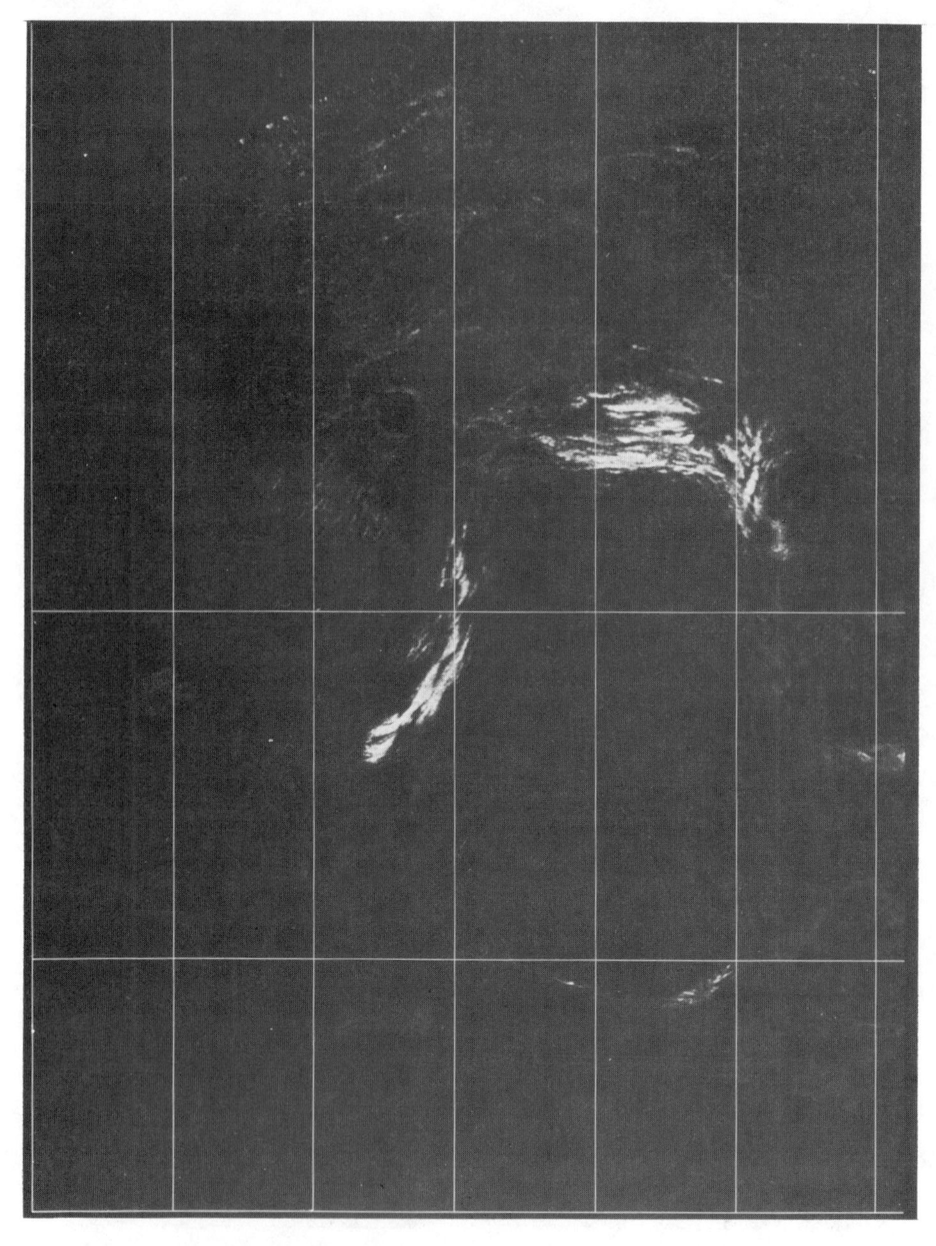

Figure 9 Radar image, taken from the Arecibo Observatory in Puerto Rico, of Lakshmi Planum. Banded structure can be seen in Akna Montes (*center*) and Freyja Montes (*top right*) (image courtesy of J. Head, Brown University, and D. B. Campbell, Arecibo Observatory; see also Campbell et al 1983).

Each of the broad bands contains numerous individual bands down to the limit of resolution. Similar banded structure is seen in Freyja and Maxwell montes.

Solomon & Head (1983) propose a tectonic origin for these mountain ranges, but conclude on the basis of mechanical models of the lithosphere that they cannot distinguish between compressional and extensional origins for the banded terrain. In either case, their simple modeling schemes predict that the elastic lithosphere was no more than a few kilometers thick at the time of tectonism. The high aspect ratio (length to width) of the shape of Akna Montes, in particular, along with the great topographic heights of the three mountain ranges, appear more consistent with a compressive stress origin. In addition, features reminiscent of V-shape fold closures are seen in the radar images. The alternative—extension comparable to that responsible for the block-faulted mountains of the Basin and Range province of the western United States—seems less likely in view of these observations and associations.

TECTONIC MODELS FOR VENUS

The limited number and poor resolution of the Venus data sets significantly hinder rigorous tests of tectonic models. It is not possible to unequivocally reject plausible tectonic models (Solomon & Head 1982b), but in the sense that the global tectonics of a terrestrial planet is a manifestation of heat loss through the lithosphere, several types of tectonics are surely operative (Phillips 1983). We cannot state, however, the relative importance as heat-loss mechanisms of any of the plausible models; the reader is forewarned that what follows falls within the realm of informed speculation.

Implicit in any discussion of tectonic models must be an assumption of bulk composition and attendant radiogenic heat sources. Cosmochemical models (e.g. those summarized in the Basaltic Volcanism Study Project 1981) suggest that Venus and Earth have similar levels of heat production because of the refractory nature of K, U, and Th in nebular condensation sequences and because of the positions of Venus and Earth in the solar system. Three different lines of evidence on Venus point to Earth-like concentrations of K, U, and Th: (*a*) Direct measurements of gamma-ray emissions by these elements with Venera 8, 9, and 10 (Surkov 1977) yield concentrations similar to those of terrestrial crustal rocks (in particular, basalt). (*b*) Venera 13 and 14 measured additional elemental compositions matching alkali and tholeiitic basalts, respectively (Surkov et al 1983). (*c*) The abundance of ^{4}He, formed by the decay of U and Th, on the surface of Venus is similar to that of the Earth (Prather & McElroy 1983).

Owing to the efficiency of fractionation of K, U, and Th within igneous melts, the heat source concentrations in the Earth's crust cannot be representative of the whole Earth. Given the apparent similarity in the abundances of these elements in the crustal materials of Venus, it is not unreasonable to assign the terrestrial bulk value to Venus, although this merely reassigns the focus of our uncertainty from Venus to Earth. Ringwood (1975) has suggested that the U and Th content of the bulk Earth must be similar to that found in chondritic meteorites; this argument is based on the abundances of other refractory elements. The measured steady-state heat flow of the Earth, if we assume crustal ratios of K/U and Th/U, yields a bulk U estimate that is in excess of the chondritic value. However, parameterized thermal history calculations (e.g. Schubert et al 1980) show that the Earth cannot be in steady state with radioactive heat production, but rather must also have a secular cooling component that could account for up to about one third of the present-day heat flux. Given these calculations, cosmochemical arguments, and the nature of the major elements in the Earth's mantle, a chondritic heat-source assumption is plausible for the Earth and presumably for Venus as well. Crustal values of the K/U ratio are distinctly subchondritic, however, for both Venus and Earth. Some nebular temperature models predict complete condensation of potassium at both Venus and Earth (e.g. Lewis 1972), and part of the "missing" potassium could reside in the core as a light alloying element (Bukowinski 1976). The deficiency of atmospheric ^{40}Ar on Venus relative to the Earth (Hoffman et al 1980) may mean that Venus has less potassium than the Earth, but it also may indicate that Venus has been less efficient in outgassing its interior. Fortunately, the uncertainty in the potassium content of Venus has only a small effect on estimates of present-day heat generation (Phillips et al 1981, Solomon & Head 1982b).

Uniform Conduction Models

In the simplest of models, Venus presently loses its interior heat by conduction through a uniform lithospheric shell, and the attendant tectonics is the result of uniform compression or tension in the lithosphere in response to the changing thermal state of the bulk planet. Regional tectonics is introduced by flexural loads on the lithosphere. The Moon apparently has been in this state for nearly 4 b.y. (Solomon & Head 1979). This likely also describes the present tectonics of Mercury and possibly Mars.

The consequences of uniform heat loss on Venus have been discussed by several authors (Phillips et al 1981, Solomon & Head 1982b, Phillips & Malin 1983). The investigation of this hypothesis involves the estimation of a heat flux for Venus and the calculation of a lithospheric thickness from which the fate of the topography can be assessed.

Solomon & Head (1982b) estimate a globally averaged heat flux on Venus of 74 mW m^{-2} by scaling the terrestrial value of 82 mW m^{-2} (Sclater et al 1980) to the mass and surface area of Venus. Phillips & Malin (1983) obtain values in the range 50–55 mW m^{-2} from chondritic thermal models. The difference arises because the Phillips-Malin models do not have any heat flux from the core (based on a strategy reflecting the deficiency of knowledge of the state of the core and the desire to work with a lower bound on heat flux for studies of isostasy). A plausible range of estimates of venusian average heat flux is 50–75 mW m^{-2}. Solomon & Head (1982b) estimate a maximum thickness of 10 km for the venusian elastic lithosphere, using a heat flux of 74 mW m^{-2}, a thermal conductivity of 3.1 W m^{-1} K^{-1}, and a base temperature of $500 \pm 150°C$ (Watts et al 1980). Phillips & Malin (1983) derive a value of $\sim$1000 mW m^{-2} km for the product of the thickness of the elastic lithosphere and the heat flux into its base. This result is based on rheological behavior that is substantially elastic for 10^8 yr and a creep law for dry olivine (Kohlstedt & Goetze 1974). With this approach, a heat flux of 74 mW m^{-2} yields a thickness of 13.5 km, essentially equivalent to the estimate of Solomon & Head (1982b); and a flux of 50 mW m^{-2} yields a reasonable upper bound of 20 km.

Given estimates of the thickness of the elastic lithosphere, arguments can then be made concerning the plausibility of isostatic support of venusian topography. The concept of isostasy implicitly assumes that the compensation zone in the interior behaves elastically; otherwise, the stresses involved will lead to creep destabilization. Alternatively, compensation is not isostatic but is maintained by dynamic processes, usually envisioned to be thermal in nature. Can venusian topography be isostatically maintained by an elastic lithosphere that is less than 20 km thick?

The stresses induced by a topographic load on an elastic lithosphere are supported by the flexural rigidity D of the lithosphere. As the ratio D/λ approaches zero, where λ is the wavelength of the load, the flexural state approaches Airy isostasy. The isostatic state is, in general, a state of minimum stress. Thus, topography that is formed as a result of internal thermomechanical processes will generally form isostatically in the absence of forces other than those due to buoyancy (Phillips & Lambeck 1980).

The isostatic configuration yields a near minimum in maximum lithospheric stress induced by a load, and this stress value is approximately ρgh, where ρ and h are the density and height of the topography, respectively. For Venus, this is about 250 bar of stress for each kilometer of topographic elevation. Thus, the topography of Aphrodite and Ishtar terrae would induce an isostatic stress of about 1 kbar, with the Maxwell Montes locally inducing stresses of 2–3 kbar.

There are two considerations in the isostatic hypothesis: (*a*) can the lithosphere support stresses in the 1–3 kbar range?; and (*b*) can the compensation be accommodated within a layer no more than about 20 km thick?

The answer to the first question depends strongly on our perceptions of the finite strength of rock materials. Phillips & Lambeck (1980) argue that finite strength in the lithosphere rarely exceeds a kilobar, while the application of laboratory experimental measurements suggests that the upper lithosphere can support stresses of several kilobars (see, for example, Byerlee 1968). On the basis of the latter view, Solomon & Head (1983) have produced an "envelope" of the maximum stress as a function of depth on Venus, assuming a strain rate of 10^{-14} s^{-1} and a temperature gradient of 20 °C km^{-1}. Use of a stress envelope diagram leads to an estimate of the elastic lithosphere thickness that is somewhat more rigorous than adopting a base temperature. In this approach, the upper part of the envelope is defined by the maximum stress that can be supported by a fractured rock, and the lower part of the envelope is governed by the maximum stress that can be supported without relaxation by ductile flow on a time scale of 1–10 m.y. (corresponding to a strain rate of 10^{-14} s^{-1}). Stresses induced by topographic loads that lie outside this envelope presumably lead to brittle failure in the upper part of the lithosphere or to significant ductile flow on geologically short time scales in the lower part of the lithosphere (i.e. these loads could not be supported isostatically).

If we assume that lithospheric rheology can be represented by the flow law for anorthosite, the stress diagram of Solomon & Head (1983) indicates that ~1-kbar stresses will be supported to maximum depths of only 2 km (horizontal compression) to 4 km (horizontal extension) over geologically long periods. The corresponding numbers for pyroxene creep are about 10 km in both cases. It is not possible to passively support the higher stresses induced by Maxwell Montes with an anorthosite flow-law rheology for time scales greater than 1 m.y. The pyroxene flow-law rheology will support the Maxwell stresses in horizontal compression to depths of about 5 km, and it will not support extensional stresses at all.

With the above limitations in mind, the second question becomes one of accommodating isostatic compensation within a layer no more than 10 km thick. Simple mass balance then requires a density contrast between the compensation zone and its surroundings to exceed 1 g cm^{-3} for Aphrodite and Ishtar terrae and 3 g cm^{-3} for Maxwell Montes. These values can be lowered slightly by adopting a lower-bound heat flux of 50 mW m^{-2} (20°C km^{-1} corresponds to ~60 mW m^{-2}). It is generally outside the range of plausibility in terrestrial experience to expect bulk density contrasts of igneous rocks to exceed 1 g cm^{-3} in the lithosphere, and thus isostatic support on geologically long time scales on Venus appears to be unlikely with a uniform conduction model.

The implications of this model (and Earth-like heat flux) and its apparent failure to provide isostatic support are as follows:

1. The topography could be very young, which severely challenges the model.
2. The heat flux is considerably less than the global average under Aphrodite and Ishtar terrae, implying a lower temperature gradient and a thicker lithosphere. The heat flux must then be abnormally high elsewhere. These considerations invalidate, by definition, the model.
3. The topography is partly supported by a lithosphere in uniform horizontal compression, as might be expected from this type of model and a cooling planet. The existence of extensional tectonic features in the Equatorial Highlands argues against the idea of uniform global compression.

In summary, a uniform conduction model provides an end member with which to test more plausible models for the tectonics of Venus. Such models must involve lateral variations in the thermal state of the mantle and in the stress state of the lithosphere. Such a tectonic style is expected for a vigorously convecting interior, and the relevant question is then, what form does the heterogeneous heat flux take?

Plate Tectonic Models

On Earth, of course, heterogeneous heat flow is manifested in plate tectonics (i.e. seafloor spreading and continental drift). Most of the heat loss in the ocean basins takes place at the spreading ridges. On Venus, the simple "equal size = equal thermal history = equal tectonics" rule predicts that some form of plate tectonics ought to be presently operative. The search for evidence of seafloor spreading on Venus has taken on two general lines of inquiry: (*a*) investigation of the topographic expression of seafloor spreading in the PV altimetry data (Arvidson & Davies 1981, Head et al 1981), and (*b*) comparison of venusian topographic profiles with quantitative models of the shape of spreading ridges (Kaula & Phillips 1981, Phillips & Malin 1983).

The search in the altimetry data for the characteristic shapes of seafloor spreading has had negative results. Thus, investigations have centered on whether or not the characteristic physiographic forms could be recognized in the *Earth's* ocean basins at the resolution of the PV altimeter, after correcting for differences in the surface environments of the two planets. The transverse topographic ridge profile of the spreading, cooling oceanic lithosphere is described theoretically by a conductively cooling, isostatically balanced lithospheric slab moving away from a ridge axis. The expression for topographic height as a function of age t is given by (Parsons & Sclater 1977)

$$h = h_0 - \delta h t^{1/2}, \tag{2a}$$

where

$$h_0 = \frac{\rho d \alpha \Delta T}{2\delta\rho}, \tag{2b}$$

and

$$\delta h = \frac{2\rho\alpha\Delta T}{\delta\rho}\left(\frac{\kappa}{\pi}\right)^{1/2}. \tag{2c}$$

Here, h_0 is the ridge height at zero age; and ρ, d, α, κ, and ΔT are the density, effective ambient thickness, coefficient of thermal expansion, thermal diffusivity, and vertical temperature change, respectively, of the lithospheric slab. The term $\delta\rho$ is the density contrast between the slab and the overlying material. In the simplest analysis, all variables are assumed the same for both planets except ΔT and $\delta\rho$. The lithospheric base (asthenospheric) temperature is assumed to be the same for both planets, so that ΔT differs by the surface temperature alone (273 K vs 740 K). The density differential $\delta\rho$ is different because the overlying material has a density of $\sim 1 \text{ g cm}^{-3}$ on Earth (ocean water) and of approximately zero on Venus.

Adjusting North Pacific bathymetry for PV resolution and sampling, and applying the ΔT–$\delta\rho$ correction, Arvidson & Davies (1981) found that many of the major seafloor-spreading forms were still discernible. They concluded that "... Venus does not display well-defined evidence for active plate spreading as that concept applies to Earth." Head et al (1981) and Solomon & Head (1982b) were not so certain of the ability of the PV altimeter to detect seafloor spreading forms in the Venus environment. The consensus of opinion (Arvidson & Davies 1981, Head et al 1981, Phillips et al 1981, Phillips & Malin 1983), however, is that the basic boundary-layer phenomenon of seafloor spreading—a decrease in elevation away from the ridge axis due to the motion of the cooling lithosphere—would be discernible for at least some of the ridge forms on Earth. When the Pacific Basin topography is modified to venusian conditions and PV altimeter sampling, for example, much of the major boundary-layer phenomena associated with the spreading centers is detectable (Figure 10; P. Morgan, personal communication, 1983).

On Venus, ridge forms *can* be seen in the PV altimetry data as well as in

Earth-based radar data. How do we know that they are *not* spreading ridges? Perhaps a more important (though somewhat philosophical) question is how do we know that the ridges seen in Figure 10 *are*? Such information cannot be determined from an image. A great deal of other geological and geophysical information was required in order to determine that the East Pacific Rise, for example, *is* a center of lithospheric spreading.

Several authors have applied simple quantitative arguments to ridge *shapes* on Venus to evaluate their potential as plate tectonic phenomena. Kaula & Phillips (1981) examined the elevated physiography of the Equatorial Highlands in a test for plate tectonics. The ridge shapes can be fit as a function of distance x from the ridge axis by

$$h(x) = e_0 - \Delta h x^n, \qquad (3)$$

where e_0 and Δh are coefficients to be determined from topographic data, as is n, an index of concavity that is theoretically one half for a cooling boundary layer with a constant horizontal velocity. Kaula & Phillips (1981)

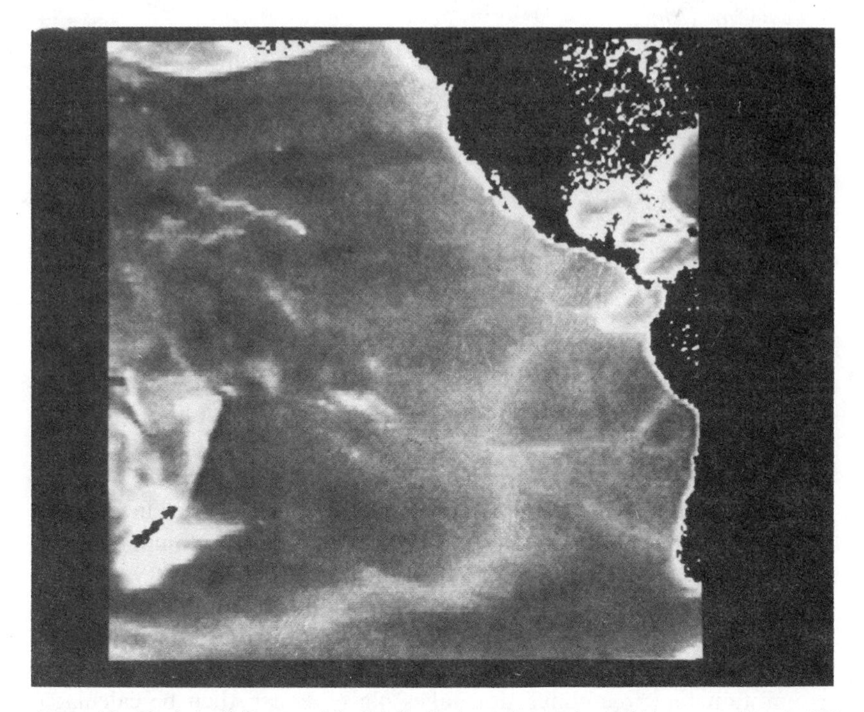

Figure 10 Shaded relief image of the Pacific Basin seafloor as it would be mapped by the PV altimeter. The data have been Venus-adjusted for the difference in surface temperature and the lack of an ocean. The brighter tones indicate greater seafloor elevation (P. Morgan, personal communication, 1983).

argue that *some* of the tests for seafloor spreading are met: (*a*) There exists on Venus a broad reference plain [the Median Plains; see Phillips et al (1981)] that has little height variation and would be analogous to the thermally mature boundary layer of the Earth's deep ocean basins. (*b*) The ridge shapes are predominantly concave ($n < 1.0$). The concavity index for Venus may be as close as the Earth is to the theoretical value of one half, both in the mean (Kaula & Phillips 1981) and individual profiles (P. Morgan, personal communication, 1983).

Topographic features on Venus differ significantly from ocean ridges on Earth in two ways (Kaula & Phillips 1981): (*a*) Unlike the Earth, the ridges of the venusian Equatorial Highlands do not form a narrow distribution about a mode in crest height above the Median Plains, and (*b*) the Venus ridges do not form as extensive an interconnected system as do the ocean ridges on Earth.

Since time and distance are related through velocity V, then from Equations (2a) and (3) we have

$$V = (\delta h)^2/(\Delta h)^2. \tag{4}$$

If we obtain δh by scaling from the Earth to Venus using boundary-layer theory, then observations of Δh constrain V under an assumption of venusian seafloor spreading. The rate of heat loss per unit length of rise is given by

$$q_{\mathrm{L}} = d\rho C_{\mathrm{p}} V \Delta T, \tag{5}$$

where C_{p} is specific heat; the total heat loss is

$$q = \int_0^B q_{\mathrm{L}} \, \mathrm{d}B, \tag{6}$$

where B is total ridge length. Kaula & Phillips (1981) estimated plate velocity from Equation (4) and measurements of Δh. Total heat loss was then estimated from Equation (6) and an estimate of B. According to Kaula & Phillips (1981), if seafloor spreading is associated with the $\sim$21,000 km of ridges identified with the Equatorial Highlands, then this mechanism would account for only 15% of venusian heat loss. Plate tectonics on Earth accounts for about 65% of the heat flux (Sclater et al 1980), with the majority of this heat loss associated with seafloor spreading.

Phillips & Malin (1983) extended the work of Kaula & Phillips (1981) by examining ridge shapes in the Equatorial Highlands, working under the assumption that free convection takes place; V can then be calculated (along with ΔT and d) from boundary-layer theory, and thus Δh can be used as an additional model parameter in the theory. The results of their calculations are shown in Figure 11, where the interior convecting temperature of Venus and the Venus-to-Earth ratio of heat-source concentrations Q are plotted against the ratio of topographic coefficients Δh for the two planets. The observed Δh ratio is about 1.0 (Kaula & Phillips 1981), and this implies, from Figure 11, that Venus is extremely depleted in heat sources relative to the Earth. However, as noted earlier, this is considered unlikely. Alternatively, with a Q ratio of unity, the predicted Δh ratio is about 0.4, or $2\frac{1}{2}$ times less than the observed value, which therefore can be interpreted in terms of a plate velocity much less than that predicted by the Rayleigh number. That is, with Earth-like heat sources, the venusian

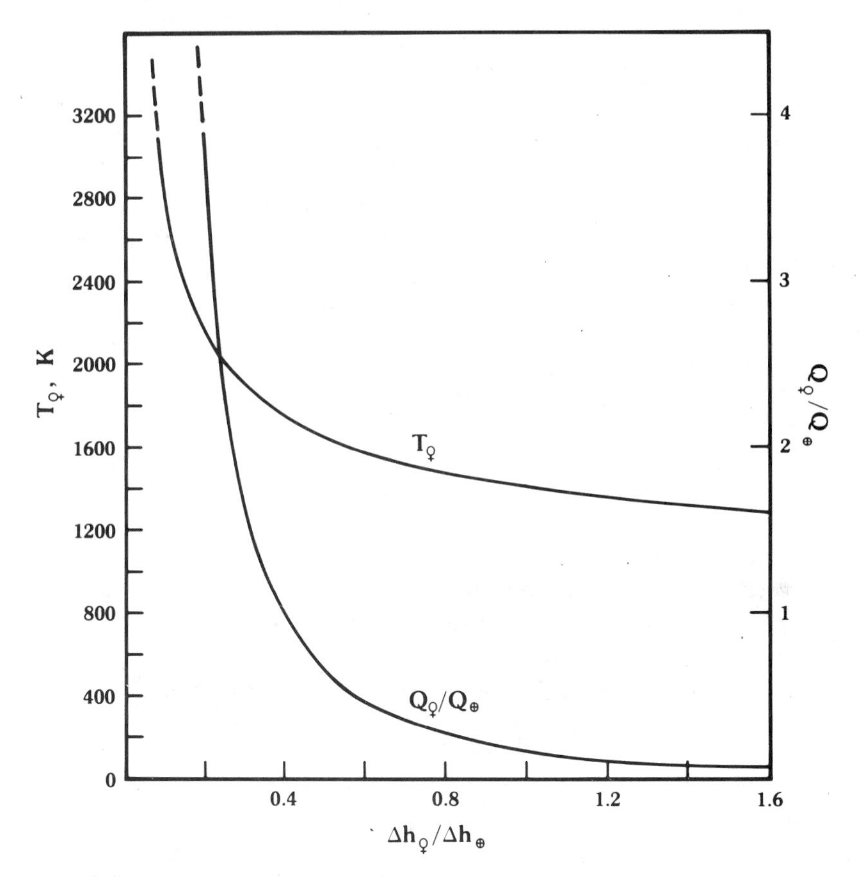

Figure 11 Boundary-layer theory results for free convection, showing the interior convecting temperature of Venus and the Venus-Earth heat-source concentration ratio as a function of the ratio of the Venus-Earth ridge-shape coefficients (from Phillips & Malin 1983).

plates could *not* be part of a freely convecting system, and plate motions and subduction would be inhibited. On Earth, inhibited motion is associated with those plates that are in part occupied by continent (e.g. Carlson 1981). This second result emphasizes the solution of Kaula & Phillips (1981), which, with unity ratio in both Δh and Q, gives a plate-velocity spectrum that is associated with lower velocities than those found on the Earth. The Earth itself possibly only approaches free-convection velocities for plates that are almost wholly oceanic with well-developed subduction boundaries.

Given the good evidence for extensional tectonism, the Equatorial Highlands could in fact be a region of slow, inhibited lithospheric spreading that does not account for heat loss in the same way as oceanic ridges on Earth. This concept is not in conflict with the hotspot hypothesis presented below. Further, if there is enough volcanism in the Equatorial Highlands to conceal Earth-like forms of plate spreading, then this volcanism would bring the major fraction of heat to the surface, making plate tectonics a secondary mechanism for heat removal (Kaula & Muradian 1982).

Solomon & Head (1982b) and Phillips & Malin (1983) both suggest that *if* Earth-like seafloor spreading and subduction and/or lithospheric delamination exist on Venus, they are most likely confined to the Median Plains. From the boundary-layer arguments, spreading forms most probably would be topographically subtle, and they could be disguised by sedimentation (Brass & Harrison 1982).

Morgan & Phillips (1983) discussed the implications for seafloor spreading from the histogram of topographic elevations (or hypsogram) for the Median Plains. The hypsogram of the Earth's ocean basins has a distinctly asymmetrical shape. This shape can be readily explained (Parsons 1982) by ocean depth increasing as a function of the square root of time or distance from the ridge axis (due to the cooling of the lithosphere), and the almost constant ocean depth for ages greater than 70–80 m.y. (due to a base-level heat flux into the lithosphere creating a "thermally mature" layer). The Median Plains hypsogram is symmetric, with about the same width as the ocean hypsogram. In terms of seafloor spreading, the possible interpretations are the following: (*a*) The entire width of the hypsogram represents seafloor spreading with preferential consumption of older seafloor. This width requires a substantially lower mantle heat flux and thicker lithosphere than the Earth. (*b*) Seafloor spreading is only represented by those elevations greater than and including the mode. This yields approximately the correct width from boundary-layer theory, but the elevations lower than the mode imply that a thermally mature base level is not reached; i.e. seafloor spreading is operative in the presence of significant topographic depressions in the boundary layer. (*c*) Seafloor spreading is operative but is topographically subtle—the hypsogram is dominated by some other type of tectonics.

CONTINENTAL ASPECTS Our discussion of plate tectonics has thus far concentrated on seafloor spreading because that provides diagnostic physiographic forms of plate tectonics. The identification and quantification of the continental aspects of plate tectonics (i.e. continental assemblage with island arcs and continental collision) are more difficult and, for Venus, have focused on the interpretation of topographic features.

Ishtar Terra has been a center of speculation because it bears the most topographic resemblance to terrestrial continents (Phillips et al 1981, Phillips & Malin 1983). Ishtar presents a well-defined boundary on much of its border; it is plateaulike, and an arcuate ridge (Ut Rupes) parallels the southwestern boundary in an island-arc-like relationship (Figures 1 and 2). High-resolution radar images show that the linear mountain belts of Akna and Freyja montes (Figure 9) are banded in radar brightness on a 20-km scale (Campbell et al 1983). It is most plausible that the banding represents ridge and valley forms indicative of horizontal stress in the lithosphere. As discussed earlier, Solomon & Head (1983) have stated that on the basis of simple mechanical models, the difference between extensional and compressional origins cannot be determined. It has been argued here that a compressional origin is more likely; a plausible speculation is that the banded terrain is analogous to structures that develop in terrestrial continental collision zones.

Physiographically, Ishtar Terra may provide the best evidence for plate tectonics on Venus. Phillips & Malin (1983) hypothesize that Ishtar may be a remnant of a plate tectonic episode that occurred on Venus prior to the formation of the atmospheric greenhouse. They also point out, however, that there is no way to assign an absolute or even relative age to this feature in order to test this hypothesis. Rheological considerations suggest that Ishtar and other elevated regions on Venus must be geologically young and/or dynamically maintained unless they are areas of abnormally low heat flow.

Hotspot Tectonics

Phillips & Malin (1983) proposed that hotspot tectonics may be an important, and possibly even the dominant, mechanism through which Venus loses its internal heat. This type of tectonism was envisioned to include concentrated regions of plumelike high heat flux from the mantle, leading to lithospheric thinning, thermal isostatic uplift, and attendant volcanism. The Hawaiian Swell provides a terrestrial analogue for this model. The thermal uplift can lead to rifting, such as on Beta Regio, and the hotspot definition can be extended to include linear regions of high heat flow—and thus could be associated with lithospheric extension as discussed above for the Equatorial Highlands. The hotspot hypothesis was proposed on the following grounds:

1. The higher surface temperature on Venus leads to a more buoyant lithosphere (Figure 12). This makes subduction less likely on Venus, which then might utilize a mechanism for heat removal that is different from seafloor spreading. It was proposed that the hotspots in the Earth's ocean basins, which now provide a secondary mechanism of heat loss, would become dominant if seafloor spreading ceased.
2. The deep (>1000 km) levels of Pratt-type compensation for Beta and Atla regione, together with basaltic volcanism and rifting associated

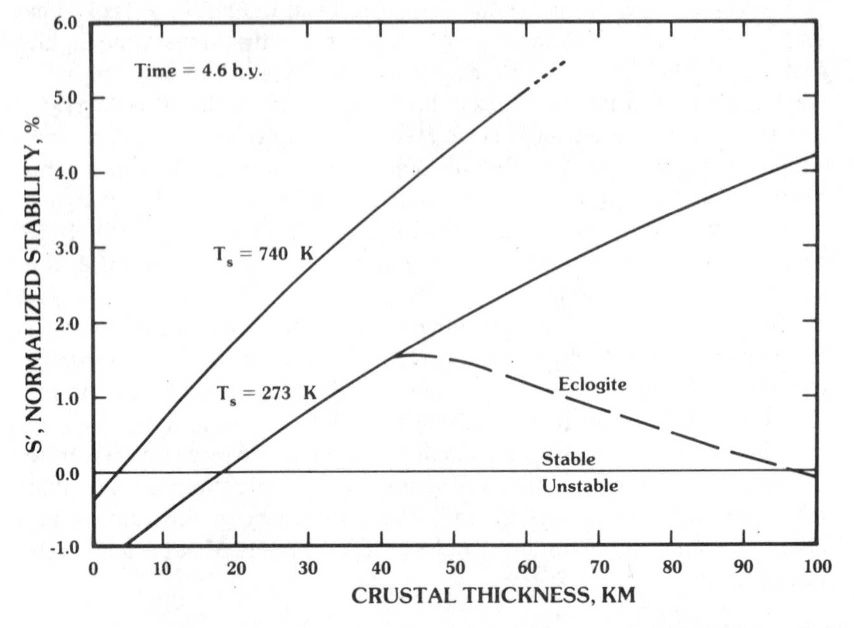

Figure 12 Present-day normalized buoyant stability of lithosphere for two different surface temperatures, as a function of crustal thickness (from Phillips & Malin 1983). The normalized stability gives the percentage deviation of the average lithospheric density from the density of the material just beneath it. Negative stability means that the lithosphere will sink, which may be a prerequisite for the initiation of subduction. The thermal profiles of the lithospheres, from which densities are calculated, are from the evolutionary models discussed by Phillips & Malin (1983). A surface temperature of 740 K (Venus) yields a much narrower range of instability than a surface temperature of 273 K (Earth). Further, if the crust is basaltic, its lower portion will enter the eclogite stability field for the lower-temperature model only.

with Beta, imply a deep thermal source and dynamic compensation beneath these features.

3. The surface of Venus displays a pervasive distribution of isolated topographic highs for which the most plausible origin is related to volcanism and thermal isostasy. The Median Plains bear a regional physiographic resemblance to the basin-and-swell topography of Africa (Burke & Wilson 1972), and the linear nature of the Equatorial Highlands is made up of a partial-to-complete coalescence of quasi-circular elevated features.

Morgan & Phillips (1983) carried out a quantitative assessment of the hotspot concept by testing the hypothesis that all of the venusian topography can be explained by thermal compensation. Adopting a mean planetary radius and a mean thermal lithospheric thickness of 6051.5 km and 100 km, respectively, they found that the thermal compensation of a lithosphere of variable thickness can account for all elevations up to 6053 km radius, where the model lithosphere thins to zero thickness. Elevations less than 6053 km makes up 93% of the mapped topography of Venus. The areas of high heat flow (thin lithosphere) account for about 5% of the surface area, or about 35 hotspots the size of the Hawaiian Swell. Additional compensation by a crust of variable thickness (perhaps created by basaltic volcanism) can account for topography greater than 6053 km. However, the high elevation occurs over areas of thin lithosphere according to the thermal compensation mechanism, and under the constraint that the crustal thickness cannot exceed the lithospheric thickness, only a limited amount of crust can be added. Further, thermal compensation must still dominate over crustal compensation because the gravity analyses indicate that the compensation is deep (Phillips & Malin 1983). If the minimum lithospheric thickness is 20 km, then elevations up to 6055 km, or 99% of the topography, can be accounted for by a combination of thermal and Airy compensation. The remaining 1% of the topography, including Maxwell Montes, must be explained by other mechanisms.

Thermal thinning of a lithosphere on the order of 100-km mean thickness can account for most of the venusian gravity anomalies. Areas such as Beta and Atla regione, where the compensation depth is on the order of 1000 km, require an additional component of compensation, most plausibly the density contrast associated with a thermal plume in the mantle.

The lithosphere model of Morgan & Phillips (1983) is presently being tested against the gravity data. For now, it is sufficient to summarize this model by stating that thermal compensation can explain most of the topography of Venus; it also serves to focus our attention on those few anomalous regions associated with inadequacies of the model.

SUMMARY

Although the data available for Venus present tantalizing glimpses of tectonic style, we are frustrated by both a lack of resolution and of any specific knowledge of the absolute, or even relative, ages of geologic features. If we assume Earth-like heat-source concentrations on Venus, it is possible, however, to build an architecture for constraining tectonic style:

1. The inferred present-day global-average heat flux points to an elastic lithosphere on the order of 10 km thick and the implausibility of topographic compensation by isostatic mechanisms for any geologically long period. This implies a spatial variability of heat flow, with isostasy possible in areas of low heat flow and topographic features in areas of high heat flow that are either very young and/or dynamically maintained.
2. Gravity modeling provides an additional constraint, indicating that compensation of at least some topography is deep (100–1000 km) and must be at least in part dynamic.
3. There is good evidence for basaltic volcanism, including direct geochemical measurements and interpretation of physiographic features.
4. The Equatorial Highlands are a region of rifting and thus extensional tectonism. Ishtar Terra does not display rift structure, though the banded terrain is strong evidence for horizontal stress in the lithosphere. The rifts of the Equatorial Highlands suggest that some spreading of the lithosphere has taken place; quantitative models of boundary-layer shape show that slow, inhibited spreading is consistent with the topographic shapes of this region.
5. Fast-spreading ridges of the type consistent with the high Rayleigh number of the venusian mantle, and their attendant well-developed subduction zones, would seem to be confined to the Median Plains if they exist. They would presently be beyond the resolution of the available altimetry and imaging.
6. Physiographically, Ishtar Terra has a terrestrial continentlike form with evidence of associated compressional plate tectonic features.
7. The hotspot hypothesis is attractive for Beta Regio because of the basaltic volcanism and also because of the great depth of compensation, which suggests dynamic compensation in the mantle. Atla Regio would also seem to fall in this category. Evidence supporting this explanation is less compelling for other regions of the Equatorial Highlands, but the depth of compensation also implies thermal mechanisms are operative beneath the high topography. It is very plausible that isolated topographic highs (as found in the Median Plains) are volcanic in origin.

Any detailed tectonic hypothesis for Venus must be highly speculative, and observations by a high-resolution radar mission may allow serious testing of models.

ACKNOWLEDGMENTS

We thank Mildred Dickey, Dory Brandt, and Rosanna Ridings for help in the preparation of the manuscript. Kevin Burke read through a draft of this work and provided helpful comments. Participants in the Second Venus Tectonics Workshop, held at Brown University in May, 1983, provided useful insight into speculations on Venus tectonics. We thank in particular Sean Solomon, Jim Head, Don Turcotte, Paul Morgan, Bill Kaula, and Norm Sleep. Head kindly provided the photographs of the Ishtar Terra radar images. This work was supported by the Planetary Geophysics and Geochemistry Program under NASA Grant NAGW-459 at Southern Methodist University, and by the Planetary Geology Program under NASA Contract NASG-1 at Arizona State University.

Literature Cited

Artyushkov, E. V. 1974. Can the Earth's crust be in a state of isostasy? *J. Geophys. Res.* 79:714–52

Arvidson, R. E., Davies, G. F. 1981. Effects of lateral resolution on the identification of volcanotectonic provinces on Earth and Venus. *Geophys. Res. Lett.* 8:741–44

Banerdt, W. B., Phillips, R. J., Sleep, N. H., Saunders, R. S. 1982. Thick shell tectonics on one-plate planets: applications to Mars. *J. Geophys. Res.* 87:9723–33

Basaltic Volcanism Study Project. 1981. *Basaltic Volcanism on the Terrestrial Planets.* New York: Pergamon. 1286 pp.

Brass, G. W., Harrison, C. G. A. 1982. On the possibility of plate tectonics on Venus. *Icarus* 49:86–96

Bukowinski, M. S. T. 1976. The effect of pressure on the physics and chemistry of potassium. *Geophys. Res. Lett.* 3:481–94

Burke, K., Wilson, J. T. 1972. Is the African Plate stationary? *Nature* 239:387–90

Byerlee, J. D. 1968. Brittle-ductile transition in rocks. *J. Geophys. Res.* 73:4741–50

Campbell, D. B., Burns, B. A. 1980. Earth-based radar imagery of Venus. *J. Geophys. Res.* 85:8271–81

Campbell, D. B., Dyce, R. B., Ingalls, R. P., Pettengill, G. H., Shapiro, I. I. 1972. Venus: topography revealed by radar data. *Science* 175:514–16

Campbell, D. B., Dyce, R. B., Pettengill, G. H. 1976. New radar image of Venus. *Science* 193:1123–24

Campbell, D. B., Burns, B. A., Boriakoff, V. 1979. Venus: further evidence of impact cratering and tectonic activity from radar observations. *Science* 204:1424–27

Campbell, D. B., Head, J. W., Harmon, J. K., Hine, A. A. 1983. Venus: identification of banded terrain in the mountains of Ishtar Terra. *Science* 221:644–47

Carlson, R. L. 1981. Boundary forces and plate velocities. *Geophys. Res. Lett.* 8:958–61

Esposito, P. B., Sjogren, W. L., Mottinger, N. A., Bills, B. G., Abbott, E. 1982. Venus gravity: analysis of Beta Regio. *Icarus* 51:448–59

Goldstein, R. M., Rumsey, H. Jr. 1970. A radar snapshot of Venus. *Science* 169:974–77

Goldstein, R. M., Green, R. R., Rumsey, H. C. 1976. Venus radar images. *J. Geophys. Res.* 81:4807–17

Goldstein, R. M., Green, R. R., Rumsey, H. C. 1978. Venus radar brightness and altitude images. *Icarus* 36:334–52

Golovkov, R. K., Kuznetsov, B. I., Petrov, G. M., Khasganov, A. F. 1976. An investigation of the topography of the equatorial zone of the venusian surface by radar observations from the U.S.S.R. Transl. in *Radio Eng. Electron. Phys.* (*USSR*) 21:1–5 (From Russian)

Head, J. W., Yuter, S. E., Solomon, S. C. 1981.

Topography of Venus and Earth: a test for the presence of plate tectonics. *Am. Sci.* 69:614–23

Hoffman, J. H., Hodges, R. R., Donahue, T. M., McElroy, M. B. 1980. Composition of the Venus lower atmosphere from the Pioneer Venus mass spectrometer. *J. Geophys. Res.* 85:7882–90

Hunten, D. M., Colin, L., Donahue, T. M., Moroz, V. I., eds. 1983. *Venus.* Tucson: Univ. Ariz. Press. 1143 pp.

Jurgens, R. F., Goldstein, R. M., Rumsey, H. C., Green, R. R. 1980. Images of Venus by three-station radar interferometry—1977 results. *J. Geophys. Res.* 85:8282–94

Kaula, W. M. 1975. The seven ages of a planet. *Icarus* 26:1–15

Kaula, W. M., Muradian, L. M. 1982. Could plate tectonics on Venus be concealed by volcanic deposits? *Geophys. Res. Lett.* 9:1021–24

Kaula, W. M., Phillips, R. J. 1981. Quantitative tests for plate tectonics on Venus. *Geophys. Res. Lett.* 8:1187–90

Kohlstedt, D. L., Goetze, C. 1974. Low-stress high-temperature creep in olivine single crystals. *J. Geophys. Res.* 79:2045–51

Ksanfomality, L. V., Scarf, F. L., Taylor, W. W. L. 1983. The electrical activity of the atmosphere of Venus. See Hunten et al 1983, pp. 565–603

Lewis, J. S. 1972. Metal silicate fractionation in the solar system. *Earth Planet. Sci. Lett.* 15:286–90

Malin, M. C., Saunders, R. S. 1977. Surface of Venus: evidence of diverse landforms from radar observations. *Science* 196:987–90

Masursky, H., Eliason, E., Ford, P. G., McGill, G. E., Pettengill, G. H., et al. 1980. Pioneer Venus radar results: geology from images and altimetry. *J. Geophys. Res.* 85:8232–60

McGill, G. E., Steenstrup, S. J., Barton, C., Ford, P. G. 1981. Continental rifting and the origin of Beta Regio, Venus. *Geophys. Res. Lett.* 8:737–40

McGill, G. E., Warner, J. L., Malin, M. C., Arvidson, R. E., Eliason, E., et al. 1983. Topography, surface properties, and tectonic evolution. See Hunten et al 1983, pp. 69–130

McKenzie, D. P., Bowin, C. 1976. The relationship between bathymetry and gravity in the Atlantic Ocean *J. Geophys. Res.* 81:1903–15

Morgan, P., Phillips, R. J. 1983. Hot spot heat transfer: its application to Venus and implications to Venus and Earth. *J. Geophys. Res.* 88:8305–17

Parsons, B. 1982. Causes and consequences of the relation between area and age of the ocean floor. *J. Geophys. Res.* 87:289–302

Parsons, B., Sclater, J. G. 1977. An analysis of the variation of ocean floor bathymetry and heat flow with age. *J. Geophys. Res.* 82:803–27

Pettengill, G. H. 1978. Physical properties of the planets and satellites from radar observations. *Ann. Rev. Astron. Astrophys.* 16:265–92

Pettengill, G. H., Ford, P. G., Brown, W. E., Kaula, W. M., Keller, C. H., et al. 1979a. Pioneer Venus radar mapper experiment. *Science* 203:806–8

Pettengill, G. H., Ford, P. G., Brown, W. E., Kaula, W. M., Masursky, H., et al. 1979b. Venus: preliminary topographic and surface imaging results from the Pioneer Orbiter. *Science* 205:91–93

Pettengill, G. H., Eliason, E., Ford, P. G., Loriot, G. B., Masursky, H., McGill, G. E. 1980. Pioneer Venus radar results: altimetry and surface properties. *J. Geophys. Res.* 85:8261–70

Phillips, R. J. 1983. Plate tectonics on Venus? *Nature* 302:655–56

Phillips, R. J., Lambeck, K. 1980. Gravity fields of the terrestrial planets: long-wavelength anomalies and tectonics. *Rev. Geophys. Space Phys.* 18:27–76

Phillips, R. J., Malin, M. C. 1983. The interior of Venus and tectonic implications. See Hunten et al 1983, pp. 159–214

Phillips, R. J., Saunders, R. S. 1975. The isostatic state of martian topography. *J. Geophys. Res.* 80:2893–97

Phillips, R. J., Sjogren, W. L., Abbott, E. A., Smith, J. C., Wimberly, R. N., Wagner, C. A. 1979. Gravity field of Venus: a preliminary analysis. *Science* 205:93–96

Phillips, R. J., Kaula, W. M., McGill, G. E., Malin, M. C. 1981. Tectonics and evolution of Venus. *Science* 212:879–87

Prather, M. J., McElroy, M. B. 1983. Helium on Venus: implications for uranium and thorium. *Science* 220:410–11

Reasenberg, R. D., Goldberg, Z. M., Shapiro, I. I. 1982. Venus: comparison of gravity and topography in the vicinity of Beta Regio. *Geophys. Res. Lett.* 9:637–40

Ringwood, A. E. 1975. *Composition and Petrology of the Earth's Mantle.* New York: McGraw-Hill. 618 pp.

Rogers, A. E. E., Ingalls, R. P. 1969. Venus: mapping the surface reflectivity by radar interferometry. *Science* 165:797–99

Rumsey, H. C., Morris, G. A., Green, R. R., Goldstein, R. M. 1974. A radar brightness and altitude image of a portion of Venus. *Icarus* 23:1–7

Russell, C. T., Elphic, R. C., Slavin, J. A. 1980. Limits on the possible intrinsic magnetic field of Venus. *J. Geophys. Res.* 85:8319–32

Saunders, R. S., Malin, M. C. 1977. Geologic interpretation of new observations of the surface of Venus. *Geophys. Res. Lett.* 4: 542–50

Schaber, G. G. 1982. Venus: limited extension and volcanism along zones of lithospheric weakness. *Geophys. Res. Lett.* 9:499–502

Schubert, G., Stevenson, D., Cassen, P. 1980. Whole planet cooling and the radiogenic heat source contents of the Earth and Moon. *J. Geophys. Res.* 85:2531–38

Sclater, J. G., Jaupart, C., Galson, D. 1980. The heat flow through oceanic and continental crust and the heat loss of the Earth. *Rev. Geophys. Space Phys.* 18:269–311

Sjogren, W. L., Phillips, R. J., Birkeland, P. W., Wimberly, R. N. 1980. Gravity anomalies on Venus. *J. Geophys. Res.* 85:8295–302

Sjogren, W. L., Bills, B. G., Birkeland, P. W., Esposito, P. B., Konopliv, A. R., et al. 1983. Venus gravity anomalies and their correlations with topography. *J. Geophys. Res.* 88:1119–28

Solomon, S. C., Head, J. W. 1979. Vertical movement in mare basins: relations to mare emplacement, basin tectonics, and lunar thermal history. *J. Geophys. Res.* 84:1667–82

Solomon, S. C., Head, J. W. 1982a. Evolution of the Tharsis Province of Mars: the importance of heterogeneous lithospheric thickness and volcanic construction. *J. Geophys. Res.* 87:9755–74

Solomon, S. C., Head, J. W. 1982b. Mechanisms for lithospheric heat transport on Venus: implications for tectonic style and volcanism. *J. Geophys. Res.* 87:9236–46

Solomon, S. C., Head, J. W. 1983. Venus banded terrain: evaluation of tectonic models for the origin of banding and implications for thermal structure. *J. Geophys. Res.* In press

Surkov, Yu. A. 1977. Geochemical studies of Venus by Venera 9 and 10 automatic interplanetary stations. *Proc. Lunar Planet. Sci. Conf., 8th*, pp. 2665–89

Surkov, Yu. A., Moskalyeva, L. P., Shcheglov, O. P., Kharyukova, V. P., Manvelyan, O. S., et al. 1983. Determination of the elemental composition of rocks on Venus by Venera 13 and Venera 14 (preliminary results). *J. Geophys. Res.* 88:A481–93 (*Proc. Lunar Planet. Sci. Conf., 13th, Part 2*)

Watts, A. B., Bodine, J. H., Steckler, M. S. 1980. Observations of flexure and the state of stress in the oceanic lithosphere. *J. Geophys. Res.* 85:6369–76

Willemann, R. J., Turcotte, D. L. 1982. The role of lithospheric stress in the support of the Tharsis Rise. *J. Geophys. Res.* 87: 9793–9801

Ann. Rev. Earth Planet. Sci. 1984. 12: 445–88

BLANCAN-HEMPHILLIAN LAND MAMMAL AGES AND LATE CENOZOIC MAMMAL DISPERSAL EVENTS

Everett H. Lindsay

Department of Geosciences, University of Arizona, Tucson, Arizona 85721

Neil D. Opdyke

Department of Geology, University of Florida, Gainesville, Florida 32611

Noye M. Johnson

Earth Sciences Department, Dartmouth College, Hanover, New Hampshire 03755

INTRODUCTION

Vertebrate fossils have produced a reliable and useful framework for interpreting geologic history during the Cenozoic Era and for unraveling the complex evolution of mammals. This framework was formalized for North America in 1941 as the North American Provincial Land Mammal Ages (Wood et al 1941). During the last 40 years, that framework has been revised, extended, and calibrated so that land mammal ages are now the prime chronologic standard for terrestrial deposits of the Cenozoic. Land mammal ages have been calibrated and resolved because of numerous studies relating mammal fossils and fossil sites to modern systematic analyses, detailed biostratigraphic analyses, new radiometric dating

0084–6597/84/0515–0445$02.00

methods, and magnetic polarity sequences. In this review, we present new paleomagnetic and radiometric data, as well as summarize and (in some cases) revise earlier interpretations of faunal correlations near the boundary between the Blancan and Hemphillian land mammal ages. Our purpose is to provide a broader and more secure foundation for interpreting geologic history and mammal evolution near the Blancan/Hemphillian boundary.

The Blancan and the Hemphillian were 2 of the original 18 land mammal ages proposed in 1941. At that time, the Blancan was considered late Pliocene and the Hemphillian was considered mid-Pliocene. The limits of late Miocene, Pliocene, and early Pleistocene have been reinterpreted many times during the past 40 years (and those limits are still subject to revision); however, the relationship of the Blancan to the Hemphillian is unchanged, although the Blancan is now considered to include most of the Pliocene, and the Miocene/Pliocene boundary is placed during the Hemphillian interval.

The Blancan can be characterized as the *Stegomastodon-Castor* faunal interval because these mammals (an elephantlike proboscidean and a beaver) commonly occur together in Blancan faunas. The Blancan is also characterized by waves of immigrants from both Eurasia and South America. The Great American Interchange reached its climax during the Blancan. The modern horse *Equus* evolved from a North American species in the Hemphillian and dispersed to Eurasia during the Blancan.

The Hemphillian can be characterized as the *Dinohippus-Osteoborus* faunal interval because these mammals (a horse and a dog) commonly occur together in Hemphillian faunas. The Hemphillian, like the Blancan, is also characterized by waves of immigrants, but these are primarily from Eurasia. The bear *Agriotherium* and the fossil wolverine *Plesiogulo*, which are common in late Hemphillian faunas, are considered to have an ancestry in Eurasia. Edentate sloths, whose ancestry was in South America, appear in North America in the early Hemphillian and mark the beginning of the Great American Interchange.

In earlier studies (Johnson et al 1975, Lindsay et al 1975, Opdyke et al 1977, Neville et al 1979), we demonstrated that Blancan faunas occur in the late Gilbert, Gauss, and early Matuyama magnetic chrons. Magnetic polarity sequences in sediments that have produced Hemphillian faunas have been documented by MacFadden (1977) in central New Mexico, by MacFadden et al (1979) in northwestern Arizona, and by Bressler & Butler (1978) in central Arizona. From these studies, Hemphillian faunas can be placed (but not restricted to) magnetic chrons 4 (Gilbert) and 5. In this article, we demonstrate that Hemphillian faunas (including some listed in the reports above) can also be correlated to magnetic chron 6.

GEOLOGIC SETTING AND MAGNETOSTRATIGRAPHY

New Sections

1. CHIHUAHUA Fossils and paleomagnetic samples were collected from unnamed sediments in the valley of Rio Papigochic near Yepómera, about 250 km west-northwest of Chihuahua city and about 320 km east of Hermosillo. Hemphillian fossils from these same deposits were most recently reviewed by Ferrusquia (1978). Jacobs & Lindsay (1981) reported new Blancan sites superposed over the Hemphillian sites in this area. Lindsay & Jacobs (1984) describe small mammal fossils of both Hemphillian and Blancan age from these sediments, placed in the stratigraphic section described below.

Sediments that yield the Hemphillian fossils are dominantly claystones and siltstones with discontinuous stringers of hard, calcareous mudstones. Blancan fossils are recovered from diatomaceous claystones that overlie and intertongue with the calcareous claystones and siltstones. The thickest paleomagnetic section (Figure 1) is from SH ravine near the Carranza School, about 5 km north of Yepómera and east of La Concha. This section is 64 m thick, with diatomaceous claystone the dominant lithology in the upper 20 m of the section. Another section was sampled from Arroyo Huachin, about 1 km west of SH ravine, that was placed stratigraphically relative to the section in SH ravine by tracing beds between the two sections. In Figure 1, the Arroyo Huachin and SH ravine partial sections are correlated paleomagnetically, as well as stratigraphically, using the reversal N_1/R_1 as the datum.

The Chihuahua section is of prime importance, since Hemphillian fossils are found in the interval of the lower normal magnetozone N_1 and directly above it, while Blancan fossils are found both above and below the upper normal magnetozone N_2. Thus, the Blancan/Hemphillian boundary can be placed in the reversed magnetozone R_2 between the two normal magnetozones.

2. REDINGTON SECTION The Redington section is in the Quiburis Formation (Fm), named by Heindl (1963) for late Cenozoic alluvial sediments in the San Pedro Valley near Mammoth, Arizona. Hemphillian fossils were discovered from the Quiburis Fm in 1941 by field parties from the American Museum of Natural History (Jacobs 1977), and subsequently both the American Museum and the University of Arizona have made large collections of Hemphillian mammals from these deposits. Jacobs (1977) described three new genera and six new species of rodents from the

Quiburis Fm near Redington, and a faunal list (compiled, in part, by R. H. Tedford and B. E. Taylor on fossils in the American Museum collection) for fossil sites near both Redington and Camel Canyon (see below) was published by Lindsay (1978).

Volcanic ashes are common in the Quiburis Fm. An initial potassium-

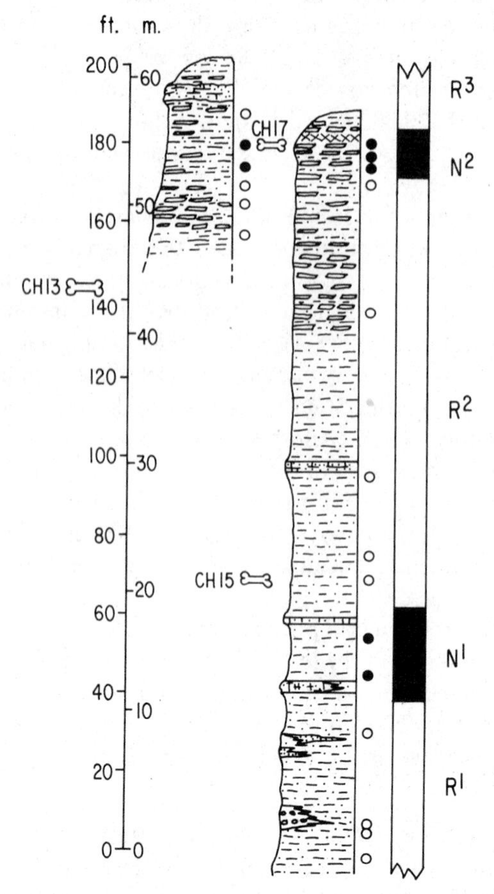

Figure 1 Chihuahua section. Stratigraphic distribution of paleomagnetic sites and fossil sites located north of Yepómera, Chihuahua. Black circles indicate normal polarity sites, while white circles indicate reversed polarity sites; N_1, N_2, etc, are normal magnetozones; R_1, R_2, etc, are reversed magnetozones. Bone symbols indicate location of fossil sites; CH 13 and CH 17 yield Blancan faunas, while CH 15 yields a Hemphillian fauna. Lithologic symbols: ...... = sandstone, –·–·– = siltstone, ——— = claystone, ▭▭▭ = diatomite, +++ = marlstone, xxx = volcanic tuff, ○○○ = conglomerate. All samples have been partially demagnetized by thermal demagnetization.

argon date of 4.6±0.4 m.y. on glass was reported from an ash in the Quiburis Fm exposed in Gust James Wash (Damon et al 1969, p. 47). Scarborough (1975) also dated ashes from the Quiburis Fm, finding a wide range of dates in the 5 and 6 m.y. range. There are no dated ashes in our Redington section.

Paleomagnetic studies in the Quiburis Fm were initiated by M. Nibbelink in 1971 as part of his graduate research (Nibbelink 1972). We have continued the work initiated by Nibbelink and gratefully acknowledge the work he contributed.

The Redington section is located in the southwest quarter of section 27, township 11 south, range 18 east. It is on the west side of the San Pedro Valley, about 6.5 km north of Redington. The section is 55 m thick, dominated by siltstones, with discontinuous and variable sandstone lenses especially in the interval 10 to 20 m above the base. Thirteen paleomagnetic sites are placed in the 55-m section. These sites define two normal magnetozones separated and limited by reversed magnetozones (see Figure 2). The lower normal magnetozone N_1 is slightly thicker (16.5 m) than the upper normal magnetozone N_2 (12 m); they are separated by a short (3 m) reversed magnetozone (R_2). The Redington fossil quarry, which yielded most of the fossils in the vicinity of the Redington section (Jacobs 1977), occurs in about the middle of the lower normal magnetozone N_1.

3. CAMEL CANYON SECTION The Camel Canyon section is also in the Quiburis Fm. Unlike the Redington section, the Quiburis Fm at Camel Canyon includes three poorly limited units of diatomaceous claystone that intertongue with fine grained clastic sediments and cherty siliceous lenses. Several volcanic ash deposits occur in the Camel Canyon section, with the two most prominent found near the bases of the lower and upper diatomaceous claystone units (see Figure 2). Zircons were separated from the lower of these ashes and dated at 6.6±0.4 m.y. (see Table 1) on the basis of fission tracks in the zircons. The zircons are euhedral and for the most part glass shrouded, which attests to their source from an airfall tuff.

The Camel Canyon section is located in section 29, township 9 south, range 18 east. It is on the east side of the San Pedro River, about 10 km east of San Manuel and about 16 km north of the Redington section. The Camel Canyon section is 40 m thick, with greater lithologic variation than the Redington section, as noted above. Thirty paleomagnetic sites in the 40-m section define two normal magnetozones that are separated and limited by reversed magnetozones (see Figure 2). Fossil sites in the Camel Canyon section occur primarily in the stratigraphic intervals near the 9- and 30-m levels, which are approximately in the middle of the lower and upper normal magnetozones.

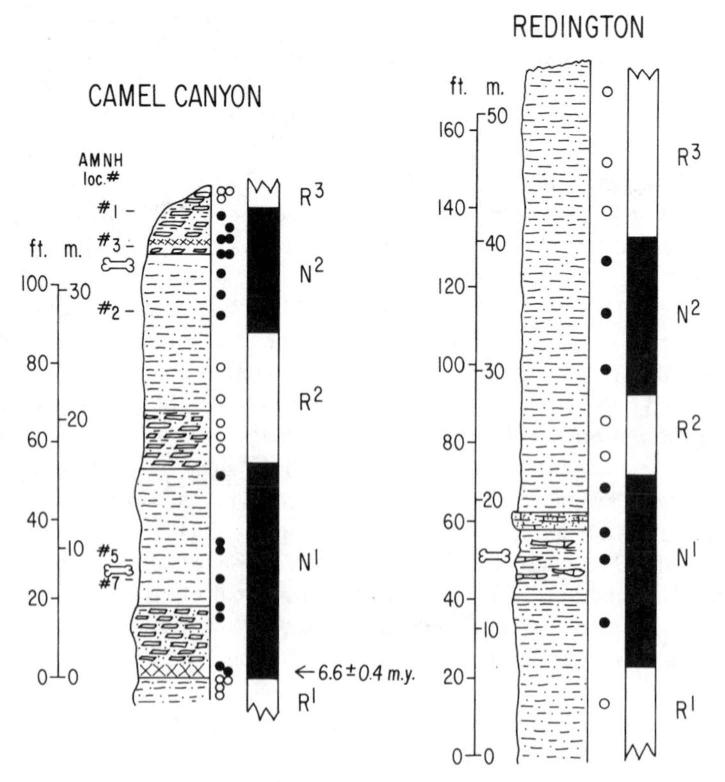

Figure 2 Quiburis sections. Camel Canyon section on the left, Redington section on the right. Symbols as given in Figure 1. Samples have been processed using alternating field magnetization. Thermal demagnetization was employed to verify the polarity.

The Redington and Camel Canyon sections are considered temporally equivalent because of similarity in fossils and magnetic reversal sequences in both sections. However, these sections cannot be correlated reliably by tracing beds or by lithologic similarity.

4. WHITE CONE SECTION The White Cone section is in the upper member of the Bidahochi Formation, a sequence of lacustrine (lower member), volcanic (middle member), and fluvial (upper member) deposits that crop out over an extensive area of northeastern Arizona and western New Mexico (Repenning et al 1958). Hemphillian fossils from White Cone were described by Stirton (1936b), with additional fossils described by Lance (1954) and Baskin (1978, 1979).

In 1964, a radiometric date of 4.1 m.y. was reported from the middle member of the Bidahochi Fm at White Cone by Evernden et al (1964). However, it was later discovered that the dated volcanic unit comes from a

Table 1 Fission track age of Camel Canyon airfall zircons

Number of zircons counted	6
Number of fossil tracks counted (N_f) ($P_s = 1.01 \times 10^6$ t cm^{-2})	177
Number of induced tracks counted (N_i) ($P_s = 9.87 \times 10^6$ t cm^{-2})	868
Correlation of N_f and N_i[a]	+0.91
Neutron dose	1.08×10^{15} cm^{-2}
U^{238} fission constant	7.03×10^{-17} yr^{-1}
Age of zircon population[b]	6.6 m.y.
Analytic error of age (calculated by the method of Johnson et al 1979)	$\sigma = 0.4$ m.y.

[a] Isochron intercept not statistically different from zero.
[b] $\lambda_f = 7.03 \times 10^{-17}$ yr^{-1}
$\lambda_d = 1.55 \times 10^{-10}$ yr^{-1}
$\gamma_n = 5.80 \times 10^{-24}$ cm^2
$I = 0.0075$
Irradiated in US Geological Survey TRIGA reactor. Dose monitored using muscovite detectors over NBS glass SRM 962 (Cu calibration).

diatreme adjacent to White Cone, and there is an airfall tuff at White Cone, at the top of the middle member of the Bidahochi Fm, that yields a date of 6.69 ± 0.16 m.y. (Scarborough et al 1974). This airfall tuff is marked on our section (Figure 3) and is used to place the section relative to the magnetic polarity time scale.

White Cone is located in sections 1 and 12, township 25 north, range 21 east. The paleomagnetic section is on the southwest exposures of the peak, in section 12. The White Cone section (Figure 3) includes exposures of both the middle and upper members of the Bidahochi Fm, with 19 m of the middle member and 51 m of the upper member. The upper member is predominantly poorly indurated siltstones and fine-grained sandstones, with friable, medium-grained sandstones becoming dominant in the upper 25 m of the section. Twenty-two paleomagnetic sites are scattered through the 70-m section. These sites have a hard magnetic overprint that is very difficult to remove by conventional methods of alternating field (a.f.) and thermal demagnetization. After intensive cleaning, it was observed that two relatively long normal magnetozones occur in the upper part of the section and one poorly defined short normal magnetozone occurs in the lower part of the section. It is plausible that a significant hiatus might also occur in the section at or below the boundary between the middle and upper members of the Bidahochi Fm. The dated volcanic ash unit is in the reversed magnetozone R_2. Hemphillian fossils are most common in the middle of the section in the magnetozone N_2.

5. RINGOLD SECTION Paleomagnetic samples were collected from three sections in the Ringold Formation exposed in bluffs along the Columbia

River north of Richland in south-central Washington. The description of these data is taken from a manuscript initiated by C. Neville and others in 1979 and is summarized below. The paleomagnetic samples were collected by Neville and Lindsay, and their polarity determinations were made by Neville and Opdyke. The Ringold section is illustrated in Figure 4.

The composite Ringold section includes samples collected from the Savage Island and Pasco Pump sections, separated by about 14 km and correlated with a datum of the White Bluffs tuff that forms a distinctive and

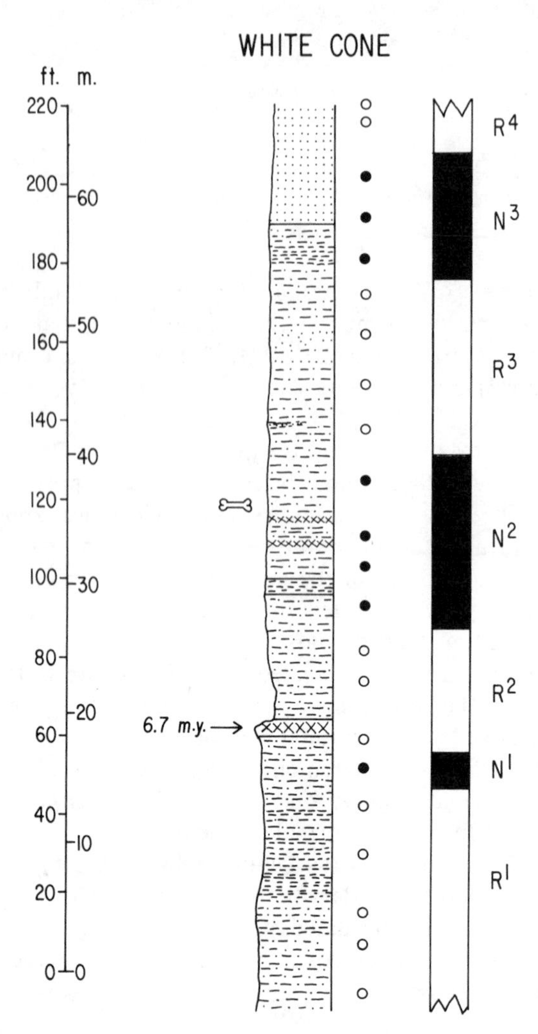

Figure 3 White Cone section. Symbols as given in Figure 1. All samples have been partially demagnetized by thermal demagnetization.

thick stratigraphic marker in the upper Ringold Formation. The upper Ringold Formation is bounded below by the Taylor Conglomerate. Thirty-three paleomagnetic sites placed in a stratigraphic interval of 149 m identify a long reversed magnetozone (R_2) in the middle of the section, capped by an unbounded normal magnetozone (N_2) at the top of the section and underlain by a short normal magnetozone (N_1) (see Figure 4). The lower normal magnetozone (N_1) is identified in both the Savage Island and Pasco

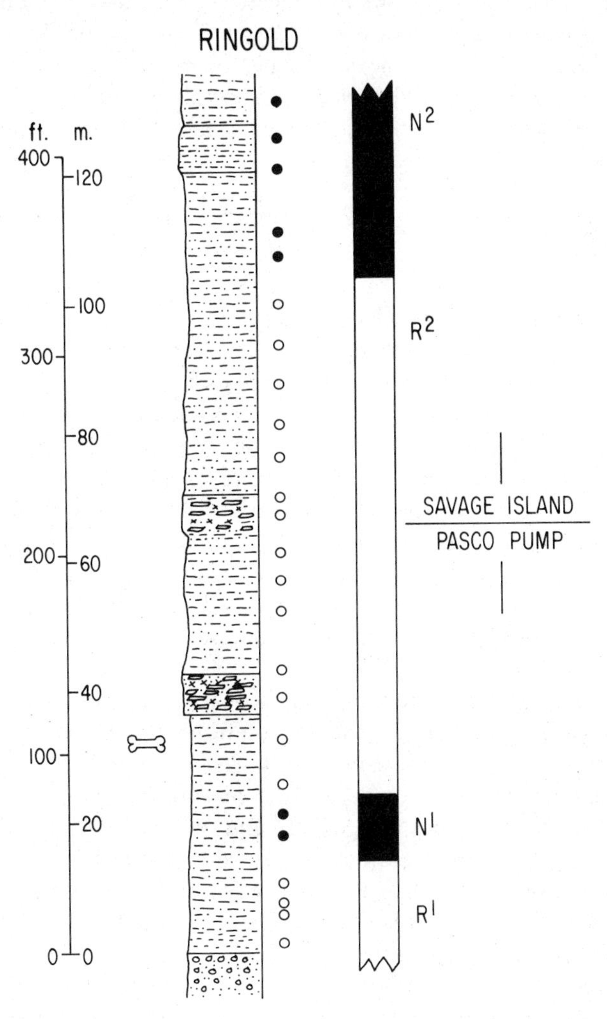

Figure 4 Ringold section. The White Bluffs tuff located in the 65–70 m interval of this composite section unites the upper Savage Island and the lower Pasco Pump parts of the section. Symbols as given in Figure 1.

Pump sections; the upper normal magnetozone is identified only in the Savage Island section, as sampling was terminated below the upper normal magnetozone in the Pasco Pump section. The long reversed magnetozone (R_2) in the middle of the Ringold section is about 57 m thick in the Savage Island section.

Fossils of the White Bluffs local fauna were described by Gustafson (1978). Most of the White Bluffs fossil mammals have been collected near (both above and below) the White Bluffs tuff, which is located about 38 m above the Taylor Conglomerate in the Pasco Pump section. Fossil locality A5927, which has yielded the microtine rodent *Ophiomys mcknighti*, is located in the lower part of the long reversed magnetozone (R_2), about 5 m below the White Bluffs tuff and about 8 m above the short normal magnetozone (N_1).

MAGNETIC POLARITY CORRELATIONS AND PLACEMENT OF FAUNAS

Correlation of magnetic polarity sequences in nine sections [the Glenns Ferry Fm (Hagerman fauna) in Idaho, the Ringold Fm in Washington, the Verde Fm in Arizona, Pliocene sediments in Chihuahua, Mexico, the Quiburis Fm in Arizona, the Big Sandy Fm (Wikieup fauna) in Arizona, the Chamita Fm in New Mexico, Pliocene sediments in Hemphill County, Texas, and the Bidahochi Fm (White Cone fauna) in Arizona] with the magnetic polarity time scale is given in Figure 5. These nine sections are scaled proportionally to the interval between the Gilbert/Gauss and the magnetic chron 5/6 boundaries. Chronologic ordering of these sequences and faunas is based primarily on (*a*) magnetic polarity sequences in the individual sections and (*b*) radiometric dating, when available, in a particular section.

Our purpose in this correlation exercise is to order the faunal data with minimal regard for biochronologic or evolutionary interpretations. This is done in order to reduce biochronologic bias from sequential placement of mammal faunas and to provide a more objective foundation for interpretation of mammal dispersal events and subtleties of faunal change. Unfortunately, there are instances considered here where faunal data have influenced the placement of faunas. These instances are clearly identified. Stratigraphic superposition is the preferred basis for ordering the faunas, but such superposition is feasible in only the Verde, Chihuahua, and Quiburis sections, and is straightforward in only the Chihuahua section.

The Glenns Ferry section studied by Neville et al (1979) includes a long normal magnetozone superposed over a longer reversed magnetozone. Radiometric dating of separate ashes in the middle of the long reversed

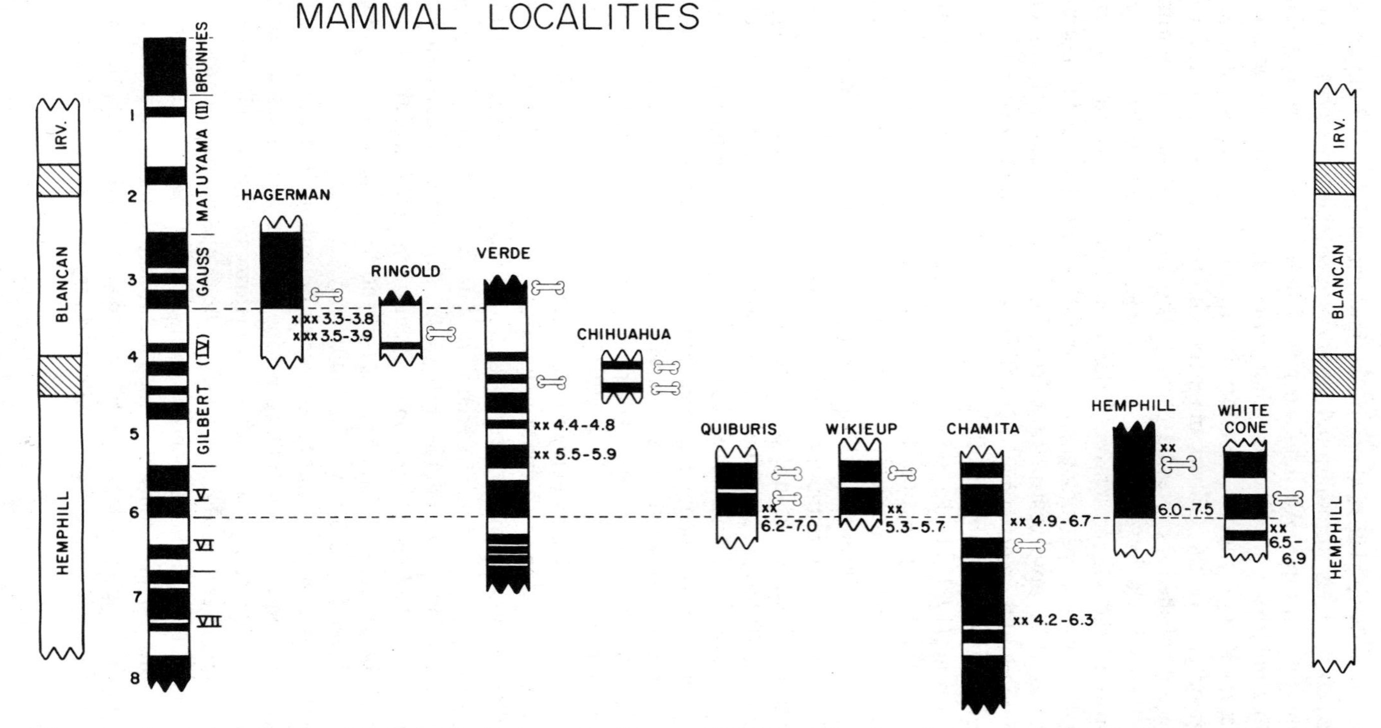

Figure 5 Correlation of sections. Magnetic polarity time scale of Ness et al (1980) for the last 8 m.y. is shown on the left. The Hagerman, Ringold, and Verde sections are correlated with the Gauss/Gilbert boundary as datum. The Verde, Quiburis, Wikieup, Chamita, Hemphill, and White Cone sections are correlated with the chron 5/6 boundary as datum. The Chihuahua section is placed relative to the Verde fauna in the Gilbert magnetic chron. Volcanic ash units that have yielded radiometric dates, and the error limits of these dates, are given in each section where located. Symbols are given in Figure 1.

magnetozone have yielded dates in the interval of 3.3 to 3.9 m.y. when referred to the latest International Union of Geological Sciences (IUGS) radiometric constants. All data we report here are so referenced. The magnetic polarity sequence and radiometric data constrain placement of the Glenns Ferry section to the interval of the Gauss magnetic chron and the upper part of the Gilbert magnetic chron. The magnetic polarity sequence published by Neville et al (1979) included a short normal magnetozone in the lower part of the sequence, interpreted as the Cochiti subchron of the Gilbert chron. Later collecting (and more thorough thermal cleaning of paleomagnetic samples) has eliminated that normal magnetozone. Its absence from the Glenns Ferry sequence also means that the faunal datum events given by Neville et al (1979) are too young by an unknown (but presumably small) factor.

Armstrong (1981) challenged our interpretation of the Glenns Ferry magnetic polarity correlation, based on older radiometric dates he and colleagues obtained from the same sequence. He concluded that the age of basalts we place in the interval 3.3 to 3.9 m.y. cannot be precisely dated, but that they should fall in the interval 4 to 5 m.y. We reject Armstrong's correlation primarily because of the sequential appearance of Blancan mammals in the lower half of the Glenns Ferry Fm. These fossils suggest a major portion (approximately two thirds) of the Blancan temporal interval is represented in the Glenns Ferry sections. If the Glenns Ferry Fm were compressed into a small portion of the Gilbert magnetic chron, as inferred by Armstrong, the entire section would represent no more than 12% of the Blancan land mammal age.

In summary, the Glenns Ferry section is relatively long and represents a major part of the temporal interval assigned to the Blancan. The presence of the Gilbert/Gauss magnetic reversal in about the middle of the Glenns Ferry sequence is a prime reference in our faunal correlations. The sequential appearance (three biostratigraphic appearance events listed by Neville et al 1979) of Blancan mammals occurs in the interval of about 3.6 to 3.8 m.y. The Hagerman fauna persists in the lower part of the Gauss magnetic chron for a duration of about 0.6 m.y.

The Ringold magnetic polarity sequence is only about half as thick as the Glenns Ferry sequence. The top of our Ringold paleomagnetic sequence is marked by a relatively long normal magnetozone. That normal magnetozone is interpreted as the lower part of the Gauss magnetic chron. The lower part of the Ringold sequence is reversed but includes a short normal magnetozone that we interpret as the Cochiti subchron. In summary, we correlate the Ringold paleomagnetic sequence as slightly older than and overlapping the lower part of the Glenns Ferry paleomagnetic sequence.

The White Bluffs fauna of the Ringold Fm (Gustafson 1978) is placed in the approximate interval 3.4 to 3.8 m.y.

Bressler & Butler (1978) established a long and detailed magnetic polarity sequence in the Verde Formation of central Arizona. Their composite sequence, equivalent to a thickness greater than 300 m, extends from the lower part of the Gauss magnetic chron through the Gilbert chron, chron 5 and 6, and into chron 7. A basalt in the Gilbert-chron part of the section yields a K-Ar date of 4.5 m.y., and another basalt in the chron 5 part of the section yields a K-Ar date of 5.5 m.y. These dated basalts attest to the validity of correlation of the Verde polarity sequence relative to the magnetic polarity time scale.

Blancan fossils have been collected from the top of the Verde Fm near Clarkdale (Breed 1962) that correlate with the Gauss magnetic chron. Bressler discovered a new fossil level while collecting paleomagnetic samples in the lower part of the Verde Fm. Fossils from that site were later collected and studied by L. Jacobs and are designated as the Verde fauna. Fossils of the Verde fauna are very near the Blancan/Hemphillian boundary; they could be placed in either an earliest Blancan or a latest Hemphillian assemblage. Additional fossils have subsequently been collected and are under study by N. Czepluski of Northern Arizona University. His analysis of the fauna, when completed, will help resolve the biochronologic assignment of the fauna or faunas. The Verde fauna occurs in the lower part of the Nunivak subchron of the Gilbert magnetic chron and indicates that the Nunivak subchron should be close to the Blancan/Hemphillian boundary. We place the Verde fauna at approximately 4.2 m.y.

Magnetostratigraphy of the Chihuahua section was described above (Figure 1). The Chihuahua section is correlated with part of the Gilbert magnetic chron because of the dominance of reversed polarity with two short normal magnetozones.

Hemphillian mammals are recorded from the stratigraphic interval of the lower normal magnetozone and are grouped as the Yepómera fauna. Blancan mammals are recorded from the stratigraphic interval surrounding the upper normal magnetozone and are grouped as the Concha fauna. The upper normal magnetozone may represent the Cochiti or the Nunivak subchron, based on correlation of the Concha fauna with either the White Bluffs fauna or the Verde fauna. Tentatively, we correlate the upper normal magnetozone of the Chihuahua section with the Nunivak subchron, which places the Blancan Concha fauna in the same polarity zone (about 4.2 m.y.) as the Verde fauna. Similarly, the Hemphillian Yepómera fauna is assigned to the interval of the Sidufjall (c_1) subchron at about 4.5 m.y. (see Figure 5).

May & Repenning (1982) correlated the short normal magnetozones in the Chihuahua section with the Sidufjall and Thvera subchrons of the Gilbert chron. We do not support their interpretation, as it would place the Verde fauna decidedly younger than the Blancan Concha fauna.

Correlation of the Chihuahua section to the interval of the Nunivak and Sidufjall subchrons of the magnetic polarity time scale is subjective; we have no compelling or unequivocal evidence that the Chihuahua section cannot be assigned to either the younger Cochiti-Nunivak or to the older Sidufjall-Thvera interval of the magnetic polarity time scale, as proposed by May & Repenning (1982). Note, however, that these alternative interpretations differ by only about 0.3 m.y.

Magnetostratigraphy of the Quiburis Formation was described above (Figure 2). Both paleomagnetic and radiometric data support correlation of the Quiburis section with magnetic chron 5. The Camel Canyon fauna is assigned an age of approximately 5.6 m.y., and the Redington fauna an age of approximately 5.9 m.y.

Magnetostratigraphy of the Big Sandy Formation was presented by MacFadden et al (1979). They assigned a mean age of 5.5 ± 0.2 m.y. for a tuff complex located slightly above the base of the lower of two relatively long normal magnetozones. The paleomagnetic and radiometric data are consistent with assignment of the Big Sandy section to the chron 5 interval of the magnetic polarity time scale. We follow the correlation given by MacFadden et al (1979) for the Big Sandy section. The age of the Wikieup fauna in the upper part of the section is placed at approximately 5.6 m.y.

Shafiqullah et al (1980) reported a K-Ar date of 9.62 ± 0.38 m.y. (UAKA-70-13) on a basalt believed to intertongue with the Big Sandy Fm southeast of Wikieup. It appears to us that this basalt underlies the Big Sandy Fm and correlates with basalts that intertongue with or cap syntectonic sediments in Burro Creek, south of and discontinuous with exposures of the Big Sandy Fm. The sediments exposed along Burro Creek are agglomerates, gravels and coarse sandstones, poorly sorted and poorly stratified, with large angular lithic fragments. They should not be confused with the well stratified, more mature sediments with interbedded tuffs that characterize the Big Sandy Fm, which are finer grained and better sorted.

Magnetostratigraphy of the Chamita Formation was presented by MacFadden (1977), who identified 12 magnetozones of variable thickness and 2 tuffaceous units with radiometric dates of 5.6 ± 0.9 (upper) and 5.2 ± 1.0 (lower) m.y. in a 500-m-thick sequence. The tuffaceous units are widely separated; they occur about 165 and 360 m above the base of the section. The Hemphillian Chamita fauna is principally from the San Juan and Rak camel quarries located between the tuffaceous units, about 320 m above the base of the section (Figure 5). Unfortunately, the chronologic range of the dated tuffaceous units does not constrain assignment of the Chamita section closer than the interval between 4 and 7 m.y., and there are several possible correlations for the Chamita section in that interval. We note also that the Chamita paleomagnetic samples were not cleaned by thermal demagnetization, so some normal sites may result from incomplete removal of a normal overprint.

MacFadden (1977) noted that the Chamita fauna is late, but not latest, Hemphillian in age. He cited a fission-track date of 5.3 ± 0.4 m.y. on glass from the Coffee Ranch section (the Hemphillian stratotype), where another late Hemphillian fauna occurs. He correlated the Chamita section to the interval between about 4.5 and 6.0 m.y., thereby placing the San Juan and Rak camel quarries in the Chamita section near the base of the Gilbert magnetic chron, slightly younger than the Coffee Ranch section that we (Lindsay et al 1975) had previously correlated to the upper part of chron 5.

We now believe the correlation given by us for the Coffee Ranch section in 1975, and that given for the Chamita section by MacFadden in 1977, should be revised. This revision is shown in Figure 5. It appears that the Chamita section correlates better in the interval from the base of the Gilbert magnetic chron to the base of chron 7, rather than to the base of chron 5. It is consistent with the radiometric date for the upper tuffaceous unit in the Chamita section (5.6 ± 0.9 m.y.) and implies the date for the lower tuffaceous unit (5.2 ± 1.0 m.y.) is unreliable. Note that we are not using the radiometric data to correlate the Chamita section; instead, we are correlating primarily on the paleomagnetic sequence. It happens that the upper radiometric date is consistent with this correlation, whereas the lower radiometric date is not. This new correlation places the San Juan-Rak camel quarries near the middle of chron 6, at about 6.5 m.y., and therefore slightly older than the Hemphillian Coffee Ranch fauna. A similar correlation was derived independently by Tedford (1981, pp. 1013–14).

Magnetostratigraphy of the Coffee Ranch section in Hemphill County, Texas, was presented by Lindsay et al (1975). The Coffee Ranch section is only about 30 m thick and is dominated by normal polarity (five sites) in the upper 25 m, with reversed polarity (one site) lower in the section. The volcanic ash that overlies the fauna occurs in the middle of the section in the normal magnetozone. Izett (1975) reported dates of 6.6 ± 0.8 m.y. (on fission tracks in zircons) and 4.7 ± 0.8 m.y. (on fission tracks in glass) from the Coffee Ranch ash. Later, Boellstorff (1976) reported a radiometric date of 5.3 ± 0.4 m.y. (on fission tracks in glass) from the same ash. We place more confidence on the fission-track date in zircons, since fission-track dates in glass consistently yield dates that are too young (Seward 1979). The one-sigma range of the zircon fission-track date (5.8 to 7.4 m.y.) would allow placement of the Hemphillian magnetostratigraphy in the lower part of chron 5 or the upper part of chron 7. We place the Hemphillian Coffee Ranch section in the lower part of chron 5 (Figure 5) and assign a younger date (about 5.9 m.y.) to the Coffee Ranch fauna than to the Chamita fauna.

Magnetostratigraphy of the White Cone section was described above (Figure 3). The radiometric date of 6.69 ± 0.16 m.y. below the White Cone fauna restricts assignment of the middle Bidahochi Formation to the interval 6.5 to 6.9 m.y., or the proximity of the chron 6/7 boundary. We place the White Cone fauna in the lower part of chron 5 (at about 5.9 m.y.) and the dated Bidahochi tuff in chron 6.

VERTEBRATE PALEONTOLOGY

A chronologic sequence of 11 mammal faunas represented in the 9 paleomagnetic sections discussed above is presented in Figure 6. The Hagerman fauna (Glenns Ferry section) and White Bluffs fauna (Ringold section) are placed in the upper part of the Gilbert magnetic chron. The Verde fauna (Verde section) and the Concha and Yepómera faunas (Chihuahua section) are placed in the middle of the Gilbert chron. The Wikieup fauna (Big Sandy section) and Camel Canyon fauna (Quiburis section) are placed in the upper part of chron 5, while the Coffee Ranch fauna (Hemphillian section), Redington fauna (Quiburis section), and White Cone fauna are placed in the lower part of chron 5. The Chamita fauna is placed in the middle of chron 6.

Faunal lists for these 11 faunas, compiled from publications, are presented in Table 2, with faunas ordered according to the sequence in Figure 6 (older faunas are to the left). The Hagerman faunal list is taken from Skinner et al (1972), Zakrzewski (1969), Hibbard (1969), Hibbard & Bjork (1971), and Bjork (1970). The White Bluffs faunal list comes from Gustafson (1978), and the Verde fauna is taken from Jacobs (1977). The faunal list for the Yepómera fauna was obtained from Ferrusquia (1978), with additions from Jacobs & Lindsay (1981) and Lindsay & Jacobs (1984); the latter is also the source of the Concha faunal list. The Wikieup fauna was reviewed by MacFadden et al (1979), and the Camel Canyon and Redington faunal lists were given in Lindsay (1978). Redington rodents were described by Jacobs (1977). The Coffee Ranch fauna was reviewed by Schultz (1977), the White Cone fauna was described and reviewed by Baskin (1978, 1979), and the Chamita fauna was reviewed by MacFadden (1977). Blancan and younger faunas of North America were extensively reviewed by Kurtén & Anderson (1980), whose work we have relied upon heavily, extending it into the Hemphillian where necessary.

Apparently, moles were widespread and diverse, if not abundant, in late Hemphillian fauna. *Domninoides* is a relatively common mole in the late

Miocene. The White Cone specimen may be the latest known record of the genus, if the isolated molar questionably identified by Baskin (1979) is really *Domninoides*. *Scapanus* (*Xeroscapheus*) is also known from the late Hemphillian Krebs Ranch, Arlington, and Little Valley faunas of Oregon (Hutchison 1968). The mole *Scalopus* is also known from the Hemphillian and Blancan of North America (Kurtén & Anderson 1980, Voorhies 1977, Schultz 1977).

Shrews are rather rare in Hemphillian faunas, but they become diverse in Blancan faunas. Four species of the shrew *Sorex* were reported from the Blancan Hagerman fauna by Hibbard & Bjork (1971). Wagner (1981) described a new species of the shrew *Cryptotis* from the late Hemphillian

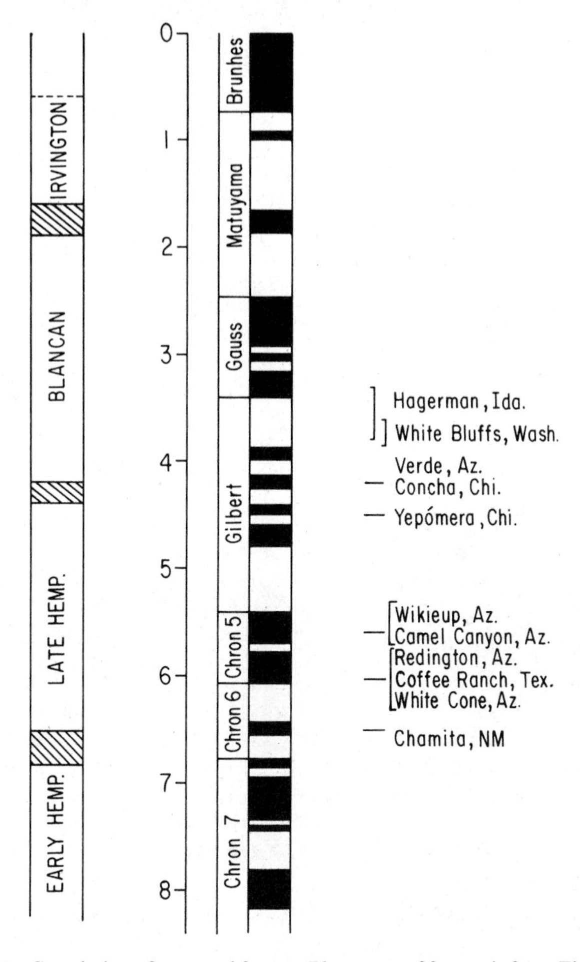

Figure 6 Correlation of mammal faunas. Placement of faunas is from Figure 5.

Table 2 Fauna from selected Blancan and Hemphillian fossil sites

	Chamita	White Cone	Coffee Ranch	Redington	Camel Canyon	Wikieup	Yepómera	Concha	Verde	White Bluffs	Hagerman
INSECTIVORA											
Talpidae			X								
Scapanus										X	X
Domninoides		?									
Soricidae			X								
Sorex											X
Paracryptotis gidleyi											X
Notiosorex		X						X			
CHIROPTERA				X							
Vespertilionidae											
Plionycteris							X				
EDENTATA											
Megalonychidae											
Pliometanastes	X										
Megalonyx										X	X
sloth			X								
LAGOMORPHA											
Leporidae											
Hypolagus	X	X	X			X				X	X
Notolagus							X				
Pratilepus											X
Nekrolagus										X	
RODENTIA											
Sciuridae											
Spermophilus		X	X	X			X			X	X
Ammospermophilus										X	?
Paenemarmota							X			?	X
Mylagaulidae											
Mylagaulus	X	X	X								
Castoridae											
Castor										X	X
Dipoides	X	X									X
Eomyidae											
Ronquillomys				X							
Geomyidae			X								
Pliogeomys							X				X
Geomys								X			
Thomomys										X	X

Table 2 (*continued*)

	—Chamita	—White Cone	—Coffee Ranch	—Redington	—Camel Canyon	—Wikieup	—Yepómera	—Concha	—Verde	—White Bluffs	—Hagerman
Heteromyidae			X								
Perognathus		X		X			X				X
Perognathoides		X									
Prodipodomys				X			X				X
Cricetidae											
Copemys		X	X	?			X				
Peromyscus										X	X
Galushamys				X							
Paronychomys		X		X							
Baiomys							X	X			X
Calomys (*Bensonomys*)		X					X	X			
Prosigmodon							X	X			
Neotoma										X	X
Mimomys (Cosomys)											X
Ophiomys										X	X
Pliopotamys											X
Nebraskomys									X		
Pliophenacomys								X	X		
CARNIVORA											
Canidae											
Canis	X		X	X		X	?			X	X
Vulpes	X		X			X	X				
Epicyon	X										
Osteoborus			X	X	?	X	?				
Borophagus										?	X
Ursidae											
Agriotherium			X	X	X	X	X				
Ursus										X	X
Tremarctus											X
Procyonidae											
Bassariscus	X				X	X					
Mustelidae											
Martes		?		X		X					
Mustela											X
Plesiogulo	X	X	X	X		X					
Ferinestrix											X
Trigonictis										X	X
Sminthosinis											X
Pliotaxidea	X		X	X		X					
Taxidea							X				X

Table 2 (*continued*)

	—Chamita	—White Cone	—Coffee Ranch	—Redington	—Camel Canyon	—Wikieup	—Yepómera	—Concha	—Verde	—White Bluffs	—Hagerman
CARNIVORA											
Mustelidae (*continued*)											
Satherium							?				X
Pliogale						X					
Buisnictis											X
Felidae											
Machairodus coloradensis			X			X					
Megantereon hesperus				?	?		X				X
Adelphailurus				X		X					
?*Homotherium*										?	X
Pseudaelurus			X			X	X				
Felis				X						X	X
Hyaenidae											?
PROBOSCIDEA											
Gomphotheriidae						X					
Stegomastodon							?				
Rhynchotherium			X								
?longirostrine gomphothere	X										
Mammutidae											
Pliomastodon					?						
Mammut										X	X
PERISSODACTYLA											
Equidae											
Neohipparion			X				X				
Nannippus			X				X				
Astrohippus	X		X				X				
Dinohippus	X		X	X	X		X	?			
Onohippidium						X					
Equus								?		X	X
Rhinoceratidae											
Aphelops	X		X								
Teleoceras	X		X				X				
ARTIODACTYLA											
Tayassuidae											
Prosthennops			X				X				
Platygonus										X	X

Table 2 (*continued*)

	—Chamita	—White Cone	—Coffee Ranch	—Redington	—Camel Canyon	—Wikieup	—Yepómera	—Concha	—Verde	—White Bluffs	—Hagerman
Camelidae											
Hemiauchenia	X		X	X	X	X				X	X
Alforjas			X								
Camelops											X
Megatylopus	X		X	X	X	X	X			X	X
Palaeomerycidae											
Pediomeryx			X								
Cervidae											X
Bretzia										X	
Antilocapridae											
Osbornoceros	X										
Ilingoceros	X										
Plioceros	X										
Sphenophalos						X					
Texoceros			X	X	X	X					
Hexobelomeryx							X				
Ceratomeryx											X
"new genus"						X					

Modesto Reservoir (= Turlock Lake) fauna of California. The White Cone occurrence of *Notiosorex* is probably the earliest known record of this relatively rare genus.

Bats (or Chiroptera) are poorly known in sedimentary deposits of North America prior to the Pleistocene. A few isolated teeth were identified as bats from the Redington fauna (Jacobs 1973), and the new genus *Plionycteris* is based on a maxilla fragment from Yepómera. Kurtén & Anderson (1980) identified the modern genus *Lasiurus* and the extinct genus *Anzanycteris* in Blancan faunas; Harrison (1978) reported the bat *Simonycteris* (which she considered inseparable from *Lasiurus*) from the Blancan Wolf Ranch fauna.

Sloths appear in North America (as immigrants from South America) approximately at the beginning of the Hemphillian land mammal age. Two genera of sloths, *Pliometanastes* and *Megalonyx*, are reported from Hemphillian faunas, and *Pliometanastes* is believed to be restricted to the early Hemphillian (Hirschfeld & Webb 1968). Hirschfeld (1981) considered *Pliometanastes protistus* from the Knight's Ferry fauna in the Mehrten

Formation of California as the most primitive ground sloth known from North America. Wagner (1981) recorded the lowest stratigraphic occurrence of *Pliometanastes protistus* in the Mehrten Fm within the Siphon Canyon local fauna (previously included in the Knight's Ferry fauna) that is overlain (4 m) by a biotite-rich white tuff that has yielded a radiometric age (K-Ar on biotite) of 8.19 ± 0.16 m.y. If this is the oldest ground sloth record in North America, the Clarendonian/Hemphillian boundary can be placed at 8.2 m.y., near the base of magnetic chron 7. MacFadden (1977) noted that ?*Pliometanastes galushai* was collected from deposits below the San Juan and Rak Camel quarries in the Chamita Formation. The Chamita ?*Pliometanastes* record is probably in magnetic chron 7, younger than the Siphon Canyon record of *Pliometanastes. Megalonyx leptostomus* was a widespread Blancan sloth, recorded from both the Hagerman and White Bluffs faunas *fide* MacDonald (1977).

Rabbits are very well represented in North American Hemphillian and Blancan deposits. *Hypolagus* is an archaeolagine rabbit (Dawson 1958) that appeared in the middle Miocene and became extinct in the Blancan. *H. vetus* is recorded from both the White Cone and Chamita faunas; a more advanced species, *H. ringoldensis*, was described by Gustafson (1978) from the White Bluffs fauna. Probably the earliest known occurrence of the archaeolagine rabbit *Notolagus* is in the Yepómera fauna. *Pratilepus* is an advanced archaeolagine or primitive leporine rabbit known only from early Blancan (Hagerman and Rexroad) faunas; it probably gave rise to *Aluralagus*, recorded from Blancan faunas of southern Arizona. *Nekrolagus* is a widespread, primitive leporine rabbit, almost completely confined to the early Blancan, whose earliest known record is in the White Bluffs fauna.

Squirrels are not very common in Hemphillian and Blancan faunas, although the genus *Paenemarmota* is restricted to the Pliocene. The Yepómera occurrence of *Paenemarmota* may be its earliest known record, although a record from the Goleta fauna of Michoacan in central Mexico may be older (Repenning 1962).

Mylagaulus is a large, beaverlike genus with a wide distribution in Hemphillian faunas (and a long history in North America) that became extinct near the end of the Hemphillian. McKay Reservoir, Oregon, or the Bone Valley, Florida, occurrences may prove the latest known record of *Mylagaulus. Dipoides* and *Castor* are Pliocene beavers, known from faunas in both North America and Europe; *Dipoides* is known throughout the late Hemphillian and into the early Blancan (Hagerman fauna, Idaho, and Rexroad fauna, Kansas), while *Castor* appears in North America in the late Hemphillian (the McKay Reservoir fauna of Oregon) and is extant.

Eomyid rodents had a long and diverse history in North America and became extinct in the Hemphillian. *Ronquillomys wilsoni*, known only from the Redington fauna, may be the latest known eomyid rodent in North America. Other Hemphillian eomyids are *Kansasimys* of the Edson fauna in Kansas (Wood 1936) and *Leptodontomys* of the McKay Reservoir and Bartlett Mt. faunas in Oregon (Shotwell 1970).

Three genera of gophers are known in Hemphillian faunas. *Pliosaccomys* is recorded from late Clarendonian and early Hemphillian faunas, *Parapliosaccomys* is known from the late Hemphillian, and *Pliogeomys* is recorded from the late Hemphillian and early Blancan. Modern genera of gophers (*Thomomys*, *Geomys*, and *Cratogeomys*) appear in the Blancan.

Heteromyid rodents (pocket mice and kangaroo rats) are well represented in Hemphillian and Blancan faunas. The modern pocket mouse, *Perognathus*, is known from the middle Miocene (Barstovian land mammal age). The White Cone record of *Perognathoides* is probably the latest known record of that genus. The genus *Prodipodomys* (*sensu stricto*) appears in the late Hemphillian and continues into the Pleistocene, giving rise to the modern kangaroo rat, *Dipodomys*, in the early Pleistocene.

Cricetid rodents underwent an explosive adaptive radiation in the Hemphillian and Blancan, especially in the subfamilies Microtinae and the Hesperomyinae. A single genus (*Copemys*) of cricetid rodent is known from the beginning of the Hemphillian. *Copemys* probably gave rise to the modern genus *Peromyscus*, plus a group of high-crowned cricetids (including *Paronychomys*, *Galushamys*, and *Repomys*) that appear in the Hemphillian (May 1981). Baskin (1978) and Jacobs & Lindsay (1981) have documented the initial radiation of Hesperomyinae in the late Hemphillian White Cone and Yepómera faunas. The North American hesperomyine cricetids entered South America in the Pliocene, where they produced an explosive radiation in the Pleistocene and Recent.

During the time of these cricetid radiations, waves of microtines were invading North America from the Arctic at irregular intervals (Repenning 1980). Repenning & Fejfar (1977) presented a framework for correlation of Pliocene and Pleistocene microtine faunas of Europe and North America, and Repenning (1980) developed a refined biochronology for late Hemphillian through Rancholabrean faunas based on successive immigrations of microtines and their endemic evolution. We should point out that the chronology for Hemphillian and Blancan presented by Repenning (1980) differs from the chronology we present here. Repenning places the Blancan/Hemphillian boundary near the base of the Thvera event, at about 4.8 m.y. The Pliocene radiation and dispersal history of cricetid rodents is still poorly documented and incompletely understood. The following faunal records are part of that history. Yepómera probably represents the latest known North American record of *Copemys*, as well as the earliest known record of both *Prosigmodon* and *Baiomys*. White Cone may represent the earliest known record of both *Paronychomys* and *Calomys* (*Bensonomys*), although both genera are also recorded from the Modesto Reservoir fauna of California (Wagner 1981). *Galushamys* is presently restricted to the Redington fauna. The earliest known record of *Pliophenacomys* is probably in the Concha or Verde fauna. *Propliophenacomys* is recorded from late Hemphillian deposits in Nebraska, including the Santee fauna (Martin 1975). The Hagerman fauna probably represents the earliest known record of *Pliopotamys*.

The Canidae have a long and rich fossil record in North America, dating back to the Oligocene. The genus *Vulpes* has been assigned to poorly known, small, gracile canids recorded from Clarendonian (late Miocene) faunas. *Vulpes stenognathus* and *Canis davisi* are both recorded from the Chamita, Coffee Ranch, Redington, and Wikieup faunas. As emphasized by Baskin (1980), the generic name *Epicyon* should be assigned to the "*Aelurodon*" *saevus* group. *Epicyon* is not known after early Hemphillian, and one of the latest records of that group is *Epicyon haydeni*, collected near the base (and about 200 m stratigraphically below the San Juan–Rak Camel quarries) of the Chamita Formation. Borophagine canids (*Osteoborus* and *Borophagus*) are very characteristic of Hemphillian and Blancan faunas. *Osteoborus* is widespread and diverse in the Hemphillian; it is replaced by *Borophagus* in the Blancan. The White Bluffs–Hagerman record of *Borophagus* may be the earliest known occurrence of that genus.

The large bear *Agriotherium* is apparently restricted to late Hemphillian deposits of North America. We consider the absence of *Agriotherium* from the White Cone and Chamita faunas significant. We correlate the Coffee Ranch fauna that records *Agriotherium* approximately temporally equivalent with the White Cone fauna, but the absence of *Agriotherium* suggests that White Cone might be slightly older than Coffee Ranch. *Ursus* is not known in North America prior to the Blancan; its occurrence in the White Bluffs fauna is considered its earliest North American record. The only known North American record of the Pliocene Panda, *Parailurus*, is from the Taunton fauna of the Ringold Formation that also produced the White Bluffs fauna (Tedford & Gustafson 1970). The Taunton fauna is considered approximately equivalent in age to the White Bluffs and Hagerman faunas. *Parailurus*, along with *Ursus* and *Agriotherium*, has an earlier record in Eurasia, and all three are interpreted as Pliocene immigrants.

Procyonids have a long and complex fossil record in North America, dating back to at least the early Miocene. However, procyonids have never been abundant or well represented in North American Cenozoic mammal faunas. The modern ring-tail, *Bassariscus*, is known from the mid-Miocene (Barstovian); its presence in Hemphillian and Blancan faunas is not especially significant. The Mustelidae, on the other hand, are diverse and well represented in Hemphillian and Blancan faunas. Several species of *Martes* have been reported from late Miocene faunas (e.g. Big Spring Canyon, South Dakota; Shansi, China; and Weze, Poland) of the Northern Hemisphere. Two species of *Martes* are presently extant in North America. The weasel, *Mustela rexroadensis*, is recorded from both the Hagerman fauna and the Fox Canyon fauna of Kansas; it is considered a Pliocene immigrant. The Pliocene wolverine *Plesiogulo* is well represented in and apparently restricted to late Hemphillian faunas in North America (Harrison 1981). Kurtén (1970) reviewed the record of *Plesiogulo* in Eurasia, noting its occurrence in the "Pontian" faunas of Tientsin, Kansu, and Shansi, China; Pavloder, Siberia; and the Dhok Pathan fauna of Pakistan. The Dhok Pathan fauna is now considered equivalent to Ruscinian (approximately equivalent to late Hemphillian); the precise ages of the other *Plesiogulo* faunas are not well established. Thus, *Plesiogulo* appears widely distributed in the Northern Hemisphere, and it may have a very restricted chronologic range. Kurtén (1970) considered *Plesiogulo* derived from a martenlike mustelid presumably from China, where it is most diverse. The Chamita record is probably the earliest known North American occurrence of *Plesiogulo*, and the Wikieup record may be its latest known North American occurrence.

The grison *Trigonictis* is a widely distributed and common Blancan carnivore. Repenning (1967) considered *Trigonictis* an Eurasian immigrant derived from either *Enhydrictis* or *Pannonictis*, both Eurasian genera. More recently, Wagner (1981) recognized the presence of grisonine mustelids in the late Hemphillian Modesto Reservoir fauna and named a new species of the grisonine *Galictis*. The Eurasian *Enhydrictis* or *Pannonictis* seem the most likely ancestor of *Trigonictis*, and the White Bluffs or Hagerman record is probably the earliest known North American occurrence of that genus.

The Pliocene badger *Pliotaxidea* is restricted to the Hemphillian and is considered ancestral to *Taxidea*, the modern badger, whose earliest known occurrence is *T. mexicana* from the latest Hemphillian Yepómera fauna (Wagner 1976). The otter *Satherium* was rather widespread in North America during the Blancan and was replaced by *Lutra*, the modern otter, in the Irvingtonian. *Satherium* is well represented in the Hagerman fauna, which is probably the earliest known occurrence of the genus. Wagner (1976, p. 107) noted that a distal humerus fragment in the Yepómera fauna identified as *Taxidea mexicana* may actually be an otter, like *Satherium*.

Systematic relationships of late Tertiary cats have recently been under study. As presently understood, the following Hemphillian and Blancan species of cats are recognized: *Nimravides thinobates*, *Nimravides catacopis*, *Nimravides galiani*, and *Machairodus coloradensis* in the Machairodontinae; *Adelphailurus kansensis* in the Metailurinae; *Megantereon hesperus* in the Smilodontinae; *Pseudaelurus hibbardi, Pratifelis martini, Felis proterolyncis*, and *Felis rexroadensis* in the Felinae (Dalquest 1969, Martin & Schultz 1975, Baskin 1981, MacFadden & Galiano 1981, Harrison 1983, Berta & Galiano 1983). If these cats were immigrants, it is believed that the nimravid *Barbourofelis* immigrated from Eurasia in the late Clarendonian, *Machairodus* dispersed from Eurasia in the early Hemphillian, and that both *Megantereon* and *Felis* dispersed from Eurasia in the late Hemphillian. It is also possible that *Megantereon* or *Felis* might have had an ancestry in North America and dispersed from North America to Eurasia in the Hemphillian (Berta & Galiano 1983). *Nimravides* is not known in North America after the early Hemphillian; and *Machairodus, Adelphailurus*, and *Barbourofelis* are not known in North America after the late Hemphillian. *Megantereon, Felis*, and the machairodont *Homotherium* are recorded from the Blancan. The record of *Machairodus coloradensis* at Wikieup may be the last known occurrence of that genus in North America. The records of *Megantereon hesperus* in the Hagerman fauna and *Adelphailurus kansensis* in the Wikieup or the late Hemphillian Edson fauna of Kansas (Hibbard 1934) may be the last known occurrences of those genera in North America. The latest known occurrence of *Pseudaelurus* in North America may be the record at Yepómera. The earliest North American record of *Felis* may be the occurrence at Redington or in the Optima fauna of Oklahoma (Savage 1941). Bjork (1970) described a fragmentary deciduous premolar (USNM 24931) from the Hagerman fauna as a questionable hyaenid. *Chasmaporthetes*, is known primarily from the Blancan; it is the only hyaenid recorded from North America. Berta (1981) reported *Chasmaporthetes* from Irvington faunas in Florida and Sonora, Mexico.

Two groups of Proboscideans, the gomphotheres and the mammutids (or mastodonts), are known from late Hemphillian and Blancan deposits (Tobien 1973). *Rhynchotherium* is the latest known long-jawed gomphothere in North America; it appears in the late Clarendonian Snake Creek fauna of Nebraska (Skinner et al 1977). It was probably derived from a North American species of *Gomphotherium* and was very common in the late Hemphillian. An important character of *Rhynchotherium* is that the lower-jaw symphysis (or chin) is strongly downturned. *Rhynchotherium* apparently gave rise to the North American short-jawed gomphotheres (*Cuvieronius, Notiomastodon*, and *Stegomastodon*), which are not known before the Blancan. *Cuvieronius* and *Notiomastodon* had an enamel band on their upper tusks (as in *Rhynchotherium*), a feature that *Stegomastodon* lacked. The longirostrine (long-jawed) gomphothere from the Chamita fauna could be *Gomphotherium* or *Rhynchotherium*.

At least nine genera of horses (*Neohipparion, Hipparion, Cormohipparion, Nannippus, Onohippidium, Hippidion, Dinohippus, Astrohippus,* and *Pliohippus*) are currently recognized in Hemphillian faunas of North America. Several of the Hemphillian species that had previously been assigned to *Pliohippus* are now assigned to *Dinohippus* (e.g. *D. leidyanus, D. interpolatus,* and *D. mexicanus*). Quinn (1955, 1958) assigned *Pliohippus* (*Pliohippus*) *mexicanus* from Yepómera to the genus *Asinus* (e.g. *Equus* (*Asinus*) *mexicanus*). However, most later workers include the Yepómera large horse in *Dinohippus* (e.g. *D. mexicanus*). *Pliohippus* had broadly expanded malar and lacrimal fossae that became deeper and more pocketed in later species (Skinner et al 1977), whereas *Dinohippus* had a shallow nasal maxillary fossa anterior to the orbit and lacked a deep malar fossa (Quinn 1955). *Equus simplicidens* had a shallow malar fossa (Stirton 1942); thus, *Dinohippus* rather than *Pliohippus* seems the best candidate for the ancestry of *Equus*. *Astrohippus* includes several other species that have previously been grouped as a subgenus of *Pliohippus*; these include *A. ansae* and *A. stocki*. *A. ansae* from the Coffee Ranch fauna had a deep, wide malar fossa (Matthew & Stirton 1930), while *A. stocki* from the Yepómera fauna had a faint, shallow malar fossa (Lance 1950). *Astrohippus* differs from *Pliohippus* and *Dinohippus* in that the upper cheek teeth are slender and straight, with small and uncomplicated fossettes; the lower cheek teeth of *Astrohippus* do not have the ectoflexid penetrating and separating the metaconid-metastylid isthmus (Quinn 1955).

As presently recognized, the genera *Neohipparion, Hipparion, Cormohipparion, Onohippidium, Dinohippus, Astrohippus,* and *Pliohippus* are not known from Blancan and younger faunas. *Hippidion* is recorded from late Pliocene and Pleistocene deposits of South America and from the early Pleistocene Vallecito Creek fauna of southern California (MacFadden & Skinner 1979). *Onohippidium* and *Hippidion* are characterized by a very deep nasal incision that unfortunately is rarely preserved in fossil skulls. *Onohippidium* also had a deep lacrimal fossa, similar to that of late-occurring species of *Pliohippus* (MacFadden & Skinner 1979). It is generally believed that *Onohippidium* and *Hippidion* are under-represented in Hemphillian and Blancan faunas because of the difficulty in distinguishing these genera from other horses in the absence of well-preserved skull material. *Nannippus* occurs in a number of Blancan faunas, but it is not recorded from either the White Bluffs or the Hagerman fauna.

The genus *Equus* is believed to be derived from a species of *Dinohippus*, with *Dinohippus mexicanus* from the Yepómera fauna of Chihuahua the most likely candidate for its ancestry. In fact, lower horse teeth from the Concha fauna are impossible to assign (with confidence) to either *D. mexicanus* or *E. simplicidens*. These specimens are designated by question marks under *Dinohippus* and *Equus* in Table 2; their proper taxonomic assignment must await the collection of more specimens, especially upper cheek teeth. Therefore, the Concha fauna of Chihuahua records either the latest known occurrence of *Dinohippus* or the earliest known occurrence of *Equus.* Yepómera records the latest known occurrence of *Neohipparion* and possibly *Dinohippus. D. mexicanus* had a shallow, subtle malar fossa, not unlike that of *E. simplicidens* (Lance 1950). Lance (1950, p. 52) noted the extreme difficulty in assigning *D. mexicanus* to the genus *Pliohippus* (or now, *Dinohippus*) rather than the genus *Plesippus* [or now, *Equus* (*Dolichohippus*)]. The assignment of the large horse in the Concha fauna to either *Dinohippus* or *Equus* is very important because the appearance of *Equus* has always been a prime datum for recognition of the beginning of the Blancan land mammal age. If the large horse of the Concha fauna can be securely identified as *Equus*, superposition of two mammal ages that are recognized, in part, on gradual evolution in autochthonous species can be demonstrated. Unfortunately, such unequivocal stratigraphic-chronologic-biologic entities are relatively rare.

When Hemphillian and Blancan land mammal ages were formalized in 1941 (Wood et al 1941), the presence of rhinoceroses in Hemphillian faunas and their absence in Blancan faunas was emphasized. Both the short-legged rhino, *Teleoceras*, and the Nebraska rhino, *Aphelops*, are recorded from the Chamita (McFadden et al 1979) and the Coffee Ranch faunas. The Coffee Ranch record of *Aphelops* is probably the latest known occurrence of that genus, while the Yepómera record is the latest known occurrence of *Teleoceras*.

Peccaries (the family Tayassuidae) are represented in North American faunas since the Oligocene, and now two species of peccaries inhabit North America. During the late Miocene and early Pliocene, the only peccary genus in North America was *Prosthennops*; the Yepómera record is probably its latest known occurrence. The genus *Platygonus* appeared in the Blancan, and either the Ringold or the Hagerman record of *Platygonus pearcei* is probably the earliest known occurrence of this genus.

Six genera of camels (*Hemiauchenia, Alforjas, Camelops, Titanocamelus, Blancocamelus,* and *Megatylopus*) inhabited North America during the Pliocene; another genus (*Palaeolama*) appeared in the Pleistocene. Webb (1974) pointed out that many Pliocene and Pleistocene long-limbed camels of North and South America previously placed in the genera *Pliauchenia, Tanupolama,* or *Palaeolama* should be grouped in the genus *Hemiauchenia,* which is most likely the ancestor of the South American llamas. Harrison (1979) described the new genus *Alforjas* as a medium-size lamine camel that is probably ancestral to the common large North American Pleistocene camel, *Camelops.*

The deerlike family Palaeomerycidae is recorded in North American faunas from the middle Miocene to the early Pliocene. *Pediomeryx hemphillensis* from the Coffee Ranch fauna is probably the latest known record of a North American paleomerycid (Stirton 1936a), although Skinner et al (1977) reported *Pediomeryx* from the late Hemphillian ZX Bar fauna of the Snake Creek Formation in Nebraska that may be about the same age. True deer (family Cervidae) were common in late Miocene and abundant in Pliocene faunas of Eurasia. They dispersed to North America in the Blancan, and *Bretzia pseudalces* of the White Bluffs fauna (Fry & Gustafson 1974) is probably the earliest known true cervid in North America. *Bretzia* is known only from the White Bluffs fauna. The extant North American deer, *Odocoileus*, is recorded from the early Blancan Fox Canyon and Rexroad faunas of Kansas (Hibbard 1954).

Pronghorns (family Antilocapridae) were common faunal elements in North America since the middle Miocene. The smooth-horned antilocaprids (subfamily Antilocaprini) with a vascular horn sheath became very diverse in the Hemphillian, as reflected in the occurrence of seven late Hemphillian genera (*Osbornoceros*, *Ilingoceros*, *Plioceros*, *Texoceros*, *Sphenophalos*, *Hexobelomeryx*, and *Hexameryx*), compared with only a single pronghorn species now living in North America. These Hemphillian pronghorns are distinguished on the basis of horns (e.g. keeled, bifurcated, spiraled, etc), which are markedly uniform in modern pronghorns (Webb 1973). It seems plausible that individual variation in horns might have been greater in the Pliocene, and that some of these Hemphillian genera will be synonymized when larger samples of fossils are available. None of the Hemphillian antilocaprids mentioned above are known from Blancan faunas. The genus *Ceratomeryx*, apparently a continuation of the late Hemphillian "blooming" of antilocaprids, is known only from the Hagerman fauna.

NORTH AMERICAN CENOZOIC MAMMAL DATUM EVENTS

By chronologically ordering the faunas discussed above, as well as reviewing the genera in each fauna, we can develop a biochronologic ordering of first or last occurrences, or faunal datum events. These faunal datum events mark the limits of local range zones. However, this review covers only a part of the known North American faunas for this interval (approximately 2 to 7 m.y.), and further collecting and study will surely identify new taxa and range extensions; thus, this biochronologic framework should be considered flexible, not final. With this caution noted, we believe the biochronologic framework we have developed is well grounded and will prove a reliable and useful tool; it includes the most diverse and well-studied faunas during the given time period. Later revisions or additions to this framework should prove minor.

This biochronologic ordering is founded on the position of a fauna placed in a stratigraphic section that has been correlated with the magnetic polarity time scale. Faunal datum events established on these local stratigraphically based criteria may be designated as lowest stratigraphic datum (LSD) and highest stratigraphic datum (HSD) to distinguish them from regionally significant biochronological faunal datum events, such as first appearance datum (FAD) and last appearance datum (LAD) events.

We identify 11 faunal datum events and note local faunal changes for the interval 1.5 to 7.0 m.y., more or less. Datum events older than 4 Ma are based on this study; datum events younger than 4 Ma are based on earlier studies (e.g., Lindsay et al 1975, Opdyke et al 1977, Neville et al 1979). These datum events are illustrated in Figure 7.

6.8 m.y.—*Pliometanastes* extinction datum

The sloth *Pliometanastes* and the canid *Epicyon* are last recorded from deposits correlated with the lowest part of magnetic chron 6. As noted above, the latest occurrence of *Epicyon* is earlier than the latest occurrence of *Pliometanastes.*

6.5 m.y.—*Plesiogulo* appearance datum

The earliest known occurrences of the wolverine *Plesiogulo* and the badger *Pliotaxidea* are at this time. This may also represent the earliest known occurrence of the horses *Dinohippus interpolatus* and *Astrohippus ansae*, the rabbit *Hypolagus vetus*, and the earliest North American occurrence of *Dipoides.*

6.0 m.y.—*Agriotherium* appearance datum

The earliest known occurrences of the bear *Agriotherium* and the cricetid rodents *Calomys* and *Paronychomys* are at this time. This is also the latest known occurrence of the heteromyid rodent *Perognathoides*, the paleomerycid *Pediomeryx*, and possibly the mole *Domninoides.*

5.6 m.y.—*Plesiogulo* extinction datum

The latest known occurrences of several characteristic Hemphillian genera, such as the wolverine *Plesiogulo*, the badger *Pliotaxidea*, the pronghorn *Sphenophalus* and the cat *Adelphailurus*, are at this time. This also corresponds with the appearance of the tapir-like horse *Onohippidium* in North America.

4.5 m.y.—*Prosigmodon* appearance datum

The earliest known records of the mice *Prosigmodon* and *Baiomys*, the squirrel *Paenemarmota*, the rabbit *Notolagus*, and the badger *Taxidea*, as well as the latest known occurrences of the bear *Agriotherium*, the rodent

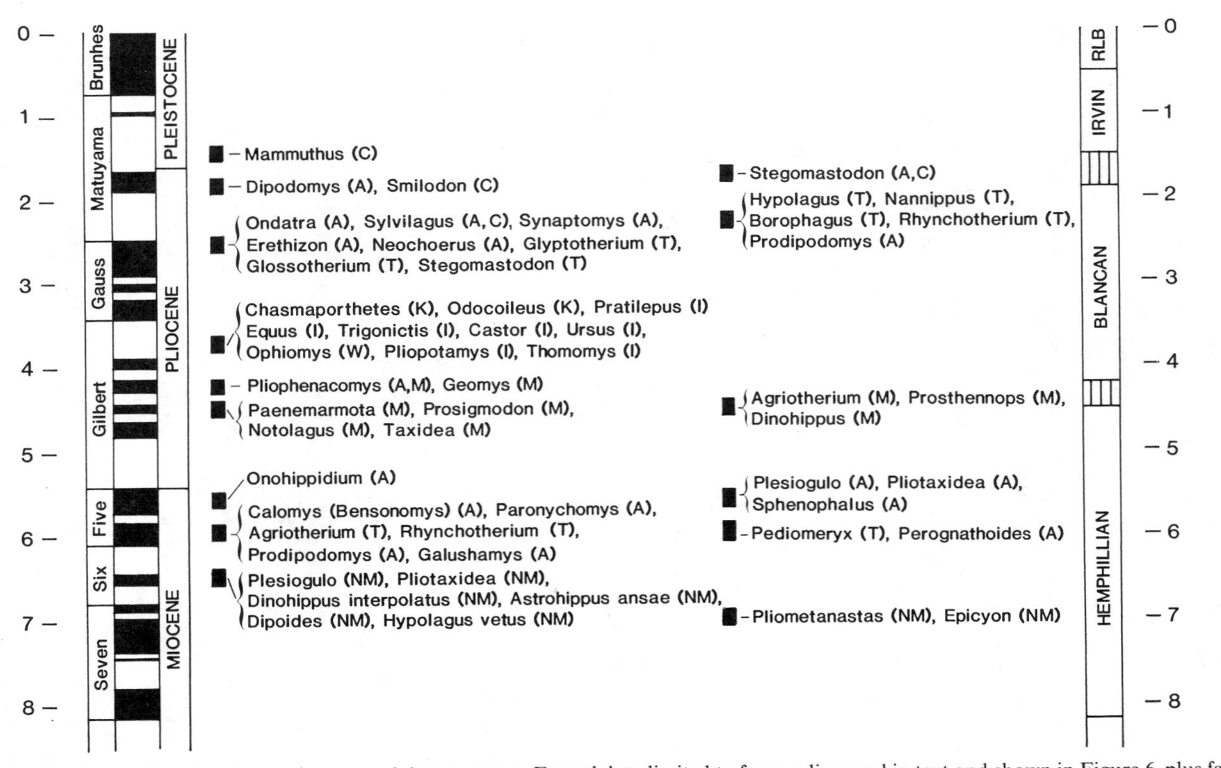

Figure 7 North American late Cenozoic mammal datum events. Faunal data limited to faunas discussed in text and shown in Figure 6, plus faunal summary of Johnson et al (1975), Lindsay et al (1975), Opdyke et al (1977), and Neville et al (1979). Letters in parentheses (A, C, I, M, NM, and T) refer, respectively, to the faunas in Arizona, California, Idaho, Mexico, New Mexico, and Texas discussed in the text or in the above references. Ma = million years, MPTS = magnetic polarity time scale, LSD = lowest stratigraphic datum, HSD = highest stratigraphic datum. Normal magnetozones have been verified by thermal demagnetization.

Copemys, the horses *Neohipparion* and *Astrohippus*, the rhino *Teleoceras*, and the peccary *Prosthennops* are at this time. As mentioned above, this time may also mark the latest record of the horse *Dinohippus*, considered ancestral to *Equus*.

4.2 m.y.—*Pliophenacomys* appearance datum

The earliest known records of the vole *Pliophenacomys* and the gopher *Geomys* occur at this time. This may represent the latest occurrence of the large cat *Machairodus* in North America.

3.7 m.y.—*Trigonictis* appearance datum

A wide variety of new appearances occur over a relatively short interval, centering on 3.7 m.y. They include the grison *Trigonictis*, the rabbits *Nekrolagus* and *Pratilepus*, the vole *Pliopotamys*, the gopher *Thomomys*, the pack rat *Neotoma*, the hyena *Chasmaporthetes*, the bear *Ursus*, the mastodont *Mammut*, the peccary *Platygonus*, the camel *Camelops*, and the deer *Bretzia*. The latest known occurrence of the large cat *Megantereon* also occurs in this interval.

2.5 m.y.—*Synaptomys* appearance datum

A number of new genera appear in North America at about this time, including the lemming *Synaptomys*, the glyptothere *Glyptotherium*, the sloth *Glossotherium*, the bunny *Sylvilagus*, the muskrat *Ondatra*, the capybara *Neochoerus*, the porcupine *Erethizon*, the bear *Tremarctos*, the gomphothere *Stegomastodon*, and the pronghorn *Tetrameryx*.

2.2 m.y.—*Nannippus* extinction datum

The latest known occurrences of several characteristic Blancan genera are recorded at this time. They include the three-toed horse *Nannippus*, the rabbit *Hypolagus*, the dog *Borophagus*, the gomphothere *Rhynchotherium*, and the kangaroo rat *Prodipodomys*.

2.0 m.y.—*Stegomastodon* extinction datum

The latest known occurrence of the gomphothere *Stegomastodon* is at this time. Also, the kangaroo rat *Dipodomys* and the saber-tooth cat *Smilodon* appear at this time.

1.6 m.y.—*Mammuthus* appearance datum

The elephant *Mammuthus* appears in North America at this time.

THE BLANCAN/HEMPHILLIAN BOUNDARY

To our knowledge, the only section with a Blancan fauna superposed over a Hemphillian fauna is that in Chihuahua with the Hemphillian Yepómera directly overlain by the Blancan Concha faunas. The Yepómera fauna includes *Dinohippus*, *Neohipparion*, *Teleoceras*, etc, that are not known in the Blancan. The Concha fauna includes *Pliophenacomys*, *Geomys*, and a horse that appears intermediate between *Dinohippus mexicanus* and *Equus simplicidens*. This horse imparts a Blancan character to the Concha fauna. As noted above, these faunas occur in strata that identify relatively short normal magnetozones, separated by a relatively long reversed magnetozone. We believe that these magnetozones represent two of the normal events in the Gilbert magnetic chron, most likely the Nunivak-Sidufjall events.

The Verde fauna of Central Arizona is very close to the Blancan/Hemphillian boundary. When the fauna is completely studied, it may be assigned either to the Hemphillian or the Blancan. It is presently considered early Blancan, primarily because it has the same species of *Pliophenacomys* (*fide* C. A. Repenning, personal communication) that occurs in the Concha fauna. The Verde fauna is rather closely associated with the Nunivak event in the Gilbert magnetic chron. Therefore, we correlate the Concha fauna, as well as the Verde fauna, with the Nunivak event. The Yepómera fauna is correlated with the Sidufjall event. If these correlations are accurate, the Blancan/Hemphillian boundary is bracketed between the Nunivak and Sidufjall events, or between 4.2 and 4.4 m.y. If the Verde fauna, after thorough analysis, is determined Hemphillian rather than Blancan, the boundary should be moved above the Nunivak event, to about 4.0 m.y.

The base of the Hemphillian land mammal age is placed at 8.2 m.y. and is recognized by the appearance of the sloth *Pliometanastes*. The early Hemphillian thus approximates the duration of magnetic chron 7 and the lower half of chron 6, or about 6.5 to 8.2 m.y. The late Hemphillian approximates the lower half of the Gilbert magnetic chron, chron 5, and the upper half of chron 6, or about 4.3 to 6.5 m.y.

The Kimballian land mammal age was proposed as a faunally distinct temporal interval between the Hemphillian and Blancan land mammal ages, based on faunas from the Ogallala Formation in Nebraska (Schultz & Stout 1961, Schultz et al 1970). The concept of Kimballian was reviewed and challenged by Breyer (1981), who noted that the Kimballian mammals occur stratigraphically high in restricted partial sections of the Ogallala Formation and that the fauna from four selected Kimballian localities are more characteristic of late Clarendonian (Kepler fauna) and early Hemphillian (Greenwood Canyon, Oshkosh, and Potter faunas) than of late Hemphillian or Blancan land mammal ages. Breyer concluded that the Kimballian is poorly conceived and temporally nonexistent.

We agree with Breyer's conclusions, and cite as evidence the absence of late Hemphillian immigrants (e.g. *Plesiogulo*, *Pliotaxidea*, and *Agriotherium*) in Kimballian faunas, as well as the presence of several taxa (e.g. *Nimravides*, *Epicyon*, and *Pseudoceras*) in Kimballian faunas that are not recorded in late Hemphillian or early Blancan faunas.

LATE CENOZOIC MAMMAL DISPERSAL EVENTS

A most intriguing aspect of vertebrate history is the realization that some groups of terrestrial mammals have on occasion dispersed from their "homeland" to become established and thriving on another continent. The dispersal of horses to Europe at four different times (*Hyracotherium* in the Eocene, *Anchitherium* in the early Miocene, *Hipparion* in the late Miocene, and *Equus* in the Pliocene) is commonly cited in historical geology texts to illustrate mammalian intercontinental dispersal. Vertebrate paleontologists utilize these intervals of "faunal mixing" to order stages of vertebrate evolution; they provide convenient means to correlate faunas on separate continents. For example, the appearance of *Hipparion* in Eurasia is correlated approximately with the appearance of that genus in North America, where the presumed ancestor of *Hipparion* (e.g., *Merychippus*) lived.

However, many dispersal events are undefined because either the phyletic history of the immigrant is poorly understood or its chronologic range is poorly known. A prime purpose of this review is to develop a chronologic "ordering" of North American Pliocene faunas and thereby identify the timing and sequence of dispersal events with greater accuracy. A recent taxonomic and biochronologic review of the stabbing cat *Megantereon* by Berta & Galiano (1983) illustrates this point. Berta & Galiano concluded that the earliest known record of *Megantereon* is from the late Hemphillian Bone Valley fauna of Florida; it is also recorded from several early to middle Blancan faunas (e.g. Hagerman of Idaho, Rexroad of Kansas, and Broadwater of Nebraska). As pointed out by Berta & Galiano, *Megantereon* is also well represented in early Villafranchian faunas (about 2.5 to 3.5 m.y.) of Europe (e.g. Les Etouaires and St. Vallier in France, Olivola and Val Darno in Italy, and La Puebla de Valverde in Spain), Asia (e.g. Nihowan in China, and possibly the upper Siwaliks of Pakistan), and Africa (e.g. Sterkfontein, Schurveberg, and Kromdraii in South Africa), all of which are considered younger than late Hemphillian (about 4.3 to 6.5 m.y.). Thus, it appears that *Megantereon* might have dispersed from North America to Eurasia and Africa during the Pliocene, and that it was more successful (as well as having a longer chronologic record) in the Old World.

Indices of faunal similarity between North America and Eurasian genera during the Pliocene, Pleistocene, and Recent should theoretically reflect the number and significance of late Cenozoic mammal dispersal events. The Simpson index ($100C/N_1$) for these intervals is Pliocene = 34, Pleistocene = 44, and Recent = 39. These data suggest that faunal interchange was probably greater in the Pleistocene and has declined since then. Such data should be used with caution, however, as the temporal divisions may not be equivalent or of equal duration, and there is an obvious "aftereffect" following the dispersal event when the occurrence of the immigrant on both continents persists an indefinite time. We suggest here that the high incidence of faunal similarity between North America and Eurasia during the Pleistocene is an "aftereffect" of the Pliocene dispersal. That is, we suggest that dispersal between North America and Eurasia was more frequent in the Pliocene than in the Pleistocene.

We recognize five significant dispersal events between North America and Eurasia in the interval between about 1.5 and 7.0 m.y. These events, illustrated in Figure 8, are the *Plesiogulo* dispersal at about 6.5 m.y., the *Agriotherium* dispersal at about 6.0 m.y., the *Trigonictis* dispersal at about 3.7 m.y., the *Synaptomys* (also *Equus*) dispersal at about 2.5 m.y., and the *Mammuthus* dispersal at about 1.6 m.y.

As noted above, *Plesiogulo* had a long and widespread history in Eurasia. Its ancestry is believed to be from Asia (Kurtén 1970), with a rather short local range zone in North America. *Agriotherium* was also widespread in the Old World, including a record from southern Africa. The North American local range zone of *Agriotherium* is also comparatively short (about 1.5 m.y.), overlapping and slightly later than the local range zone of *Plesiogulo*. Both genera were very characteristic members of late Hemphillian faunas, reflecting their short but successful existence in North America. The 3.7 m.y. dispersal event marks a number of Eurasian immigrants that are characteristic of Blancan faunas. In addition to the grison *Trigonictis*, the bear *Ursus*, the mastodont *Mammut*, and the deer *Bretzia* are probably immigrants of this dispersal event. The 2.5 m.y. dispersal event marks the dispersal of *Equus* from North America to Eurasia (and the beginning of the European E-L-E fauna). Immigrants to North America at that time include the lemming *Synaptomys* and the bear *Tremarctos*. This dispersal event also approximates the Great American Interchange (North America–South America), since several South American immigrants, such as the glyptodont *Glyptotherium*, the sloth *Glossotherium*, the capybara *Neochoerus*, and the porcupine *Erethizon*, appear in North America between 2.5 and 2.8 m.y. Based on the lower stratigraphic occurrence of *Glyptotherium* and *Neochoerus* relative to the appearance of *Synaptomys* in the 111 Ranch section of southeastern Arizona, it seems that the Great American Interchange (N.A.–S.A.) was probably slightly earlier than the 2.5 m.y. dispersal of *Equus* to Eurasia. However, the dispersal center for South American immigrants (Mexico) is much closer to southeastern Arizona than is the Eurasian dispersal center (Alaska), so the 2.5 m.y. Eurasian dispersal might actually be coincident with the Great American Interchange. If this is true, the time of this dispersal event must actually be closer to 2.8 m.y. than the 2.5 m.y. "timing" we have indicated. The 1.6 m.y. dispersal event marks the appearance of the elephant *Mammuthus* in North America, following a record in Eurasia for about 1 m.y. Apparently very few other mammals invaded North America from Eurasia at this time.

We believe there may be another poorly defined and incompletely identified dispersal event approximately in the interval of 4.9 to 5.2 m.y. This event is marked by the appearance in North America of the microtine rodent *Promimomys* and the beaver *Castor*. The first known record of these immigrants is in the McKay Reservoir fauna of Oregon. The McKay reservoir fauna includes the rodents *Mylagaulus* and *Dipoides*, the wolverine *Plesiogulo*, the badger *Pliotaxidea*, and the cat *Machairodus*, in addition to *Promimomys* and *Castor*, which firmly establish it as a late Hemphillian fauna. However, the McKay Reservoir fauna lacks the bear *Agriotherium*, and thus it seems that it should be placed younger than the

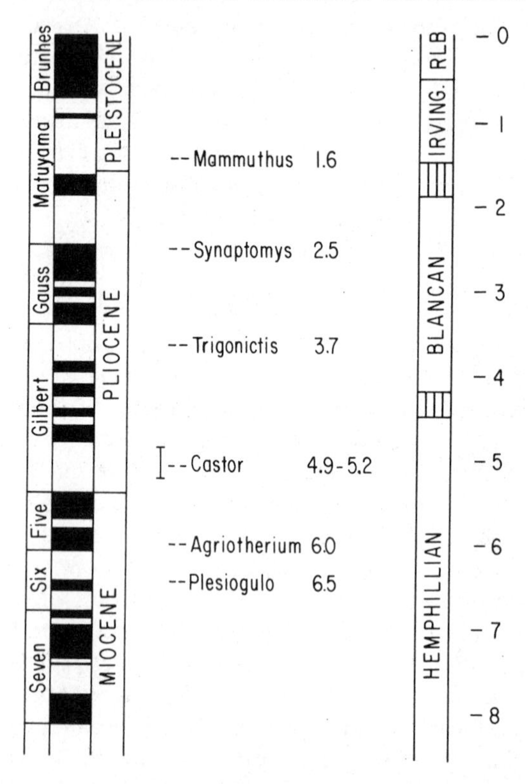

Figure 8 Nearctic-Palearctic mammal dispersal events.

chron 5 faunas (e.g. Wikieup), since none of those faunas record *Agriotherium* with either *Promimomys* or *Castor*. The absence of *Agriotherium* in the McKay Reservoir fauna and its presence in the younger Yepómera fauna is anomalous. Another late Hemphillian fauna that may reflect a "*Promimomys-Castor*" dispersal event is the Santee fauna of Nebraska (Martin 1975, Voorhies 1977). The primitive microtine rodent *Propliophenacomys* is recorded from this fauna. Boellstorff (1976) reported a date of 5.0 ± 0.2 m.y. on fission tracks in glass from an ash that overlies the Santee fauna (Voorhies 1977). Fission-track dates on glass are usually slightly younger than other radiometric dates taken on the same volcanic rock, which suggests that the Santee fauna might be placed very near the Gilbert/chron 5 boundary.

Note that if the McKay Reservoir fauna can be placed with confidence in the lower part of the Gilbert magnetic chron, the extinctions of *Plesiogulo* and *Pliotaxidae* (which are now placed in the upper part of chron 5) would be moved up a notch with the McKay Reservoir fauna.

Microtine rodents underwent an explosive radiation in the Pliocene after their appearance in the late Miocene. *Promimomys* is recorded in only two North American late Hemphillian faunas, the McKay Reservoir and Christmas Valley faunas of Oregon (Repenning 1968), but it is recorded from a number of Ruscinian faunas of Europe [e.g. Podlesice, Poland; Malusteni, Romania; Kardia and Ptolemais, Greece; Terrats and Vendargue, southern France (Michaux 1976, Repenning & Fejfar 1977)]; it is considered the stem of much of the European microtine radiation and was very probably responsible for much of the North American microtine radiation as well.

Repenning (1978) recently studied the North American microtine radiation and identified seven microtine dispersal events during the interval of the late Hemphillian through the Rancholabrean (i.e. Pliocene and Pleistocene). He divided the Pliocene and Pleistocene faunas of North America into 10 discrete and sequential microtine faunas, based on the dispersal and subsequent evolution of microtines. Some of Repenning's microtine dispersal events (at 5.4, 4.9, 3.8, 2.6, 1.8, 1.2, and 0.48 m.y.) nearly coincide with our more general dispersal events (e.g. 3.7, 2.5, and 1.6 m.y.), which is not surprising, since we are both working (more or less) with the same data set. However, his correlation of several Pliocene faunas differs from our correlations, leading to alternate interpretations.

Repenning (personal communication, 1980) would place *Promimomys* in the 5.4 m.y. dispersal event, whereas we would place it in the 4.9 m.y. dispersal event. Repenning's 4.9 m.y. dispersal event marks the appearance by immigration of the *Mimomys* (including *Ophiomys*) group of microtines with three alternating prisms on the M_1. One member of this group, *Mimomys mcknighti*, occurs stratigraphically below the White Bluffs tuff and above a normal magnetic event we interpret as the Cochiti event in the Ringold-White Bluffs section. *Mimomys mcknighti* also occurs in the Alturas fauna of California, where it is placed stratigraphically below a basalt that has repeatedly yielded a K-Ar date of 4.7 ± 0.5 m.y. (= 4.2 to 5.2 m.y.) (Repenning, personal communication, 1980). Repenning places the White Bluffs fauna, along with the Concha, Verde, and Alturas faunas, in the middle of the Gilbert magnetic chron, and he interprets the normal event in the White Bluffs section below the locality of *M. mcknighti* as the Sidufjall event, rather than the Cochiti event in the Gilbert magnetic chron. Based on this interpretation, Repenning places several other Blancan faunas (e.g. Rexroad, Bender, and Fox Canyon) in the upper Gilbert magnetic chron, below the Gauss magnetic chron and the 3.8 m.y. dispersal event. He also places the McKay Reservoir, Christmas Valley, and Santee faunas near the base of the Gilbert magnetic chron, later than (above) the 5.4 m.y. dispersal event. Apparently, Repenning considers that the large mammal immigrants in the White Bluffs fauna (e.g. *Trigonictis* and *Ursus*) entered North America in the 4.9 m.y. dispersal event, along with *Mimomys*, whereas we identify these immigrants with the 3.7 m.y. dispersal event. Moreover, we feel that the absence of *Mimomys* in the Verde, Concha, and Yepómera faunas, all of which have been screen washed and yield abundant small mammals (which we place in the middle Gilbert magnetic chron), argues against Repenning's concept of a widespread radiation of *Mimomys* in North America after the 4.9 m.y. dispersal event. We believe a more reasonable, if not more parsimonious, interpretation is that *Promimomys* and *Propliophenacomys* immigrated to North America during a poorly defined dispersal event in the early part of the Gilbert magnetic chron (the *Promimomys-Castor* immigration event at about 4.9 to 5.2 m.y.), and that by the middle Gilbert they had given rise to *Mimomys* and *Pliophenacomys*, but species of these genera were not as widespread and abundant as later species of microtines were.

The microtine *Pliophenacomys* that occurs in the Concha and Verde faunas had five alternating prisms on M_1, as did *Propliophenacomys*. Another microtine, *Nebraskomys*, that is recorded from the Verde fauna had three alternating prisms on M_1, similar to *Mimomys*. We consider that both *Nebraskomys* and *Mimomys* were probably derived from *Promimomys* in the early Gilbert magnetic chron. And, it is not necessary to have a second dispersal event in the 4.5 to 4.9 m.y. interval to account for the appearance of *Pliophenacomys*, *Nebraskomys*, and *Mimomys* at about 4.2 m.y.

However, successive stages of development of *Mimomys* (plus *Dolomys*, *Pliomys*, *Germanomys*, etc) are well documented in Villafranchian faunas of

Europe, and their presumed counterpart can frequently be found in North American Blancan faunas. If North American species of *Mimomys* evolved from a different species of *Promimomys* than did the Eurasian species of *Mimomys*, as we propose, then separate North American and European lineages of *Mimomys* should be distinct. Recognition of these lineages is very difficult because of parallelism, repeated mixing of faunas on both continents during later dispersal events, and the possibility of (more or less) simultaneous invasion of both continents by similar species of microtines from a high polar center of microtine radiation. Extensive comparison of North American and European microtine lineages is presently underway by Repenning and O. Fejfar.

The results of our analysis suggest that the appearance of *Promimomys* (and *Castor*) in North America, represented in the McKay Reservoir fauna, should be close to 5.0 m.y., slightly younger than the characteristic "chron 5" faunas with *Agriotherium* and *Plesiogulo*. We find that this single microtine dispersal event is sufficient to explain the explosive radiation of microtines in North America prior to 3.7 m.y. Our results also suggest the "*Promimomys* faunas" of Europe (equal MN zone 14) should occur in the interval close to 5.0 m.y.

SUMMARY

This review illustrates the application of biostratigraphy and magnetostratigraphy to better understand and resolve problems in Earth history. We have evaluated and utilized much data generated by others and have also contributed new data for this synthesis. Our efforts have focused on chronologic resolution of late Cenozoic events in the history of mammals. We have provided a framework for interpreting these and other events in Earth history, but this framework is fragile and must not be accepted as final. Our sincere hope is that it will spark interest in the chronologic resolution of Earth history and will provide a better foundation for interpreting biologic and geologic evolution.

Specific points we wish to emphasize are the following:

1. A number of Hemphillian (Chamita, White Cone, Coffee Ranch, Redington, Camel Canyon, Wikieup, and Yepómera) as well as Blancan (Concha, Verde, White Bluffs, and Hagerman) faunas have been "ordered" by correlation with the magnetic polarity time scale. This ordering, in addition to earlier studies (e.g. the San Pedro Valley and Anza-Borrego faunal sequences), permits chronologic resolution of local range zones for numerous North American fossil mammals.

2. These local range zones have been analyzed to identify 11 terrestrial

faunal datum events (e.g. *Pliometanastes* extinction, *Plesiogulo* appearance, *Agriotherium* appearance, *Plesiogulo* extinction, *Prosigmodon* appearance, *Pliophenacomys* appearance, *Trigonictis* appearance, *Synaptomys* appearance, *Nannippus* extinction, *Stegomastodon* extinction, and *Mammuthus* appearance) in North America during the interval from about 1.5 to 7.0 m.y. These faunal datum events are tied to (and calibrated with) the magnetic polarity time scale; they are called "highest or lowest stratigraphic datum" events because they are based on local stratigraphic ranges of the taxon in question.

3. This framework permits chronologic resolution and division of North American Pliocene land mammal ages: Early Hemphillian (approximates the joint occurrence of *Epicyon* and *Pliometanastes*) from about 6.8 to 8.2 m.y. BP; late Hemphillian (approximates the joint occurrence of *Plesiogulo* and *Agriotherium*) from about 4.3 to 6.5 m.y. BP; early Blancan (approximates the joint occurrence of *Nannippus* and *Equus*) from about 2.2 to 4.3 m.y. BP; late Blancan (approximates the joint occurrence of *Sylvilagus* and *Stegomastodon*) from about 1.6 to 2.2 m.y. BP; and early Irvingtonian (approximates the joint occurrence of *Dipodomys* and *Mammuthus*) from about 1.0 to 1.6 m.y. BP. The Kimballian land mammal age, proposed as an interval between the Hemphillian and Blancan land mammal ages, is not necessary or justified.

4. Mammal dispersal events are important factors in the history of terrestrial evolution. They have always been important criteria for correlating and limiting the units of mammalian biochronology. This framework has identified five significant mammal dispersal events between North America and Eurasia during the interval 1.5 to 7.0 m.y. These are the *Plesiogulo* dispersal (6.5 m.y.), *Agriotherium* dispersal (6.0 m.y.), *Trigonictis* dispersal (3.7 m.y.), *Synaptomys* or *Equus* dispersal (2.5 m.y.), and *Mammuthus* dispersal (1.6 m.y.). The climax of the Great American Interchange approximates, and may be coincident with, the *Equus* dispersal event. Improved chronologic resolution of mammalian history on all continents is needed in order to clarify additional mammal dispersal events and the directions of these dispersals.

Acknowledgments

Results presented here reflect the efforts of many students and colleagues, in addition to those of the authors. We gratefully acknowledge these efforts, especially those of former students Mark Nibbelink, Colleen Neville, Louis Jacobs, Jon Baskin, Jessica Harrison, and Hugh Wagner, who have gone on to take a position in their chosen profession. This work was supported by National Science Foundation Grants DES 74-13860 and EAR-8206184.

Literature Cited

Armstrong, R. L. 1981. Discussion of "Magnetic stratigraphy of Pliocene deposits of the Glenns Ferry Formation, Idaho, and its implications for North American mammalian biostratigraphy" by Neville, Opdyke, Lindsay and Johnson. *Am. J. Sci.* 282:730–33

Baskin, J. A. 1978. *Bensonomys*, *Calomys*, and the origin of the phyllotine group of Neotropical cricetines (Rodentia: Cricetidae). *J. Mammal.* 59:125–35

Baskin, J. A. 1979. Small mammals of the Hemphillian age White Cone local fauna, northeastern Arizona. *J. Paleontol.* 53: 695–708

Baskin, J. A. 1980. The generic status of *Aelurodon* and *Epicyon* (Carnivora, Canidae). *J. Paleontol.* 54(6):1349–51

Baskin, J. A. 1981. *Barbourofelis* (Nimravidae) and *Nimravides* (Felidae), with a description of two new species from the late Miocene of Florida. *J. Mammal.* 62(1):122–39

Berta, A. 1981. The Plio-Pleistocene hyaena *Chasmaporthetes ossifragus* from Florida. *J. Vertebr. Paleontol.* 1:341–56

Berta, A., Galiano, H. 1983. *Megantereon hesperus* from the late Hemphillian of Florida with remarks on the phylogenetic relationships of machairodonts (Mammalia, Felidae, Machairodontinae). *J. Paleontol.* 57:892–99

Bjork, P. R. 1970. The Carnivora of the Hagerman local fauna (late Pliocene) of southwestern Idaho. *Trans. Am. Philos. Soc.* 60(7):1–54

Boellstorff, J. D. 1976. The succession of late Cenozoic volcanic ashes in the Great Plains: a progress report. *Midwest. Friends Pleistocene Guideb., Ann. Meet., 24th, Meade County, Kans.*, pp. 37–71

Breed, W. J. 1962. Road log—Globe to Flagstaff, Arizona. *N. Mex. Geol. Soc. Guideb., Field Conf., 13th, Mogollon Rim Reg.*, pp. 31–49

Bressler, S. L., Butler, R. F. 1978. Magnetostratigraphy of the late Tertiary Verde Formation, central Arizona. *Earth Planet. Sci. Lett.* 38:319–30

Breyer, J. A. 1981. The Kimballian land-mammal age: Mene, Mene, Tekel, Upharsin (Dan. 5:25). *J. Paleontol.* 55:1207–16

Dalquest, W. W. 1969. Pliocene carnivores of the Coffee Ranch. *Bull Tex. Mem. Mus. No. 15*, pp. 1–43

Damon, P. E., Lovering, T. S., Livingston, D. E., Laughlin, A. W., Palmer, R. A., Pushkar, P. D. 1969. Correlation and chronology of ore deposits and volcanic rocks. *US At. Energy Comm. Prog. Rep. No. C00-689-120, Contract AT(11-1)-689*, pp. 1–90

Dawson, M. R. 1958. Later Tertiary Leporidae of North America. *Univ. Kans. Paleontol. Contrib. Vertebrata No. 6*, pp. 1–75

Evernden, J. E., Savage, D. E., Curtis, G. H., James, G. T. 1964. Potassium-argon dates and the Cenozoic mammalian chronology of North America. *Am. J. Sci.* 262:145–98

Ferrusquia, I. 1978. Conexiones terrestres entre norte y sudamerica. XIII. Distribution of Cenozoic vertebrate faunas in middle America and problems of migration between North and South America. *Bol. Inst. Geol. Univ. Nat. Autón. Méx.* 101:193–329

Fry, W. E., Gustafson, E. P. 1974. Cervids from the Pliocene and Pleistocene of central Washington. *J. Paleontol.* 48:375–86

Gustafson, E. P. 1978. The vertebrate faunas of the Pliocene Ringold Formation, south-central Washington. *Bull. Mus. Nat. Hist. Univ. Oreg. No. 23*, pp. 1–62

Harrison, J. A. 1978. Mammals of the Wolf Ranch local fauna, Pliocene of the San Pedro Valley, Arizona. *Occas. Pap. Mus. Nat. Hist. Univ. Kans. No. 73*, pp. 1–18

Harrison, J. A. 1979. Revision of the Camelinae (Artiodactyla, Tylopoda) and description of the new genus *Alforjas*. *Univ. Kans. Paleontol. Contrib. No. 95*, pp. 1–20

Harrison, J. A. 1981. A review of the extinct wolverine *Plesiogulo* (Carnivora: Mustelidae) from North America. *Smithson. Contrib. Paleobiol. No. 46*, pp. 1–27

Harrison, J. A. 1983. The Carnivora of the Edson local fauna (late Hemphillian), Kansas. *Smithson. Contrib. Paleobiol. No. 54*, pp. 1–42

Heindl, L. A. 1963. Cenozoic geology of the Mammoth area, Pinal County, Arizona. *US Geol. Surv. Bull. No. 1141-E*, pp. 1–41

Hibbard, C. W. 1934. Two new genera of Felidae from the middle Pliocene of Kansas. *Trans. Kans. Acad. Sci.* 37:239–55

Hibbard, C. W. 1954. Second contribution to the Rexroad fauna. *Trans. Kans. Acad. Sci.* 57:221–37

Hibbard, C. W. 1969. The rabbits (*Hypolagus* and *Pratilepus*) from the upper Pliocene, Hagerman local fauna of Idaho. *Pap. Mich. Acad. Sci., Arts Lett.* 1:81–97

Hibbard, C. W., Bjork, P. R. 1971. The insectivores of the Hagerman local fauna, upper Pliocene of Idaho. *Contrib. Mus. Paleobiol. Univ. Mich.* 23:171–80

Hirschfeld, S. E. 1981. *Pliometanastes protistus* (Edentata, Megalonychidae) from

Knights Ferry, California with discussion of early Hemphillian megalonychids. *Paleobios. No. 31*, pp. 1–17

Hirschfeld, S. E., Webb, S. D. 1968. Plio-Pleistocene megalonychid sloths of North America. *Bull. Fla. State Mus.* 12:213–96

Hutchison, J. H. 1968. Fossil Talpidae (Insectivora, Mammalia) from the later Tertiary of Oregon. *Bull. Mus. Nat. Hist. Univ. Oreg. No. 11*, pp. 1–117

Izett, G. A. 1975. Late Cenozoic sedimentation and deformation in northern Colorado and adjacent areas. *Geol. Soc. Am. Mem. 144*, pp. 179–209

Jacobs, L. L. 1973. *Small mammals of the Quiburis formation, southeastern Arizona.* MSc thesis. Univ. Ariz., Tucson. 77 pp.

Jacobs, L. L. 1977. Rodents of the Hemphillian age Redington local fauna, San Pedro Valley, Arizona. *J. Paleontol.* 51: 505–19

Jacobs, L. L., Lindsay, E. H. 1981. *Prosigmodon oroscoi*, a new sigmodont rodent from the late Tertiary of Mexico. *J. Paleontol.* 55:425–30

Johnson, N. M., Opdyke, N. D., Lindsay, E. H. 1975. Magnetic polarity stratigraphy of Pliocene-Pleistocene terrestrial deposits and vertebrate faunas, San Pedro Valley, Arizona. *Geol. Soc. Am. Bull.* 86:5–12

Johnson, N. M., McGee, V. E., Naeser, C. W. 1979. A practical method of estimating standard error of age in the fission track dating method. *Nucl. Tracks* 3:93–99

Kurtén, B. 1970. The Neogene *Plesiogulo* and the origin of *Gulo* (Carnivora, Mammalia). *Acta Zool. Fenn.* 131:1–22

Kurtén, B., Anderson, E. 1980. *Pleistocene Mammals of North America.* New York: Columbia Univ. Press. 442 pp.

Lance, J. F. 1950. Paleontolgia y estratigrafia del Plioceno de Yepómera, estado de Chihuahua. 1. Equidos, excepto Neohipparion. *Bol. Inst. Geol. Univ. Nac. Autón. Méx.* 54:1–81

Lance, J. F. 1954. Age of the Bidahochi Formation, Arizona. *Geol. Soc. Am. Bull.* 65:1276 (Abstr.)

Lindsay, E. H. 1978. Late Cenozoic vertebrate faunas, southeastern Arizona. *N. Mex. Geol. Soc. Guideb., Field Conf., 29th, Land of Cochise*, pp. 269–75

Lindsay, E. H., Jacobs, L. L. 1984. Pliocene small mammal fossils from Chihuahua, Mexico. *Bol. Inst. Geol. Univ. Nac. Autón. Méx.* In press

Lindsay, E. H., Johnson, N. M., Opdyke, N. D. 1975. Preliminary correlation of North American land mammal ages and geomagnetic chronology. *Studies on Cenozoic Paleontology and Stratigraphy in Honor of C. W. Hibbard, Univ. Mich. Pap. Paleontol. No. 12*, pp. 111–19

MacDonald, H. G. 1977. *Description of the osteology of the extinct gravigrade edentate* Megalonyx, *with observations on its ontogeny, phylogeny, and functional anatomy.* MSc thesis. Univ. Fla., Gainesville

MacFadden, B. J. 1977. Magnetic polarity stratigraphy of the Chamita Formation Stratotype (Mio-Pliocene) of north-central New Mexico. *Am. J. Sci.* 277:769–800

MacFadden, B. J., Galiano, H. 1981. Late Hemphillian cat (Mammalia, Felidae) from the Bone Valley Formation of central Florida. *J. Paleontol.* 55(1):218–26

MacFadden, B. J., Skinner, M. F. 1979. Diversification and biogeography of the one-toed horse *Onohippidium* and *Hippidion. Postilla No. 175*, pp. 1–10

MacFadden, B. J., Johnson, N. M., Opdyke, N. D. 1979. Magnetic polarity stratigraphy of the Mio-Pliocene mammal-bearing Big Sandy Formation of western Arizona. *Earth Planet. Sci. Lett.* 44:349–64

Martin, L. D. 1975. Microtine rodents from the Ogallala Pliocene of Nebraska and the early evolution of the Microtinae in North America. *Univ. Mich. Pap. Paleontol. No. 12*, pp. 101–10

Martin, L. D., Schultz, C. B. 1975. Scimitar-toothed cats, *Machairodus* and *Nimravides*, from the Pliocene of Kansas and Nebraska. *Bull. Univ. Nebr. State Mus.* 10(1):55–63

Matthew, W. D., Stirton, R. A. 1930. Equidae from the Pliocene of Texas. *Univ. Calif. Publ. Geol. Sci.* 19(17):349–96

May, S. R. 1981. *Repomys* (Mammalia: Rodentia gen. nov.) from the late Neogene of California and Nevada. *J. Vertebr. Paleontol.* 1:219–30

May, S. R., Repenning, C. A. 1982. New evidence for the age of the Mount Eden fauna, southern California. *J. Vertebr. Paleontol.* 2:109–13

Michaux, J. 1976. Découverte d'une faune de petits mammifére dans le Pliocene continental de la vallee de la Canterrane (Rousillon); ses conséquences stratigraphiques. *Bull. Soc. Geol. Fr. Ser. 7* 18:165–70

Ness, G., Levi, S., Couch, R. 1980. Marine magnetic anomaly timescales for the Cenozoic and late Cretaceous: a précis, critique, and synthesis. *Rev. Geophys. Space Phys.* 18:753–70

Neville, C., Opdyke, N. D., Lindsay, E. H., Johnson, N. M. 1979. Magnetic stratigraphy of Pliocene deposits of the Glenns Ferry Formation, Idaho, and its implications for North American mammalian biostratigraphy. *Am. J. Sci.* 279:503–26

Nibbelink, M. 1972. *The paleomagnetic stra-*

tigraphy of the Pliocene age Quiburis Formation near Mammoth, Arizona. MSc thesis. Dartmouth Coll., Hanover, N.H.

Opdyke, N. D., Lindsay, E. H., Johnson, N. M., Downs, T. 1977. The paleomagnetism and magnetic polarity stratigraphy of the mammal-bearing section of Anza-Borrego State Park, California. *Quat. Res.* 7:316–29

Quinn, J. H. 1955. Miocene Equidae of the Texas Gulf Coastal Plain. *Bur. Econ. Geol. Univ. Texas No. 5516*, pp. 1–102

Quinn, J. H. 1958. New Pleistocene *Asinus* from southwestern Arizona. *J. Paleontol.* 32:603–10

Repenning, C. A. 1962. The giant ground squirrel *Paenemarmota. J. Paleontol.* 36: 540–56

Repenning, C. A. 1967. Palearctic-Nearctic mammalian dispersal in the late Cenozoic. In *The Bering Land Bridge*, ed. D. M. Hopkins, pp. 288–311. Stanford, Calif: Stanford Univ. Press

Repenning, C. A. 1968. Mandibular musculature and the origin of the subfamily Arvicolinae (Rodentia). *Acta Zool. Cracov.* 13:29–72

Repenning, C. A. 1980. Faunal exchanges between Siberia and North America. In *The Ice-Free Corridor and Peopling of the New World, Can. J. Anthropol. Spec. Issue* 1:37–44

Repenning, C. A., Fejfar, O. 1977. Holarctic correlations of microtid rodents. In *Quaternary Glaciations in the Northern Hemisphere*, ed. V. Sibrava, pp. 234–50. *IGCP Proj. 73/1/24, Rep. 4*

Repenning, C. A., Lance, J. F., Irwin, J. H. 1958. Tertiary stratigraphy of the Navajo Country. *N. Mex. Geol. Soc. Guideb., Field Conf., 9th, Black Mesa Basin*, pp. 123–29

Savage, D. E. 1941. Two new middle Pliocene carnivores from Oklahoma with notes on the Optima fauna. *Am. Midl. Nat.* 25:692–710

Scarborough, R. B. 1975. *Chemistry and age of late Cenozoic air-fall ashes in southeastern Arizona.* MSc thesis. Univ. Ariz., Tucson

Scarborough, R., Damon, P., Shafiqullah, M. 1974. K-Ar age for a basalt from the volcanic member (unit 5) of the Bidahochi Formation. *Geol. Soc. Am. Abstr. with Programs* 6:472

Schultz, C. B., Stout, T. M. 1961. Field conference on the Tertiary and Pleistocene of western Nebraska. *Guideb. Field Conf. Soc. Vertebr. Paleontol., 9th, Spec. Publ. Univ. Nebr. State Mus.* 2:1–55

Schultz, C. B., Schultz, M. R., Martin, L. D. 1970. A new tribe of the saber-toothed cats (Barbourofelini) from the Pliocene of North America. *Bull. Univ. Nebr. State Mus.* 9:1–31

Schultz, G. E. 1977. *Guidebook for Field Conference on Late Cenozoic Biostratigraphy of the Texas Panhandle and Adjacent Oklahoma, Kilgore Res. Cent., West Tex. State Univ., Spec. Publ. No. 1.* 160 pp.

Seward, D. 1979. Comparison of zircon and glass fission-track ages from tephra horizons. *Geology* 7:479–82

Shafiqullah, M., Damon, P. E., Lynch, D. J., Reynolds, S. J., Rehrig, W. A., Raymond, R. H. 1980. K-Ar geochronology and geologic history of southwestern Arizona and adjacent areas. *Ariz. Geol. Soc. Dig.* 12:201–60

Shotwell, J. A. 1970. Pliocene mammals of southeastern Oregon and adjacent Idaho. *Bull. Mus. Nat. Hist. Univ. Oreg. No. 17*, pp. 1–103

Skinner, M. F., Hibbard, C. W., Gutentag, E. D., Smith, G. R., Lundberg, J. G., et al. 1972. Early Pleistocene preglacial and glacial rocks and faunas of north-central Nebraska. *Bull. Am. Mus. Nat. Hist.* 148: 1–148

Skinner, M. F., Skinner, S. M., Gooris, R. J. 1977. Stratigraphy and biostratigraphy of late Cenozoic deposits in central Sioux County, western Nebraska. *Bull. Am. Mus. Nat. Hist.* 158:263–370

Stirton, R. A. 1936a. A new ruminant from the Hemphill middle Pliocene of Texas. *J. Paleontol.* 10:644–47

Stirton, R. A. 1936b. A new beaver from the Pliocene of Arizona with notes on the species of *Dipoides. J. Mammal.* 17:279–81

Stirton, R. A. 1942. Comments on the origin and generic status of *Equus. J. Paleontol.* 16:627–37

Tedford, R. H. 1981. Mammalian biochronology of the late Cenozoic basins of New Mexico. *Geol. Soc. Am. Bull.* 92: 1008–22

Tedford, R. H., Gustafson, E. P. 1977. First North American record of the extinct panda *Parailurus. Nature* 265:621–23

Tobien, H. 1973. On the evolution of mastodonts (Proboscidea, Mammalia). I. The bunodont trilophodont groups. *Notizbl. Hess. Landesamtes Bodenforsh., Wiesbaden* 101:202–76

Voorhies, M. R. 1977. Fossil moles of late Hemphillian age from northwestern Nebraska. *Trans. Nebr. Acad. Sci.* 4:129–38

Wagner, H. M. 1976. A new species of *Pliotaxidea* (Mustelidae: Carnivora) from California. *J. Paleontol.* 50:107–27

Wagner, H. M. 1981. *Geochronology of the Mehrten Formation, Stanislaus County, California.* PhD thesis. Univ. Calif., Riverside. 342 pp.

Webb, S. D. 1973. Pliocene pronghorns of Florida. *J. Mammal.* 54:203–21

Webb, S. D. 1974. Pleistocene llamas of Florida, with a brief review of the Lamini. In *Pleistocene Mammals of Florida*, ed. S. D. Webb, pp. 170–213. Gainesville: Univ. Presses Fla.

Wood, A. E. 1936. A new rodent from the Pliocene of Kansas. *J. Paleontol.* 10:392–94

Wood, H. E., Chaney, R. W., Clark, J., Colbert, E. H., Jepsen, G. L., et al. 1941. Nomenclature and correlation of the North American continental Tertiary. *Geol. Soc. Am. Bull.* 52:1–48

Zakrzewski, R. J. 1969. The rodents from the Hagerman local fauna, upper Pliocene of Idaho. *Contrib. Mus. Paleontol. Univ. Mich.* 23:1–36

Ann. Rev. Earth Planet. Sci. 1984. 12: 489–518

STRUCTURE AND TECTONICS OF THE HIMALAYA: Constraints and Implications of Geophysical Data

Peter Molnar

Department of Earth and Planetary Sciences, Massachusetts Institute of Technology, Cambridge, Massachusetts 02139

INTRODUCTION

The Himalayan range is a product of continental collision between India and Eurasia, and for several reasons it is a logical place to examine the physical causes and mechanisms of mountain building. The exceptionally high mountains and the great length of the range alone single the Himalaya out as a major structural feature. The convergence rate between India and Eurasia of about 50 mm yr^{-1} (Minster & Jordan 1978) requires rapid rates of slip, and the occurrence of four major earthquakes with magnitudes in excess of 8.4 in the last 100 years attests to a high level of seismicity and a rapid rate of active deformation (e.g. Richter 1958, Seeber & Armbruster 1981). The exposures of high-grade metamorphic rocks in the Greater Himalaya may suggest large vertical transport of lower crustal rocks (e.g. Valdiya 1980a,b), and the occurrences of young, postcollision granites imply recent, large perturbations of the temperature structure of the range (e.g. LeFort 1981). Moreover, as sediments derived from the Himalaya fill the broad Ganga Basin south of the range, the mountain belt simultaneously seems to override the basin (e.g. G. D. Johnson et al 1979, N. M. Johnson et al 1982). As such, the processes occurring now and since the collision began in the early Tertiary, as well as the present structure of the Himalayan belt, are probably similar to those at different stages of collision and mountain building in other, older orogenic belts, such as the Appalachians, the Scottish and Scandinavian Caledonides, the Alps, and

0084–6597/84/0515–0489$02.00

the Urals. Correspondingly, an understanding of the Himalaya is likely to be helpful in explaining some geologic aspects of these other ranges.

A thorough understanding of orogenesis will no doubt draw from both similarities and differences among various belts, both active and inactive, and eroded to different levels. The size and the high level of activity in the Himalaya make it a logical place to examine many of the large-scale aspects of mountain building, but the reconnaissance nature of most of the geological and geophysical work done so far in the Himalaya makes it difficult at present to understand many of the equally, or more, important details of orogenesis that must be studied before we will have a quantitative understanding of how mountain belts evolve. In particular, the details of timing remain uncertain in many portions of the Himalaya, and the throws on the major faults are controversial; moreover, numerous small-scale processes and the mechanisms of deformation are probably better studied in more accessible and better-mapped belts, such as the Alps, or where deeper levels can be examined, such as the Scandinavian Caledonides. Partly for this reason and partly because of my greater familiarity with the geophysical aspects of the Himalaya, this review concentrates on the large-scale structure and tectonics, particularly those in which geophysical data offer strong constraints. This review also focuses on the simpler part of the Himalaya, east of Pakistan. I urge readers interested in learning more about the overall geology to consult Gansser (1964), Hashimoto et al (1973), LeFort (1975), Stöcklin (1980), and the series *Himalayan Geology*, published by the Wadia Institute of Himalayan Geology in Dehra Dun, India; those interested in an area particularly well studied geologically to read Valdiya (1977, 1980a,b, 1981) on the Kumaon Himalaya; those interested in the principal tectonic feature, the Main Central thrust (MCT), to read Pêcher (1978); those interested in the Siwalik sequence and details of recent uplift and sedimentation to consult G. D. Johnson et al (1982) and companion papers; those interested in metamorphism to read Hashimoto et al (1973); those interested in a thorough discussion of the major earthquakes to read Seeber & Armbruster (1981); those interested in observations made on the Tibetan side of the Himalaya to consult papers in *Geological and Ecological Studies of Qinghai-Xizang Plateau* (1981), Tapponnier et al (1981a), or Nicolas et al (1981); and those interested in observations made in Pakistan to consult Farah & DeJong (1979) or Tahirkheli (1982).

GEOLOGIC BACKGROUND

It is convenient and customary to divide the geologic structure of the Himalaya into a series of terrains that trend parallel to the range (e.g. Gansser 1964). Although variations along strike are clear, these terrains can be traced for most of the length of the range. Consequently, cross sections at different parts of the range are quite similar to one another, and a simple description of the gross features of the geology can be given using one such cross section as an example (Figure 1).

Transhimalayan (or Kangdese) Granites

In southern Tibet and Ladakh, just north of the Himalaya, a chain of granodioritic batholiths, the Transhimalayan granites, extends virtually continuously from one end of the range to the other with a width of about 50 km (Figure 2). Most reported radiometric ages have been obtained from K-Ar decay (Bally et al 1980, Zhang et al 1981, Sharma et al 1978a, Zhou et al 1981), but a few from $^{39}Ar/^{40}Ar$ (Brookfield & Reynolds 1981, Maluski et al 1982), Rb-Sr (Honegger et al 1982, Tu et al 1981), and U-Pb (Honegger et al 1982, U-Pb Dating Group 1981) techniques are consistent with intrusion from about 110 to 40 m.y. ago, with some ages as old as 120 m.y. and with the younger period represented also by lavas. A few younger ages may suggest a waning of activity from about 40 to even 10 m.y. ago (e.g. Sharma et al 1978a, Tu et al 1981), but too little information has been given to evaluate the significance of the dates. The Transhimalayan granites are generally ascribed to subduction of oceanic lithosphere beneath southern Tibet, which appears to have been the southern margin of Eurasia about 100 m.y. ago.

Indus-Tsangpo Suture Zone

A belt of ophiolites, the Indus-Tsangpo suture zone, lies just south of the Kangdese granites and appears to mark the collision zone between the Indian subcontinent and Eurasia (e.g. Bally et al 1980, Frank et al 1977, Gansser 1964, 1980, Honegger et al 1982, Nicolas et al 1981, Tapponnier et al 1981a, Thakur 1981, Wu 1981). Therefore, virtually all of the material comprising the Himalayan range was once part of the Indian subcontinent and not part of Asia. The ophiolites of the Indus-Tsangpo suture zone include all members of a classical ophiolite suite, and the sequence in most areas is not especially dismembered (Nicolas et al 1981, Tapponnier et al 1981a). The suture zone does not form a continuous belt of ophiolites (Figure 2), and in some areas sedimentary sequences (e.g. the Shigatse formation; Bally et al 1980, Cao 1981, Tapponnier et al 1981a, Wang & Wang 1981) apparently typical of a fore-arc setting are directly in (tectonic) contact with the sedimentary sequences that apparently once formed the northern margin of India. Hence, the vast majority of the ocean floor that once lay between India and Eurasia has been subducted, and only a few discontinuous bodies record the history of that ocean floor. Moreover, the close proximity of the ophiolites to the Kangdese granites and the

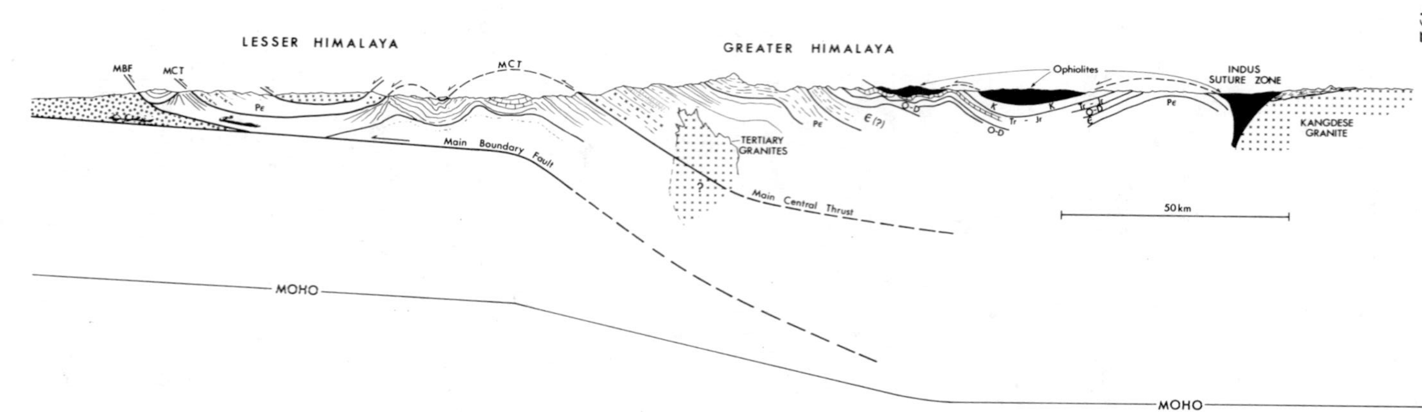

Figure 1 Geologic cross section of the Himalaya (from Molnar et al 1983). Top 3 km are taken from Gansser's (1964) cross section across the Kumaon Himalaya. The lower parts are redrawn, in part, using results from an analysis of gravity anomalies by Lyon-Caen & Molnar (1983) from the Mount Everest region. That study showed that the Moho dips gently at a few degrees beneath the Lesser Himalaya. As Seeber et al (1981) inferred, and as fault-plane solutions and depths of foci of eathquakes imply, the Indian plate is shown as a coherent plate beneath the Lesser Himalaya. However, the gravity anomalies from the Mount Everest region require that the dip of the Moho steepen to about 15° beneath the Greater Himalaya in that area, presumably because the Indian plate is weaker and can bend more there than it can beneath the Lesser Himalaya (Lyon-Caen & Molnar 1983). The Indian plate is shown here with a sliver of crust having been detached from the rest of the lithosphere when slip began on the Main Boundary Fault (MBF). Now following some 100–150 km of slip on that fault, the Lesser Himalaya has ramped up onto the Indian plate, causing the steepening of the now-presumed-inactive Main Central thrust (MCT) and rapid erosion of the portion that has already ramped onto the coherent portion of the Indian plate. In this scheme, developed in more detail by Lyon-Caen & Molnar (1983), the uplift and erosion of the Greater Himalaya where the MCT outcrops would be very recent [less than 10 m.y. (and maybe only a couple of million years)]. The possibility that the granites in the Greater Himalaya are derived from below the MCT is shown as questionable; this idea is discussed in Molnar et al (1983), but has not been inferred from geologic mapping.

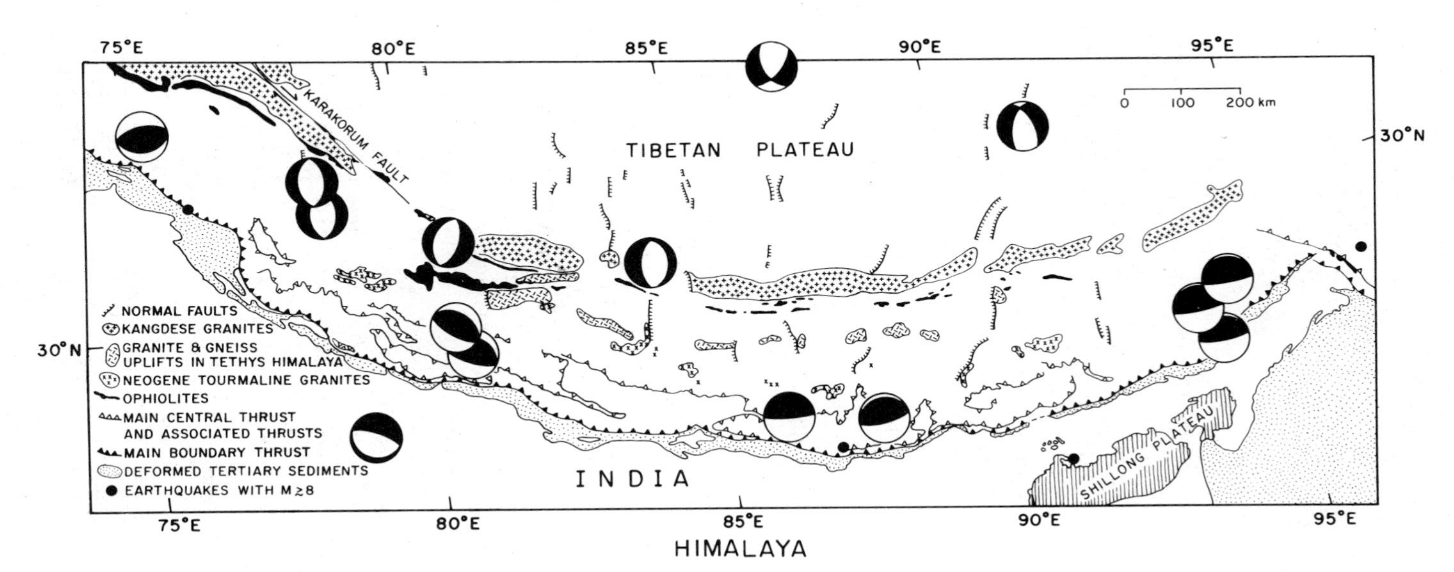

Figure 2 Neotectonic map and fault-plane solutions in the Himalaya (from Molnar & Chen 1982). Basic geologic features (the Kangdese granites and late Tertiary granites, the Indus-Tsangpo ophiolites, the Main Central thrust, and the Main Boundary thrust) are shown (from Gansser 1977). Lower hemisphere projections of fault-plane solutions (from Baranowski et al 1984, Molnar & Tapponnier 1978, Molnar et al 1973, 1977) are plotted at the epicenters of the corresponding earthquakes. In cases where two events occurred with the nearby epicenters, only one solution is shown. Blackened areas show quadrants with compressional first motions. Locations of normal faults in Tibet and the Tethys Himalaya are from P. Tapponnier's analysis of *Landsat* imagery in Molnar & Tapponnier (1978). Black dots show epicenters of large earthquakes ($M > 8$) in the Himalaya in 1905, 1934, 1897, and 1950, from west to east. Rupture zones of all probably exceeded 200 km parallel to the range (Seeber & Armbruster 1981).

predominant northward vergence of thrusts in the ophiolites imply substantial postcollisional shortening of the region (Tapponnier et al 1981a).

Tethys Himalaya

South of the Indus-Tsangpo suture, virtually all of the sedimentary rock appears to have been part of the Indian subcontinent or its northern continental margin from the time of its deposition until the collision with Eurasia. The sedimentary sequence is nearly continuous from the early Paleozoic until the early Tertiary. The sequence of Paleozoic and, in parts, Mesozoic sedimentary rocks records a stable, shallow-water environment with flora and fauna typical of the southern continents and the Tethyan realm (e.g. Gansser 1964, LeFort 1975, Sinha 1981). In the northern part of the region, however, Mesozoic sedimentary rocks contain abundant flysch with basic dikes and sills (Stöcklin 1980, Tapponnier et al 1981a, Wang & Wang 1981), suggestive of continental rifting and the formation of a continental margin (e.g. Bally et al 1980, Sengör 1981). Marine sedimentation continued into Eocene time, and its termination provides an upper bound on the date when India and Eurasia collided.

The timing of the collision is not tightly constrained. Too little detailed work has been done to disprove that at least in some localities marine sediments were deposited since early Eocene time and subsequently were eroded or overthrust and buried. Moreover, the likelihood that the two continental margins did not meet flush implies that a precise date of the collision at one locality could be in error by several m.y. elsewhere. I consider it likely that the subduction of oceanic lithosphere was not complete until the latter part of the Eocene. First, mammals apparently did not appear on India until after middle Eocene time; deposits of more recent age, however, contain abundant fossils of mammals similar to those that had evolved in Mongolia (Sahni & Kumar 1974), suggesting a land bridge only after middle Eocene time. Second, the rate of convergence between India and Eurasia slowed abruptly from more than 100 mm yr^{-1} to about 50 mm yr^{-1} between 50 and 40 m.y. ago (Molnar & Tapponnier 1975). Such a decrease is to be expected when oceanic lithosphere has been subducted and continents collide. These arguments, however, clearly do not provide tight constraints on the timing of the collision at the Himalaya.

Greater Himalaya

The Paleozoic sedimentary rocks of the Tethys Himalaya apparently were deposited on metamorphic rock of Precambrian age. These metamorphic rocks are exposed in the Greater Himalaya where the peaks are highest and accessibility is most difficult. Whereas rivers in the Tethys Himalaya are often braided and even meander without eroding, torrential rivers have rapidly cut deep gorges in the Greater Himalaya. With the difficulties of travel in the Greater Himalaya, it is not surprising that the geology there is controversial. Nevertheless, a few facts seem to be widely accepted. First, the metamorphic grade increases downward through the section and attains a high grade near the base of the Greater Himalaya (e.g. LeFort 1975). Second, the rocks are highly sheared and have been thrust over the Lesser Himalaya on a major zone of thrust faulting, the Main Central thrust (e.g. Pêcher 1978). Finally, within the Greater Himalaya, and parts of the Tethys Himalaya as well, there are Tertiary two-mica granitic intrusions. Analyses of K-Ar (LeFort 1975, 1981, Wang et al 1981) and Rb-Sr (Vidal 1978) isotopes give mineral ages of 10 to 20 m.y., and high $^{87}Sr/^{86}Sr$ ratios imply anatectic melting (Hamet & Allègre 1976, Vidal 1978, Vidal et al 1982, Wang et al 1981). A close relationship between these granites and the MCT is usually assumed, with the full range of plausible causal connections between slip on the fault and the granites represented. Although most authors agree that a large throw on the MCT probably placed rocks from the middle, or even lower, crust onto the Lesser Himalaya and that crustal material melted to form these granites, controversy surrounds the timing and duration of slip on the MCT, the precise origin of the granites, and the source of heat for melting.

Lesser Himalaya

The Lesser Himalaya is comprised largely of low-grade metasediments and klippen of older metamorphic rocks and granitic intrusions (e.g. Bhanot et al 1977, 1981, Jäger et al 1971, Valdiya 1980b, Virdi 1980). Because of disagreements about where the MCT lies and probably because of limited mapping, authors disagree about whether or not the MCT forms a sharp boundary between the high-grade terrains above it and the lower-grade terrain below it (e.g. Valdiya 1980a,b, 1981) and about where the highest metamorphic grade lies with respect to the MCT (e.g. Bordet et al 1981, LeFort 1975, 1981). The ages of most of the metasediments are poorly constrained, and these formations have not been correlated and apparently *do not* correlate with the sedimentary sequence of the Tethys Himalaya. As a result, placing tight constraints on the throw on the MCT has been an elusive task. Klippen of crystalline rock similar to that of the Greater Himalaya, however, have been recognized in parts of the Himalaya and have been interpreted as evidence for at least 100 km of throw on the MCT (e.g. Brunel & Andrieux 1977, Evans 1964, Fuchs & Sinha 1978, Gansser 1966, Stöcklin 1980, Virdi 1980). The rocks of the Lesser Himalaya, in turn, are cut by numerous thrust faults (e.g. Valdiya 1980b, 1981), but most authors infer that the entire terrain has been thrust as a package over the

India shield on another major thrust fault, the Main Boundary Fault (MBF).

Ganga Basin

The main strand of the Main Boundary Fault crops out as a steeply northward-dipping fault within sediments of Miocene age—the Siwalik sequence (e.g. G. D. Johnson et al 1982, N. M. Johnson et al 1982). This sequence consists of terrigenous sediments derived by erosion of the Himalaya. Current opinion holds that the MBF consists of several strands and forms a schuppenzone (a closely spaced series of thrust faults) within the Siwaliks and younger sediments. The sedimentary fill of the Ganga Basin is thus a continuation of the Siwaliks. The late Cenozoic sediments of the Ganga Basin lap onto the Indian shield to the south, and they apparently increase in thickness to reach a maximum of 4–6 km at the foot of the Himalaya (Sastri et al 1971, Stöcklin 1980). Presumably the entire Ganga Basin, as well as part of the Lesser Himalaya, is underlain by the Indian shield.

SEISMICITY AND FAULT-PLANE SOLUTIONS

The Himalaya are seismically active, not only with moderate earthquakes ($M \sim 6$), but also with great earthquakes ($M > 8$). The distribution of activity is not tightly constrained because most of what we know is based on both a short sample of time (since about 1960), with recordings made on large distances ($\Delta > 30°$), or on intensity distributions of earthquakes occurring in the last 160 yrs. Only in the northwest Himalaya has there been extensive study using a local network of stations (Armbruster et al 1978, Seeber & Armbruster 1979, Seeber & Jacob 1977), but since that area seems to be very complicated, it is not discussed further here. Most of the best-located events throughout the rest of the range appear to have occurred beneath the Lesser Himalaya between the MCT and MBF (e.g. Molnar et al 1973, 1977), but there is only a small number of well-located events. Given the northward dips of the MBF and MCT, it appears that the MCT is not very active, if at all, and that if these events occurred on one major fault, it was the MBF.

Fault-Plane Solutions

Most fault-plane solutions are consistent with an active MBF. They show steeply southward-dipping nodal planes and gently northward-dipping nodal planes (Figure 2) (Baranowski et al 1984, Chandra 1978, Fitch 1970, Molnar et al 1973, 1977, Rastogi 1974). Given the preponderance of northward-dipping thrust faults in the Himalaya, the northward-dipping nodal planes are almost certainly the fault planes. Slip vectors plunge gently beneath the range and perpendicular to its local trend. In the eastern part of the range, the plunges typically are only a few degrees ($<10°$), but in the west they are steeper (20–25°) (Baranowski et al 1984, Molnar et al 1977). Whether this difference reflects a variation in the dip of the MBF from east to west (or from south to north) or some other complexity is, to my knowledge, still an open question. In any case, the seismicity clearly implies that convergence continues.

Fault-plane solutions of earthquakes in the Tethys Himalaya show normal faulting and reflect east-west extension, like those for events in Tibet (Molnar & Chen 1983, Molnar & Tapponnier 1975, 1978, Ni & York 1978). Hence, although the rocks of the Tethys Himalaya were originally part of India and Gondwana, they now participate in the active tectonic regime of Tibet and not that of the Himalaya (see also Tapponnier et al 1981b). Apparently style of deformation changed from thrust faulting to normal faulting when elevations and crustal thicknesses increased sufficiently that the vertical stress passed from being the least to the most compressive stress (Tapponnier & Molnar 1976). Note also that the present extension in Tibet is reflected by the variation in azimuth of slip vectors along the trend of the arc; northward underthrusting in the eastern Himalaya and northeastward underthrusting in the western Himalaya require east-west extension of southern Tibet, if we assume that India behaves rigidly (Armijo et al 1982, Baranowski et al 1984, Molnar & Chen 1982, 1983).

The fault-plane solution of one earthquake (15 August 1967) in Figure 2 deserves particular attention. This event occurred beneath the Indian shield south of the Himalaya. The fault-plane solution shows normal faulting, with its T-axis perpendicular to the Himalaya (Molnar et al 1973, 1977). It, therefore, probably reflects extension of the top of the shield as it bends down and plunges beneath the Himalaya.

Hence, the fault-plane solutions are broadly consistent with the Indian plate bending down in front of the Himalaya and sliding beneath the range, but with the present activity of the Tethys Himalaya being part of another regime.

Focal Depths

The absence of local networks of stations makes reliable routine determination of focal depths impossible, but by matching synthetic and recorded seismograms from earthquakes with known fault-plane solutions, it is possible to obtain depths for those few events large enough to be well recorded by the World Wide Standardized Seismograph Network (WWSSN) (Baranowski et al 1984, Ni & Barazangi, in preparation). The depths of these events are about 15 km ($\pm$3–5 km) below the surface of the Lesser Himalaya, and epicenters of these earthquakes are usually 80–100 km north of the front of the range. Therefore, these earthquakes occur at or close to the top surface of the Indian plate as it slides beneath the Himalaya (Baranowski et al 1984, Molnar & Chen 1982, Ni & Barazangi, in preparation). Recall that the elevation of the Lesser Himalaya is about 2 km and that the maximum depth of the sediments in the Ganga Basin is about 4 km. Were the top of the Indian plate to dip north and reach a depth of 15 km, 80 to 100 km north of the Himalayan front, its dip would be arctan $[(15-6\,\text{km})/(90\,\text{km}) = 0.1] \approx 6°$, which accords well with the dips of nodal planes ($\sim 5°$), at least in the eastern Himalaya. The steeper dips ($\sim 25°$) of the planes in the western part of the Himalaya might instead reflect deformation of the overriding material or possibly a steepening of the MBF, and therefore of the top surface of India.

Seeber et al (1981) suggested that Indian lithosphere slides coherently under the Lesser Himalaya along a gently dipping fault, which they call "the Detachment" and assume to extend far south of the Himalayan front beneath the Ganga Basin. They deduce this in part from the distribution of isoseismals of four great earthquakes in 1897, 1905, 1934, and 1950 (Seeber & Armbruster 1981). I disagree with their conclusions about the extent of faulting during two of the events (1897 and 1934), but the fault-plane solutions and focal depths clearly support their inferences about the gentle dip of the major active thrust fault beneath the Himalaya and about the underthrusting of India intact beneath the Lesser Himalaya.

GRAVITY ANOMALIES AND THE DEEP STRUCTURE OF THE HIMALAYA

Virtually every study of gravity anomalies in the Himalaya has concluded that the crust must thicken toward the north (e.g. Choudhury 1975, Kono 1974, Qureshy 1969, Warsi & Molnar 1977). Bouguer anomalies become increasingly negative toward the north, implying at least partial compensation for the increasing elevation. Most of these studies, however, have sought fits to observed gravity anomalies using somewhat arbitrarily chosen blocks of different densities, and consequently the literature includes a spectrum of opinion on whether or not (and if so, to what extent) isostatic equilibrium prevails.

Given the inference that India slides coherently beneath the Lesser Himalaya, Hélène Lyon-Caen and I examined gravity anomalies under the assumption that the Indian plate was flexed down as an elastic plate under the load of the overthrust Himalayan mountain range (Lyon-Caen & Molnar 1983). By making such an assumption, we calculated the shape of the top surface of the Indian plate as well as a profile of gravity anomalies with only a small number of free parameters—the flexural rigidity of the plate, the density differences among the sediments, crust, and mantle, the position of the end of the elastic plate, and the force and bending moment applied to the end. Here, I give only a brief summary of the results.

First, the calculated Bouguer anomalies are not very sensitive to the plausible density differences among Ganga Basin sediments, crustal rock, and the mantle. The flexural rigidity of the Indian plate is within a factor of 3 of 7×10^{24} N m, corresponding to equivalent elastic plate thicknesses within a factor of 1.5 of 80 km. The Lesser Himalaya seems to be supported by the elastic plate, and beneath this part of the range, the plate dips gently north ($\sim 3°$). A slight improvement of the fit of calculated to observed gravity anomalies can be obtained if light material (sediments subducted with the Indian plate) is assumed to underlie part of the Lesser Himalaya (Figure 3). Beneath the Greater Himalaya, however, the Moho must steepen to about 15° (see also Kono 1974, Warsi & Molnar 1977).

The steepening of the Moho implies a sharp bending of the plate. We infer that between 100 and 150 km north of the Himalayan front, the Indian plate must weaken so that it can be bent to dip 15°. A smaller flexural rigidity of the northern part of the plate is not surprising. First, as the Indian plate underthrusts beneath the Himalaya, it should warm and weaken. Moreover, and perhaps more important, the detachment of large thrust sheets between the Main Boundary thrust (MBT) and the MCT and between the MCT and the Indus-Tsangpo suture zone must have thinned the leading edge of the Indian lithosphere. It seems possible that the detachment of the thrust sheet of the MBF may have occurred when India was 100 to 150 km farther to the south of its present position with respect to the Himalaya. Hence, this detachment would be responsible for the weak, thin leading edge of the Indian plate, and if this is so, then the throw on the Main Boundary Fault would be 100–150 km.

Inclusion of a weaker segment of plate alone, however, does not allow an acceptable fit to the gravity anomalies. The heavy load of the Himalaya flexes the plate down too much, and some additional force or set of forces must be found to help support this load (Lyon-Caen & Molnar 1983). We showed that a bending moment applied to the end of the plate can flex part of the plate up and contribute to the support of the heavy load of the Greater Himalaya (Figure 4). The value of the bending moment depends on the value of the flexural rigidity of the northern segment of the plate (Figure 4), and it therefore is unlikely that we can reliably deduce one or the other.

An origin for the prescribed bending moment requires more faith in the simple physical model than does a mechanism for weakening the plate. One possible source for the moment could be a horizontal force acting on the

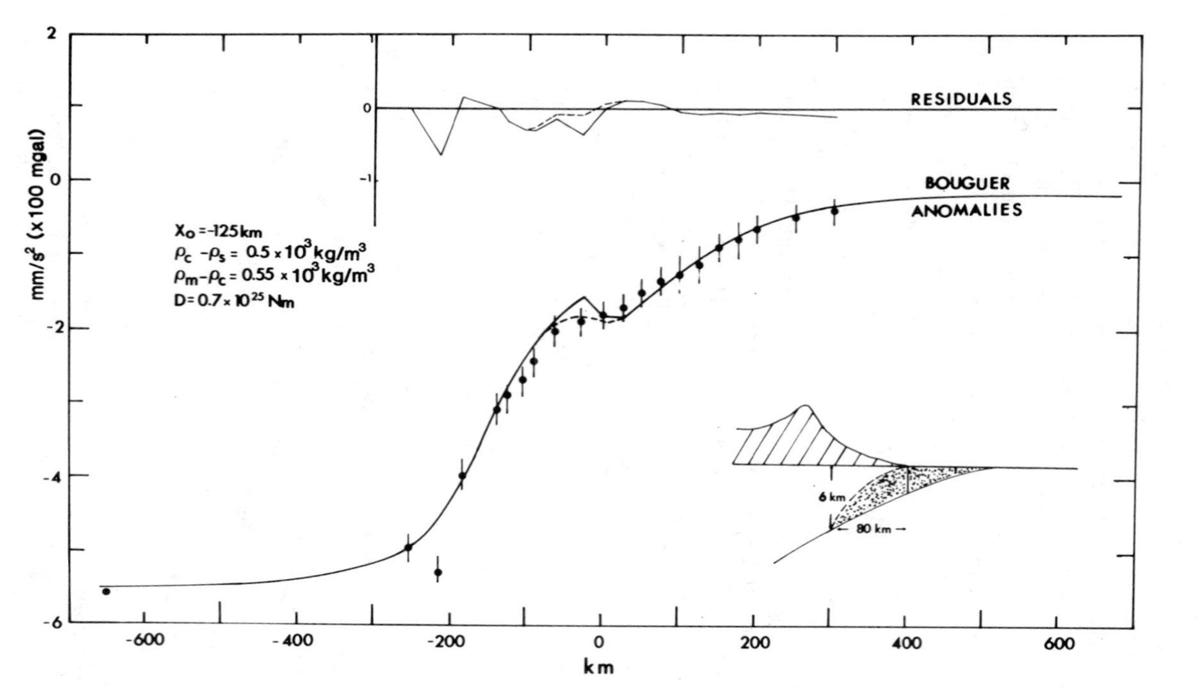

Figure 3 Comparison of observed and calculated Bouguer gravity anomalies along a north-south profile through Mount Everest (from Lyon-Caen & Molnar 1983). Black dots show observed gravity anomalies with assumed uncertainties. To calculate the anomalies, we assumed that a two-dimensional plate underlies southern India, the Ganga Basin, and the Himalaya for a distance (X_0) of 125 km from the front of the range. Using the weight of the Himalaya, and using density differences of 0.5 gm cm^{-3} for the sediments in the Ganga Basin and crustal rock and 0.55 gm cm^{-3} for the crust and mantle, we calculated the shape of the elastic plate. We assumed that Tibet, beginning 250 km north of the Himalayan front, is in isostatic equilibrium with the Moho 60 km below sea level. We then assumed that the Moho dips smoothly from the end of the plate ($X_0 = -125$ km) to southern Tibet ($X = -200$ km). In the top of the figure, residuals (observed minus calculated anomalies) are shown with an expanded scale. We found that the peak in the calculated anomalies at the foot of the Lesser Himalaya ($X \approx -20$) could be eliminated by the inclusion of light material beneath the range (see insert), which we infer to be sediments.

leading edge of the downflexed plate. To generate an adequate bending moment, however, requires a horizontal compressive stress of 5–10 kbar acting over a thickness of 100 km on a plate flexed down 30 km (Lyon-Caen & Molnar 1983). This large force seems unlikely. Alternatively, a torque could act on the end of the elastic plate; gravity acting on cold material north of, but attatched to, the effectively elastic plate could generate such a torque.

In this scenario, as the Indian plate underthrusts the Himalaya, slivers of crust would be detached from the Indian lithosphere deep in the crust and possibly at the Moho. These slivers would become part of the "overthrust plate" of the Himalaya as India continues to underthrust beneath the range. The thinning of the Indian lithosphere would weaken it and allow it to bend more easily than where the full thickness remains intact. At the same time, gravity acting on the portion of the Indian lithosphere stripped of much of its crust would help draw India toward Eurasia and would apply a torque or bending moment to the part of the Indian plate farther south.

Proof that such a scenario is appropriate will require a variety of investigations, but perhaps the most obvious is to determine the velocity structure of the upper mantle beneath southern Tibet. A cold region of downwelling might reveal itself by high velocities.

A BALANCED LITHOSPHERIC CROSS SECTION ACROSS THE HIMALAYA

The surface geology in the cross section in Figure 1 is based on Gansser's (1964) profile across the Kumaon Himalaya. The results from fault-plane solutions and focal depths and from gravity anomalies described above were used to infer the deeper structure. The gentle dip of the MCT *within* the Lesser Himalaya (Figure 1; Pêcher 1978, pp. 314–21) suggests that when it was active, it too may have defined the interface between the coherent Indian plate and the overriding sliver that now constitutes the Greater Himalaya and the Tethys Himalaya. With this in mind, Hélène Lyon-Caen and I constructed a series of idealized, but balanced, cross sections across the Himalaya since the collision. The full list of assumptions and the rationale for them are given in Lyon-Caen & Molnar (1983).

Following the collision, we arbitrarily allowed 100 km of underthrusting of India beneath southern Tibet (Figure 5*a*). More could have occurred, especially if Tibet's crust were thin before the collision. In any case, after a certain amount of subduction of Indian crust, the Main Central thrust formed, detaching the upper crust and permitting the entire, coherent Indian lithosphere to slide beneath this detatched sliver (Figure 5*b*). We assumed that the thrust dipped at a gentle angle where India's lithosphere

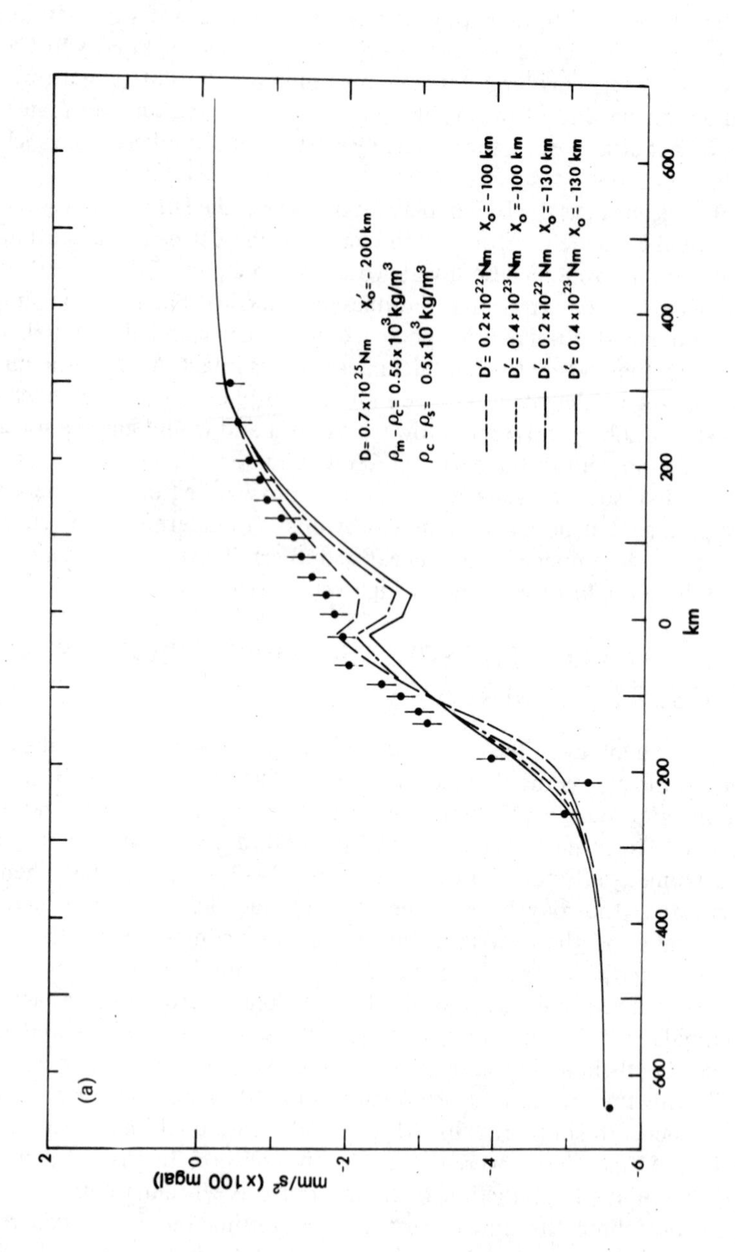
(a)
D= 0.7 x 10^25 Nm X'o= - 200 km
ρm - ρc = 0.55 x 10^3 kg/m^3
ρc - ρs = 0.5 x 10^3 kg/m^3
D' = 0.2 x 10^22 Nm Xo = -100 km
D' = 0.4 x 10^23 Nm Xo = -100 km
D' = 0.2 x 10^22 Nm Xo = -130 km
D' = 0.4 x 10^23 Nm Xo = -130 km
mm/s^2 (x 100 mgal)
2
0
-2
-4
-6
-600
-400
-200
0
200
400
600
km

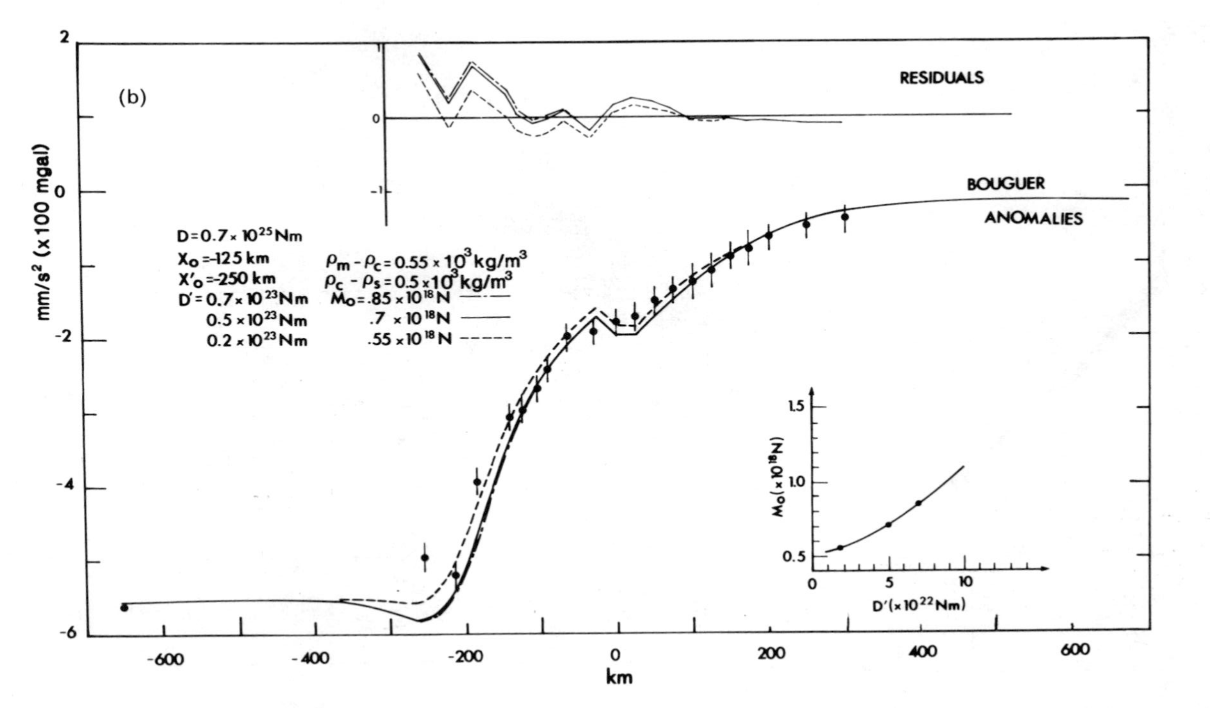

Figure 4 Comparison of observed and calculated Bouguer anomalies for structures in which the Indian plate is intact from $X \to \infty$ to $X = X_0$ (≈ -100 to -150 km), and with a plate with a lower flexural rigidity (D') than that beneath the Lesser Himalaya and Ganga Basin (D) from X_0 to $X'_0 = -200$ km (from Lyon-Caen & Molnar 1983). Layout is as in Figure 3. (*a*) Poor fit of data and calculations because the weight of the Himalaya is too great and forces the elastic plate to bend down too far. (*b*) Acceptable fits, for which a bending moment (M_0) has been applied to the end of the plate at $X'_0 = -250$ km. This moment flexes the end of the plate down but helps elevate the portion beneath the Greater and Lesser Himalaya. Note that three different combinations of M_0 and flexural rigidity D' yield similar Bouguer anomalies. Insert in lower right shows the interdependence between the bending moment and the flexural rigidity of the portion of plate between $X_0 = -125$ km and $X'_0 = -250$ km.

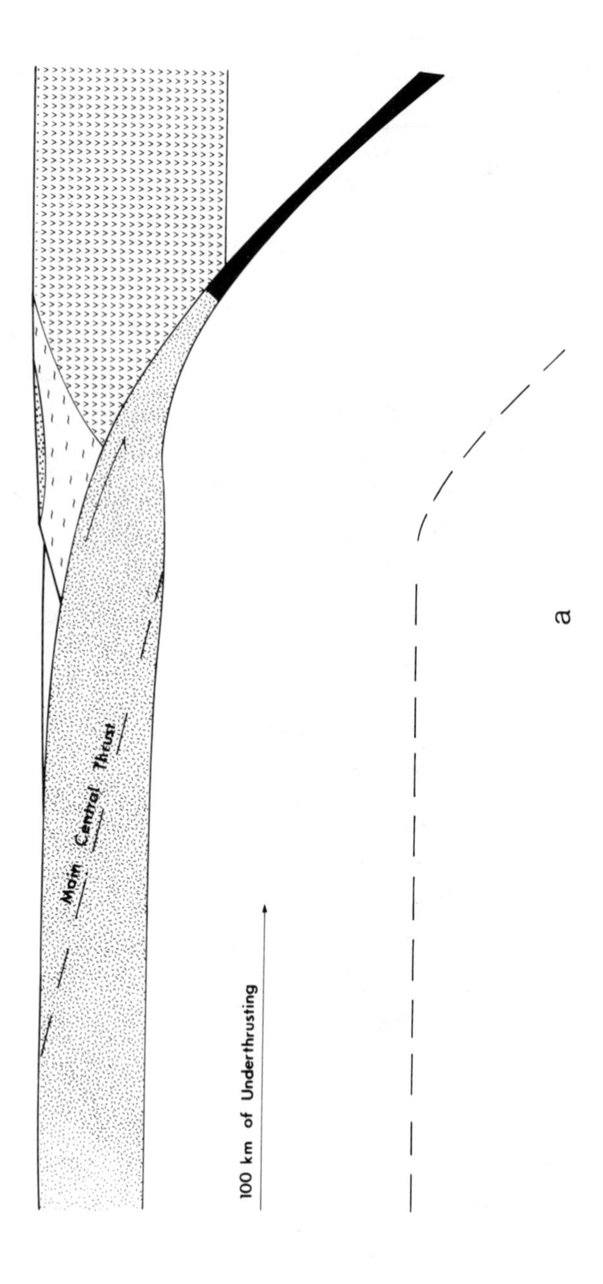
Collision and Formation of The Main Central Thrust
Main Central Thrust
100 km of Underthrusting
a

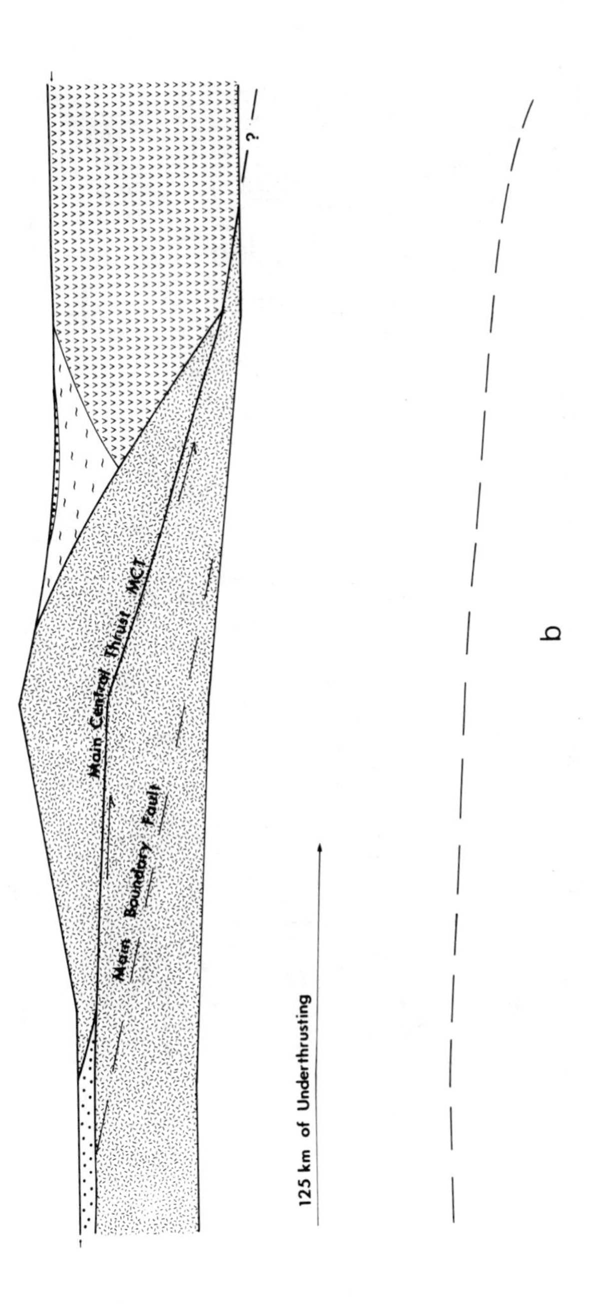

Underthrusting Along The MCT And Formation of The Main Boundary Fault

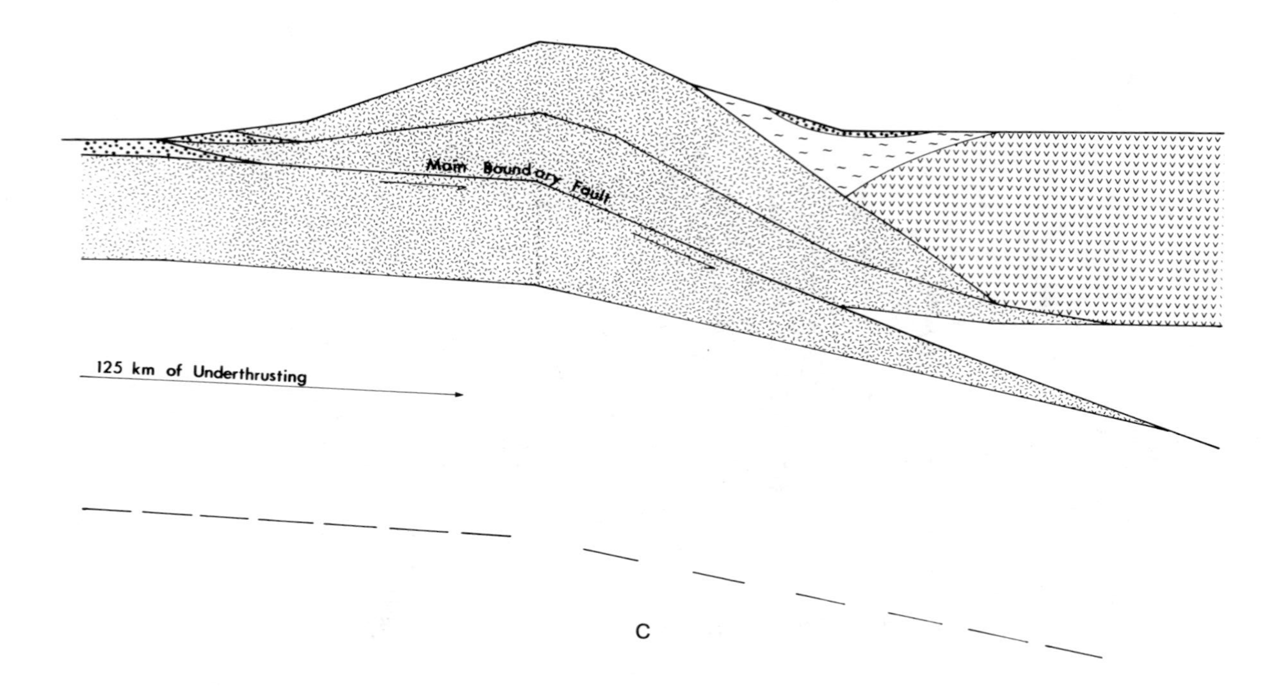
Underthrusting Along The Main Boundary Fault
Main Boundary Fault
125 km of Underthrusting
C

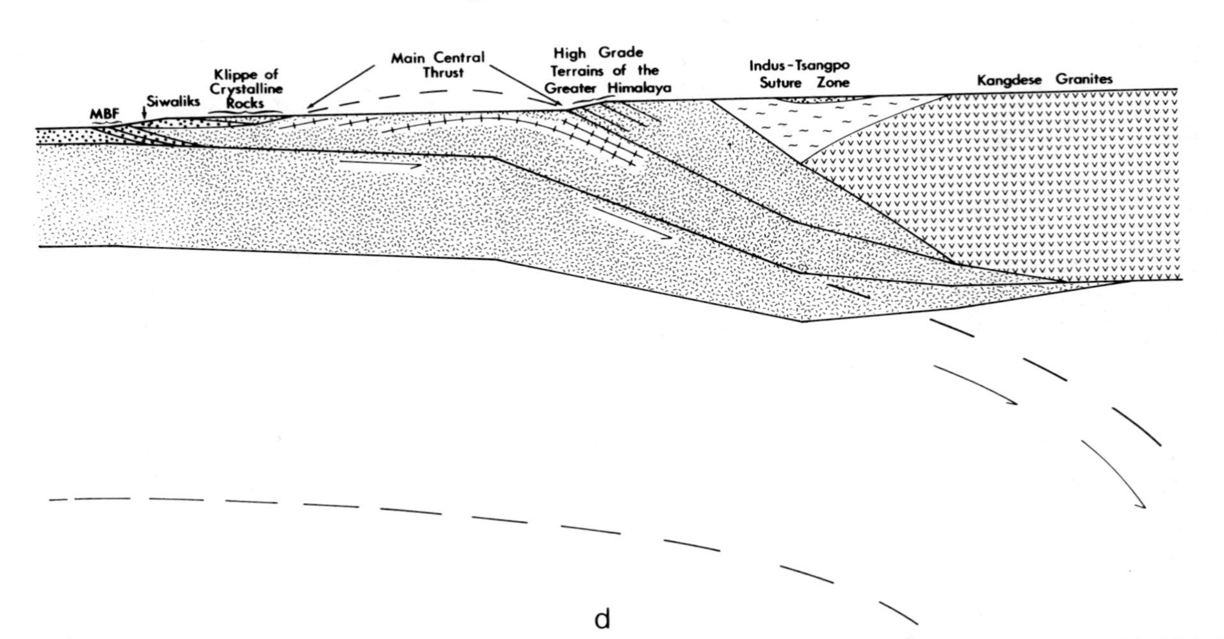

Figure 5 Sequence of idealized cross sections from the collision to the present (from Lyon-Caen & Molnar 1983). (*a*) Formation of the Main Central thrust after (an arbitrarily assumed amount of) 100 km of subduction of part of the northern margin of India. (*b*) Formation of the Main Boundary Fault after (an assumed) 125 km of underthrusting of India along the MCT. Note the marked uplift of material over the MCT. (*c*) Underthrusting of (an assumed) 125 km of India along the MBF. Note that again pronounced uplift occurs where the MBF changes dip. (*d*) Same as (*c*), but with material eroded to the level of the present topography. Note that many features of the present Himalaya are present (see Figure 1): the overthrust sediments are analogous to the Siwalik sequence of Miocene terrigenous sediments at the Himalayan front, the klippe of crystalline rocks transported by the MCT to the south is present, the MCT dips at a gentle angle to the south in the Lesser Himalaya but more steeply to the north beneath the Greater Himalaya, the metamorphosed sediments in the Lesser Himalaya are domed slightly, and high-grade metamorphic rocks are present above the MCT.

would have been strong. We also assumed that a basin analogous to the present Ganga Basin formed by the flexure. It would have been filled by the predecessor of the Siwalik Series that was eroded from the Himalaya in the Oligocene and early Miocene, but we temporarily ignore erosion. Once again, after further convergence and subduction, a new fault (the Main Boundary Fault) formed, and again we allowed 125 km of slip. The Moho in Figure 5*c* thus assumes approximately the shape deduced from gravity anomalies, briefly described above. The high mountains in Figure 5*c*, of course, would not and do not exist; their inclusion is to show how much material would have been eroded and to show where the rock now at the surface may have once lain with respect to the overlying material.

When the excess elevation in the profile in Figure 5*c* above that in a typical profile is removed, the gross structure of the predicted cross section (Figure 5*d*) looks very similar to that in Figure 1. There is a klippe of crystalline rock overthrust onto molasse (pre-Siwalik sediments) and separated from the low-grade terrain of the Lesser Himalaya by a nearly flat thrust (MCT). The low-grade terrain probably would not have achieved a high-grade because it never would have been buried deeply. The MCT beneath the Greater Himalaya is shown to dip steeply to the north, and the high-grade terrain above it would have once been lower crustal material later transported over the shallower, lower-grade Lesser Himalaya along the MCT. The high-grade metamorphism might therefore be due to burial in the lower curst for hundreds of million of years before the collision started, and not to heating associated with slip on the Main Central thrust. The predicted structure shown in Figure 5*d*, however, does not match that observed near the Indus-Tsangpo suture, but as noted above, the southward dips of thrust and the close proximity of the ophiolites to the Kangdese granites imply some late-stage back thrusting in that area (Tapponnier et al 1981a), something ignored in the construction of Figure 5.

THERMAL HISTORY OF THE HIMALAYA

Although the high-grade metamorphic terrains of the Greater Himalaya may not imply unusually high crustal temperatures, the occurrences of late Tertiary, two-mica, tourmaline granites clearly require melting since the collision with India. As noted above, there is little agreement about the source of heat for the melting. Among possible principal sources, probably frictional heating on the Main Central thrust still has the greatest number of adherents (see Bird et al 1975, LeFort 1975, Scholz 1980, Sinha-Roy 1981). Some advocates, however, have defected toward other sources of heat. For instance, Bird (1978) concluded that no reasonable combination of frictional and conductive heating is adequate, and he appealed to "delamination" of the crust from the mantle to expose the crust to asthenospheric temperatures. In contrast, LeFort (1981) concluded that the addition of water into an already hot upper thrust plate would reduce the melting temperature of it and allow it to melt without the need for additional heat sources. Bird (1978) had already rejected this idea because a colder lower thrust plate would draw heat out of the upper plate too rapidly to allow it to melt. In this respect, I concur with Bird, unless the melting temperature is very low, less than 600°C. I disagree with Bird, however, that other sources of heat could not cause the heating and that delamination is required.

Molnar et al (1983) made a series of simple calculations of temperatures in overthrust terrains and concluded that a wide variety of plausible combinations of physical parameters could account for the melting of crustal rocks to form the two-mica granites, assuming a melting temperature of 650°C. If the granites originated sufficiently deep in the crust (10–20 km below the MCT) and if radioactive heating and the heat supplied to the base of the lithosphere were somewhat higher than is typical of shields (0.6 instead of 0.3 Heat Generation Units, and 0.9 instead of 0.6 Heat Flow Units), then no frictional heating would be necessary unless melting occurred only a short time after the initiation of slip. On the other hand, if melting occured within 10 m.y. after slip began or if the lower plate were as cold as shields are typically thought to be (e.g. Sclater et al 1980), then several kilobars of stress or an alternative process such as delamination would be needed to raise the temperature of crustal material to 650°C. Scattered heat-flow measurements in India are somewhat higher than normal shields (Rao et al 1976), but the absence of measurements within the Himalaya and uncertainties in the depth of melting, in the duration of slip on the MCT, and in other relevant physical parameters, such as the coefficient of thermal conductivity, make it impossible to decide which combinations of parameters are most probable. Only a range of already implausible values can be shown to be impossible. Thus, in my opinion the source of heat for the granites remains open, but Figure 1 is drawn according to my prejudice that melting began below the MCT. The reader, however, should know that this view is not widely shared.

RATES OF CONVERGENCE AT THE HIMALAYA

Several facts, inferences, and prejudices combine to suggest a rate of convergence of the order of 10–20 mm yr^{-1} across the Himalaya.

Average Long-Term Rates

An upper bound on the convergence rate is given by the present rate between India and Eurasia of about 50 mm yr^{-1} (Minster & Jordan 1978), but the abundant evidence for active and Tertiary deformation north of the

Himalaya implies that less convergence occurs in the Himalaya (e.g. Molnar & Tapponnier 1975). A more reasonable upper bound is obtained by assuming that Tibet was and continues to be underthrust by India; taking 1000 km for the amount of underthrusting and 40–50 m.y. for its duration, we obtain a rate of 20–25 mm yr^{-1}. If only 300–500 km of underthrusting actually took place (e.g. Gansser 1966, Mattauer 1975, Warsi & Molnar 1977, or Figure 5), then the average rate is about 10 mm yr^{-1} or less. Finally, readers who share my prejudice that the late Tertiary granites originate below the Main Central thrust might note that the apparent young age of these granites ($\sim$15 m.y.) puts a lower bound on when slip on the MCT would have ceased and, hence, when the MBF formed. If 100–150 km of the slip has occurred on it, as argued above, then again the average rate of slip is 10 mm yr^{-1} or less. Thus, the average rate of slip has probably been less than 25 mm yr^{-1} and may have been less than 10 mm yr^{-1}.

Short-Term Rates

Estimates of average rates over shorter periods of time also are not very definitive. Sinvhal et al (1973) operated a strain meter for several months across one major fault in the Lesser Himalaya. They obtained a relatively constant rate of slip of about 10 mm yr^{-1}, but the measured direction of slip was approximately parallel to the range, not perpendicular to it or parallel to the relative motion between India and Eurasia. Moreover, given the possible slip on other fault strands, it is perhaps premature to speculate on the significance of this result.

The seismic moments of earthquakes offer another possible way to estimate the rate slip during the last 80 yr (Brune 1968). When combined, the values give average rates of underthrusting of 10 to 20 mm yr^{-1} (Chen & Molnar 1977, Molnar et al 1977), but the uncertainties are at least a factor of two.

Although active and recently active faults have been recognized in the Himalaya, I am not aware of estimates of amounts of slip on these faults. Armijo et al (1982), however, estimated a rate of extension in Tibet and used the difference in directions of underthrusting in the Himalaya (discussed above) to infer 10 to 20 mm yr^{-1} of underthrusting. Since the rate of extension is based on data from one basin, this rate, too, cannot be more certain than an additional factor of two. Therefore, its value is more in corroborating a reasonable range of extension rates in Tibet (e.g. Baranowski et al 1984) than in deducing a rate of underthrusting at the Himalaya.

Taken together, I think that these estimates of short-term rates of 10–20 mm yr^{-1} support the supposition that part of India's convergence with

Eurasia is absorbed north of the Himalaya, but none of them offers any preference for fast (25 mm yr^{-1}) or slow (10 mm yr^{-1}) rates.

RATES OF UPLIFT OF THE HIMALAYA

Studies of fission-track ages in different minerals of the same rock bodies indicate rates of uplift, averaged over durations of tens of millions of years since the collision, of less than 1 mm yr^{-1} and typically about 0.3 mm yr^{-1} (e.g. Mehta 1980, Sharma et al 1978b, Zeitler et al 1982b). Note that such rates, if constant, would imply 10 to 30 km of uplift in a 30 m.y. period. Fission-track ages, however, also imply higher rates in some portions of the Himalaya for more recent periods of time; for instance, the area around Nanga Parbat has been rising at nearly 10 mm yr^{-1} for the last few million years (Zeitler et al 1982a), and some of the smaller ranges of the foothills of Pakistan seem to have risen at a rate of 1 to 4 mm yr^{-1} in the last 1 or 2 m.y. (G. D. Johnson et al 1979, Opdyke et al 1979). Similarly, there are scattered reports from the Greater Himalaya of pollen and spores in Quaternary sediments at high elevations, but from plants typically found at much lower elevations (e.g. Li et al 1981, Song & Liu 1981, Xu Ren 1981), and they have been used to infer rates of uplift of several millimeters per year.

I am not aware of data that allow a quantitative estimate for the rate of uplift of the Lesser Himalaya. One repeated leveling profile across the front of the Himalaya suggests as much as 6 mm yr^{-1} of uplift in a 12-yr period of the front with respect to the Ganga Basin, but the profile does not extend far into the Himalaya [work of Chugh (1974), reproduced in Molnar et al 1977]. Thus, there are several data sources suggesting very rapid uplift (several millimeters per year) in some areas in recent times, but with lower average rates when longer durations are considered.

These data can be used to infer that the Himalaya has risen recently, possibly at an accelerating rate. Although this inference is clearly allowed by the data, I do not consider it likely. The evidence for high rates of uplift comes from the front of the range and from the Greater Himalaya. If the Indian plate remains intact, it presumably will flex down and slide beneath the Lesser Himalaya, pushing the Lesser Himalaya up only at its front, where splays of listric thrust faults reach the surface. If the fault changes dip beneath the Greater Himalaya (Figure 5), but the motion of India with respect to the overriding Himalaya is northward underthrusting on a gently dipping plane, then the material above the more steeply dipping thrust must rise (Lyon-Caen & Molnar 1983). If India underthrusts the Lesser Himalaya at a rate v, along a plane of negligible dip ($\lesssim 5°$), and if the dip, ϕ, is much greater beneath the Greater Himalaya, then the rate of uplift there should be $v \tan \phi$. For $v = 10$ or 20 mm yr^{-1} and $\phi = 20$ or $30°$, we obtain rates of uplift of 3.6 or 7.2 mm yr^{-1} (for $\phi = 20°$) and 5.8 and 11.6 mm yr^{-1} (for $\phi = 30°$). Hence, my inclination is to believe that the evidence for rapid and recent uplift applies more to the areas where it is observed—at the front of the Lesser Himalaya and in the Greater Himalaya—than to the most recent part of geologic time. The gentle grades of rivers in the Tethys Himalaya, the rapid downcutting in the Greater Himalaya, the more mature topography of the Lesser Himalaya, and the abrupt rise of the Himalaya at its front are consistent with this view.

SUMMARY

Several basic controveries underscore our ignorance of the Himalaya and its evolution. Accordingly, any summary of the structure and tectonics will be either somewhat prejudiced or long. I choose the former, but try to direct the reader to references with views different from mine when I have not already done so.

The Himalaya appears to be one of the products of the collision between India and Eurasia, which began in or near Eocene time. The rocks exposed at the Earth's surface in the Himalaya consist of material that was part of the Indian subcontinent since being deposited or consolidated. The geologic structure seems to be divided into slivers separated by major thrust faults that, when active, dipped north [see comments by W. Hamilton in Bally et al (1980, p. 38) for a different opinion]. There seems to have been a steady southward progression of thrust faulting, beginning at the Indus-Tsangpo suture zone when collision occurred in Eocene time. Subsequently, the Main Central thrust formed, and slip of at least 100 km, and perhaps several hundred kilometers, occurred. More recently, slip on the Main Boundary Fault began, probably with a cessation of activity on the MCT. Again, probably 100–150 km of slip has occurred on the MBF (Lyon-Caen & Molnar 1983), and others have inferred as much as 1000 km (e.g. Powell & Conaghan 1973, Seeber et al 1981).

The southward migration of thrust faulting probably occurred because it is energetically more favorable than continued slip on one fault. As the elevation increases and the crust thickens, the gravitational potential energy stored in the elevated areas and in the crustal root increases as the square of the elevation and of the excess crustal thickness, respectively (e.g. Molnar et al 1977). Thus, each increment of slip on a thrust fault stores an increasing amount of potential energy in the mountain belt and its root. Probably, less work is eventually required to start a new fault than to maintain slip on an existing fault (Molnar et al 1977).

The present structure and tectonics accord with India continuing to penetrate into Eurasia at about 10 to 20 mm yr^{-1}, and with the same processes that built the range continuing unabated. Fault-plane solutions indicate continued underthrusting of India beneath the Himalaya, and focal depths imply that many of these events occur on the top surfaces of the Indian plate as it slides intact beneath the Lesser Himalaya. The fault-plane solution of one earthquake beneath the Ganga Basin indicates normal faulting, presumably in response to a bending of the Indian plate down beneath the mountains. The wedge of sediments in the Ganga Basin fill the trough created by that flexure. Gravity anomalies corroborate the inference of flexure and accord with the view that the Indian plate has slid intact for at least 100 to 150 km beneath the range (Lyon-Caen & Molnar 1983). Elastic stress in the flexed Indian plate would support the weight of the Lesser Himalaya. I presume that these processes and this structure are not unique to the present, but that they prevailed throughout most of the history of the range and probably for other ranges too.

One of the major controversies about Himalayan tectonics concerns the amount of underthrusting beneath Tibet. A wealth of literature favors wholesale underthrusting of India beneath Tibet, as Argand (1924) and later Barazangi & Ni (1982), Powell & Conaghan (1973), Seeber & Armbruster (1981), and others have advocated. I dissent from this view for several reasons. One is that *S*-wave travel times from earthquakes in Tibet arrive too late to be underlain by a shield structure (Chen & Molnar 1981, Molnar & Chen 1984). Another is that the gravity anomalies in the Himalaya imply that the Moho steepens beneath the Greater Himalaya, suggesting a weakening of the plate there and that an additional force is necessary to support the weight of the Himalaya (Lyon-Caen & Molnar 1983). The inferred weakening of the Indian plate could result either from a heating of it or by an earlier thinning of it by detachment of crust when the Main Boundary Fault formed. I favor the latter view and consider that the change in dip of the Moho provides a measure of the throw on the MBT (100–150 km). Gravity acting on cold, sinking material and applying a bending moment to the Indian plate seems to me to be the force that helps to support the Himalaya. If India were to underthrust the entire Tibetan plateau, then one would have to find another method for supporting the enormous weight of the mountains. Perhaps one can be found, but until it is, I will consider these gravity anomalies to be a more persuasive indication of India's fate than the data used to infer its presence beneath Tibet.

My prejudices are illustrated by Figure 5. As India underthrusts the Himalaya, slices of the crust on its leading edge are detached and become the material constituting the mountain range that we see. The remaining underlying mantle lithosphere, possibly with the lowermost crust, plunges back into the asthenosphere. Among the forces driving India into Eurasia, the push from the ridge crests and the pull of sinking slabs beneath the

Sunda arc are likely to be significant. Unlike those that infer wholesale underthrusting of India beneath Tibet, however, I think that the sinking of cold, Indian mantle lithosphere also contributes to India's motion. The detachment of crust from the top of the Indian lithosphere should leave the mantle lithosphere sufficiently negatively buoyant that it can be expected to sink and to help draw India northward (e.g. Molnar & Gray 1979).

Finally, the occurrences of young, apparently anatectic granites in the Himalaya require melting of crustal materials since the collision began. As noted above, the source of the heat is controversial and, in my opinion, not resolvable with present data. The cross section in Figure 1 shows my prejudice that the granites originated below the MCT and not at or above the MCT as inferred by others (e.g. LeFort 1975, 1981). This prejudice depends on the portion of the Indian shield that underthrust the Himalaya being hotter than typical shields. Although scattered heat-flow measurements from India show substantially higher values than those of most shields (Rao et al 1976), the gravity anomalies and the inferred flexural rigidity imply a thick plate, comparable with estimates from other shields (Walcott 1970). This inconsistency between an apparently relatively thick elastic plate and an apparently relatively thin thermally conducting plate suggests that at least one of the prejudices expressed above may be wrong. Given the uncertainties in measurements and inferences, however, I am loath to abandon any specific one yet. Moreover, if only one is wrong, then I will not be ashamed of this summary. Regardless of whether or not my personal opinions are wrong, I hope that the reader will be able to extract the relevant facts and references from this review to gain an improved understanding of the Himalaya.

ACKNOWLEDGEMENTS

I thank H. Lyon-Caen for permission to use illustrations in advance of their publication, K. Hodges and P. LeFort for critically reading an early draft of this paper, D. Frank for typing it, and K. S. Valdiya and V. C. Thakur for very instructive field trips in the Himalaya. Part of the work described here is a result of continued support from the National Science Foundation (Grant EAR 81 21184) and NASA (Grant NAG-141).

Literature Cited

Argand, E. 1924. *La Tectonique de l'Asie, Rep. Int. Geol. Congr., 13th*, 1 : 170–372

Armbruster, J., Seeber, L., Jacob, K. H. 1978. The northwestern termination of the Himalayan mountain front: active tectonics from microearthquakes. *J. Geophys. Res.* 83 : 269–82

Armijo, R., Tapponnier, P., Mercier, J. L., Han, T. 1982. A field study of Pleistocene rifts in Tibet. *Eos, Trans. Am. Geophys. Union* 63 : 1093 (Abstr.)

Bally, A. W., Allen, C. R., Geyer, R. E., Hamilton, W. B., Hopson, C. A., et al. 1980. Notes on the geology of Tibet and adjacent areas—report of the American Plate Tectonics delegation to the People's

Republic of China. *US Geol. Surv. Open-File Rep. 80-501*. 100 pp.

Baranowski, J., Armbruster, J., Seeber, L., Molnar, P. 1984. Focal depths and fault plane solutions of earthquakes and active tectonics of the Himalaya. *J. Geophys. Res.* In press

Barazangi, M., Ni, J. 1982. Velocities and propagation characteristics of Pn and Sn beneath the Himalayan arc and Tibetan plateau: possible evidence for underthrusting of Indian continental lithosphere beneath Tibet. *Geology* 10:179–85

Bhanot, V. B., Pandey, B. R., Singh, V. P., Thakur, V. C. 1977. Rb-Sr whole-rock age of the granite gneiss from the Askot area, eastern Kumaun and its implications on tectonic interpretation of the area. In *Himalayan Geology*, 7:118–22. Dehra Dun, India: Wadia Inst. Himalayan Geol.

Bhanot, V. B., Singh, V. P., Pandey, B. K., Singh, R. 1981. Rb-Sr isochron age for the gneissic rocks of Askot Crystallines, Kumaun Himalaya (U.P.). In *Contemporary Scientific Researches in Himalaya*, ed. A. K. Sinha, 1:117–19. Dehra Dun, India: Bishen Singh Mahendra Pal Singh

Bird, P. 1978. Initiation of intracontinental subduction in the Himalaya. *J. Geophys. Res.* 83:4975–87

Bird, P., Toksöz, M. N., Sleep, N. H. 1975. Thermal and mechanical models of continent-continent convergence zones. *J. Geophys. Res.* 80:4405–16

Bordet, P., Colchen, M., LeFort, P., Pêcher, A. 1981. The geodynamic evolution of the Himalaya—ten years of research in central Nepal Himalaya and some other regions. In *Zagros, Hindu Kush, Himalaya, Geodynamic Evolution, Geodyn. Ser.* 3:149–68. Washington DC: Am. Geophys. Union

Brookfield, M. E., Reynolds, P. H. 1981. Late Cretaceous emplacement of the Indus Suture Zone ophiolite mélanges and an Eocene-Oligocene magmatic arc on the northern edge of the Indian plate. *Earth Planet. Sci. Lett.* 55:157–62

Brune, J. N. 1968. Seismic moment, seismicity, and rate of slip along major fault zones. *J. Geophys. Res.* 72:777–84

Brunel, M., Andrieux, J. 1977. Déformations superposées et mécanismes associés au chevauchement central Himalayan "M.C.T.": Nepal oriental. In *Himalaya: Sciences de la Terre*, pp. 69–84. Paris: Ed. Cent. Natl. Rech. Sci.

Cao, R. 1981. Lithological features and geological significance of Yarlung Zangbo river ophiolite belt and trench sediments in Xizang plateau. In *Geological and Ecological Studies of Quinghai-Xizang Plateau*, 1:611–20. Beijing: Science Press

Chandra, U. 1978. Seismicity, earthquake mechanisms and tectonics along the Himalayan mountain range and vicinity. *Phys. Earth Planet. Inter.* 16:109–31

Chen, W. P., Molnar, P. 1977. Seismic moments of major earthquakes and the average rate of slip in Central Asia. *J. Geophys. Res.* 82:2945–69

Chen, W. P., Molnar, P. 1981. Constraints on the seismic wave velocity structure beneath the Tibetan plateau and their tectonic implications. *J. Geophys. Res.* 86: 5937–62

Choudhury, S. K. 1975. Gravity and crustal thickness in the Indo-Gangetic Plains and Himalayan region, India. *Geophys. J. R. Astron. Soc.* 40:441–52.

Chugh, R. S. 1974. *Study of recent crustal movements in India and future programs.* Presented at Int. Symp. Recent Crustal Movements, Zürich

Evans, P. 1964. The tectonic framework of Assam. *J. Geol. Soc. India* 5:80–96

Farah, A., DeJong, K. A. 1979. *Geodynamics of Pakistan.* Quetta: Geol. Surv. Pak. 361 pp.

Fitch, T. J. 1970. Earthquake mechanisms in the Himalayan, Burmese and Andaman regions and continental tectonics in Central Asia. *J. Geophys. Res.* 75:2699–2709

Frank, W., Gansser, A., Trommsdorff, V. 1977. Geological observations in the Ladakh area (Himalayas)—a preliminary report. *Schweiz. Mineral. Petrogr. Mitt.* 57:89–133

Fuchs, G., Sinha, A. K. 1978. The tectonics of the Garhwal-Kumaun Lesser Himalaya. *Jahrb. Geol. Bundesanst. (Austria)* 121(2): 219–41

Gansser, A. 1964. *Geology of the Himalayas.* London: Wiley-Interscience. 289 pp.

Gansser, A. 1966. The Indian Ocean and the Himalayan: a geologic interpretation. *Eclogae Geol. Helv.* 59:832–48

Gansser, A. 1977. The great suture zone between Himalaya and Tibet. In *Himalaya: Sciences de la Terre*, pp. 181–91. Paris: Ed. Cent. Natl. Rech. Sci.

Gansser, A. 1980. The significance of the Himalayan suture zone. *Tectonophysics* 62:37–52

Hamet, J., Allègre, C. J. 1976. Rb-Sr systematics in granite from central Nepal (Manaslu): significance of the Oligocene age and high $^{87}Sr/^{86}Sr$ ratio in Himalayan orogeny. *Geology* 4:470–72

Hashimoto, S., Ohta, Y., Akiba, C., eds. 1973. *Geology of the Nepal Himalayas.* Sapporo, Jpn: Saikon Publ. Co. 292 pp.

Honegger, K., Dietrich, V., Frank, W., Gansser, A., Thöni, M., Trommsdorff, V.

1982. Magmatism and metamorphism in the Ladakh Himalayas (the Indus-Tsangpo suture zone). *Earth Planet. Sci. Lett.* 60:253–92

Jäger, E., Bhandari, A. K., Bhanot, V. B. 1971. Rb-Sr age determinations on biotites and whole-rock samples from the Mandi and Chor granites. Himachal Pradesh, India. *Eclogae Geol. Helv.* 64:521–27

Johnson, G. D., Johnson, N. M., Opdyke, N. D., Tahirkheli, R. A. K. 1979. Magnetic reversal stratigraphy and sedimentary tectonic history of the upper Siwalik group, eastern Salt Range and southwestern Kashmir. In *Geodynamics of Pakistan*, pp. 149–66. Quetta: Geol. Surv. Pak.

Johnson, G. D., Johnson, N. M., Opdyke, N. D., Tahirkheli, R. A. K. 1982. The occurrence and fission-track ages of late Neogene and Quaternary volcanic sediments, Siwalik group, northern Pakistan. *Palaegeogr. Palaeoclimatol. Palaeoecol.* 37:63–93

Johnson, N. M., Opdyke, N. D., Johnson, G. D., Lindsay, E. H., Tahirkheli, R. A. K. 1982. Magnetic polarity stratigraphy and ages of Siwalik group rocks of the Potwar Plateau, Pakistan. *Palaegeogr. Palaeoclimatol. Palaeoecol.* 37:17–42

Kono, M. 1974. Gravity anomalies in east Nepal and their implications to the crustal structure of the Himalayas. *Geophys. J. R. Astron. Soc.* 39:283–300

LeFort, P. 1975. Himalayas: the collided range. Present knowledge of the continental arc. *Am. J. Sci.* 275-A:1–44

LeFort, P. 1981. Manaslu leucogranite: a collision signature of the Himalaya, a model for its genesis and emplacement. *J. Geophys. Res.* 86:10545–68

Li, J., Li, B., Wang, F., Zhang, Q., Wen, S., Zheng, B. 1981. The process of the uplift of the Qinghai-Xizang plateau. In *Geological and Ecological Studies of Qinghai-Xizang Plateau*, 1:111–18. Beijing: Science Press

Lyon-Caen, H., Molnar, P. 1983. Constraints on the structure of the Himalaya from an analysis of gravity anomalies and a flexural model of the lithosphere. *J. Geophys. Res.* 88:8171–91

Maluski, M., Proust, F., Xiao, X. C. 1982. First results of $^{39}Ar/^{40}Ar$ dating of the Transhimalaya, calc-alkaline magnetism of southern Tibet. *Nature* 298:152–54

Mattauer, M. 1975. Sur le mecanisme de formation de la schistocité dans l'Himalaya. *Earth Planet. Sci. Lett.* 28:144–54

Mehta, P. K. 1980. Tectonic significance of the young mineral dates and rates of cooling and uplift in the Himalaya. *Tectonophysics* 62:205–17

Minster, B., Jordan, T. 1978. The present-day plate motions. *J. Geophys. Res.* 83:5331–54

Molnar, P., Chen, W. P. 1982. Seismicity and mountain building. In *Mountain Building Processes*, ed. U. Briegel, K. J. Hsü, pp. 41–56. London: Academic

Molnar, P., Chen, W. P. 1983. Focal depths and fault plane solutions of earthquakes under the Tibetan plateau. *J. Geophys. Res.* 88:1180–96

Molnar, P., Chen, W. P. 1984. *S-P* wave residuals and lateral inhomogeneity in the mantle beneath Tibet and the Himalaya. *J. Geophys. Res.* In press

Molnar, P., Gray, D. 1979. Subduction of continental crust: some constraints and uncertainties. *Geology* 7:58–62

Molnar, P., Tapponnier, P. 1975. Cenozoic tectonics of Asia: effects of a continental collision. *Science* 189:419–26

Molnar, P., Tapponnier, P. 1978. Active tectonics of Tibet. *J. Geophys. Res.* 83:5361–74

Molnar, P., Fitch, T. J., Wu, F. T. 1973. Fault plane solutions of shallow earthquakes and contemporary tectonics in Asia. *Earth Planet. Sci. Lett.* 19:101–12

Molnar, P., Chen, W. P., Fitch, T. J., Tapponnier, P., Warsi, W. E. K., Wu, F. T. 1977. Structure and tectonics of the Himalaya: a brief summary of relevant geophysical observations. In *Himalaya: Sciences de la Terre*, pp. 267–94. Paris: Ed. Cent. Natl. Rech. Sci.

Molnar, P., Chen, W. P., Padovani, E. 1983. Calculated temperatures in overthrust terrains and possible combinations of heat sources responsible for the Tertiary granites in the Greater Himalaya. *J. Geophys. Res.* 88:6415–29

Ni, J., York, J. 1978. Late Cenozoic tectonics of Tibetan Plateau. *J. Geophys. Res.* 83:5375–84

Nicolas, A., Girardeau, J., Marcoux, J., Dupre, B., Wang, X., et al. 1981. The Xigaze ophiolite (Tibet): a peculiar oceanic lithosphere. *Nature* 294:414–17

Opdyke, N. D., Lindsay, E., Johnson, G. D., Johnson, N., Tahirkheli, R. A. K., Mirza, M. A. 1979. Magnetic polarity stratigraphy and vertebrate paleontology of the Upper Siwalik group of northern Pakistan. *Palaegeogr. Palaeoclimatol. Palaeoecol.* 27:1–34

Pêcher, A. 1978. *Deformations et metamorphisme associes a une zone de cisaillement. Exemple du grand chevauchement central Himalayen (M.C.T.), transversale des Annapurnas et du Manaslu, Nepal.* These d'Etat. Univ. Sci. Méd. Grenoble, France. 354 pp.

Powell, C. M., Conaghan, P. 1973. Plate

tectonics and the Himalayas. *Earth Planet. Sci. Lett.* 20:1–12

Qureshy, M. N. 1969. Thickening of the basaltic layer as a possible cause for the uplift of the Himalayas—a suggestion based on gravity data. *Tectonophysics* 7:137–57

Rao, R. U. M., Rao, G. V., Narain, H. 1976. Radioactive heat generation and heat flow in the Indian shield. *Earth Planet. Sci. Lett.* 30:57–64

Rastogi, B. K. 1974. Earthquake mechanisms and plate tectonics in the Himalayan region. *Tectonophysics* 21:47–56

Richter, C. F. 1958. *Elementary Seismology.* San Francisco: Freeman. 768 pp.

Sahni, A., Kumar, V. 1974. Palaeogene palaeobiogeography of the Indian subcontinent. *Palaegeogr. Palaeoclimatol. Palaeoecol.* 15:209–26

Sastri, V. V., Bhandari, L. L., Rasu, A. T. R., Datta, A. K. 1971. Tectonic framework and subsurface stratigraphy of the Ganga Basin. *J. Geol. Soc. India* 12:222–33

Scholz, C. H. 1980. Shear heating and the state of stress on faults. *J. Geophys. Res.* 85:6174–84

Sclater, J. G., Jaupart, C., Galson, D. 1980. The heat flow through oceanic and continental crust and the heat loss of the earth. *Rev. Geophys. Space Phys.* 18:269–311

Seeber, L., Armbruster, J. 1979. Seismicity of the Hazara arc in northern Pakistan; Decollement vs. basement faulting. In *Geodynamics of Pakistan*, pp. 131–42. Quetta: Geol. Surv. Pak.

Seeber, L., Armbruster, J. 1981. Great detachment earthquakes along the Himalayan arc and long-term forecasting. In *Earthquake Prediction: An International Review, Maurice Ewing Ser.*, 4:259–79. Washington DC: Am. Geophys. Union

Seeber, L., Jacob, K. H. 1977. Microearthquake survey of northern Pakistan: preliminary results and tectonic implications. In *Himalaya: Sciences de la Terre*, pp. 347–60. Paris: Ed. Cent. Natl. Rech. Sci.

Seeber, L., Armbruster, J. G., Quittmeyer, R. C. 1981. Seismicity and continental subduction in the Himalayan arc. In *Zagros, Hindu-Kush Himalaya, Geodynamic Evolution, Geodyn. Ser.*, 3: 215–42. Washington DC: Am. Geophys. Union

Sengör, A. M. C. 1981. The evolution of Palaeo-Tethys in the Tibetan segment of the Alpides. In *Geological and Ecological Studies of Qinghai-Xizang Plateau*, 1:51–56. Beijing: Science Press

Sharma, K. K., Sinha, A. K., Bagdasarian, G. P., Gukasian, R. Ch. 1978a. Potassium-argon dating of Dras volcanics, Shyok volcanics and Ladakh granite, Ladakh, northwest Himalaya. In *Himalayan Geology*, 8(1):288–95. Dehra Dun, India: Wadia Inst. Himalayan Geol.

Sharma, K. K., Saini, H. S., Nagpaul, K. K. 1978b. Fission track annealing, ages of apatites from Mandi granite and their application to tectonic problems. In *Himalayan Geology*, 8(1):296–312. Dehra Dun, India: Wadia Inst. Himalayan Geol.

Sinha, A. K. 1981. Geology and tectonics of the Himalayan region of Ladakh, Himachal, Garwhal-Kumaun and Arunachal Pradesh: a review. In *Zagros, Hindu-Kush, Himalaya, Geodynamic Evolution, Geodyn. Ser.*, 3:122–48. Washington DC: Am. Geophys. Union

Sinha-Roy, S. 1981. Comments on "Initiation of intracontinental subduction in the Himalaya," by Peter Bird. *J. Geophys. Res.* 86:9320–22

Sinvhal, H., Agrawal, P. N., King, G. C. P., Gaur, V. K. 1973. Interpretation of measured movement at a Himalayan (Nahan) thrust. *Geophys. J. R. Astron. Soc.* 34:203–10

Song, Z., Liu, G. 1981. Tertiary palynological assemblages from Xizang with reference to their paleogeographical significance. In *Geological and Ecological Studies of Qinghai-Xizang Plateau*, 1:207–14. Beijing: Science Press

Stöcklin, J. 1980. Geology of Nepal and its regional frame. *J. Geol. Soc. London* 137:1–34

Tahirkheli, R. A. K. 1982. *Geology of the Himalaya, Karakorum, and Hindukush in Pakistan, Geol. Bull., Univ. Peshawar*, Vol. 15. 51 pp.

Tapponnier, P., Molnar, P. 1976. Slip-line field theory and large scale continental tectonics. *Nature* 264:319–24

Tapponnier, P., Mercier, J. L., Proust, F., Andrieux, J., Armijo, R., et al. 1981a. The Tibetan side of the India-Eurasia collision. *Nature* 294:405–10

Tapponnier, P., Mercier, J. L., Armijo, R., Han, T., Zhou, J. 1981b. Field evidence for active normal faulting in Tibet. *Nature* 294:410–14

Thakur, V. C. 1981. Regional framework and geodynamic evolution of the Indus-Tsangpo suture zone in the Ladakh Himalayas. *Trans. R. Soc. Edinburgh: Earth Sci.* 72:89–97

Tu, G. z., Zhang, Y. q., Zhao, Zh. l., Wang, Zh. q. 1981. Characteristics and evolution of granitoids of southern Xizang. In *Geological and Ecological Studies of Qinghai-Xizang Plateau*, 1:353–62. Beijing: Science Press

U-Pb Dating Group. 1981. A discussion on zircon U-Pb dating of intermediate-acid rocks in southern Xizang. In *Geological and Ecological Studies of Qinghai-Xizang Plateau*, 1:497–506. Beijing: Science Press

Valdiya, K. S. 1977. Structural set-up of the Kumaun Lesser Himalaya. In *Himalaya: Sciences de la Terre*, pp. 449–62. Paris: Ed. Cent. Natl. Rech. Sci.

Valdiya, K. S. 1980a. The two intracrustal boundary thrusts of the Himalaya. *Tectonophysics* 66:323–48

Valdiya, K. S. 1980b. *Geology of the Kumaun Lesser Himalaya.* Dehra Dun, India: Wadia Inst. Himalayan Geol. 291 pp.

Valdiya, K. S. 1981. Tectonics of the central sector of the Himalaya. In *Zagros, Hindu-Kush, Himalaya, Geodynamic Evolution, Geodyn. Ser.*, 3:87–110. Washington DC: Am. Geophys. Union

Vidal, Ph. 1978. Rb-Sr systematics in granite from central Nepal (Manaslu): significance of the Oligocene age and high $^{87}Sr/^{86}Sr$ ratio in Himalayan orogeny: comment. *Geology* 6:196

Vidal, Ph., Cocherie, A., LeFort, P. 1982. Geochemical investigations of the origin of the Manaslu leucogranite (Himalaya, Nepal). *Geochim. Cosmochim. Acta* 46: 2279–92

Virdi, N. S. 1980. Problem of the root-zone of nappes in the western Himalaya—a critical review. *Himalayan Geology*, 10:55–77. Dehra Dun, India: Wadia Inst. Himalayan Geol.

Walcott, R. I. 1970. Flexural rigidity, thickness, and viscosity of the lithosphere. *J. Geophys. Res.* 75:3941–54

Wang, J., Chen, Z., Gui, X., Xu, R., Zhang, Y. 1981. Rb-Sr isotopic studies on some intermediate-acid plutons in southern Xizang. In *Geological and Ecological Studies of Qinghai-Xizang Plateau*, 1:515–20. Beijing: Science Press

Wang, L. C., Wang, D. A. 1981. Character of the sedimentary facies-belts and the sedimentary model of the subsidence belt in the Yarlung Zangbo-Xiangquan River. In *Geological and Ecological Studies of Qinghai-Xizang Plateau*, 1:599–610. Beijing: Science Press

Warsi, W. E. K., Molnar, P. 1977. Gravity anomalies and plate tectonics in the Himalaya. In *Himalaya: Sciences de la Terre*, pp. 463–78. Paris: Ed. Cent. Natl. Rech. Sci.

Wu, H. 1981. Gravity anomalies and plate tectonics in the Himalaya. In *Geological and Ecological Studies of Qinghai-Xizang Plateau*, 1:567–78. Beijing: Science Press

Xu, R. 1981. Vegetational changes in the past and the uplift of Qinghai-Xizang plateau. In *Geological and Ecological Studies of Qinghai-Xizang Plateau*, 1:139–44. Beijing: Science Press

Zeitler, P. K., Johnson, N. M., Naeser, C. W., Tahirkheli, A. K. 1982a. Fission-track evidence for Quaternary uplift of the Nanga Parbat region, Pakistan. *Nature* 298:255–57

Zeitler, P. K., Tahirkheli, R. A. K., Naeser, C. W., Johnson, N. M. 1982b. Unroofing history of a suture zone in the Himalaya of Pakistan by means of fission-track annealing ages. *Earth Planet. Sci. Lett.* 57:227–40

Zhang, Y. q., Dai, T. m., Hong, A. s. 1981. Isotopic geochronology of granitoid rocks in southern Xizang plateau. In *Geological and Ecological Studies of Qinghai-Xizang Plateau*, 1:483–96. Beijing: Science Press

Zhou, Y. s., Zhang, Q., Jin, C., Deng, W. m. 1981. The migration and evolution of magmatism and metamorphism in Xizang since Cretaceous and their relation to the Indian Plate motion—a possible model for the uplift of Qinghai-Xizang plateau. In *Geological and Ecological Studies of Qinghai-Xizang Plateau*, 1:363–78. Beijing: Science Press

SUBJECT INDEX

O

P

Q

R

S

CUMULATIVE INDEXES

CONTRIBUTING AUTHORS VOLUMES 1–12

CHAPTER TITLES VOLUMES 1–12

TECTONOPHYSICS AND REGIONAL GEOLOGY

ORDER FORM

se list the volumes you wish to order by volume number. If you wish a standing order (the latest volume to you automatically each year), indicate volume number to begin order. Volumes not yet published will shipped in month and year indicated. All prices subject to change without notice. Prepayment required n individuals. Telephone orders charged to VISA, MasterCard, American Express, welcomed.

ANNUAL REVIEW SERIES

		Prices Postpaid per volume USA/elsewhere	Regular Order Please send: Vol. number	Standing Order Begin with: Vol. number
ual Review of ANTHROPOLOGY				
Vols. 1-10	(1972-1981)	$20.00/$21.00		
Vol. 11	(1982)	$22.00/$25.00		
Vol. 12	(1983)	$27.00/$30.00		
Vol. 13	(avail. Oct. 1984)	$27.00/$30.00	Vol(s). ______	Vol. ______
nual Review of ASTRONOMY AND ASTROPHYSICS				
Vols. 1-19	(1963-1981)	$20.00/$21.00		
Vol. 20	(1982)	$22.00/$25.00		
Vol. 21	(1983)	$44.00/$47.00		
Vol. 22	(avail. Sept. 1984)	$44.00/$47.00	Vol(s). ______	Vol. ______
nual Review of BIOCHEMISTRY				
Vols. 29-50	(1960-1981)	$21.00/$22.00		
Vol. 51	(1982)	$23.00/$26.00		
Vol. 52	(1983)	$29.00/$32.00		
Vol. 53	(avail. July 1984)	$29.00/$32.00	Vol(s). ______	Vol. ______
nnual Review of BIOPHYSICS AND BIOENGINEERING				
Vols. 1-10	(1972-1981)	$20.00/$21.00		
Vol. 11	(1982)	$22.00/$25.00		
Vol. 12	(1983)	$47.00/$50.00		
Vol. 13	(avail. June 1984)	$47.00/$50.00	Vol(s). ______	Vol. ______
nnual Review of EARTH AND PLANETARY SCIENCES				
Vols. 1-9	(1973-1981)	$20.00/$21.00		
Vol. 10	(1982)	$22.00/$25.00		
Vol. 11	(1983)	$44.00/$47.00		
Vol. 12	(avail. May 1984)	$44.00/$47.00	Vol(s). ______	Vol. ______
Annual Review of ECOLOGY AND SYSTEMATICS				
Vols. 1-12	(1970-1981)	$20.00/$21.00		
Vol. 13	(1982)	$22.00/$25.00		
Vol. 14	(1983)	$27.00/$30.00		
Vol. 15	(avail. Nov. 1984)	$27.00/$30.00	Vol(s). ______	Vol. ______

SEE ORDERING INFORMATION ON PAGE 4.

		Prices Postpaid per volume USA/elsewhere	Regular Order Please send: Vol. number	Standing O Begin wi Vol. numb

Annual Review of ENERGY

Vols.	Years	Price	Regular Order	Standing Order
Vols. 1-6	(1976-1981)	$20.00/$21.00		
Vol. 7	(1982)	$22.00/$25.00		
Vol. 8	(1983)	$56.00/$59.00		
Vol. 9	(avail. Oct. 1984)	$56.00/$59.00	Vol(s). ______	Vol. ______

Annual Review of ENTOMOLOGY

Vols.	Years	Price	Regular Order	Standing Order
Vols. 7-16, 18-26	(1962-1971; 1973-1981)	$20.00/$21.00		
Vol. 27	(1982)	$22.00/$25.00		
Vol. 28	(1983)	$27.00/$30.00		
Vol. 29	(avail. Jan. 1984)	$27.00/$30.00	Vol(s). ______	Vol. ______

Annual Review of FLUID MECHANICS

Vols.	Years	Price	Regular Order	Standing Order
Vols. 1-13	(1969-1981)	$20.00/$21.00		
Vol. 14	(1982)	$22.00/$25.00		
Vol. 15	(1983)	$28.00/$31.00		
Vol. 16	(avail. Jan. 1984)	$28.00/$31.00	Vol(s). ______	Vol. ______

Annual Review of GENETICS

Vols.	Years	Price	Regular Order	Standing Order
Vols. 1-15	(1967-1981)	$20.00/$21.00		
Vol. 16	(1982)	$22.00/$25.00		
Vol. 17	(1983)	$27.00/$30.00		
Vol. 18	(avail. Dec. 1984)	$27.00/$30.00	Vol(s). ______	Vol. ______

Annual Review of IMMUNOLOGY

Vols.	Years	Price	Regular Order	Standing Order
Vol. 1	(1983)	$27.00/$30.00		
Vol. 2	(avail. April 1984)	$27.00/$30.00	Vol(s). ______	Vol. ______

Annual Review of MATERIALS SCIENCE

Vols.	Years	Price	Regular Order	Standing Order
Vols. 1-11	(1971-1981)	$20.00/$21.00		
Vol. 12	(1982)	$22.00/$25.00		
Vol. 13	(1983)	$64.00/$67.00		
Vol. 14	(avail. Aug. 1984)	$64.00/$67.00	Vol(s). ______	Vol. ______

Annual Review of MEDICINE: Selected Topics in the Clinical Sciences

Vols.	Years	Price	Regular Order	Standing Order
Vols. 1-3, 5-15	(1950-1952; 1954-1964)	$20.00/$21.00		
Vols. 17-32	(1966-1981)	$20.00/$21.00		
Vol. 33	(1982)	$22.00/$25.00		
Vol. 34	(1983)	$27.00/$30.00		
Vol. 35	(avail. April 1984)	$27.00/$30.00	Vol(s). ______	Vol. ______

Annual Review of MICROBIOLOGY

Vols.	Years	Price	Regular Order	Standing Order
Vols. 17-35	(1963-1981)	$20.00/$21.00		
Vol. 36	(1982)	$22.00/$25.00		
Vol. 37	(1983)	$27.00/$30.00		
Vol. 38	(avail. Oct. 1984)	$27.00/$30.00	Vol(s). ______	Vol. ______

Annual Review of NEUROSCIENCE

Vols.	Years	Price	Regular Order	Standing Order
Vols. 1-4	(1978-1981)	$20.00/$21.00		
Vol. 5	(1982)	$22.00/$25.00		
Vol. 6	(1983)	$27.00/$30.00		
Vol. 7	(avail. March 1984)	$27.00/$30.00	Vol(s). ______	Vol. ______

Annual Review of NUCLEAR AND PARTICLE SCIENCE

Vols.	Years	Price	Regular Order	Standing Order
Vols. 12-31	(1962-1981)	$22.50/$23.50		
Vol. 32	(1982)	$25.00/$28.00		
Vol. 33	(1983)	$30.00/$33.00		
Vol. 34	(avail. Dec. 1984)	$30.00/$33.00	Vol(s). ______	Vol. ______

'G INFORMATION ON PAGE 4.